Springer Complexity

Springer Complexity is an interdisciplinary program publishing the best research and academic-level teaching on both fundamental and applied aspects of complex systems—cutting across all traditional disciplines of the natural and life sciences, engineering, economics, medicine, neuroscience, social and computer science.

Complex Systems are systems that comprise many interacting parts with the ability to generate a new quality of macroscopic collective behavior the manifestations of which are the spontaneous formation of distinctive temporal, spatial or functional structures. Models of such systems can be successfully mapped onto quite diverse "real-life" situations like the climate, the coherent emission of light from lasers, chemical reaction-diffusion systems, biological cellular networks, the dynamics of stock markets and of the Internet, earthquake statistics and prediction, freeway traffic, the human brain, or the formation of opinions in social systems, to name just some of the popular applications.

Although their scope and methodologies overlap somewhat, one can distinguish the following main concepts and tools: self-organization, nonlinear dynamics, synergetics, turbulence, dynamical systems, catastrophes, instabilities, stochastic processes, chaos, graphs and networks, cellular automata, adaptive systems, genetic algorithms and computational intelligence.

The three major book publication platforms of the Springer Complexity program are the monograph series "Understanding Complex Systems" focusing on the various applications of complexity, the "Springer Series in Synergetics", which is devoted to the quantitative theoretical and methodological foundations, and the "Springer Briefs in Complexity" which are concise and topical working reports, case studies, surveys, essays and lecture notes of relevance to the field. In addition to the books in these two core series, the program also incorporates individual titles ranging from textbooks to major reference works.

Understanding Complex Systems

Founding Editor: S. Kelso

Future scientific and technological developments in many fields will necessarily depend upon coming to grips with complex systems. Such systems are complex in both their composition – typically many different kinds of components interacting simultaneously and nonlinearly with each other and their environments on multiple levels – and in the rich diversity of behavior of which they are capable.

The Springer Series in Understanding Complex Systems series (UCS) promotes new strategies and paradigms for understanding and realizing applications of complex systems research in a wide variety of fields and endeavors. UCS is explicitly transdisciplinary. It has three main goals: First, to elaborate the concepts, methods and tools of complex systems at all levels of description and in all scientific fields, especially newly emerging areas within the life, social, behavioral, economic, neuro- and cognitive sciences (and derivatives thereof); second, to encourage novel applications of these ideas in various fields of engineering and computation such as robotics, nano-technology, and informatics; third, to provide a single forum within which commonalities and differences in the workings of complex systems may be discerned, hence leading to deeper insight and understanding.

UCS will publish monographs, lecture notes, and selected edited contributions aimed at communicating new findings to a large multidisciplinary audience.

More information about this series at http://www.springer.com/series/5394

Eckehard Schöll · Sabine H.L. Klapp
Philipp Hövel

Editors

Control of Self-Organizing Nonlinear Systems

 Springer

Editors
Eckehard Schöll
Institut für Theoretische Physik
Technische Universität Berlin
Berlin
Germany

Philipp Hövel
Institut für Theoretische Physik
Technische Universität Berlin
Berlin
Germany

Sabine H.L. Klapp
Institut für Theoretische Physik
Technische Universität Berlin
Berlin
Germany

ISSN 1860-0832
Understanding Complex Systems
ISBN 978-3-319-28027-1
DOI 10.1007/978-3-319-28028-8

ISSN 1860-0840 (electronic)

ISBN 978-3-319-28028-8 (eBook)

Library of Congress Control Number: 2015958534

Printed on acid-free paper

This Springer imprint is published by SpringerNature
The registered company is Springer International Publishing AG Switzerland

Preface

This book summarizes state-of-the-art research on the control of self-organizing nonlinear systems by a selection of contributions from leading international experts in this new emerging field. The first focus concerns recent methodological developments from both the physical and the mathematical side, including control of networks and of noisy and time-delayed systems. As a second focus, the book features novel innovative concepts of application including control of quantum systems, soft condensed matter, biological systems, and complex networks. Special topics reflecting the active research in the field are the analysis and control of chimera states in classical networks and in quantum systems, the mathematical treatment of multiscale systems, the control of colloidal and quantum transport, and the control of epidemics and of neural network dynamics.

The International Conference on Control of Self-Organizing Nonlinear Systems held from August 25–28, August 2014 in Rostock-Warnemünde, Germany, was organized by the Collaborative Research Center (Sonderforschungsbereich) SFB 910 *Control of Self-Organizing Nonlinear System—Theoretical Methods and Concepts of Application*, Berlin, to provide a forum for such topics. We took this opportunity to assemble a list of world-leading experts which now enables us to present perspectives of the cutting-edge-research in this field. The book covers mathematical foundations as well as applications. The individual contributions summarize recent research results and also address the broader context. Thus, the presentation is kept accessible for a large audience. The 24 chapters cover various aspects, ranging from fundamental aspects like synchronization and control of complex networks, time-delayed feedback control, interplay of noise and delay, optimal control, effective models, and multiscale systems, to applications of feedback control and chimera patterns in quantum transport and photonics, colloidal systems and liquid films, neuroscience, epidemiology, and evolutionary dynamics. The chapters are grouped into two parts: I Theoretical Methods and II Concepts of Applications.

The first part addresses fundamental issues of controlling nonlinear dynamical systems. The contribution by Zakharova et al. discusses the interplay of structure, noise, and delay in the control of chimera patterns, in particular amplitude chimeras and chimera death, in networks of Stuart–Landau oscillators. Olmi and Torcini analyze the synchronization transition of a globally coupled network of phase oscillators with inertia, whose natural frequencies are unimodally or bimodally distributed. Hövel et al. present an adaptive control scheme, based on the speed-gradient method, for in-phase and cluster synchronization in delay-coupled networks of Stuart–Landau oscillators, adapting the topology by changing the link weights. Atay investigates the problem of controlling oscillations in nonlinear systems by delayed feedback, thus driving the system to a stable limit cycle with prescribed amplitude and frequency. Purewal et al. treat global effects of time-delayed feedback control applied to the Lorenz system, and demonstrate the stabilization of one of its two saddle periodic orbits. Schneider and Fiedler study the symmetry-breaking stabilization of rotating waves, i.e., unstable periodic orbits in ring networks of Stuart–Landau oscillators, by time-delayed feedback control. D'Huys et al. focus on the interplay of noise and delay in delay-coupled oscillators, in particular on delay-induced multistability and noise-induced switching between different periodic orbits. Just et al. develop some basic analytical perturbation schemes for noisy dynamical systems with time delay, and apply this to time-delayed feedback control, coherence resonance, and the computation of power spectra. Li et al. develop a computationally efficient numerical method for the study of noise-induced bifurcations in nonautonomous dynamical systems, and apply it to explosive and dangerous stochastic bifurcations, characterized by sudden jumps of the response probability distribution. Ryll et al. deal with optimal control of traveling wave solutions of reaction–diffusion systems, in particular with the position control of selected spatiotemporal patterns. Curran et al. report recent rigorous results on reaction–diffusion equations with discontinuous hysteretic nonlinearities and treat the pattern formation mechanism of rattling as an application. Mielke develops mathematical tools for deriving effective models for multi-scale systems via evolutionary Γ-convergence and applies them to perturbed gradient systems, e.g., the homogenization of reaction–diffusion systems. Kuehn presents a review of moment closure methods which enable the derivation of a closed hierarchy of coupled differential equations in the modeling of complex systems.

The second part discusses a number of recent innovative applications, starting with feedback control on the quantum scale. The contribution of Emary summarizes theoretical strategies to manipulate the properties of electron flows and states in quantum transport devices, covering both, measurement-based, and coherent control. Strasberg et al. present two measurement-based feedback schemes for a paradigmatic nonlinear quantum system and discuss their application on stabilization of steady states. The contribution of Bastidas et al. extends the phenomenon of chimera states, i.e., partially synchronized patterns, to the quantum regime, and uncovers intriguing quantum signatures of these states in the quantum correlations and the quantum information. Moving towards classical systems, Weicker et al. present a combined theoretical and experimental study of a time-delayed FitzHugh–Nagumo system exhibiting a threshold nonlinearity related to multirhythmicity. The theoretical observations are supported by parallel experiments involving an electrical circuit. Based on a semiclassical approach, Böhm and Lüdge investigate small networks of semiconductor lasers with optical all-to-all coupling, focusing on synchronization patterns and the occurrence of chimera states. As a first application to soft condensed matter, Gernert et al. discuss feedback control strategies to manipulate transport properties and transport efficiency in colloidal systems, including systems with non-negligible interactions. Another type of soft-matter system, namely thin films containing self-propelled particles is discussed by Pototsky et al., focusing on the interplay of self-propulsion and linear stability of the film. On a more methodological level, Kraft and Gurevich address the formation and control of spatiotemporal patterns in a Swift–Hohenberg equation subject to time-delayed feedback. The contribution of Belik et al. involves control of a large-scale hospital network. Based on a large number of real, patient referral patterns they propose an agent-based computational model of hospital-related infections and analyze the model predictions including the effect of various control strategies. Ladenbauer et al. present an overview of control strategies with time-delayed feedback for neural networks, focusing on the impact of plasticity of synaptic coupling strengths and changes of neuronal adaptation properties. Finally, Claussen discusses examples of macroscopic evolutionary dynamics, particularly the stabilization of steady states of coexistence via payoff and global feedback.

Owing to the cross-disciplinary nature of the topic, we hope that this book will have substantial impact across field boundaries. It is aimed to bring together the nonlinear dynamics control concepts, the classical mathematical control theory, and quantum control. In particular, we envisage to stimulate future developments and interactions in the areas of control theory, functional differential equations, dynamical network science, hard and soft condensed matter, nonlinear optics, neuroscience, and socio-economic systems. This book thus provides a snapshot of the vibrant research related to controlling nonlinear systems from across different fields. It will not only be of great interest to specialists working on related problems, but also provide a valuable resource for other scientists and newcomers to the field.

Berlin   Eckehard Schöll
November 2015   Sabine H.L. Klapp
                                                                                                    Philipp Hövel

Contents

Contributors

Fatihcan M. Atay Max Planck Institute for Mathematics in the Sciences, Leipzig, Germany

Moritz Augustin Department of Software Engineering and Theoretical Computer Science, Technische Universität Berlin, Berlin, Germany; Bernstein Center for Computational Neuroscience Berlin, Berlin, Germany

Victor Manuel Bastidas Institut für Theoretische Physik, Technische Universität Berlin, Berlin, Germany; Centre for Quantum Technologies, National University of Singapore, Singapore, Singapore

Vitaly Belik Institut für Theoretische Physik, Technische Universität Berlin, Berlin, Germany; Helmholtz Centre for Infection Research, Braunschweig, Germany

Fabian Böhm Institut für Theoretische Physik, Technische Universität Berlin, Berlin, Germany

Tobias Brandes Institut für Theoretische Physik, Technische Universität Berlin, Berlin, Germany

Jens Christian Claussen Computational Systems Biology Lab, Jacobs University Bremen, Bremen, Germany

Mark Curran Institute of Mathematics I, Free University of Berlin, Berlin, Germany

Jan Danckaert Applied Physics Research Group (APHY), Vrije Universiteit Brussel, Brussels, Belgium

Otti D'Huys Universität Würzburg, Würzburg, Germany; Duke University, NC, USA

Clive Emary Department of Physics and Mathematics, University of Hull, Hull, UK; School of Mathematics and Statistics, Newcastle University, Tyne and Wear, UK

Harald Engel Institut für Theoretische Physik, Technische Universität Berlin, Berlin, Germany

Thomas Erneux Optique Nonlinéaire Théorique, Université Libre de Bruxelles, Bruxelles, Belgium

Bernold Fiedler Institut für Mathematik, Freie Universität Berlin, Berlin, Germany

Alexander Fradkov Department of Theoretical Cybernetics, Saint-Petersburg State University, Saint-Petersburg, Russia; Institute for Problems of Mechanical Engineering, Russian Academy of Sciences, Saint-Petersburg, Russia

Gaetan Friart Optique Nonlinéaire Théorique, Université Libre de Bruxelles, Bruxelles, Belgium

Paul M. Geffert School of Mathematical Sciences, Queen Mary University of London, London, UK; Institut für Theoretische Physik, Technische Universität Berlin, Berlin, Germany

Robert Gernert Institut für Theoretische Physik, Technische Universität Berlin, Berlin, Germany

Aleksandar Gjurchinovski Faculty of Natural Sciences and Mathematics, Institute of Physics, Sts. Cyril and Methodius University, Skopje, Macedonia

Kongming Guo School of Electromechanical Engineering, Xidian University, Xi'an, China

Pavel Gurevich Institute of Mathematics I, Free University of Berlin, Berlin, Germany; Peoples' Friendship University of Russia, Moscow, Russia

Svetlana V. Gurevich Institut für Theoretische Physik, Westfälische Wilhelms-Universität Münster, Münster, Germany

Ling Hong State Key Laboratory for Strength and Vibration, Xi'an Jiaotong University, Xi'an, China

Philipp Hövel Institut für Theoretische Physik, Technische Universität Berlin, Berlin, Germany; Bernstein Center for Computational Neuroscience Berlin, Berlin, Germany

Jun Jiang State Key Laboratory for Strength and Vibration, Xi'an Jiaotong University, Xi'an, China

Wolfram Just School of Mathematical Sciences, Queen Mary University of London, London, UK; Institut für Theoretische Physik, Technische Universität Berlin, Berlin, Germany

Thomas Jüngling Universitat de les Illes Balears IFISC (CSIC-UIB), Campus UIB, Palma de Mallorca, Spain

Lars Keuninckx Applied Physics Research Group (APHY), Vrije Universiteit Brussel, Brussels, Belgium

Wolfgang Kinzel Universität Würzburg, Würzburg, Germany

Sabine H.L. Klapp Institut für Theoretische Physik, Technische Universität Berlin, Berlin, Germany

Alexander Kraft Institut für Theoretische Physik, Technische Universität Berlin, Berlin, Germany

Bernd Krauskopf Department of Mathematics, University of Auckland, Auckland, New Zealand

Christian Kuehn Institute for Analysis and Scientific Computing, Vienna University of Technology, Vienna, Austria

Josef Ladenbauer Department of Software Engineering and Theoretical Computer Science, Technische Universität Berlin, Berlin, Germany; Bernstein Center for Computational Neuroscience Berlin, Berlin, Germany

Judith Lehnert Institut für Theoretische Physik, Technische Universität Berlin, Berlin, Germany

Zigang Li State Key Laboratory for Strength and Vibration, Xi'an Jiaotong University, Xi'an, China

Ken Lichtner Institut für Theoretische Physik, Technische Universität Berlin, Berlin, Germany

Jakob Löber Institut für Theoretische Physik, Technische Universität Berlin, Berlin, Germany

Sarah A.M. Loos Institut für Theoretische Physik, Technische Universität Berlin, Berlin, Germany

Kathy Lüdge Institut für Theoretische Physik, Technische Universität Berlin, Berlin, Germany

Steffen Martens Institut für Theoretische Physik, Technische Universität Berlin, Berlin, Germany

Alexander Mielke Weierstraß-Institut für Angewandte Analysis und Stochastik, Berlin, Germany

Rafael Mikolajczyk Helmholtz Centre for Infection Research, Braunschweig, Germany

Klaus Obermayer Department of Software Engineering and Theoretical Computer Science, Technische Universität Berlin, Berlin, Germany; Bernstein Center for Computational Neuroscience Berlin, Berlin, Germany

Simona Olmi CNR—Consiglio Nazionale delle Ricerche, Istituto dei Sistemi Complessi, Sesto Fiorentino, Italy; INFN Sez. Firenze, Sesto Fiorentino, Italy

Iryna Omelchenko Institut für Theoretische Physik, Technische Universität Berlin, Berlin, Germany

Claire M. Postlethwaite Department of Mathematics, University of Auckland, Auckland, New Zealand

Andrey Pototsky Department of Mathematics, Faculty of Science Engineering and Technology, Swinburne University of Technology, VIC, Australia

Anup S. Purewal Department of Mathematics, University of Auckland, Auckland, New Zealand

Christopher Ryll Institut für Mathematik, Technische Universität Berlin, Berlin, Germany

Gernot Schaller Institut für Theoretische Physik, Technische Universität Berlin, Berlin, Germany

Isabelle Schneider Institut für Mathematik, Freie Universität Berlin, Berlin, Germany

Eckehard Schöll Institut für Theoretische Physik, Technische Universität Berlin, Berlin, Germany

Anton Selivanov School of Electrical Engineering, Tel Aviv University, Tel Aviv, Israel; Department of Theoretical Cybernetics, Saint-Petersburg State University, Saint-Petersburg, Russia

Julien Siebert Institut für Theoretische Physik, Technische Universität Berlin, Berlin, Germany

Holger Stark Institut für Theoretische Physik, Technische Universität Berlin, Berlin, Germany

Philipp Strasberg Institut für Theoretische Physik, Technische Universität Berlin, Berlin, Germany

Uwe Thiele Institut für Theoretische Physik, Westfälische Wilhelms-Universität Münster, Münster, Germany

Sergey Tikhomirov Max Planck Institute for Mathematics in the Sciences, Leipzig, Germany; Chebyshev Laboratory, Saint-Petersburg State University, Saint-Petersburg, Russia

Alessandro Torcini CNR—Consiglio Nazionale delle Ricerche, Istituto dei Sistemi Complessi, Sesto Fiorentino, Italy; INFN Sez. Firenze, Sesto Fiorentino, Italy; Aix-Marseille Université, Inserm, INMED UMR 901 and Institut de Neurosciences des Systèmes UMR 1106, Marseille, France; Aix-Marseille Université, Université de Toulon, CNRS, CPT, UMR 7332, Marseille, France

Fredi Tröltzsch Institut für Mathematik, Technische Universität Berlin, Berlin, Germany

Lionel Weicker Optique Nonlinéaire Théorique, Université Libre de Bruxelles, Bruxelles, Belgium; Applied Physics Research Group (APHY), Vrije Universiteit Brussel, Brussels, Belgium; LMOPS, CentraleSupélec, Université Paris-Saclay, Metz, France; LMOPS, CentraleSupélec, Université de Lorraine, Metz, France

Anna Zakharova Institut für Theoretische Physik, Technische Universität Berlin, Berlin, Germany

Part I
Theoretical Methods

Chapter 1
Controlling Chimera Patterns in Networks: Interplay of Structure, Noise, and Delay

Anna Zakharova, Sarah A.M. Loos, Julien Siebert,
Aleksandar Gjurchinovski, Jens Christian Claussen and
Eckehard Schöll

Abstract We investigate partially coherent and partially incoherent patterns (chimera states) in networks of Stuart-Landau oscillators with symmetry-breaking coupling. In particular, we study two types of chimera states, amplitude chimeras and chimera death, under the influence of time delay and noise. We find that amplitude chimeras are long-living transients, whose lifetime can be controlled by varying the noise intensity and the value of time delay.

1.1 Introduction

Collective behavior of coupled nonlinear dynamical systems can take diverse forms, ranging from various synchronization patterns and oscillation suppression to chimera states, which have been receiving growing interest of researchers from different fields during the past decade [1]. Originally found for the model of phase oscillators [2, 3], chimera states imply spatial coexistence of coherent (synchronized) and incoherent (desynchronized) domains in a dynamical network and have been found in a large variety of different systems [4–29]. The most intriguing feature of chimera states is that they appear for networks of identical elements and symmetric coupling configurations.

Numerous experimental reports on chimera states [30–40] have stimulated further investigations in the field. Additionally, the burst of activity in chimera research is motivated by the wide range of its possible applications. In neural networks, for

A. Zakharova (✉) · S.A.M. Loos · J. Siebert · E. Schöll
Institut für Theoretische Physik, Technische Universität Berlin, Hardenbergstr. 36, 10623 Berlin, Germany
e-mail: anna.zakharova@tu-berlin.de

A. Gjurchinovski
Institute of Physics, Faculty of Natural Sciences and Mathematics,
Sts. Cyril and Methodius University, P. O. Box 162, 1000 Skopje, Macedonia

J.C. Claussen
Computational Systems Biology Lab, Jacobs University Bremen, Campus Ring 1, 28759 Bremen, Germany

© Springer International Publishing Switzerland 2016
E. Schöll et al. (eds.), *Control of Self-Organizing Nonlinear Systems*,
Understanding Complex Systems, DOI 10.1007/978-3-319-28028-8_1

3

example, chimeras can be associated with bump states [41] and in the dynamics of
the heart they may be used to model ventricular fibrillation [42]. In the investigation
of power grids it is important to understand how to avoid chimera states, since they
may initiate a blackout—partial or full desynchronization of the power network [43].
For social systems chimeras may be linked to the situation of partial consensus in
the two-population network of social agents [44]. Unihemispheric sleep of some sea
mammals and birds can be related to chimera behavior [45]. Chimeras have also
been suggested as a mechanism for the termination of epileptic seizure [46].

In recent studies chimera states have been extended to systems which involve not
only phase but also amplitude dynamics and named amplitude-mediated chimeras
in the case when both amplitude and phase are characterized by chimera behavior
[20, 21]. More complicated patterns in which chimera structures are formed with
respect to the amplitudes while the phases remain correlated for the whole network
have been first reported in [47]. This particular type of chimera states, *amplitude
chimeras*, is investigated in the present work.

While modelling real-world systems it is important to take stochasticity and time
delay into account. Arising naturally, these two factors lead to a plethora of complex
phenomena with applications to various fields. Moreover, both may result in opposite
effects and can be exploited for control purposes. Our objective is to establish efficient
control mechanisms based on noise and time delay. In particular, we address the question of how time delay and noise influence the behavior of amplitude chimera states
in networks of Stuart-Landau oscillators. Additionally, we study another recently
discovered type of chimera states, *chimera death* [47], which, through death of the
oscillations, generalizes the chimera feature of coexistence of spatially coherent and
incoherent domains to steady states.

1.2   Model

We consider a network of N Stuart-Landau oscillators [2, 47–50] under the impact of
external white noise $\xi_j(t)$ and in the presence of time delay τ. The local deterministic
dynamics of each node $j \in \{1, \ldots, N\}$ is given by $\dot{z}_j = f(z_j)$, with the normal form
of a supercritical Hopf bifurcation

$$f(z_j) = (\lambda + i\omega - |z_j|^2)z_j, \qquad (1.1)$$

where $z_j = x_j + i\,y_j = r_j e^{i\phi_j} \in \mathbb{C}$, with $x_j, y_j, r_j, \phi_j \in \mathbb{R}$, and $\lambda, \omega > 0$. At $\lambda =
0$ a Hopf bifurcation occurs, so that for $\lambda > 0$ the single Stuart-Landau oscillator
exhibits self-sustained oscillations with frequency ω and radius $r_j = \sqrt{\lambda}$, and the
unique fixed point $x_j = 0$, $y_j = 0$ is unstable.

We investigate a ring of N non-locally coupled Stuart-Landau oscillators, where
each node is coupled to its P nearest neighbors in both directions with the strength
$\sigma > 0$, and is subject to noise of intensity $D > 0$:

$$\dot{z}_j = f(z_j) + \frac{\sigma}{2P} \sum_{k=j-P}^{j+P} (\text{Re}z_k - \text{Re}z_j) + \sqrt{2D}\xi_j(t) \qquad (1.2)$$

where $j = 1, 2, \ldots, N$ and all indices are modulo N. The normalized number of nearest neighbors P/N is denoted as coupling range. The coupling and the noise are only applied to the real parts, and $\xi_j(t) \in \mathbb{R}$ is additive Gaussian white noise [51], i.e., $\langle \xi_j(t) \rangle = 0$, $\forall j$, and $\langle \xi_i(t)\xi_j(t') \rangle = \delta_{ij}\delta(t - t')$, $\forall i, j$, where δ_{ij} denotes the Kronecker-Delta and $\delta(t - t')$ denotes the Delta-distribution. Hence the noise is spatially uncorrelated.

Further we study the impact of time delay using the following model:

$$\dot{z}_j = f(z_j) + \frac{\sigma}{2P} \sum_{k=j-P}^{j+P} (\text{Re}z_k(t - \tau) - \text{Re}z_j(t)), \qquad (1.3)$$

where τ is time delay.

1.3 Deterministic Amplitude Chimera and Chimera Death

For a deterministic network with instantaneous coupling as demonstrated in [47], various different states can be found in the network given by Eq. (1.2). Which particular state actually arises, depends on the specific values of the coupling parameters and the initial conditions, as Eq. (1.2) describes a multistable system. Among the possible states, two different types of asymptotically stable states can be found, on the one hand oscillatory states, and on the other hand steady state patterns which are related to oscillation death. The latter are represented by completely coherent or completely incoherent oscillation death patterns, as well as by chimera death patterns consisting of coexisting domains of coherent and incoherent steady states. The asymptotically stable oscillatory states appear in two different spatio-temporal patterns: in-phase synchronized oscillations and coherent traveling waves. Besides these, long lasting oscillatory transients with interesting features occur, i.e., amplitude chimera states. In this work we demonstrate that all these states can also be observed under the influence of noise and time delay. Before an asymptotic oscillatory state is approached, *amplitude chimera* states can appear as long transients, potentially lasting for hundreds or even thousands of oscillation periods. In contrast to classical phase chimeras, all nodes (including the ones within the incoherent domains) oscillate with the same period, $T = \frac{2\pi}{\omega}$, and a spatially correlated phase, but they show spatially incoherent behavior with respect to the *amplitudes* in part of the system. Figure 1.1 shows an exemplary amplitude chimera configuration. The nodes within the two coherent domains (here $13 \le j \le 85$ and $113 \le j \le 185$) perform synchronized oscillations, all with the same amplitudes. The coherent domains always appear pairwise, such that for every time t, all nodes within one coherent domain have a phase lag of π with

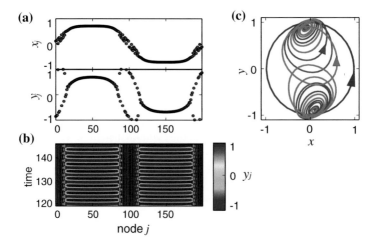

Fig. 1.1 Amplitude chimera state in system (1.2) with $N = 200$ nodes, for coupling range $P/N = 0.04$ and coupling strength $\sigma = 18$: **a** snapshot (*top* x_j, *bottom* y_j), **b** space-time plot, **c** phase plot in the complex plane: trajectories of 12 nodes of the incoherent (*red* and *green*), and 12 nodes of the coherent (*blue*) domains, the *arrows* indicate the direction of the motion. *Initial condition*: See Sect. 1.4. *Other parameters*: $D = 0$, $\lambda = 1$, $\omega = 2$

respect to all nodes of the other, antipodal domain. Hence they always fulfill the "anti-phase partner" condition $z_j = -z_{j+N/2}$, j mod N, assuming even N. As visible in Fig. 1.1c, the trajectories in the complex plane of all nodes are cycles, illustrating that all nodes have periodic dynamics in time. This is a fundamental difference between the classical phase chimera states where a part of the network demonstrates chaotic temporal behavior. The nodes of the coherent domains all oscillate on a perfect circle around the origin. Both coherent domains are represented by one single blue line in Fig. 1.1c, which as well represents time the trajectory of all nodes when the completely in-phase synchronized oscillatory solution is approached. The two antipodal coherent domains are separated by incoherent domains. There, neighboring nodes can be in completely different states at a given time t. Their trajectories are deformed circles, whose centers are shifted from the origin. The *completely arbitrary sequence* of nodes that oscillate around centers in the upper and in the lower half-plane reflects the incoherent nature. Transient amplitude chimeras with very narrow incoherent domains can be observed, as well as with broad ones.

If the coupling strength and coupling range exceeds certain values, the oscillations of the Stuart-Landau nodes can be suppressed due to the stabilization of a new inhomogeneous steady state created by the coupling. Instead of performing oscillations, each node approaches a fixed point close to one of the following two branches: $(x^{*1}, y^{*1}) \approx (-0.1, +0.85)$ or $(x^{*2}, y^{*2}) \approx (+0.1, -0.85)$ (for $\lambda = 1$), and remains there for all times. The oscillation death states exhibit a huge variety of spatial patterns, including multiple coherent and multiple incoherent oscillation death states [47, 50, 52, 53]. Two exemplary configurations of completely coherent oscilla-

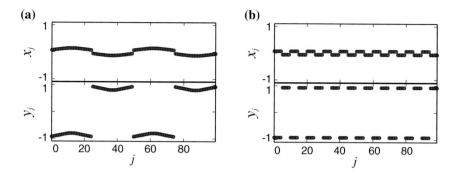

Fig. 1.2 Snapshots of coherent oscillation death states: **a** coupling strength $\sigma = 18$ and coupling range $P/N = 0.14$ (2-cluster), **b** $\sigma = 8, P/N = 0.04$ (10-cluster). *Initial condition*: Nodes $0 \leq j \leq 24$ and $50 \leq j \leq 74$ are set to $(x_j, y_j) = (0.1, -1)$, all other nodes are set to $(x_j, y_j) = (-0.1, +1)$. *Other parameters*: $N = 100, D = 0, \lambda = 1, \omega = 2$

tion death patterns are shown in Fig. 1.2 (2-cluster and 10-cluster oscillation death). The oscillation death regime is characterized by very high multistability. Among the oscillation death states, *chimera death* patterns can be found, which combine the characteristics of both phenomena: chimera state and oscillation death. These patterns consist of coexisting domains of coherent and incoherent populations of the inhomogeneous steady state branches. Within the incoherent domains, the population of the two branches (upper and lower) follows a random sequence, as for example visible in Fig. 1.3. Within the coherent domains, the number of clusters of neighboring nodes that populate the same branch of the inhomogeneous steady state can vary. An m-cluster chimera death state (m-CD), with $m \in \{1, 3, 5, 7, 9, \ldots\}$, is characterized by the occurrence of m clusters i.e., sets of neighboring nodes that populate the same branch of the inhomogeneous steady state within each coherent domain. The coherent domains always appear pairwise with anti-phase symmetry $z_j = -z_{j+N/2}$, similarly to the coherent domains of the amplitude chimera configurations. Our numerical results confirm that the stable oscillation death patterns fulfill the "anti-phase partner" condition.

1.4 The Impact of Initial Conditions

For $D = 0$, Eq. (1.2) is known to describe a multistable system [47]. Both types of chimera states appear in coupling parameter regimes, where other oscillation death patterns and coherent oscillatory states can be found as well. In order to increase the probability of finding chimera states, we use specially prepared initial conditions [54]. A very simple initial condition that produces transient amplitude chimeras in a certain parameter regime (of about $0.01 < P/N < 0.05, \sigma < 33$), is when all nodes of one half of the network ($1 \leq j \leq \frac{N}{2}$) are set to the same value $(x_j, y_j) = (x_0^1, y_0^1)$

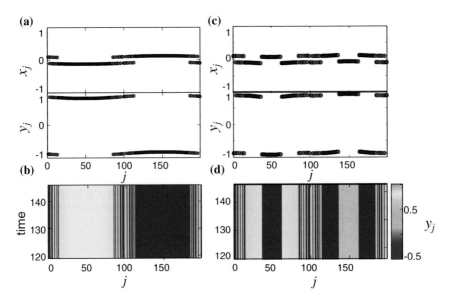

Fig. 1.3 **a, b** 1-cluster chimera death (1-CD) for coupling range $P/N = 0.4$, **c, d** 3-cluster chimera death (3-CD) for $P/N = 0.2$. Snapshots x_j, y_j are shown in panels (**a, c**), space-time plots in panels (**b, d**). *Initial condition*: See Sect. 1.4. *Parameters*: $N = 200$, $\sigma = 18$, $D = 0$, $\lambda = 1$, $\omega = 2$

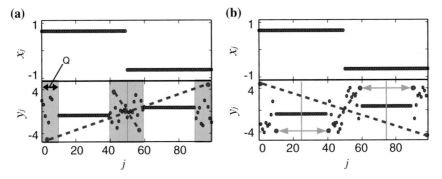

Fig. 1.4 Specially prepared initial conditions for amplitude chimeras: **a** point-symmetric type, **b** fully-symmetric type. *Top panels* x_j, *bottom panels* y_j. *Dashed lines* indicate the point-symmetry about the center, *solid green lines* indicate the axial symmetry within both network halves. *System size* $N = 100$

(excluding the choice $(0, 0)$), and the rest is set to $(x_0^2, y_0^2) = (-x_0^1, -y_0^1)$. Hence, amplitude chimera states can evolve out of initial configurations that only consist of two completely coherent parts. We choose the values $(x_0^1, y_0^1) = (\sqrt{0.5}, -\sqrt{0.5})$, so that the nodes start on the limit cycle with $r = \sqrt{\lambda} = 1$, which is the solution for the in-phase synchronized oscillation. The amplitude chimera lifetime nevertheless appears to be of the same order for other values (e. g. $(x_0^1, y_0^1) = (1, -1)$).

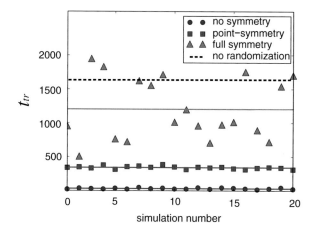

Fig. 1.5 Transient times of amplitude chimera states t_{tr} for 20 realizations of specially prepared random initial conditions (no symmetry, point-symmetry, full symmetry, depicted by *different symbols*) with $Q = 10$, $V = 2$. *Horizontal solid lines* Mean values. *Dashed line* t_{tr} for the initial condition with $V = 0$ (no randomization). *System parameters* $N = 100$, $\sigma = 14$, $P/N = 0.04$, $D = 0, \lambda = 1, \omega = 2$

By adding random numbers to y_j, we construct a more general class of specially prepared random initial conditions for amplitude chimeras. In particular, we add a random number drawn from a Gaussian distribution with variance V to y_j of the Q nodes on the left and on the right side of the borders between both halves (at $j = \frac{N}{2}$ and $j = N$), as indicated in Fig. 1.4a, with $Q \in \mathbb{N}$ and $0 < Q \leq \frac{N}{4}$. Besides the range Q of incoherence, we also vary $V \geq 0$. For a proper choice of the two initial condition parameters (Q and V), we obtain amplitude chimeras. Using the achieved amplitude chimera lifetime as a quality measure for the initial condition, we compare multiple realizations of the specially prepared random initial conditions for the deterministic system with $P/N = 0.04$ and $N = 100$. We observe that among all considered kinds of initial conditions (different choices for Q and V, symmetry conditions, x_j randomized as well, a different underlying distribution for the random numbers), the applied symmetry of the initial condition has the greatest effect upon the transient time.

Figure 1.5 shows the transient times and their mean value (solid lines) for multiple realizations of the initial conditions following three different symmetry schemes. For the particular choice ($Q = 10$, $V = 2$) all symmetry types lead on average to shorter lifetimes than an initial condition without random component (black dashed line). For the initial configurations without symmetry, a random number is chosen independently for each node within the four incoherent intervals. These configurations clearly create the shortest amplitude chimera lifetimes, lasting at most for a couple of oscillation periods. This symmetry type also leads to the shortest transients in other regimes of Q, and V (not shown here). In contrast, for the point-symmetric initial conditions, we mirror the random numbers used for $j \in \{1, \ldots, \frac{N}{2}\}$ with respect to the center $j = 0$, $y_j = 0$, and use their negative counterparts for the second half.

We hence only generate $2Q$ random numbers in total. The initial configurations are point-symmetric with respect to the center, see Fig. 1.4a. The lifetimes of the occurring amplitude chimeras are much longer than in the non-symmetric case. However, the symmetry type which leads to the longest lifetimes is the one referred to as full symmetry; the initial conditions fulfill two symmetries: The randomly chosen values of the positions of the nodes within the first incoherent interval $1 \leq j \leq Q$ are mirrored to the nodes $\frac{N}{2} - Q \leq j \leq \frac{N}{2}$, by setting: $z_j = z_{\frac{N}{2}+1-j}$. To obtain the positions of the second network half, then a phase shift of π is applied, such that the "anti-phase partner" condition is fulfilled ($z_j = -z_{j+\frac{N}{2}}$ and j mod N). Thus, we only generate Q different random numbers in total. The configurations are again point-symmetric with respect to the center, and have an additional axial symmetry with orthogonal axes through $j = \frac{N}{4}$ and $j = \frac{3N}{4}$, as indicated in Fig. 1.4b. Of course, the simple initial condition with no randomization also fulfills these symmetry conditions and can therefore be regarded as one special type of the fully-symmetric specially prepared initial conditions (with $V = 0$). We have further tested another type of initial condition that solely fulfills the anti-phase partner condition: $z_j = -z_{j+\frac{N}{2}}$, but has no other symmetries. This type of initial condition also certainly leads to transient amplitude chimeras, but only within very narrow ranges of Q and V. For $Q = 10$, $V = 2$, the mean lifetime (of about $t_{tr} \approx 49$) is only slightly increased compared to the non-symmetric initial condition (not shown here). Since the symmetry which is applied to the initial conditions remains preserved during the dynamic evolution, this observation means that the fully-symmetric amplitude chimeras are most stable and have the longest lifetimes.

By decreasing the variance in the interval $0.1 \leq V \leq 2$, the mean amplitude chimera lifetimes increase. In the range of small variances of about $V < 0.5$, amplitude chimeras occur for all choices of the incoherence range Q, and the particular choice of Q does not influence the transient time much. For $Q = \frac{N}{4}$, all nodes are randomized, see Fig. 1.6a, which appears to be a natural choice. Figure 1.6b shows the corresponding transient times belonging to a set of 40 realizations of the specially prepared random initial condition with $Q = \frac{N}{4}$ and $V = 0.5$, and for a set with $V = 0.1$. The mean transient times are much longer than for $V = 2$ (cf. Fig. 1.5). They are at least of the same order (and can be larger) as the transient time for the simple initial condition with no randomization, $V = 0$ (dashed black line). For the choice $V = 0.1$, the transient times are increased as compared to $V = 0.5$. We use the fully-symmetric initial conditions with $Q = \frac{N}{4}$ and $V = 0.1$ for all investigations presented in this chapter.

Besides oscillatory states, oscillation death states can occur in a large variety of different spatial patterns. Our numerical results suggest that in the appropriate parameter regime every amplitude chimera snapshot can be used as initial condition to certainly produce a chimera death state. How many clusters in the coherent domain of the chimera death pattern occur, depends on the initial condition as well as on the parameter choice (see Sect. 1.5.2).

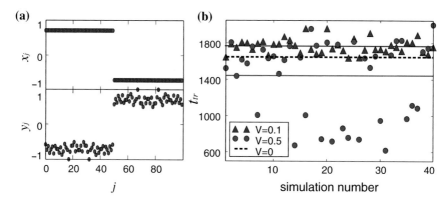

Fig. 1.6 a Fully-symmetric random initial condition with range of incoherence $Q = \frac{N}{4}$ and variance $V = 0.1$. **b** Transient times of amplitude chimeras t_{tr} for 40 realizations of the initial condition shown in (**a**): *circles* $V = 0.5$, *triangles* $V = 0.1$. *Horizontal solid lines*: Mean values, *dashed line*: t_{tr} for $V = 0$. *System parameters*: $N = 100$, $\sigma = 14$, $P/N = 0.04$, $D = 0$, $\lambda = 1$, $\omega = 2$

1.5 Stochastic Case

1.5.1 Control of Amplitude Chimera Lifetime by Noise

In this section we will study the role of noise for chimera patterns [54]. By using the same initial conditions which lead to amplitude chimera states and chimera death in the deterministic case, we also observe these states in Eq. (1.2) in the presence of noise in a wide range of the coupling parameters. Figure 1.7 shows one exemplary configuration for an amplitude chimera which occurs in a system under the impact of noise of intensity $D = 5 \cdot 10^{-3}$.

In general, the transient times of amplitude chimeras decrease with increasing noise intensity. Figure 1.8 shows the average transient times and the corresponding standard deviations in dependence of the noise intensity D, for three choices of the coupling strength σ, in a semi-logarithmic plot. The average is over 50 different fully symmetric initial conditions (with $Q = \frac{N}{4}$, see Sect. 1.4) drawn from different realizations of the associated random distribution. For each one of those realizations of the initial conditions, a different realization of the Gaussian white noise $\xi_j(t)$ is considered. The average transient times show a clear linear decrease as a function of the logarithmic noise intensity. This behavior is found throughout the range $6 \leq \sigma \leq 24$, i.e., $t_{tr} = -\frac{1}{\mu} ln(D) + \eta$ with slope $-\frac{1}{\mu}$ and axis intercept η. This gives the scaling law

$$D \sim e^{-\mu t_{tr}}. \tag{1.4}$$

The lines in Fig. 1.8 show the linear fits, and the inset depicts the slope in dependence on the coupling strength σ. For the same set of 50 initial conditions, Fig. 1.9a depicts the mean transient time in dependence of the coupling strength for four different

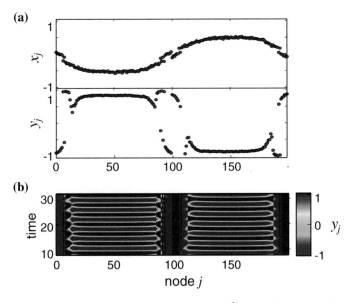

Fig. 1.7 Amplitude chimera for noise intensity $D = 5 \cdot 10^{-3}$: **a** snapshot (*top* x_j, *bottom* y_j), **b** space-time plot. *Parameters*: $N = 200$, $P/N = 0.04$, $\sigma = 19$, $\lambda = 1$, $\omega = 2$. *Initial condition*: See Sect. 1.4

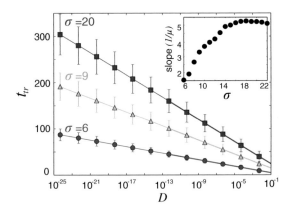

Fig. 1.8 Transient times of amplitude chimeras t_{tr} versus noise strength D (log-scaled) for different values of coupling strength σ. *Symbols*: Average over 50 fully symmetric initial conditions (with $Q = \frac{N}{4}$), each associated with a different realization of the random force $\xi(t)$; *error bars*: standard deviations; *lines*: linear fits from Eq. (1.4). *Inset*: Slope versus σ. *Parameters*: $N = 100$, $P/N = 0.04$, $\lambda = 1$, $\omega = 2$

noise intensities D, and Fig. 1.9b shows a color-coded density plot of the mean transient times of amplitude chimeras in the (σ, D)-plane. The transient times generally decrease with increasing noise, and increase with increasing coupling strength up to a saturation value at about $\sigma \approx 15$. The vertical error bars in panel (a) show that

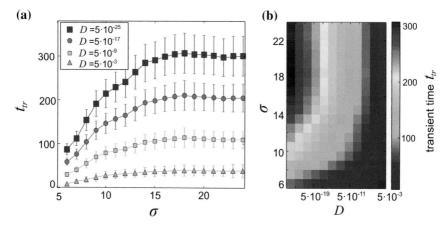

Fig. 1.9 Transient times of amplitude chimeras t_{tr} averaged over 50 initial conditions and noise realizations (the same set of initial conditions as used in Fig. 1.8): **a** t_{tr} versus coupling strength σ for different noise intensities. *Symbols*: Mean transient times; *error bars*: standard deviations. The *lines* serve as a guide to the eye. **b** t_{tr} in the plane of coupling strength σ and noise intensity D. *Other parameters*: $N = 100$, $P/N = 0.04$, $\lambda = 1$, $\omega = 2$

the transient times are less sensitive to the initial condition, the larger the noise is. We generally find that the spread of the amplitude chimera lifetimes for different initial conditions (and different noise realizations), is smaller with increasing noise strength.

Transient amplitude chimeras can last for thousands of oscillation periods until they disappear. Even under disturbance by external noise they persist for a significant time. Noise does not essentially change their spatial configuration. If noise throws the system onto an adjacent trajectory in the underlying high-dimensional phase space of the network, this does not normally lead to a flow into a completely different direction in phase space. Geometrically speaking, this shows that there are some attracting directions in phase space along which the system dynamics is pushed towards the amplitude chimera. Furthermore, amplitude chimeras can evolve out of initial configurations that do not show the characteristic coexistence of coherent and incoherent domains (see Sect. 1.4). In fact, they can be observed when completely incoherent initial configurations are used, as well as when the initial condition consists of two completely coherent parts. These dynamical properties indicate that the flow within a certain volume of the phase space is directed towards the amplitude chimera state. From the perspective of the amplitude chimera, there must exist some associated "stable directions". However, even in the absence of any external perturbation, for all system sizes, the amplitude chimera states disappear after some time, and the system approaches a coherent oscillatory state. Accordingly, there must also exists at least one "unstable direction" in phase space. These findings can be explained by the structure of the phase space, which is schematically depicted in Fig. 1.10.

Fig. 1.10 Schematic phase-space structure of an amplitude chimera (AC) as a saddle-point. *Thick solid lines*: Stable directions, *thick dashed lines*: unstable directions, *thin solid lines*: different trajectories, with arrows denoting the direction of time evolution. *Grey shaded region*: Scheme of amplitude chimera configuration, *green disk*: set of initial conditions (IC), *yellow area*: impact of Gaussian white noise

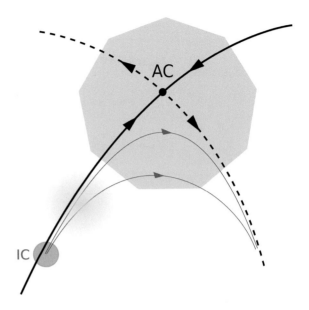

The lifetime of amplitude chimeras in the deterministic case strongly depends on the choice of initial conditions as discussed in Sect. 1.4. In general the sensitivity of chimera states to the initial configurations is explained by the fact that classical chimera states typically coexist with the completely synchronized state, for which the basin of attraction is significantly larger. For amplitude chimeras, all our numerical results support the idea that amplitude chimera patterns can be seen as a saddle state composed of stable (solid lines in Fig. 1.10) and unstable (dashed lines Fig. 1.10) manifolds. The set of initial conditions leading to amplitude chimeras can be represented as a volume restricted in phase space (green disk in Fig. 1.10). The observed amplitude chimera corresponds to trajectories starting from this set and passing the saddle-point from the stable direction towards the unstable manifold. The lifetime of an amplitude chimera, therefore, depends on the chosen trajectory: the closer to the saddle-point it gets, the longer is the lifetime. In other words, the transient time is determined by the time the system spends in the vicinity of the saddle-point where coherent and incoherent oscillating domains coexist before it escapes to the in-phase synchronized regime along the direction of the unstable manifold. Such a phase space scenario explains the sensitivity of transient times to initial conditions since they determine the particular path the system takes.

Our numerical investigations of the stochastic model Eq. (1.2) show that Gaussian white noise dramatically reduces the impact of initial conditions on the lifetime of amplitude chimeras. In more detail, we have tested a set of realizations of initial conditions which lead to significantly different lifetimes of amplitude chimeras without random forcing. In the presence of relatively weak noise $D = 5 \cdot 10^{-13}$ all realizations result in amplitude chimeras with similar lifetime. This again supports our view of the amplitude chimera as a saddle-point, and allows for the following explanation.

The stochastic force, which continuously perturbs the system, makes it randomly switch between different trajectories close to the saddle-point. Therefore, the system's dynamics is not determined by a single trajectory anymore, but rather affected by a set of trajectories belonging to the N-dimensional hyper-sphere. This reduces the sensitivity of the amplitude chimera lifetime to specific initial conditions. In Fig. 1.10 the impact of noise is illustrated by yellow shading, denoting the stochastic forces applied to the system at one instant of time.

1.5.2 Maps of Dynamic Regimes: Interplay of Noise and Coupling

For a large range of the coupling parameters σ and P we calculate the asymptotically stable state and the transient time of amplitude chimeras for $N = 100$. For each choice of (σ, P) we start with the same amplitude chimera configuration as initial condition. For an exemplary initial condition, the results belonging to four different noise intensities are shown in Fig. 1.11. For very small coupling strength σ or very small coupling range P/N the asymptotic states are coherent oscillatory

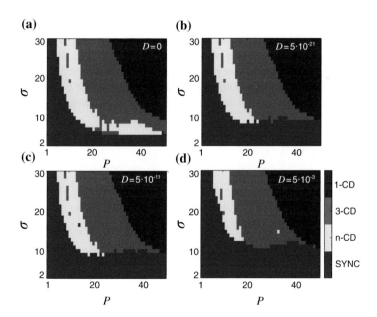

Fig. 1.11 Map of dynamic regimes in the plane of coupling strength σ and number of nearest neighbors P for noise intensities: **a** $D = 0$, **b** $D = 5 \cdot 10^{-21}$, **c** $D = 5 \cdot 10^{-11}$, **d** $D = 5 \cdot 10^{-3}$. *Color code*: 1-cluster chimera death (1-CD), 3-cluster chimera death (3-CD), multi-cluster chimera death (n-CD, $n > 3$), in-phase synchronized oscillations and coherent traveling waves (SYNC). *Initial condition*: Snapshot of an amplitude chimera calculated for $D = 0$, $P = 4$, $\sigma = 14$, $t = 150$. *Maximum simulation time*: $t = 5000$. *Parameters*: $N = 100$, $\lambda = 1$, $\omega = 2$

states, either in-phase synchronized oscillations or traveling waves (dark blue region, labeled SYNC). For very small coupling range, we observe amplitude chimeras as transients. For larger σ and P, we find chimera death states (yellow, orange, and red regions) with one coherent domain (1-CD), or for slightly smaller P, with three (3-CD), or more (n-CD with $n > 3$) coherent domains. For all noise intensities, there exists a chimera death regime (1-CD, 3-CD, n-CD), as well as a coherent oscillatory regime (SYNC).

The regime of chimera death states is characterized by high multistability. The boundary between the oscillatory regime and the chimera death regime is roughly independent of the particular amplitude chimera snapshot used as initial condition. In contrast, for many values of (σ, P), the particular type of chimera death depends on the realization of the initial condition. Note that there is nevertheless a clear tendency that the m-CD patterns (with $m < k$) are generally found for larger coupling ranges than the k-CD states ($k, m \in \{1, 3, n\}$). This tendency is especially pronounced for large coupling strengths. Noise influences the dynamic regimes in different ways. First, the boundaries between the different cluster types of chimera death appear to be almost unaffected by the applied external noise. We do not observe any noise-induced switching between the different types of chimera death. The applied noise does not influence the asymptotic chimera death state. Second, with increasing noise intensity, the boundary between the oscillatory regime and the oscillation death regime is shifted towards higher coupling strengths. This means that the stochastic force pushes the system out of the deterministic inhomogeneous steady state into the basin of attraction of the stable coherent oscillatory state, and induces oscillations in a parameter regime where in the absence of noise the steady state is a stable asymptotic solution. The size of this parameter regime depends on the applied noise intensity. In order to facilitate the comparison, the boundaries between the oscillatory regime and the chimera death regime are depicted for different noise intensities in Fig. 1.12.

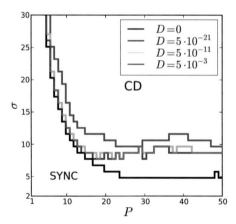

Fig. 1.12 Boundary between the oscillatory regime and the chimera death regime for different noise intensities D, extracted from the maps of dynamic regimes shown in Fig. 1.11

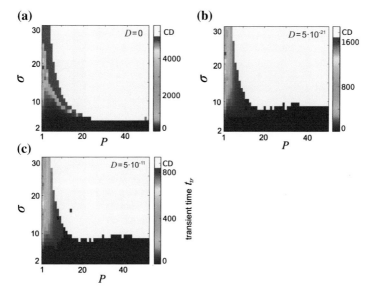

Fig. 1.13 Transient times of amplitude chimeras t_{tr} in the plane of coupling strength σ and number of nearest neighbors P, for the noise intensities: **a** $D = 0$, **b** $D = 5 \cdot 10^{-21}$, **c** $D = 5 \cdot 10^{-11}$. System parameters, initial condition and simulation time as in Fig. 1.11

Generally, the stronger the applied noise is, the smaller is the regime of chimera death states.

In the oscillatory regime, we observe transient amplitude chimeras. In Fig. 1.13 their lifetimes are depicted, obtained from the same simulations described above. One can see that generally the transient time decreases with decreasing coupling strength, and with increasing noise intensity, as shown already in Figs. 1.8 and 1.9 for a restricted range of coupling parameters. Note that Fig. 1.13b ($D = 5 \cdot 10^{-11}$) and Fig. 1.13c ($D = 5 \cdot 10^{-21}$) look very similar up to rescaling of the transient times. This illustrates that the impact of the applied noise upon the dynamics is rather independent of the strength and range of the coupling.

In the deterministic case (Fig. 1.13a) there is a regime of high values of the coupling strength, at the border between the oscillatory regime and the chimera death regime, where the transient amplitude chimeras last longer than the maximum simulation time of $t = 5000$ (bright orange). For several values of (σ, P) in this region, we have simulated much longer time series until $t = 40,000$ (more than 12,700 oscillation periods T), and have found that the amplitude chimeras persist. However, they disappear much earlier as soon as a tiny amount of external noise is applied. This indicates that the amplitude chimera states are also unstable in this region. The extremely long transient times might simply be related to our choice of initial conditions in the deterministic system.

(a)

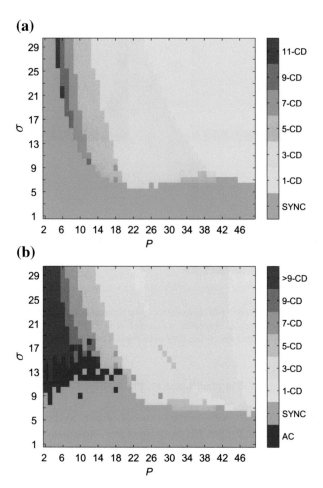

Fig. 1.14 Map of dynamic regimes in the plane of number of nearest neighbors P and coupling strength σ **a** for $\tau = 0$ and **b** $\tau = \pi$. *Color scale*: *1-CD* 1-cluster chimera death; *3-CD* 3-cluster chimera death; *n-CD* n-cluster chimera death. *SYNC* coherent states (synchronized oscillations, traveling waves). *AC* amplitude chimera. *Parameters*: $N = 100, \lambda = 1, \omega = 2$

1.6 Control of Chimeras by Time Delay

We have shown that amplitude chimeras are preserved in the stochastic case and their lifetime can be decreased by tuning the noise intensity. Next we demonstrate that amplitude chimeras are also observed for the time-delayed coupling and their lifetime can be significantly enlarged by an appropriate choice of delay time.

As initial condition for the simulation of the system Eq. (1.3) we choose a snapshot of an amplitude chimera. The corresponding phase diagram for $\tau = 0$ is illustrated in Fig. 1.14a. Since the integration time used for this plot is rather large ($t = 5000$),

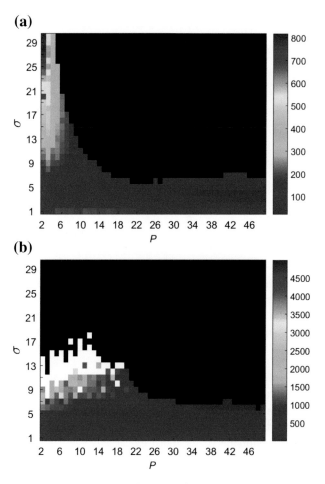

Fig. 1.15 Transient times t_{tr} in the plane of number of nearest neighbors P and coupling strength σ: *color code* indicates the time of transition from incoherent states (amplitude chimera) to coherent states for **a** $\tau = 0$ and **b** $\tau = \pi$. The *black region marks* chimera death states. Integration time until t = 5000. The *white dots* are amplitude chimeras and related structures that are stable in the simulation timespan. *Parameters*: $\lambda = 1$, $\omega = 2$, $N = 100$

amplitude chimeras do not survive for that long in the absence of time-delayed coupling and transform into the in-phase synchronized regime (green region in Fig. 1.14a). Therefore, for $\tau = 0$ the phase diagram contains only chimera death states with different number of clusters and asymptotic coherent states (in-phase synchronized oscillations and coherent travelling waves).

In the presence of delay, however, amplitude chimeras live significantly longer. In particular, for $\tau = \pi$, which corresponds to the period of the single Stuart-Landau oscillator, they still exist at $t = 5000$ for a certain range of coupling strength $7 < \sigma < 19$ and number of nearest neighbors $2 < P < 20$, see Fig. 1.14b. Moreover, chimera

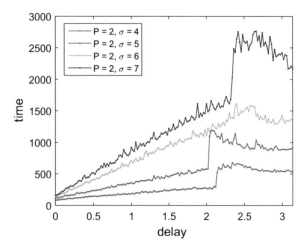

Fig. 1.16 Transient times t_{tr} of amplitude chimeras in dependence on time delay τ for different values of coupling strength $\sigma = 4, 5, 6, 7$. *Other parameters*: $N = 100$, $\omega = 2, \lambda = 1, P = 2$

death patterns with large number of clusters now dominate while the region of 1-cluster chimera death is strongly reduced. Additionally, the in-phase synchronized state observed in the deterministic system for small number of nearest neighbors ($P < 6$) and large coupling strength ($\sigma > 15$) in the presence of time delay is replaced by chimera death patterns with the number of clusters exceeding 9 (red region in Fig. 1.14b).

To compare directly the results for the lifetime of amplitude chimeras we calculate transient times t_{tr} in the plane of number of nearest neighbors P and coupling strength σ for $\tau = 0$ and $\tau = \pi$ (Fig. 1.15a, b, respectively). In both cases chimera death states are the dominating patterns in the phase diagram (black color in Fig. 1.15). The long-living amplitude chimeras ($t_{tr} \gtrsim 5000$) appear only when the links between the nodes include time delay (white region in Fig. 1.15b), while for $\tau = 0$ the lifetime of amplitude chimera is relatively short ($t_{tr} < 900$).

In order to quantitatively characterize the impact of the time-delayed coupling we calculate the lifetime of amplitude chimeras in dependence on time delay for different values of coupling strength $\sigma = 4, 5, 6, 7$ (Fig. 1.16). The transient times increase with time delay for all considered values of σ. Therefore, by appropriately choosing the value of time delay one can realize a desired lifetime of amplitude chimeras.

1.7 Conclusions

We have investigated two types of chimera states for a paradigmatic network of oscillators under the influence of noise and time delay. We have presented numerical results demonstrating that transient amplitude chimeras and chimera death states in a ring network of identical Stuart-Landau oscillators with symmetry-breaking coupling continue to exist in the presence of Gaussian white noise or if time delay is introduced to the coupling.

In the presence of external noise, transient amplitude chimeras occur in the same range of coupling parameters as in the deterministic case. The key quantity we use to characterize them is the transient time. The latter decreases logarithmically with the applied noise intensity. For a constant noise intensity, the transient times increase with the coupling strength up to a saturation value. The amplitude chimera lifetimes depend sensitively on the particular realization of the randomized initial condition. We have introduced a class of specially prepared random initial conditions that produce long lasting amplitude chimeras. We have shown that initial configurations that fulfill a symmetry conditions which is also found in oscillation death patterns result in the longest living amplitude chimera transients.

The chimera death patterns also persist under the impact of stochastic forces. However, the coupling parameter regime where they occur is reduced with increasing noise intensity. The boundary between the coherent oscillatory regime and the chimera death regime is shifted towards higher values of the coupling strength. That means that the system favors oscillatory behavior for a larger coupling parameter regime. In contrast, this boundary appears to be independent on the particular realization of the initial condition. The number of clusters within the coherent domains appears to be unaffected by the external noise, but depends on the particular initial condition.

In the presence of time delay the lifetime of amplitude chimera patterns is essentially enlarged. Moreover, time delay induces amplitude chimeras for the coupling parameter values for which in the absence of time delay no chimera patterns are observed.

Thus, the lifetime of amplitude chimeras can be controlled by tuning the noise intensity and the value of time delay, which, therefore, play the role of control parameters. Noise allows one to decrease the lifetime of amplitude chimeras, while time delay can significantly increases it.

Our numerical findings can be explained by the underlying phase space structure. More specifically, we propose that amplitude chimera states can be represented by saddle states in the phase space of the network. This elucidates the behavior of their lifetime, and explains that generally the initial conditions become less important under the influence of noise.

Acknowledgments This work was supported by DFG in the framework of SFB 910.

References

1. M.J. Panaggio, D.M. Abrams, Chimera states: Coexistence of coherence and incoherence in networks of coupled oscillators. Nonlinearity **28**, R67 (2015)
2. Y. Kuramoto, D. Battogtokh, Coexistence of coherence and incoherence in nonlocally coupled phase oscillators. Nonlin. Phen. in Complex Sys. **5**(4), 380–385 (2002)
3. D.M. Abrams, S.H. Strogatz, Chimera states for coupled oscillators. Phys. Rev. Lett. **93**(17), 174102 (2004)

22   A. Zakharova et al.

4. D.M. Abrams, R.E. Mirollo, S.H. Strogatz, D.A. Wiley, Solvable model for chimera states of coupled oscillators. Phys. Rev. Lett. **101**(8), 084103 (2008)
5. G.C. Sethia, A. Sen, F.M. Atay, Clustered chimera states in delay-coupled oscillator systems. Phys. Rev. Lett. **100**(14), 144102 (2008)
6. C.R. Laing, The dynamics of chimera states in heterogeneous Kuramoto networks. Physica D **238**(16), 1569–1588 (2009)
7. A.E. Motter, Nonlinear dynamics: spontaneous synchrony breaking. Nat. Phys. **6**(3), 164–165 (2010)
8. E.A. Martens, C.R. Laing, S.H. Strogatz, Solvable model of spiral wave chimeras. Phys. Rev. Lett. **104**(4), 044101 (2010)
9. S. Olmi, A. Politi, A. Torcini, Collective chaos in pulse-coupled neural networks. Europhys. Lett. **92**, 60007 (2010)
10. G. Bordyugov, A. Pikovsky, M. Rosenblum, Self-emerging and turbulent chimeras in oscillator chains. Phys. Rev. E **82**(3), 035205 (2010)
11. J.H. Sheeba, V.K. Chandrasekar, M. Lakshmanan, Chimera and globally clustered chimera: impact of time delay. Phys. Rev. E **81**, 046203 (2010)
12. A. Sen, R. Dodla, G. Johnston, G.C. Sethia, Amplitude death, synchrony, and chimera states in delay coupled limit cycle oscillators, in *Complex Time-Delay Systems*, vol. 16, Understanding Complex Systems, ed. by F.M. Atay (Springer, Berlin, 2010), pp. 1–43
13. M. Wolfrum, O.E. Omel'chenko, Chimera states are chaotic transients. Phys. Rev. E **84**(1), 015201 (2011)
14. C.R. Laing, Fronts and bumps in spatially extended Kuramoto networks. Phys. D **240**(24), 1960–1971 (2011)
15. I. Omelchenko, Y. Maistrenko, P. Hövel, E. Schöll, Loss of coherence in dynamical networks: spatial chaos and chimera states. Phys. Rev. Lett. **106**, 234102 (2011)
16. I. Omelchenko, B. Riemenschneider, P. Hövel, Y. Maistrenko, E. Schöll, Transition from spatial coherence to incoherence in coupled chaotic systems. Phys. Rev. E **85**, 026212 (2012)
17. I. Omelchenko, O.E. Omel'chenko, P. Hövel, E. Schöll, When nonlocal coupling between oscillators becomes stronger: patched synchrony or multichimera states. Phys. Rev. Lett. **110**, 224101 (2013)
18. S. Nkomo, M.R. Tinsley, K. Showalter, Chimera states in populations of nonlocally coupled chemical oscillators. Phys. Rev. Lett. **110**, 244102 (2013)
19. J. Hizanidis, V. Kanas, A. Bezerianos, T. Bountis, Chimera states in networks of nonlocally coupled hindmarsh-rose neuron models. Int. J. Bifurcat. Chaos **24**(03), 1450030 (2014)
20. G.C. Sethia, A. Sen, G.L. Johnston, Amplitude-mediated chimera states. Phys. Rev. E **88**(4), 042917 (2013)
21. G.C. Sethia, A. Sen, Chimera states: the existence criteria revisited. Phys. Rev. Lett. **112**, 144101 (2014)
22. A. Yeldesbay, A. Pikovsky, M. Rosenblum, Chimeralike states in an ensemble of globally coupled oscillators. Phys. Rev. Lett. **112**, 144103 (2014)
23. F. Böhm, A. Zakharova, E. Schöll, K. Lüdge, Amplitude-phase coupling drives chimera states in globally coupled laser networks. Phys. Rev. E **91**(4):040901 (R) (2015)
24. A. Buscarino, M. Frasca, L.V. Gambuzza, P. Hövel, Chimera states in time-varying complex networks. Phys. Rev. E **91**(2), 022817 (2015)
25. I. Omelchenko, A. Provata, J. Hizanidis, E. Schöll, P. Hövel, Robustness of chimera states for coupled FitzHugh-Nagumo oscillators. Phys. Rev. E **91**, 022917 (2015)
26. I. Omelchenko, A. Zakharova, P. Hövel, J. Siebert, E. Schöll, Nonlinearity of local dynamics promotes multi-chimeras. Chaos **25**, 083104 (2015)
27. P. Ashwin, O. Burylko, Weak chimeras in minimal networks of coupled phase oscillators. Chaos **25**, 013106 (2015)
28. V. Bastidas, I. Omelchenko, A. Zakharova, E. Schöll, T. Brandes, Quantum signatures of chimera states. Phys. Rev. E **92**, 062924 (2015)
29. N. Semenova, A. Zakharova, E. Schöll, V.S. Anishchenko. Does hyperbolicity impede emergence of chimera states in networks of nonlocally coupled chaotic oscillators? Europhys. Lett. **112**, 40002 (2015)

30. A.M. Hagerstrom, T.E. Murphy, R. Roy, P. Hövel, I. Omelchenko, E. Schöll, Experimental observation of chimeras in coupled-map lattices. Nat. Phys. **8**, 658–661 (2012)
31. M.R. Tinsley, S. Nkomo, K. Showalter, Chimera and phase cluster states in populations of coupled chemical oscillators. Nat. Phys. **8**, 662–665 (2012)
32. E.A. Martens, S. Thutupalli, A. Fourriere, O. Hallatschek, Chimera states in mechanical oscillator networks. Proc. Nat. Acad. Sci. **110**, 10563 (2013)
33. L. Larger, B. Penkovsky, Y. Maistrenko, Virtual chimera states for delayed-feedback systems. Phys. Rev. Lett. **111**, 054103 (2013)
34. T. Kapitaniak, P. Kuzma, J. Wojewoda, K. Czolczynski, Y. Maistrenko, Imperfect chimera states for coupled pendula. Sci. Rep. **4**, 6379 (2014)
35. M. Wickramasinghe, I.Z. Kiss, Spatially organized dynamical states in chemical oscillator networks: synchronization, dynamical differentiation, and chimera patterns. PLoS ONE **8**(11), e80586 (2013)
36. M. Wickramasinghe, I.Z. Kiss, Spatially organized partial synchronization through the chimera mechanism in a network of electrochemical reactions. Phys. Chem. Chem. Phys. **16**, 18360–18369 (2014)
37. L. Schmidt, K. Schönleber, K. Krischer, V. Garcia-Morales, Coexistence of synchrony and incoherence in oscillatory media under nonlinear global coupling. Chaos **24**(1), 013102 (2014)
38. L.V. Gambuzza, A. Buscarino, S. Chessari, L. Fortuna, R. Meucci, M. Frasca, Experimental investigation of chimera states with quiescent and synchronous domains in coupled electronic oscillators. Phys. Rev. E **90**, 032905 (2014)
39. D.P. Rosin, D. Rontani, N.D. Haynes, E. Schöll, D.J. Gauthier, Transient scaling and resurgence of chimera states in coupled Boolean phase oscillators. Phys. Rev. E **90**, 030902(R) (2014)
40. L. Larger, B. Penkovsky, Y. Maistrenko, Laser chimeras as a paradigm for multistable patterns in complex systems. Nat. Comm. **6**, 7752 (2015)
41. C.R. Laing, C.C. Chow, Stationary bumps in networks of spiking neurons. Neural Comput. **13**(7), 1473–1494 (2001)
42. J.M. Davidenko, A.M. Pertsov, R. Salomonsz, W. Baxter, J. Jalife, Stationary and drifting spiral waves of excitation in isolated cardiac muscle. Nature **355**, 349 (1992)
43. A.E. Motter, S.A. Myers, M. Anghel, T. Nishikawa, Spontaneous synchrony in power-grid networks. Nat. Phys. **9**, 191–197 (2013)
44. J.C. Gonzalez-Avella, M.G. Cosenza, M.S. Miguel, Localized coherence in two interacting populations of social agents. Phys. A **399**, 24–30 (2014)
45. N.C. Rattenborg, C.J. Amlaner, S.L. Lima, Behavioral, neurophysiological and evolutionary perspectives on unihemispheric sleep. Neurosci. Biobehav. Rev. **24**, 817–842 (2000)
46. A. Rothkegel, K. Lehnertz, Irregular macroscopic dynamics due to chimera states in small-world networks of pulse-coupled oscillators. New J. Phys. **16**, 055006 (2014)
47. A. Zakharova, M. Kapeller, E. Schöll, Chimera death: symmetry breaking in dynamical networks. Phys. Rev. Lett. **112**, 154101 (2014)
48. F.M. Atay, Distributed delays facilitate amplitude death of coupled oscillators. Phys. Rev. Lett. **91**, 094101 (2003)
49. A. Zakharova, I. Schneider, Y.N. Kyrychko, K.B. Blyuss, A. Koseska, B. Fiedler, E. Schöll, Time delay control of symmetry-breaking primary and secondary oscillation death. Europhys. Lett. **104**, 50004 (2013)
50. A. Zakharova, M. Kapeller, E. Schöll, Amplitude chimeras and chimera death in dynamical networks. J. Phys. Conf. Ser. (2015). arXiv:1503.03371
51. H. Risken, *The Fokker-Planck Equation*, 2nd edn. (Springer, Berlin, 1996)
52. I. Schneider, M. Kapeller, S. Loos, A. Zakharova, B. Fiedler, E. Schöll, Stable and transient multi-cluster oscillation death in nonlocally coupled networks. Phys. Rev. E **92**, 052915 (2015)
53. T. Banerjee, Mean-field diffusion-induced chimera death state. Europhys. Lett. **110**, 60003 (2015)
54. S. Loos, A. Zakharova, J.C. Claussen, E. Schöll. Chimera patterns under the impact of noise. Phys. Rev. E (2016). arXiv:1508.04010v2

Chapter 2
Dynamics of Fully Coupled Rotators with Unimodal and Bimodal Frequency Distribution

Simona Olmi and Alessandro Torcini

Abstract We analyze the synchronization transition of a globally coupled network of N phase oscillators with inertia (rotators) whose natural frequencies are unimodally or bimodally distributed. In the unimodal case, the system exhibits a discontinuous hysteretic transition from an incoherent to a partially synchronized (PS) state. For sufficiently large inertia, the system reveals the coexistence of a PS state and of a standing wave (SW) solution. In the bimodal case, the hysteretic synchronization transition involves several states. Namely, the system becomes coherent passing through traveling waves (TWs), SWs and finally arriving to a PS regime. The transition to the PS state from the SW occurs always at the same coupling, independently of the system size, while its value increases linearly with the inertia. On the other hand the critical coupling required to observe TWs and SWs increases with N suggesting that in the thermodynamic limit the transition from incoherence to PS will occur without any intermediate states. Finally a linear stability analysis reveals that the system is hysteretic not only at the level of macroscopic indicators, but also microscopically as verified by measuring the maximal Lyapunov exponent.

S. Olmi (✉) · A. Torcini
CNR - Consiglio Nazionale delle Ricerche - Istituto dei Sistemi Complessi,
via Madonna del Piano 10, I-50019 Sesto Fiorentino, Italy
e-mail: simona.olmi@fi.isc.cnr.it

S. Olmi · A. Torcini
INFN Sez. Firenze, via Sansone, 1, I-50019 Sesto Fiorentino, Italy
e-mail: alessandro.torcini@cnr.it

A. Torcini
Aix-Marseille Université, Inserm, INMED UMR 901 and Institut de Neurosciences
des Systèmes UMR 1106, 13000 Marseille, France

A. Torcini
Aix-Marseille Université, Université de Toulon, CNRS, CPT, UMR 7332,
13288 Marseille, France

© Springer International Publishing Switzerland 2016
E. Schöll et al. (eds.), *Control of Self-Organizing Nonlinear Systems*,
Understanding Complex Systems, DOI 10.1007/978-3-319-28028-8_2

25

2.1   Introduction

The renowned Kuramoto model [1] for phase oscillators was generalized in 1997 by Tanaka, Lichtenberg and Oishi (TLO) [2, 3] by including an additional inertial term. The TLO model revealed, at variance with the usual Kuramoto model, first order synchronization transitions even for unimodal distributions of the natural frequencies. TLO have been inspired in their extension by a work of Ermentrout published in 1991 [4]; in this paper Ermentrout has introduced a pulse coupled phase oscillator model with inertia to mimic the *perfect synchrony* achieved by a specific type of fireflies, the *Pteroptix Malaccae* (but also by certain species of crickets and humans). The peculiarity of these fireflies is that they are able to synchronize their flashing activity to some forcing frequency (even quite distinct from their own intrinsic flashing frequency) with an almost zero phase lag. This happens because they adapt their period of oscillation to that of the driving oscillator. After its introduction, the Kuramoto model with inertia has been employed to describe synchronization phenomena in crowd synchrony on Londons Millennium bridge [5], as well as in Huygens pendulum clocks [6]. Furthermore, phase oscillators with inertia (rotators) have recently found application in the study of self-synchronization in power and smart grids [7–11], as well as in the analysis of disordered arrays of underdamped Josephson junctions [12]. Cluster explosive synchronization has been reported for an adaptive network of Kuramoto oscillators with inertia, where the natural frequency of each oscillator is assumed to be proportional to the degree of the corresponding node [13]. Rotators arranged in two symmetrically coupled populations have recently revealed the emergence of intermittent chaotic chimeras [14], imperfect chimera states have been found in a ring with nonlocal coupling [15], and transient waves have been observed in regular lattices [16].

There is a wide literature devoted to coupled rotators with an unimodal frequency distribution, however only a really limited number of studies have been devoted to this model with a bimodal distribution, despite the subject being extremely relevant for the modeling of the power grids [8, 9, 17]. To our knowledge the synchronization transition in populations of globally coupled rotators with bimodal distribution has been previously analyzed only by Acebrón et al. in [18]. More specifically, the authors considered a model with white noise and a distribution composed by two δ-functions localized at $\pm\Omega_0$. As suggested in [19], the presence of noise blurs the δ-functions in bell-shaped functions analogous to Gaussian distributions. Therefore one expects a similar phenomenology to the one observable for deterministic systems with bimodal Gaussian distributions of the frequencies. A multiscale analysis of the model, in the limit of sufficiently large Ω_0, reveals the emergence, from the incoherent state, of stable standing wave solutions (SWs) and of unstable traveling wave solutions (TWs) via supercritical bifurcations, while partially synchronized stationary states (PSs) bifurcates subcritically from incoherence. However, the authors affirm that in the considered limit the bifurcation diagram coincides with that of the usual Kuramoto model without inertia [20].

In this article we analyze the synchronization transitions observable for unimodal and bimodal frequency distributions for a population of globally coupled rotators in a fully deterministic system. In particular, we will analyze the influence of inertia and of finite size effects on the synchronization transitions. Moreover, we will study the macroscopic and microscopic characteristics of the different regimes emerging during adiabatic increase and decrease of the coupling among the rotators. In particular, Sect. 2.1 will be devoted to the introduction of the model, of the indicators used to characterize the synchronization transition, and of the different protocols employed to perform adiabatic simulations. The Lyapunov linear stability analysis is introduced in Sect. 2.2.1. The results for unimodal distributions are reported in Sect. 2.3, with a particular emphasis to the TLO mean field theory and its extension to any generic state observable within the hysteretic region (Sect. 2.3.2). The emergence of clusters of locked and whirling oscillators is described in details in Sect. 2.3.3. The dynamics of the network for bimodal distributions is analyzed in Sect. 2.4. In particular, Sect. 2.4.1 is devoted to two non overlapping distributions and Sect. 2.4.2 to largely overlapping Gaussian distributions. Section 2.5 report the result of linear stability analysis for the considered distributions. Finally, in Sect. 2.6 the reported results are briefly summarized and discussed.

2.2 Model and Indicators

By following Refs. [2, 3], we study the following version of the Kuramoto model with inertia for N fully coupled rotators :

$$m\ddot{\theta}_i + \dot{\theta}_i = \Omega_i + \frac{K}{N} \sum_j \sin(\theta_j - \theta_i), \qquad (2.1)$$

where θ_i and Ω_i are, respectively, the instantaneous phase and the natural frequency of the i-th oscillator, K is the coupling. In the following we will consider random natural frequencies Ω_i Gaussian distributed according to: an unimodal distribution $g(\Omega) = \frac{1}{\sqrt{2\pi}} e^{-\frac{\Omega^2}{2}}$ with zero average and an unitary standard deviation or a bimodal symmetric distribution $g(\Omega) = \frac{1}{2\sqrt{2\pi}} \left[e^{-\frac{(\Omega - \Omega_0)^2}{2}} + e^{-\frac{(\Omega + \Omega_0)^2}{2}} \right]$, which is the overlap of two Gaussians with unitary standard deviation and with the peaks located at a distance $2\Omega_0$ (see Fig. 2.4).

To measure the level of coherence between the oscillators, we employ the complex order parameter [21]

$$r(t) e^{i\phi(t)} = \frac{1}{N} \sum_j e^{i\theta_j}, \qquad (2.2)$$

where $r(t) \in [0 : 1]$ is the modulus and $\phi(t)$ the phase of the macroscopic indicator. An asynchronous state, in a finite network, is characterized by $r \simeq \frac{1}{\sqrt{N}}$, while for

$r \equiv 1$ the oscillators are fully synchronized and intermediate r-values correspond to partial synchronization.

Another relevant indicator for the state of the rotator population is the number of locked oscillators N_L, characterized by a vanishingly small average phase velocity $\bar{\omega}_i \equiv \frac{d\bar{\theta}_i}{dt}$, and the maximal locking frequency Ω_M, which corresponds to the maximal natural frequency $|\Omega_i|$ of the locked oscillators.

In general we will perform sequences of simulations by sweeping up/down adiabatically the coupling parameter K with two different protocols. Namely, for the first protocol (I) the series of simulations are initialized for the decoupled system by considering random initial conditions for $\{\theta_i\}$ and $\{\omega_i\}$. Afterwards the coupling is increased in steps ΔK until a maximal coupling K_M is reached. For each value of K, apart the very first one, the simulations is initialized by employing the last configuration of the previous simulation in the sequence. For the second protocol (II), starting from the final coupling K_M achieved by employing the protocol (I), the coupling is swept down in steps ΔK until $K = 0$ is recovered. At each step the system is simulated for a transient time T_R followed by a period T_W during which the average value of the order parameter $\bar{r}$ and of the velocities $\{\bar{\omega}_i\}$, as well as Ω_M, are estimated.

2.2.1 Lyapunov Analysis

The stability of Eq. (2.1) can be analyzed by following the evolution of infinitesimal perturbations $\mathcal{T} = (\delta\dot{\theta}_1, \ldots, \delta\dot{\theta}_N, \delta\theta_1, \ldots, \delta\theta_N)$ in the tangent space, whose dynamics is ruled by the linearization of Eq. (2.1) as follows:

$$m\,\delta\ddot{\theta}_i + \delta\dot{\theta}_i = \frac{K}{N}\sum_{j=1}^{N} \cos\left(\theta_j - \theta_i\right)(\delta\theta_j - \delta\theta_i). \qquad (2.3)$$

We will limit to estimate the maximal Lyapunov exponent λ_M, by employing the method developed by Benettin et al. [22]. This amounts to follow the dynamical evolution of the orbit and of the tangent vector $\mathcal{T}$ for a time lapse T_W by normalizing at fixed time intervals Δt, after discarding an initial transient evolution T_R.

Furthermore, the values of the components of the maximal Lyapunov vector $\mathcal{T}$ can give important information about the oscillators that are more actively contributing to the chaotic dynamics. It is useful to introduce the following squared amplitude component of the normalized vector for each rotator [14, 23]

$$\xi_i(t) = [\delta\dot{\theta}_i(t)]^2 + [\delta\theta_i(t)]^2 , \qquad i = 1, \ldots, N. \qquad (2.4)$$

The time average $\bar{\xi}_i$ of this quantity gives a measure of the contribution of each oscillator to the chaotic dynamics.

2.3 Unimodal Frequency Distribution

2.3.1 Hysteretic Synchronization Transitions

In Fig. 2.1 the results for a sequence of simulations obtained by following protocol (I) and (II) are reported for a not too small inertia (namely, $m = 2$) and unimodal frequency distribution. For the first protocol the system remains incoherent up to a critical value $K = K_1^c \simeq 2$, where $\bar{r}$ jumps to a finite value and then increases with K reaching $\bar{r} \simeq 1$ for sufficiently large coupling. Starting from the last state and by reducing K one notices that $\bar{r}$ assumes larger values than during protocol (I) and the system becomes incoherent at a smaller coupling, namely $K_2^c < K_1^c$. This is a clear indication of the hysteretic nature of the synchronization transition in this case.

For the chosen values of the inertia, we observe the creation of an unique cluster of N_L locked oscillators with $\bar{\omega}_i \simeq 0$, for larger m the things will be more complex, as we will discuss in the following. The maximal locking frequency Ω_M becomes finite for $K > K_1^c$ and increases with K. The frequency Ω_M attains a maximal value when $\bar{r} \simeq 1$, no more oscillators can be recruited in the large locked cluster. Once reached this value, even if K is reduced following the protocol (II), Ω_M remains constant for a wide K interval. Then Ω_M shows a rapid decrease towards zero by approaching K_2^c. This behavior will be explained in the following two sub-sections.

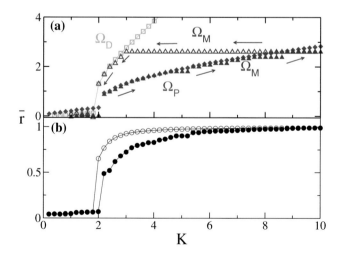

Fig. 2.1 Unimodal frequency distribution. **a** Maximal locking frequency Ω_M (*blue triangles*) and **b** time averaged order parameter $\bar{r}$ (*black circles*) as a function of the coupling K for two series of simulations performed following the protocol (I) (*filled symbols*) and the protocol (II) (*empty symbols*). The data refer to inertia: $m = 2$, for which we set $\Delta K = 0.2$ and $K_M = 10$; moreover $N = 500$, $T_R = 5{,}000$ and $T_W = 200$. The (*magenta*) diamonds indicate $\Omega_P = \frac{4}{\pi}\sqrt{\frac{K\bar{r}}{m}}$ for protocol (I) and the (*green*) squares $\Omega_D = K\bar{r}$ for protocol (II)

2.3.2 Mean-Field Theory

In order to derive a mean field description of the dynamics of each single rotator, we
can rewrite Eq. (2.1) by employing the order parameter definition (2.2) as follows

$$m\ddot{\theta}_i + \dot{\theta}_i = \Omega_i - Kr\sin(\theta_i - \phi), \tag{2.5}$$

this corresponds to the evolution equation for a damped driven pendulum. Equation
(2.5) admits, for sufficiently small forcing frequency Ω_i, two fixed points: a stable
node and a saddle. At larger frequencies $\Omega_i > \Omega_P \simeq \frac{4}{\pi}\sqrt{\frac{Kr}{m}}$ the saddle, via a homoclinic bifurcation, gives rise to a limit cycle. The stable limit cycle and a stable fixed
point coexist until a saddle node bifurcation, taking place at $\Omega_i = \Omega_D = Kr$, leads
to the disappearance of the fixed points and for $\Omega_i > \Omega_D$ only the limit cycle persists. This scenario is correct for sufficiently large inertia; at small m one has a direct
transition from a stable node to a periodic oscillating orbit at $\Omega_i = \Omega_D = Kr$ [24].
Therefore for sufficiently large m there is a coexistence regime where, depending on
the initial conditions, the single oscillator can rotate or stay quiet. The fixed point
(limit cycle) solution corresponds to locked (drifting) rotators.

The TLO theory [2, 3] has explained the origin of the first order hysteretic transitions by considering two opposite initial states for the network: (I) the completely
incoherent phase ($r = 0$) and (II) the completely synchronized one ($r \equiv 1$). In case
(I) the oscillators are all initially drifting with finite velocities ω_i; by increasing K
the oscillators with smaller natural frequencies $|\Omega_i| < \Omega_P$ begin to lock ($\bar{\omega}_i = 0$),
while the other continue to drift. This is confirmed by the data reported in Fig. 2.1,
where it is clear that the locking frequency Ω_M is well approximated by Ω_P. The
process continues until all the oscillators are finally locked, leading to $r = 1$ and to
a plateau in Ω_M.

In the second case, initially all the oscillators are already locked, with an associated
order parameter $r \equiv 1$. Therefore, the oscillators can start to drift only when the
stable fixed point solution will disappear, leaving the system only with the limit
cycle solution. This happens, by decreasing K, whenever $|\Omega_i| \geq \Omega_D = Kr$. This is
numerically verified, indeed, as shown in Fig. 2.1, where it is clear that the maximal
locked frequency Ω_M remains constant until, by decreasing K, it encounters the
curve Ω_D and then Ω_M follows this latter curve down towards the asynchronous
state. The case (II) corresponds to the situation observable for the usual Kuramoto
model, where there is no bistability [1].

In both the considered cases there is a group of desynchronized oscillators and
one of locked oscillators separated by a frequency, Ω_P (Ω_D) in case (I) (case (II)).
At variance with the usual Kuramoto model, both these groups contribute to the total
level of synchronization, namely

$$r = r_L + r_D \tag{2.6}$$

where r_L (r_D) is the contribution of the locked (drifting) population.

The contribution of the locked population is simply given by

$$r_L^{I,II} = Kr \int_{-\theta_{P,D}}^{\theta_{P,D}} \cos^2\theta g(Kr\sin\theta)d\theta, \tag{2.7}$$

where $\theta_P = \sin^{-1}(\Omega_P/Kr)$ and $\theta_D = \sin^{-1}(\Omega_D/Kr) \equiv \pi/2$.

The contribution r_D of the drifting rotators is negative and it has been estimated analytically by TLO by performing a perturbative expansion to the fourth order in $1/(mK)$ and $1/(m\Omega)$. The obtained expression, valid for sufficiently large inertia, reads as

$$r_D^{I,II} \simeq -mKr \int_{\Omega_{P,D}}^{\infty} \frac{1}{(m\Omega)^3}g(\Omega)d\Omega, \tag{2.8}$$

with $g(\Omega) = g(-\Omega)$.

By considering an initially desynchronized (fully synchronized) system and by increasing (decreasing) K one can get a theoretical approximation for the level of synchronization in the system by employing the mean-field expression (2.7), (2.8) and (2.6) for case I (II). In this way, two curves are obtained in the phase plane (K, r), namely $r^I(K)$ and $r^{II}(K)$. For a certain coupling K the system can attain all the possible levels of synchronization between $r^I(K)$ and $r^{II}(K)$.

Let us notice that the expression for r_L and r_D reported in Eqs. (2.7) and (2.8) are the same for case (I) and (II), only the integration extrema change in the two cases. These are defined by the frequency which discriminates locked from drifting oscillator, that in case (I) is Ω_P and in case (II) Ω_D. It should be noticed that the value of these frequencies is a function of the order parameter r and of the coupling constant K, therefore one should solve implicit integrals to obtain r.

However, one could also fix the discriminating frequency to some arbitrary value Ω_0 and solve self-consistently the equations Eqs. (2.6), (2.7), and (2.8) for different values of the coupling K. This corresponds to solve the equation

$$\int_{-\theta_0}^{\theta_0} \cos^2\theta g(Kr^0\sin\theta)d\theta - m\int_{\Omega_0}^{\infty} \frac{1}{(m\Omega)^3}g(\Omega)d\Omega = \frac{1}{K}, \tag{2.9}$$

with $\theta_0 = \sin^{-1}(\Omega_0/Kr^0)$. A solution $r^0 = r^0(K, \Omega_0)$ exists provided that $\Omega_0 \le \Omega_D(K) = r^0 K$. Therefore the part of the plane delimited by the curve $r^{II}(K)$, will be filled with the curves $r_0(K)$ obtained for different Ω_0 values (as shown in Fig. 2.2a). These solutions represent clusters of N_L oscillators for which the maximal locking frequency and N_L do not vary upon changing the coupling strength. In particular, for $K > K_2^c$ these states can be observed in numerical simulations in the portion of the phase space delimited by the two curves $r^I(K)$ and $r^{II}(K)$ (see Fig. 2.2b).

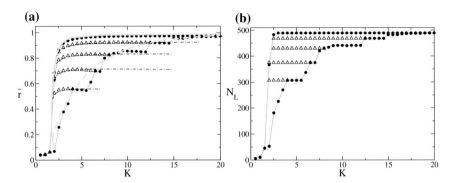

Fig. 2.2 Unimodal Distribution. Panel **a**: Average order parameter $\bar{r}$ versus the coupling constant K. Theoretical mean field estimates: the *dashed* (*solid*) *grey curves* refer to $r^I = r_L^I + r_D^I$ ($r^{II} = r_L^{II} + r_D^{II}$) as obtained by employing Eqs. (2.7) and (2.8) following protocol (I) (protocol (II)); the (*black*) *dot-dashed curves* are the solutions $r^0(K, \Omega_0)$ of Eq. (2.9) for different Ω_0 values. The employed values from *bottom* to *top* are: $\Omega_0 = 0.79$, 1.09, 1.31 and 1.79. *Numerical simulations* (*black*) filled circles have been obtained by following protocol (I) and then (II) starting from $K = 0$ until $K_M = 20$ with steps $\Delta K = 0.5$; (*black*) *empty triangles* refer to simulations performed by starting from a final configuration obtained during protocol (I) and by decreasing the coupling from such initial configurations. The Panel (**b**) displays N_L versus K for the numerical simulations reported in (**a**). The numerical data refer to $m = 6$, $N = 500$, $T_R = 5000$, and $T_W = 200$

2.3.3 Clusters of Locked and Whirling Oscillators

By observing the results reported in Fig. 2.2a for $m = 6$, it is evident that the numerical data obtained by following the procedure (II) are quite well reproduced from the mean field approximation r^{II} (solid grey curve). This is not the case for the theoretical estimation r^I (dashed grey curve), which does not reproduce the step-wise structure revealed for the data corresponding to protocol (I). This step-wise structure emerges only for sufficiently large inertia (as it is clear from Fig. 2.1b, where it is absent for $m = 2$); this is due to the break down of the independence of the whirling oscillators: namely, to the formation of clusters of drifting oscillators moving coherently at the same non zero velocity [2]. Oscillators join in small groups to the locked stationary cluster and not individually as it happens for smaller inertia; this is clearly revealed by the behavior of N_L versus the coupling K as reported in Fig. 2.2b.

Furthermore, once formed, these stationary locked clusters are particularly robust, as it can be appreciated by considering as initial condition a partially synchronized state obtained following protocol (I) for a certain coupling $K_S > K_1$. This state is characterized by a cluster of N_L locked; if now we reduce the coupling K, the number of locked oscillators remain constant until we do not reach the descending curve obtained with protocol (II), see the black empty triangles in Fig. 2.2. On the other hand $\bar{r}$ decreases slightly with K, this behavior is well reproduced by the mean field solutions of Eq. (2.9), namely $r^0(K, \Omega_0)$ with $\Omega_0 = \Omega_P(K_s, r^I(K_S)) = \frac{4}{\pi}\sqrt{\frac{K_s r^I}{m}}$, these are shown in Fig. 2.2a as black dot-dashed lines. As soon as, by decreasing

K, the frequency Ω_0 becomes equal or smaller than Ω_D, the order parameter has a rapid drop towards zero following the upper limit curve r^{II}. These results indicate that hysteretic loops of any size are possible within the region delimited by the two curves $\bar{r}$ obtained by following protocol I and II respectively, as shown in [17].

For sufficiently small m, the synchronization occurs starting from the incoherent state via the formation of an unique cluster of locked oscillators, and the size of this cluster increases with K, as evident from Fig. 2.3a for $m = 2$. At the same time the value of r also increases with K and its evolution is characterized only by finite size fluctuations vanishing in the thermodynamic limit (see the inset of Fig. 2.3a). As already mentioned, the situation is different for sufficiently large inertia, now the partially synchronized phase is characterized by the coexistence of the main cluster of locked oscillators with $\bar{\omega}_i \simeq 0$, but also by the emergence of clusters composed by drifting oscillators with common finite velocities, see the data for $\bar{\omega}_i$ reported in Fig. 2.3b for $m = 6$. In particular, the clusters of whirling oscillators emerge always in couple and they are characterized by the same average velocity but opposite sign. These states are indicated as standing waves (SWs), therefore we have a SW coexisting with a partially synchronized stationary state (PS) (as shown in Fig. 2.3b).

The effect of these extra clusters on the collective dynamics is to induce oscillations in the temporal evolution of the order parameter, as one can see from the inset of Fig. 2.3b. In presence of drifting clusters characterized by the same average velocity (in absolute value), as for $m = 6$ and $K = 5$ in Fig. 2.3b, r exhibits almost regular oscillations and the period of these oscillations corresponds to the one associated to the oscillators in the drifting cluster.

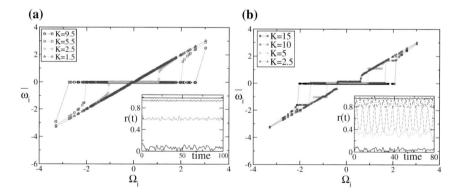

Fig. 2.3 Unimodal Distribution. Average phase velocity $\bar{\omega}_i$ of the rotators versus their natural frequencies Ω_i for $N = 500$ and inertia $m = 2$ (**a**) and $m = 6$ (**b**). In panel (**a**) (panel (**b**)) *magenta triangles* refer to $K = 1.5$ ($K = 2.5$), *green diamonds* to $K = 2.5$ ($K = 5$), *red squares* to $K = 5.5$ ($K = 10$) and *black circles* to $K = 9.5$ ($K = 15$). The *insets* report the time evolution of the order parameters $r(t)$ for the corresponding coupling constants, apart for the extra *blue line* shown in the inset in (**b**) which refers to $K = 1$. For each simulation an initial transient $T_R \simeq 5,500$ has been discarded and the time averages have been estimated over a window $T_W = 5,000$

2.4 Bimodal Distribution

In this section we consider a bimodal distribution, we will initially focus on two
almost non overlapping Gaussians (see Fig. 2.4c), namely we consider $\Omega_0 = 2$,
while Sect. 2.4.2 is devoted to overlapping Gaussians (see Fig. 2.4b), examined for
$\Omega_0 = 0.2$.

2.4.1 Non-overlapping Gaussians

For $\Omega_0 = 2$ and sufficiently small inertia ($m = 1$ and 2), we observe a very rich synchronization transition, as shown in Fig. 2.5a. In particular, by following protocol
(I) we observe that the system leaves the incoherent state abruptly by exhibiting a
jump to a finite $\bar{r}$ value at K^{TW}; above such value in the network emerges a single
cluster of oscillators, drifting together with a finite velocity $\simeq \Omega_0$, this corresponds
to Traveling Wave (TW) solution. By further increasing K a second finite jump of
the order parameter at K^{SW} denotes the passage to a Standing Wave (SW) solution, corresponding to two clusters of drifting oscillators with symmetric opposite
velocities $\simeq \pm\Omega_0$. A final jump at K^{PS} leads the system to a Partially Synchronized
(PS) phase, characterized by an unique cluster of locked rotators with zero average
velocity. By increasing the coupling the PS state smoothly approaches the fully synchronized regime. Starting from this final state the return sequence of simulations,
following protocol (II), displays a simpler phenomenology. The network stays in the
PS regime, characterized by an order parameter larger than that measured during
protocol (I) simulations, until $K^{DS} < K^{TW}$. For smaller coupling, the system leaves
the PS state; however, depending on the realization of the natural frequencies and on

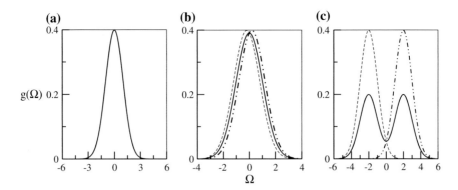

Fig. 2.4 Panel (**a**): Unimodal frequency distribution centered in $\Omega = 0$. Panel (**b**): The *solid line*
represents a bimodal frequency distribution for two overlapping Gaussians (*dashed and dot-dashed
lines*) and $\Omega_0 = 0.2$. Panel (**c**): The *solid line* represents bimodal frequency distribution for non
overlapping Gaussians (*dashed and dot-dashed lines*) and $\Omega_0 = 2$

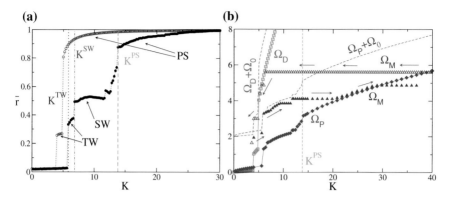

Fig. 2.5 Bimodal frequency distribution. Panel (**a**): Average order parameter $\bar{r}$ versus K for two series of simulations performed following the protocol (I) (*filled symbols*) and (II) (*empty symbols*). The *dotted vertical blue line* refers to K^{TW}; the *dashed-dotted magenta line* to K^{SW}. Panel (**b**): Maximal locking frequency Ω_M (*blue triangles*) versus K for simulations reported in (a) for protocol (I) (*filled symbols*) and (II) (*empty symbols*). The *magenta diamonds* indicate $\Omega_P = \frac{4}{\pi}\sqrt{\frac{K\bar{r}}{m}}$ for protocol (I) and the (*green*) *squares* $\Omega_D = K\bar{r}$ for protocol (II). The *dashed magenta line* represents the *curve* $\Omega_P + \Omega_0$ and the *dashed green curve* $\Omega_D + \Omega_0$. In both panels the *dashed orange vertical line* denotes the critical value K^{PS}. The data refer to $m = 2$, $\Omega_0 = 2$, $N = 2000$, $T_R = 10,000$ and $T_W = 200$, for the sequence of simulations we employed $\Delta K = 0.4$ until $K = 14.8$ and $\Delta K = 0.4$ for $14.8 < K < 39.8$

the initial conditions, it can ends up or in a TW (most of the cases) or in a SW, or it can even reach directly the incoherent state (as shown in Figs. 2.5a and 2.7a). This first analysis clearly shows hysteretic effects and coexistence of macroscopic states with different level of synchronization for a wide range of couplings.

Let us now try to examine the observed transitions in terms of the maximal locking frequency, Ω_M. This frequency is now defined in a different way with respect to the unimodal distribution, in the present case Ω_M represents the maximal absolute value of the natural frequencies of the oscillators belonging to the main clusters present in the system, therefore in this estimation are considered both stationary and drifting clusters. As shown in Fig. 2.5b, Ω_M increases with K for protocol (I) simulations. In particular Ω_M shows a finite jump in correspondence of $K = K^{TW}$, and then its evolution is reasonably well approximated by the curve $\Omega_P + \Omega_0$, where $\Omega_P = \frac{4}{\pi}\sqrt{\frac{K\bar{r}}{m}}$. By approaching K^{PS} the maximal frequency displays a constant plateau which extends beyond K^{PS}, this indicates that the two symmetric drifting clusters merge at $K = K^{PS}$ giving rise to an unique locked cluster with zero average velocity, however no other oscillators join this cluster up to a larger coupling. Whenever this happens, Ω_M starts again to increase, but this time it follows the curve $\Omega_P = \frac{4}{\pi}\sqrt{\frac{K\bar{r}}{m}}$. Finally, for $\bar{r} \simeq 1$ the maximal locking frequency attains a maximal value. Moreover, by reducing the coupling, following now protocol (II), Ω_M remains stacked to such a value for a wide K interval. The fully synchronized cluster is difficult to break down due to the inertia effects. Finally, Ω_M reveals a rapid decrease towards zero whenever it encounters the curve $\Omega_D = K\bar{r}$, initially it follows this curve, however as soon as the system desynchronizes towards a TW the decrease of Ω_M is better described by the curve $\Omega_D + \Omega_0$. The observed behavior can be explained by the fact that for $K < K^{PS}$ ($K < K^{(DS)}$) for protocol (I) (protocol (II)), the network behaves as two independent sub-networks each characterized by an unimodal frequency distribution, one centered at Ω_0 and the other one at $-\Omega_0$. The extension of the analysis reported in Sect. 2.3.2 for an unimodal distribution not centered around zero simply amounts to shift the limiting curves Ω_P and Ω_D by Ω_0. However, for sufficiently large coupling constant, once the system exhibits only one large cluster with zero velocity, the network behaves as a single entity and Ω_M closely follows Ω_P or Ω_D as for a single unimodal distribution centered in zero.

Let us now describe the TW and SW states in more details with the help of the examples reported in Fig. 2.6 for inertia $m = 2$ and $N = 2000$. The TW is an asymmetrical cluster of whirling oscillators with a finite velocity $\bar{\omega}_i \simeq \Omega_0$, in particular in Fig. 2.6a the oscillators have natural frequencies in a range around Ω_0, namely $0.67 \leq \Omega_i \leq 3.34$. The effect of this cluster on the collective dynamics is to increase the average value of the order parameter without inducing any clear oscillating behavior in $r(t)$. However, oscillators with positive natural frequencies are much more synchronized with respect to the ones with negative frequencies, as can be inferred by observing the order parameter r_p (r_n) estimated only on the sub-population of oscillators with positive (negative) natural frequencies and reported in Fig. 2.6b. We believe that the emergence of the TW state is related to the finite sampling of the distribution of the natural frequencies, which due to finite size effects

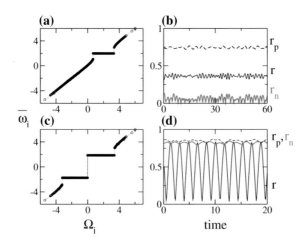

Fig. 2.6 Bimodal frequency distributions. Average phase velocity $\bar{\omega}_i$ of the rotators versus their natural frequencies Ω_i for coupling strength K$=$6.2 (**a**) and K$=$6.8 (**c**). Panels (**b**) and (**d**) display the order parameter $r(t)$ (*black line*) versus time for the same coupling constants as in (**a**) and (**c**), respectively. The *dashed black* (*continuous grey*) line denotes the time evolution of r_p (r_n). For each simulation an initial transient time $T_R = 1000$ has been discarded and the average are estimated over a time interval $T_W = 200$. In both cases $m = 2$, $N = 2000$ and $\Omega_0 = 2$

can be non perfectly symmetric. The asymmetric cluster will emerge around $+\Omega_0$ ($-\Omega_0$) depending on the positive (negative) sign of the average natural frequency.

The SWs are observable at larger coupling constants, this is characterized by two symmetrical clusters with opposite average velocities $\simeq \pm\Omega_0$, as shown in Fig. 2.6c. The presence of these two clusters now induces clear periodic oscillations in the order parameter $r(t)$, as observable in Fig. 2.6d. The period of the oscillations is related to $|\Omega_0|$, i.e. the average frequency of the clustered oscillators. However, at variance with the results reported in Fig. 2.3b for the unimodal distribution the two symmetric clusters do not coexist with a cluster of locked oscillators with zero average velocity. By examining separately r_p and r_n reported in Fig. 2.3d, we notice that each sub-population is much more synchronized than the global one, in fact r_p and r_n have a higher average value than r with superimposed irregular oscillations.

The data reported so far refer to a single system size, however finite size effects are quite relevant for this model, as shown in [17] for unimodal distributions. In Fig. 2.7a we report the synchronization transition for several system sizes, namely $1,000 \leq N \leq 50,000$ for a small inertia value ($m = 1$). We observe that K^{TW} and K^{SW} increase with the size N; in particular, the incoherent state is observable on a wider coupling interval by increasing N (similarly to what reported in [17] for unimodal distributions). Finite size fluctuations induce transitions from the incoherent branch to the TW branch and from this to the SW branch. The fact that we do not observe transition back to the original states indicates that the energy barriers are higher from these sides. A quite astonishing result is the fact that the transition value K^{PS} and K^{DS} seem completely independent from N. The combination of these results seem to suggest that in the thermodynamic limit the incoherent state will loose stability at K^{PS} and therefore the two branches corresponding to TW and SW will be no more visited, at least by following protocol (I).

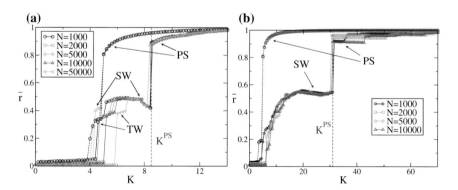

Fig. 2.7 Bimodal frequency distribution with $\Omega_0 = 2$. Average order parameter $\bar{r}$ versus K for various system sizes N: **a** $m = 1$, **b** $m = 6$. The numerical data have been obtained by following protocol (I) and then protocol (II) from $K = 0$ up to $K_M = 20$ ($K_M = 200$) for inertia m $=1$ (m$=6$) with $\Delta K = 0.2$ ($\Delta K = 0.5$). The *vertical dashed blue line* refers to K^{PS}. The average has been performed over a time window $T_W = 200$, after discarding a transient time $T_R = 5,000-50,000$ depending on the system size; the larger T_R have been employed for the larger N

By observing all the data reported in Fig. 2.7a for various N and for protocol (I) and (II), it seems that there are clear indications that the two branches of solutions, corresponding to TW and SW, emerge via a supercritical bifurcation at the same coupling, namely $K \simeq 3.8$, while the transition to PS is clearly subcritical. These results confirm the analysis reported in [18] for a system with noise (in particular, see Fig. 17 in that paper). However, Acebrón et al. affirm that the SW is stable, while the TW is unstable. From our results, both branches seem to become inaccessible (in absence of noise) from the incoherent state, while at least a part of these branches appear to be reachable from the PS state by decreasing K below K^{DS} following protocol (II). Another important difference with respect to the results reported in [18] is that the PS regime is clearly hysteretic revealing two coexisting branches of PS states visited by following protocol (I) or (II).

As shown in Fig. 2.7b, for larger inertia (namely, $m = 6$), the transition from the incoherent state following protocol (I) occurs via the emergence of many small clusters leading finally to a SW state. In this case the critical value at which the incoherent state looses stability seems to saturate to a constant value already for $N \geq 2,000$. The value of K^{PS} is also in this case insensible to the system size. For large inertia values, the TWs seem no more observable.

As a further aspect, we will report the numerical results of the dependence on the inertia of the critical coupling constant K^{PS}, while the value of $K^{DS} \simeq 4.9$ is independent not only by N, but also by the inertia. As shown in Fig. 2.8, K^{PS} increases linearly with the inertia and this scaling is already valid for not too large inertia values. The linear scaling with the inertia is analogous to the scaling recently found within a theoretical mean-field analysis for the coupling K_1^{MF}, which delimits the range of linear stability of the asynchronous state [18, 25]. In particular, the authors in [17] have shown for a Gaussian unimodal distribution of width σ that $K_1^{MF} \simeq$

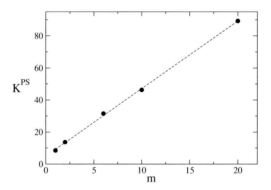

Fig. 2.8 Critical value K^{PS} as a function of the inertia. The *dashed line* represents the fit of the numerical data and indicates a linear increasing of K^{PS} as a function of the inertia being the fit $K^{PS} = 4(1.245 + 1.0525m)$. For all cases $N = 2000$, $\Omega_0 = 2$. The data have been obtained by employing protocol (I) and for each simulation an initial transient time $T_R = 5000$ has been discarded and data are averaged over a time $T_W = 200$

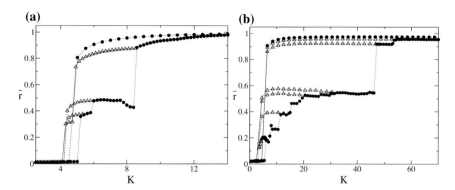

Fig. 2.9 Bimodal frequency distribution. Average order parameter $\bar{r}$ versus the coupling constant K for $m = 1$ and $N = 10{,}000$ (panel (**a**)) and for $m = 10$ and $N = 2{,}000$ (panel (**b**)). The *filled circles* have been obtained by following protocol (I) and then (II) starting from $K = 0$ until $K_M = 20$ ($K_M = 100$) with steps $\Delta K = 0.2$ ($\Delta K = 0.5$); the *empty triangles* refer to simulations performed by starting from a final configuration obtained during protocol (I) and by decreasing the coupling from such initial configurations. The numerical data refer to $\Omega_0 = 2$, $T_R = 50{,}000$ ($T_R = 5000$), and $T_W = 2000$

$2\sigma(0.64 + m\sigma)$, which shows a linear dependence on the inertia and a quadratic dependence on the variance of the frequency distribution.

In the final part of this sub-section we perform an analysis analogous to that reported in Sect. 2.3.3, in particular starting from states with a finite level of synchronization obtained by following protocol (I) we decrease the coupling and observe how these states evolve. In Fig. 2.9, we report the results of these simulations (shown as empty triangles) for two different inertia values, namely $m = 1$ and $m = 10$. Starting from PS states we observe that the cluster survives until the descending curve obtained with protocol (II) is encountered, analogously to the results reported in Fig. 2.3 for the unimodal distribution. Therefore any part of the hysteretic portion of the (K, r)-plane delimited by the PS curves obtained via protocol (I) or (II) is accessible. However, if one starts for $m = 1$ from a TW or a SW state, one observes only two curves (corresponding to the TW and SW branches previously discussed) which seem to end up at the same critical coupling which is smaller than K^{DS}. Therefore it seems that there are no evidences of hysteresis for this small inertia for SW and TW solutions (as shown in Fig. 2.9a). For large inertia values $m = 10$, since now, apart the SW solutions, there are solutions with many small clusters, the situations is much more complex. By starting from different values of $K < K^{PS}$ and by decreasing K, these curves seem all to end up at the same critical coupling smaller than K^{PS}, see Fig. 2.9b. These results suggest that for large inertia values is possible to observe a continuum of possible states even starting from states characterized by (many) drifting clusters and that these states coexist in a wide range of coupling.

2.4.2 Overlapping Gaussians

In this sub-section, we analyze a bimodal distribution, where the two Gaussians are largely overlapping, since $\Omega_0 = 0.2$. In this case we expect to observe a phenomenology of the synchronization transition quite similar to the one seen for the unimodal case. In Fig. 2.10a is reported the average order parameter $\bar{r}$ versus the coupling constant K estimated by following protocol (I) and (II) for various inertia values and for a fixed system size, namely $N = 10,000$. We observe that all the curves obtained for protocol (II) almost overlap irrespectively of the used inertia, while the protocol (I) curves reveal a strong dependence on m. In particular, the hysteretic region widens with m. For small inertia values, namely $m = 1$ and 2, there is a sudden transition from the asynchronous state to a PS state at K^{PS} and neither traveling waves nor standing waves are observable: a single cluster at zero velocity emerges in correspondence of K^{PS} and the order parameter never shows oscillating behavior in time.

For $m = 6$, it is possible to observe a scenario similar to the one reported in Fig. 2.3b, where not only a cluster at zero velocity is present, but also two symmetrical clusters at finite velocities emerge. In particular, following protocol (I) for $K > 2.4$ a small cluster of locked oscillators emerges; at larger coupling, namely $K \geq 3$, two symmetrical clusters of whirling oscillators emerge and coexist with the zero velocity cluster. Finally, at $K = 10.8$ the PS regime arises, corresponding to a single large cluster of locked oscillators. Furthermore, in the range $3 \leq K < 10.8$ the order parameter reveals irregular oscillations. A more detailed analysis is needed to understand the origin of these oscillations as done in the next section.

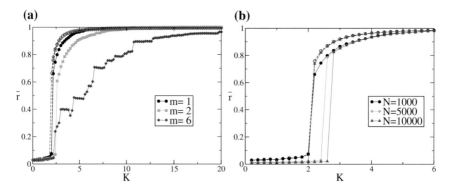

Fig. 2.10 Bimodal frequency distribution for $\Omega_0 = 0.2$. Panel (**a**): Average order parameter $\bar{r}$ versus K for various inertia values and N = 1000. The numerical data have been obtained by following protocol (I) and then protocol (II) from $K = 0$ up to $K_M = 20$ for all inertia values with $\Delta K = 0.2$. Panel (**b**): Average order parameter $\bar{r}$ versus the coupling constant K for various system sizes N and $m = 1$. Data have been obtained by averaging the order parameter over a time window $T_W = 200$, after discarding a transient time $T_R = 5,000$–50,000 depending on the system size

Finally, we examine the influence of the system size on the studied transitions: for $m = 1$ the results for the protocol (I) [protocol (II)] simulations are reported in Fig. 2.10b for sizes ranging from $N = 1,000$ to $N = 10,000$. It is immediately evident that the transition from synchronized state to the asynchronous state, following protocol (II), does not depend on N: for all considered sizes the transition happens in correspondence of $K \simeq 2$, analogously to what reported for unimodal distributions [17]. Starting from the incoherent regime and following protocol (I) the system reveals a jump to a finite $\bar{r}$ value for critical couplings increasing with N, quite similar once more to the results reported for unimodal distributions. We can conclude this sub-section by affirming that the phenomenology seen for bimodal, but largely overlapping, distributions should not differ much from the one observed for unimodal distributions.

2.5 Linear Stability Analysis

To better characterize the synchronization transitions and the stability of the observed states it is worth estimating the maximal Lyapunov exponent λ_M following protocol (I) and (II) for an unimodal and a bimodal distributions. This quite time consuming analysis has been performed for a single inertia value $m = 6$ and a single system size $N = 1,000$, the scaling of λ_M with N will be discussed in the following for specific coupling constant values.

In general, we observe that once the system fully synchronizes, λ_M vanishes; therefore for most of the simulations associated to protocol (II) corresponding to fully synchronized cluster down to the desynchronization transition, λ_M is zero. This is not the case for protocol (I) simulations which reveal a positive λ_M as soon as $\bar{r}$ is non zero. Thus indicating that not only the dynamics characterized in terms of the macroscopic order parameter $\bar{r}$ is hysteretic, but also at the level of the microscopic dynamics, investigated via λ_M, the system has a clear hysteretic behavior.

The behavior of λ_M with K exhibits chaotic dynamical states with windows of regularity for both unimodal and bimodal distribution with $\Omega = 2$, as shown in Fig. 2.11a, b. As a general aspect, we observe the maximal level of chaoticity immediately after the transition from the incoherent state to partially coherence, where small clusters of synchronized oscillators and drifting oscillators coexist. The increase of $\bar{r}$ is accompanied by a trend of λ_M to decrease and finally to vanish for $\bar{r} \to 1$.

An important aspect to understand is if this dynamics is *weakly* chaotic or not, in particular this amounts to verify if, in the thermodynamic limit, λ_M will vanish or will remain finite. In order to test for this aspect, we have considered a configuration obtained by following protocol (I) for a specific coupling and analyzed λ_M versus the system size for $200 \leq N \leq 32,000$. The results for unimodal distributions, as well as for bimodal ones with $\Omega_0 = 2$ and $\Omega_0 = 0.2$ are shown in Fig. 2.11e. It is clear for all the considered cases that the system remains chaotic for diverging system sizes.

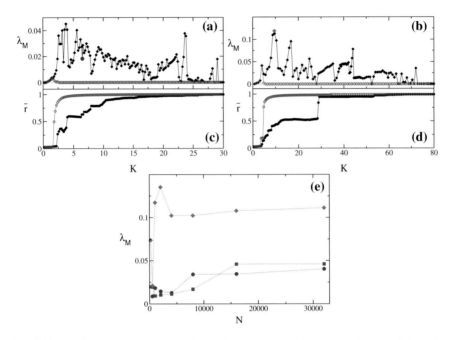

Fig. 2.11 Maximal Lyapunov exponent λ_M and the average order parameter $\bar{r}$ versus K for unimodal (bimodal with $\Omega_0 = 2$) are shown in panel (**a**) (panel (**b**)) and in panel (**c**) (panel (**d**)), respectively. The numerical data have been obtained by following protocol (I) (*black circles*) and then protocol (II) (*red diamonds*) from $K = 0$ up to $K_M = 30$ ($K_M = 80$) with $\Delta K = 0.2$ ($\Delta K = 0.8$). For panels (**a**), (**c**) $T_R = 500$, and $T_W = 50{,}000$; for panels (**b**), (**d**) $T_R = 500$, and $T_W = 400{,}000$. The *different symbols* in (**a**) and (**b**) denote the value for which the further analysis reported in panel (**e**) has been done. Panel (**e**): λ_M versus N for different couplings and frequency distributions. *Blue circles* refer to unimodal distribution and coupling constant $K = 6.5$; *magenta squares* (*green diamonds*) refer to binomial distributions with $\Omega_0 = 0.2$ and $K = 6.7$ ($\Omega_0 = 2$ and $K = 9.5$). λ_M has been averaged over a time window $T_W = 4{,}000$–$400{,}000$, after discarding a transient time $T_R = 1{,}000$–$10{,}000$ depending on the system size. For all panels $m = 6$ and $N = 1{,}000$.

As a final aspect we would like to understand which oscillators contribute more to the chaotic dynamics of the system; this can be understood by measuring the average squared amplitude of the components of the maximal Lyapunov vector $\bar{\xi}_i$ (see the definition reported in Eq. 2.4). In particular, we consider the three cases analyzed in Fig. 2.11e for $N = 1{,}000$. The corresponding results are shown in Fig. 2.12. From panel (a) and (b) of the figure it is clear that for the unimodal distribution, as well as for the largely overlapping bimodal distributions, the chaotic activity is associated almost exclusively to the rotators which are outside the large clusters of locked oscillators with $\bar{\omega}_i \simeq 0$. Thus confirming recent results reported for two coupled populations of rotators with identical natural frequencies [14].

However, the situation for the bimodal distribution with $\Omega_0 = 2$ is different; in particular, as shown in Fig. 2.12c, the network for this large value of the inertia and the considered coupling does not exhibit a cluster of locked oscillators with $\bar{\omega}_i \simeq 0$,

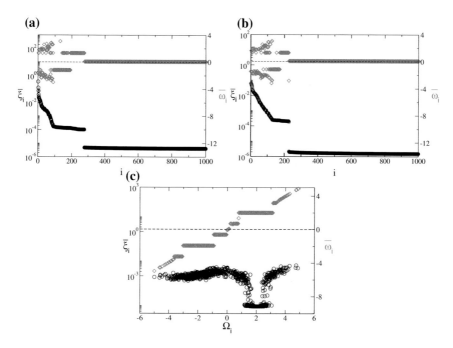

Fig. 2.12 Average squared amplitude of the components of the maximal Lyapunov vector $\bar{\xi}_i$ (*black circles*) and average frequencies $\bar{\omega}_i$ (*grey diamonds*) versus the rotator index for unimodal (**a**) and bimodal distribution with $\Omega_0 = 0.2$ (**b**). In both panels the oscillators are rdered according to the values of $\bar{\xi}_i$. Panel (**c**): $\bar{\xi}_i$ (*black circles*) and average frequencies $\bar{\omega}_i$ (*grey diamonds*) of the oscillators as a function of their natural frequency for bimodal distribution with $\Omega_0 = 2$. Oscillators are ordered according to the values of $\bar{\omega}_i$ and not to the values of $\bar{\xi}_i$ as in panels (**a**), (**b**). Panel (**a**) refers to coupling constant $K = 6.5$, panel (**b**) to $K = 6.7$ and panel (**c**) to $K = 9.5$. For all panels inertia $m = 6$ and $N = 1,000$, $T_R = 500$, and $T_W = 400,000$

but only drifting clusters. In this case the rotators outside and inside the clusters seem to contribute equally to the maximal Lyapunov vector, with the possible exclusion of a group of rotators with $\Omega_i \simeq \Omega_0$.

2.6 Conclusions

We have studied the synchronization transition for a globally coupled Kuramoto model with inertia for different frequency distributions. For the unimodal frequency distribution we have shown that clusters of locked oscillators of any size coexist within the hysteretic region. This region is delimited by two curves in the plane defined by the coupling and the average value of the order parameter. Each curve corresponds to the *synchronization* (*desynchronization*) profile obtained starting from the fully desynchronized (synchronized) state. For sufficiently large inertia values, clusters composed by drifting oscillators with opposite velocities (standing wave state) emerge in addition to the locked oscillators clusters. The presence of clusters of whirling rotators induces oscillatory behavior in the order parameter.

For bimodal frequency distribution the scenario can become more complex since it is possible to play with an extra parameter: the distance between the peaks of the distributions. For simplicity we have analyzed only two cases: largely overlapping distributions ($\Omega_0 = 0.2$), and almost not overlapping distributions ($\Omega_0 = 2$). The phenomenology observed for $\Omega_0 = 0.2$ resembles strongly that found for the unimodal distribution. The analysis of the non overlapping case reveals new interesting features. In particular, the transition from incoherence to coherence occurs via several states: namely, traveling waves, standing waves and finally partial synchronization. This scenario resembles that reported for the usual Kuramoto model for a bimodal distribution [19, 26, 27]. However, in our case the transition is always largely hysteretic, and for non overlapping distributions, traveling waves are clearly observable at variance, not only with the results for the Kuramoto model [19, 26, 27], but also with the theoretical phase diagram reported in [18] for oscillators with inertia. A peculiar aspect is that in the thermodynamic limit we expect a direct discontinuous jump from the incoherent to the coherent phase, without passing through any intermediate state. The critical coupling K^{PS} required to pass from incoherence to partial synchronization is independent of the system size and grows linearly with inertia, while the partially synchronized state looses its stability at a smaller coupling $K^{DS} < K^{PS}$ which is the same for any inertia value and system size.

Furthermore, by performing a linear stability analysis we have been able to show that the hysteretic behavior is not limited to macroscopic observables, as the level of synchronization, but it is revealed also by microscopic indicators as the maximal Lyapunov exponent. In particular, we expect that in a large interval of coupling values chaotic and non chaotic states will coexist.

Finally, it would be challenging to apply optimal control schemes to increase stability and/or to enhance synchronization in a network of oscillators. As a first step time-delay in the coupling can be used to control the collective dynamics, and in particular to stabilize the synchronization of clusters or groups of oscillators, similarly to what done in [28, 29]. In order to control the synchronization of a whole globally coupled system, a time delayed mean field term can be fed-back into the ensemble [30] or an adaptive control strategy can be implemented with non-stationary, time-varying parameters [31]; in particular tuning the amplification and the delay of the feed-back loop will be possible to change the critical point at which the system synchronizes.

However, the onset of synchronization is mostly desired in realistic systems, like power grids, characterized by a sparse connectivity and a bimodal frequency distribution [9]. If a realistic, diluted network is considered, it will be worth investigating the presence of possible mesoscale symmetries, since mesoscale subgraphs can have precise and distinct consequences for the system-level dynamics [32]. In particular, if mesoscale symmetries are present, dynamical instabilities associated with these subgraphs can be analyzed without considering the topology of the embedding network, even though the instabilities generally do not remain confined to the subgraphs.

Therefore the application of adaptive control to engineer mesoscale structure will play a central role in influencing the entire network behavior, thus resulting fundamental also to control decentralized, widely distributed power grids.

Acknowledgments We would like to thank E.A. Martens, D. Pazó, E. Montbrió, M. Wolfrum for useful discussions. We acknowledge partial financial support from the Italian Ministry of University and Research within the project CRISIS LAB PNR 2011-2013. This work is part of of the activity of the Marie Curie Initial Training Network 'NETT' project # 289146 financed by the European Commission. A.T. has also been supported by the A*MIDEX grant (No. ANR-11-IDEX-0001-02) funded by the French Government "program Investissements d'Avenir".

References

1. Y. Kuramoto, *Chemical Oscillations, Waves, and Turbulence* (Courier Dover Publications, Mineola, 2003)
2. H.A. Tanaka, A.J. Lichtenberg, S. Oishi, Phys. Rev. Lett. **78**(11), 2104 (1997)
3. H.A. Tanaka, A.J. Lichtenberg, S. Oishi, Phys. D: Nonlinear Phenom. **100**(3), 279 (1997)
4. B. Ermentrout, J. Math. Biol. **29**(6), 571 (1991)
5. S.H. Strogatz, D.M. Abrams, A. McRobie, B. Eckhardt, E. Ott, Nature **438**(7064), 43 (2005)
6. M. Bennett, M.F. Schatz, H. Rockwood, K. Wiesenfeld, Proc. Math. Phys. Eng. Sci. **458**, 563 (2002)
7. F. Salam, J.E. Marsden, P.P. Varaiya, IEEE Trans. Circ. Syst. **31**(8), 673 (1984)
8. G. Filatrella, A.H. Nielsen, N.F. Pedersen, Eur. Phys. J. B **61**(4), 485 (2008)
9. M. Rohden, A. Sorge, M. Timme, D. Witthaut, Phys. Rev. Lett. **109**(6), 064101 (2012)
10. F. Dörfler, M. Chertkov, F. Bullo, Proc. Nat. Acad. Sci. **110**(6), 2005 (2013)
11. T. Nishikawa, A.E. Motter, J. Phys. **17**(1), 015012 (2015)
12. B. Trees, V. Saranathan, D. Stroud, Phys. Rev. E **71**(1), 016215 (2005)
13. P. Ji, T.K.D. Peron, P.J. Menck, F.A. Rodrigues, J. Kurths, Phys. Rev. Lett. **110**, 218701 (2013)
14. S. Olmi, E.A. Martens, S. Thutupalli, A. Torcini, Phys. Rev. E **92**(3), 030901 (2015)
15. P. Jaros, Y. Maistrenko, T. Kapitaniak, Phys. Rev. E **91**(2), 022907 (2015)
16. D.J. Jörg, Chaos: an interdisciplinary. J. Nonlinear Sci. **25**(5), 053106 (2015)
17. S. Olmi, A. Navas, S. Boccaletti, A. Torcini, Phys. Rev. E **90**(4), 042905 (2014)
18. J. Acebrón, L. Bonilla, R. Spigler, Phys. Rev. E **62**(3), 3437 (2000)
19. E.A. Martens, E. Barreto, S. Strogatz, E. Ott, P. So, T. Antonsen, Phys. Rev. E **79**(2), 026204 (2009)
20. J. Acebrón, R. Spigler, Phys. Rev. Lett. **81**(11), 2229 (1998)
21. A. Winfree, *The Geometry of Biological Time* (Springer, New York, 1980)
22. G. Benettin, L. Galgani, A. Giorgilli, J.M. Strelcyn, Meccanica **15**(1), 9 (1980)
23. F. Ginelli, K.A. Takeuchi, H. Chaté, A. Politi, A. Torcini, Phys. Rev. E **84**(6), 066211 (2011)
24. S.H. Strogatz, *Nonlinear Dynamics and Chaos: with Applications to Physics, Biology, Chemistry, and Engineering* (Westview press, Boulder, 2014)
25. S. Gupta, A. Campa, S. Ruffo, Phys. Rev. E **89**(2), 022123 (2014)
26. J.D. Crawford, J. Stat. Phys. **74**(5–6), 1047 (1994)
27. D. Pazó, E. Montbrió, Phys. Rev. E **80**(4), 046215 (2009)
28. C.U. Choe, T. Dahms, P. Hövel, E. Schöll, Phys. Rev. E **81**(2), 025205 (2010)
29. T. Dahms, J. Lehnert, E. Schöll, Phys. Rev. E **86**(1), 016202 (2012)
30. M.G. Rosenblum, A.S. Pikovsky, Phys. Rev. Lett. **92**(11), 114102 (2004)
31. G. Montaseri, M.J. Yazdanpanah, A. Pikovsky, M. Rosenblum, Chaos: an interdisciplinary. J. Nonlinear Sci. **23**(3), 033122 (2013)
32. A.L. Do, J. Höfener, T. Gross, New J. Phys. **14**(11), 115022 (2012)

Chapter 3
Adaptively Controlled Synchronization of Delay-Coupled Networks

**Philipp Hövel, Judith Lehnert, Anton Selivanov,
Alexander Fradkov and Eckehard Schöll**

Abstract We discuss an adaptive control delay-coupled networks of Stuart-Landau oscillators, an expansion of systems close to a Hopf bifurcation. Based on the considered, automated control scheme, the speed-gradient method, the topology of a network adjusts itself by changing the link weights in a self-organized manner such that the target state is realized. We find that the emerging topology of the network is modulated by the coupling delay. If the delay time is a multiple of the system's eigenperiod, the coupling within a cluster and to neighboring clusters is on average positive (excitatory), while the coupling to clusters with a phase lag close to π is negative (inhibitory). For delay times equal to odd multiples of half of the eigenperiod, we find the opposite: Nodes within one cluster and of neighboring clusters are coupled by inhibitory links, while the coupling to clusters distant in phase state is excitatory. In addition, the control scheme is able to construct networks such that they exhibit not only a given cluster state, but also oscillate with a prescribed frequency. Finally, we demonstrate the efficiency of the speed-gradient method in cases where only part of the network is accessible.

P. Hövel (✉) · J. Lehnert · E. Schöll
Institut für Theoretische Physik, Technische Universität Berlin,
Hardenbergstraße 36, 10623 Berlin, Germany
e-mail: phoevel@physik.tu-berlin.de

P. Hövel
Bernstein Center for Computational Neuroscience Berlin,
Philippstraße 13, 10115 Berlin, Germany

A. Selivanov
School of Electrical Engineering, Tel Aviv University, Tel Aviv, Israel

A. Selivanov · A. Fradkov
Department of Theoretical Cybernetics,
Saint-Petersburg State University, Saint-Petersburg, Russia

A. Fradkov
Institute for Problems of Mechanical Engineering, Russian Academy of Sciences,
Bolshoy Ave, 61, V. O., 199178 St. Petersburg, Russia

© Springer International Publishing Switzerland 2016
E. Schöll et al. (eds.), *Control of Self-Organizing Nonlinear Systems*,
Understanding Complex Systems, DOI 10.1007/978-3-319-28028-8_3

3.1   Introduction

Networks are ubiquitous. They can be found in a large variety of different research areas such as social science, economics, psychology, biology, physics, and mathematics [1–3], where networks are used to model the interactions of coupled systems or large number of agents. Two important lines of research have formed: (i) investigations of network topologies including data-mining and constructive models of their generation [1–6] and (ii) studies of dynamics on networks with fixed topology [7–16]. The concept of adaptive networks aims to bring these two directions together by considering topologies that evolve according to the states of the network nodes, which are in turn influenced by the topology [17].

In the wide spectrum of dynamical scenarios of coupled systems, zero-lag synchronization, which is also known as in-phase or complete synchronization, has been at the center of attention for a long time. Within the last decade, other, more complex synchronization patterns have moved into the focus of increasing research activities. These include cluster and group synchronization, which was studied in theory [11, 18–21] and realized in experiments [22, 23]. Prominent examples of these types of synchrony have been reported in many biological systems including dynamics of neurons [24], central pattern generation in animal locomotion [25], or population dynamics [26]. The difference between cluster and group synchronization can be described as follows: Group synchronization corresponds to the case where each cluster potentially exhibits different local dynamics. This dynamical state is more general than an M-cluster state, for which the compound system exhibits M clusters with zero-lag synchronization between the nodes within one cluster, but—in the case of oscillating dynamics—with a constant phase lag of $2\pi/M$ between the clusters.

If the network dynamics does not settle in the desired cluster state in a self-organized way, control methods can help to adaptively change the topology of the network in order to realize the target state. This has previously been investigated, to our knowledge, only by a few researchers: Lu et al., for instance, considered the control of cluster synchronization by means of changing topology. As a limiting restriction for the applicability, their method requires a-priori knowledge to which cluster each node should belong in the final state [27]. Furthermore, the majority of algorithms, which have been developed to control synchrony by adaptation of the network topology, take advantage of local mechanisms. A large number of these control schemes can be related to Hebb's rule: *Cells that fire together, wire together* [28]. The method that we propose below, however, uses a global goal function to realize self-organized control. It is hence a powerful alternative and complements existing control schemes.

In short, we will present an algorithm that adapts the weights of the links in the network such that a desired cluster state is reached. We will show that our method is robust towards different initial conditions and also works for a large parameter range. This includes the potential adding of new links, if the initial weight had been zero, or the removal of links, if the respective weight is set to zero. The adaptation algorithm for the network structure is based on the *speed-gradient method* [29, 30]. The goal function, which we employ, has a strong advantage compared to other methods: it does not rely on a-priori ordering of nodes, i.e., it is not necessary to assign each node to a specific cluster in advance.

As a proof of concept, we will consider the normal form of a Hopf bifurcation, which is also known as Stuart-Landau oscillator. This model is generic for many oscillating systems present in nature and technological applications. In addition, we take into account time delays in the coupling between the nodes because delays naturally arise in many applications. Note that our scheme also works for instantaneous coupling. Furthermore, we will demonstrate that our method does not require to control all links of a network. The control scheme is still successful if only a subset of links is accessible, which we will explicitly illustrate for random networks [31]. We will also show how networks can be constructed in which cluster states oscillate with a prescribed frequency. This includes zero frequency and gives rise to a freezing of the dynamical motion. The final topology, i.e., distribution of link weights, of these controlled networks will contain some randomness because we start with random initial conditions for the state of the nodes. Despite this randomness, we will show that on average the topology is characterized by common features. As a crucial parameter shaping a topology, which enables synchronization, we identify the delay time.

Delay is an ubiquitous phenomenon in nature and technology and arises whenever time in the propagation or the processing of a signal is needed [32, 33]. For example, in laser networks the finite speed of light gives rise to a propagation delay [34–36]. Time delay in neural networks emanates from the finite speed of the transmission of an action potential between two neurons where the propagation velocity of an action potential varies between 1 and 100 m/s depending on the diameter of the axon and whether the fibers are myelinated or not [37]. The influence of delay on the dynamics on networks has been investigated by several authors [13, 16, 19, 38–57]. Depending on the context, delay can play a constructive or a destructive role [58–69].

The rest of this chapter is organized as follows: In Sect. 3.2 we introduce the model of Stuart-Landau oscillator and discuss the application of the speed-gradient method on the coupling matrix. In Sect. 3.3, we present the main results including applications of the control scheme to select a frequency of the ensemble of oscillators and restricted accessibility of the controller. We wrap up in Sect. 3.4 and finish with an outlook for future research directions and additional questions.

3.2 Model Equation and Control Scheme

In this chapter, we will first introduce the model equations of the Hopf normal form. Then, we will show the application of the speed-gradient method for an automated adjustment of the network topology by changing the weights of the links. The control scheme is based on a goal function that will be designed such that it becomes minimal for the desired M-cluster state.

3.2.1 Stuart-Landau Oscillator

We consider the Stuart-Landau oscillator given by the following equations

$$\dot{z} = \left[\lambda + i\omega - |z|^2\right] z \qquad (3.1)$$

with the complex variable $z \in \mathbb{C}$ and parameters $\lambda, \omega \in \mathbb{R}$ [70]. This system arises generically in a center manifold expansion close to a supercritical Hopf bifurcation with λ as the bifurcation parameter. Below the bifurcation, i.e., for negative λ, the system exhibits a stable focus at the origin, which becomes unstable at the bifurcation point $\lambda = 0$. Above the bifurcation, a stable limit cycle with radius $r = \sqrt{\lambda}$ coexists with the unstable focus. The parameter ω is the frequency of the limit cycle and determines the intrinsic timescale.

Throughout this chapter, we discuss networks of N delay-coupled Stuart-Landau oscillators z_j, $j = 1, \ldots, N$, described by

$$\dot{z}_j(t) = [\lambda + i\omega - |z_j|^2]z_j + K \sum_{n=1}^{N} G_{jn}(t)[z_n(t-\tau) - z_j(t)] \qquad (3.2)$$

with a real coupling strength K and coupling delay τ. For notational convenience, we use in the following the abbreviation $z_{n,\tau} \equiv z_n(t-\tau)$. The matrix $\{G_{jn}(t)\}_{j,n=1,\ldots,N}$ describes the topology of the network. Its elements might change over time, because it is subject to the adaptive control as discussed in Sect. 3.2.2 below.

In order to investigate the amplitude and phase dynamics of the complex variable z, it is convenient to rewrite Eq. (3.1) using $r_j = |z_j|$ and $\varphi_j = \arg(z_j)$:

$$\dot{r}_j(t) = \left[\lambda - r_j^2\right]r_j + K \sum_{n=1}^{N} G_{jn} \left\{r_{n,\tau} \cos\left[\varphi_{n,\tau} - \varphi_j\right] - r_j\right\}, \qquad (3.3a)$$

$$\dot{\varphi}_j(t) = \omega + K \sum_{n=1}^{N} G_{jn} \left\{\frac{r_{n,\tau}}{r_j} \sin\left[\varphi_{n,\tau} - \varphi_j\right]\right\}. \qquad (3.3b)$$

One class of solutions of Eqs. (3.3) are M-cluster states that exhibit a common amplitude $r_j \equiv r_0$. The phases of the oscillators in an M-cluster state are given by $\varphi_j = \Omega_M t + j2\pi/M$, where Ω_M is the collective frequency. A special cluster state is complete, in-phase, or zero-lag synchronization, i.e., $M = 1$, where all nodes are in one cluster. The other extreme case are splay states with $M = N$, where each cluster consists of a single node only. In the continuum limit, the splay state on a unidirectionally coupled ring corresponds to a rotating wave. For a schematic diagram of (a) in-phase synchronization, (b) a 3-cluster state, and (c) a splay state, see Fig. 3.1.

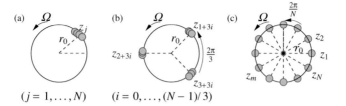

Fig. 3.1 Schematic examples of **a** in-phase synchronization ($M = 1$), **b** a 3-cluster ($M = 3$), and **c** a splay state ($M = N$). Each cluster consists of the same number of nodes

3.2.2 Speed-Gradient Method

For a dynamical system in general notation

$$\dot{x}(t) = F(x, u, t),$$ (3.4)

an additional set of equations for the control vector u can be derived using the gradient (with respect to the accessible parameters) of the speed (temporal change) of an appropriately chosen goal function Q [29]:

$$\frac{du}{dt} = -\Gamma \nabla_u \dot{Q}(x, u, t)$$ (3.5)

with a positive definite gain matrix Γ. Intuitively, the control scheme works as follows: The speed $\dot{Q}$ may decrease along the direction of its negative gradient. As $\dot{Q}$ becomes negative, the control function Q will decrease as well and will finally reach its minimum indicating that the control goal is realized. For details and conditions, see Refs. [71, 72].

In the following, we will apply this *speed-gradient control* to the elements of the coupling matrix $\{G_{jn}(t)\}_{j,n=1,\dots,N}$ of Eq. (3.2), i.e.,

$$\dot{G}_{jn} = -\gamma \frac{\partial}{\partial G_{jn}} \dot{Q}_M$$

with $\gamma > 0$ and choose the goal function Q_M to realize the M-cluster state as [73]:

$$Q_M = 1 - \underbrace{\frac{1}{N^2} \sum_{j=1}^{N} e^{Mi\varphi_j} \sum_{k=1}^{N} e^{-Mi\varphi_k}}_{I} + \underbrace{\frac{1}{2} \sum_{p=1}^{M-1} \sum_{j=1}^{N} e^{pi\varphi_j} \sum_{k=1}^{N} e^{-pi\varphi_k}}_{II}$$

$$+ \underbrace{\frac{1}{2} \sum_{i,k=1}^{N} (r_i - r_k)^2}_{III} + \underbrace{\frac{c}{2} \int_0^t \sum_{k=1}^{N} \left(\sum_{i=1}^{N} G_{ki} - 1 \right)^2 dt}_{IV}$$ (3.6)

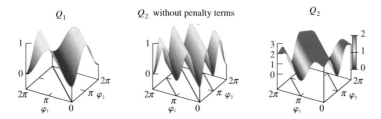

Fig. 3.2 Effects of the penalty term (II) in Eq. (3.6). Both Q_1 and Q_2 without penalty terms have a minimum for $\varphi_1 = \varphi_2$. Taking the penalty terms for $p < M = 2$ into account, this minimum vanishes and Q_2 becomes zero only for anti-phase synchronization

with $c > 0$. The goal function becomes minimized, once an M-cluster state is reached, and consists of the following terms: (I) a Kuramoto-type order parameter generalized for an M-cluster state, (II) a penalty term for all p-cluster states with $p < M$, (III) a penalty term to realize identical amplitudes of all oscillators, and (IV) a term added to guarantee a constant row-sum of $\{G_{jn}(t)\}_{j,n=1,\dots,N}$ designed such that $\sum_{n=1}^{N} G_{jn} = 1$. Figure 3.2 illustrates the effects of the penalty term (II) for in-phase and anti-phase synchronization of a network motif of two coupled nodes. The penalty term (IV) takes into account all deviations from the unity row-sum during the growth of the network, such that Q_M will not vanish completely in the goal state. Thus, we define $q_M \equiv Q_M - \frac{c}{2}\int_0^t \sum_k \left(\sum_i G_{ki} - 1\right)^2 dt$, i.e., the sum over the terms (I)–(III), as better measure for the quality of synchronization.

Calculating the derivation of Q_M and the gradient with respect to the matrix elements G_{jn}, we obtain the following N^2 equations after some algebraic manipulation [73]

$$
\begin{aligned}
\dot{G}_{jn} = &-\gamma K \left[\frac{r_{n,\tau}}{r_j} \sin(\varphi_{n,\tau} - \varphi_j)\right] \\
&\times \sum_{k=1}^{N} \left\{\sum_{1 \le p < M} p \sin[p(\varphi_k - \varphi_j)] - \frac{2M}{N^2}\sin[M(\varphi_k - \varphi_j)]\right\} \\
&- 2\gamma K \sum_{k=1}^{N} (r_j - r_k)\left[r_{n,\tau}\cos(\varphi_{n,\tau} - \varphi_j) - r_j\right] - \gamma c \left(\sum_{i=1}^{N} G_{ji} - 1\right).
\end{aligned}
\tag{3.7}
$$

The case, when not all N^2 elements of the coupling matrix are accessible, will be discussed in Sect. 3.3.3.

Next, we will present some results on the generation of various cluster states, that is, different combinations of number of elements N and number of clusters M.

3.3   Results

In this chapter, we will present the main findings of our study. At first, we will demonstrate the success of the proposed control method along the lines of an exemplary 8-cluster state. Then, we will address the impact of the time delay on the distribution of the coupling weights. Finally, we will discuss two applications of the controller: (i) a frequency selection of the cluster state and (ii) targeted control, when only a fraction of the network is accessible.

3.3.1   Automated Adjustment of Network Topology

Figure 3.3 presents a successful realization of an 8-cluster state and depicts the time series of the radii, the phase differences with respect to the reference node 0, the weights of the coupling matrix, and the goal function with (Q_8) and without (q_8) the unity row-sum term IV of Eq. (3.6) during the growth of the network. The simulations starts from a unidirectional ring as initial topology and random initial conditions $z_j(-50) = r_j(-50)e^{i\varphi_j(-50)}$, $j = 1, \ldots, N$. The control is switched on at time $t = 0$. One can see that after a short transient, the radii and phases rapidly converge to the desired 8-cluster state, and Q_8 and q_8 approach their minimum. Once the target state is reached, the coupling weights do not change anymore. The specific choice of the parameter γ influences the transient times to reach the final network that supports the desired cluster state. Note that the generated network contains excitatory links, i.e., $G_{jn} > 0$, and inhibitory ones with $G_{jn} < 0$. The distribution of the weights in the final topology and the perturbation of the network (marked by the dotted line at $t = 80$) will be discussed in detail in Sect. 3.3.2.

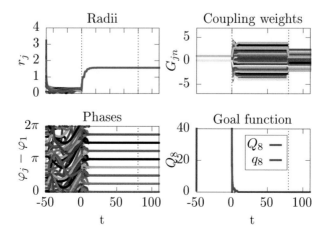

Fig. 3.3   Control of an 8-cluster state for $N = 40$. Parameters: $\lambda = 0.1, \omega = 1, c = 0.01, K = 0.1,$ $\tau = \pi, \gamma = 10$

We also study the fraction of successful realizations f_c and the time to reach the target state t_c in dependence on the coupling strength K and the delay time τ. We find that the fraction of successfully controlled networks f_c is close to 1 and t_c is roughly 10 units in the considered range $0.1 < K \leq 5, 0 \leq \tau \leq 3\pi$ (cf. Fig. 10.4 in Ref. [74]), demonstrating that our method works very reliably independently of the coupling parameters. The quantities f_c and t_c will be helpful in Sect. 3.3.3, where we will apply the control only to a fraction of the links in the network.

3.3.2 Dependence on Time Delay

In the following, we discuss the structural properties of the networks after successful control for different coupling delays. For this purpose, we consider the coupling weights of the final topology as a function of the final phase difference between all pairs of oscillators. This will allow us to investigate the influence of delay on networks that enable synchronization in the prescribed cluster state.

Figure 3.4 shows the weights G_{jn} of an 8-cluster state as an average over 100 realizations. This ensemble average $\langle G_{jn} \rangle$ is presented in dependence on the final phase difference $\Delta_{jn} \equiv \lim_{t \to \infty} [\varphi_j(t) - \varphi_n(t)]$, where the different colors correspond to different coupling delays. The network of the exemplary case shown in Fig. 3.3 is included in the dark-blue curve for $\tau = \pi$.

It can be seen that the curves have the form of a shifted cosine, i.e., $\langle G_{jn} \rangle \propto \cos\left(\frac{2\pi(j-n)}{M} - \tau\right)$. Focusing on the 8-cluster states of Fig. 3.3, we find a negative coupling between nodes with a small phase difference and a positive coupling between nodes with a phase equal or close to π.

For further insight into the network structure, we consider a row-wise discrete Fourier transform of the coupling matrix $\{G_{jn}\}$ after successful control. For this

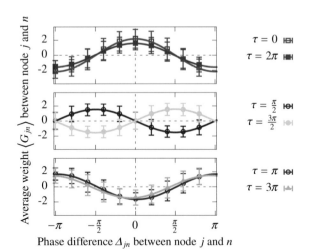

Fig. 3.4 Dependence of the elements of coupling matrix averaged over 100 realizations on the phase difference $\Delta_{jn} = \lim_{t \to \infty} [\varphi_j(t) - \varphi_n(t)]$ for $N = 30$, $M = 10$, and different time delays τ. Other parameters as in Fig. 3.3

purpose, we introduce an auxiliary $N \times M$ matrix $\Gamma_{jk} = \sum_{l=0}^{m-1} \tilde{G}_{j,k+lM}$, where m is the number of nodes in one cluster, i.e., $m = N/M$, and the final topology of the network is given by $\tilde{G} = G(t_\infty)$. For notational convenience, we label the nodes such that the synchronized state can be described by $r_j \equiv r_{0,M}$ and $\varphi_j \equiv \Omega_M t + j\frac{2\pi}{M}$, $j = 1, \ldots, N$, where $r_{0,M}$ and Ω_M denote the common radius and the common frequency, respectively. In other words, Γ_{jn} represents the total input which node j receives from all nodes in cluster n. Representing each row of Γ as a discrete Fourier series, the corresponding Fourier coefficients are given by

$$a_l^j = \frac{2}{M} \sum_{k=1}^{M} \Gamma_{jk} \sin\left(\frac{l(k-j)2\pi}{M} - \Omega_M \tau\right)$$
$$= \frac{2}{N} \sum_{k=1}^{N} \tilde{G}_{jk} \sin\left(\frac{l(k-j)2\pi}{M} - \Omega_M \tau\right), \tag{3.8a}$$

$$b_l^j = \frac{2}{M} \sum_{k=1}^{M} \Gamma_{jk} \cos\left(\frac{l(k-j)2\pi}{M} - \Omega_M \tau\right)$$
$$= \frac{2}{N} \sum_{k=1}^{N} \tilde{G}_{jk} \cos\left(\frac{l(k-j)2\pi}{M} - \Omega_M \tau\right), \tag{3.8b}$$

where l labels the lth coefficient in the Fourier series of the jth row. Note that the coefficients a_0^j are equal to zero and due to the constant row-sum condition of G_{jn}, we have $b_0^j = \frac{1}{2M\cos(\Omega_M \tau)}$.

It is straight-forward, but lengthy to perform a linear stability analysis to compute the impact of perturbation on the radii, phases, and Fourier coefficients on the desired cluster state. One will finally derive a characteristic equation, whose infinite number of roots are the Floquet exponents. We are only interested in the one with the largest real part, which we denote by Re Λ. If this quantity is negative, the cluster solution will be stable, otherwise the solution is unstable. For details, on the derivation see Sect. 10.6.2 of Ref. [74].

Figure 3.5 shows the result of this stability analysis for (a) all higher Fourier coefficients being zero, i.e., $a_l^j = b_l^j = 0$ for $l > 1$ and $j = 1, \ldots, N$, (b) random higher Fourier coefficients, and (c) constant higher Fourier coefficients, i.e., $a_l^j = b_l^j = 10$ for $a_l^j = b_l^j = 10$ and $j = 1, \ldots, N$, in dependence on the common radius $r_{0,M}$ and the common frequency Ω_M. Random means that for each value of $r_{0,M}$ and Ω_M the coefficients are drawn from a uniform distribution on the interval $[-10, 10]$. We find that the stability is only affected by the higher Fourier coefficients if $r_{0,M}$ is small: For small $r_{0,M}$ the unstable regions (yellow to orange color code) have a qualitatively different form in panels (a), (b), and (c), while for large $r_{0,M}$ stability is found in all three cases.

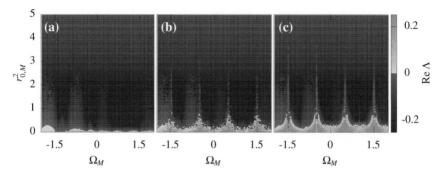

Fig. 3.5 Stability as a function of common frequency Ω_M and radius $r^2_{0,M}$ for **a** vanishing higher Fourier coefficients, i.e., $a_l^j = b_l^j = 0$ for $l > 1$, **b** random higher Fourier coefficients, **c** constant higher Fourier coefficients, i.e., $a_l^j = b_l^j = 10$ for $l > 1$. $N = 8$, $M = 4$. Other parameters as in Fig. 3.3

Another possibility to test the influence of the higher coefficients is to disturb them during or after the course of the adaptation process. This is shown for the 8-cluster state in Fig. 3.3, where at $t = 80$, we set each of the higher Fourier coefficients to a random value in the interval $[-3, 3]$. One can see that the common frequency and radius do not change as a result of this perturbation.

In summary, these results can be seen as evidence that the higher Fourier coefficients do not affect the stability of the desired cluster state. Analyzing the first Fourier coefficients, one can derive the following equations for the common radius $r_{0,M}$ and frequency Ω_M [73, 74]:

$$r^2_{0,M} = \lambda + K \left[\frac{b_1^j N}{2} - 1 \right], \qquad (3.9a)$$

$$\Omega_M = \omega + K \left[\frac{a_1^j N}{2} \right]. \qquad (3.9b)$$

Considering that these equations have to be satisfied for all $j = 1, \ldots, N$, we conclude that a solution with a common radius and a common frequency exists, only if $a_1^1 = a_1^2 = \ldots = a_1^N \equiv a$ and $b_1^1 = b_1^2 = \ldots = b_1^N \equiv b$. In fact, the average topology is mainly given by b, because a and b_0 are typically small and the higher coefficients average out as discussed above. Therefore, we obtain $G_{jn} \approx b \cos \left(\frac{2\pi(j-n)}{M} - \Omega_M \tau \right)$, which explains the cosine form of the curves shown in Fig. 3.4.

3.3.3 Applications of Controller

3.3.3.1 Frequency Selection

In the following, we will exploit Eq. (3.9b) to select a common frequency via constructing an appropriate matrix. For this purpose, we set $a = \frac{2}{N}\left(\frac{\Omega_M - \omega}{K}\right)$. To demonstrate the effect of this choice, we consider the case of a stationary cluster with a common frequency Ω_M. Figure 3.6 depicts the corresponding time series of the radii, phases, coupling weights, goal function, and Ω_M. At $t = 0$ (first dotted line), we start the adaptive control with $M = N$, that is, with the goal function leading to a splay state. Then, at $t = 40$ (second dotted line), the adaptive control is switched off and a is set to $a = \frac{2\omega}{NK}$ forcing Ω_M to approach zero.

3.3.3.2 Targeted Control

In the previous sections, we have assumed that every link of the network is subject to the control scheme, that is, all elements of the coupling matrix $\{G_{jn}(t)\}_{j,n=1,\dots,N}$ are accessible. This might be not realistic for applications. We will show in the following that it suffices to control a subset of links, while the other links are left unchanged. This will be demonstrated for the example of a directed random network that consists of P links chosen from the $L = N(N-1)$ possible links excluding self-coupling. From this set of P links, we select, again randomly, A links which are subject to adaptation as given by Eq. (3.7).

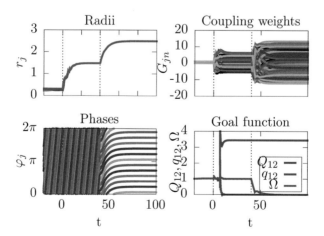

Fig. 3.6 Freezing of the motion of a splay state with $N = 12 = M$. Parameters: $\lambda = 0.1$, $\omega = 1$, $c = 0.01$, $K = 0.1$, $\tau = \pi$, $\gamma = 10$

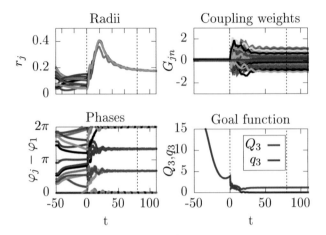

Fig. 3.7 Control of $A/L = 30\%$ of the links for $N = 15$, $M = 3$, and $P/L = 0.4$. Parameters: $\lambda = 0.1$, $\omega = 1$, $c = 0.01$, $K = 0.1$, $\tau = \pi$, $\gamma = 10$

Figure 3.7 depicts the realization of 3-cluster state in a network of 15 nodes with the time series of the radii, the phase differences, the elements of the coupling matrix, and the goal function shown in the different panels, respectively. The nodes are coupled on a random network with density 0.4 and with 30 % of the links accessible, i.e., $A/L = 0.3$. It can be seen that using the goal function Q_3 the network consists of 3 equally sized clusters after successful control.

Next, we explore the performance of our method with respect to the links present in the networks and the fraction of these links subject to adaptation. Figure 3.8a depicts the fraction f_c of successfully controlled networks as a function of P/L and

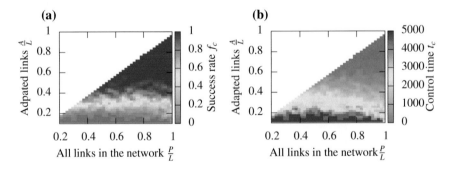

Fig. 3.8 **a** Fraction f_c of successfully controlled networks and **b** times t_c needed to reach the control goal as a function of number of random links P and number of controlled links A, normalized by $L = N(N-1)$. We simulated 10 realizations for each parameter combination. $K = 0.2$. Other parameters as in Fig. 3.7

A/L. We define a network as successfully controlled in an M-cluster state at time t_c if it was in this state for $t \in [t_c - 1, t_c]$. Figure 3.8b shows the corresponding control time t_c.

One can see that the success rate f_c does not depend strongly on the total number of links P in the network and is rather constant for fixed A. The rate, however, depends on the ratio of adapted links A/L. We conclude that the links additionally present in the network, but not subject to control, have very little effect on the synchronizability of the network. For example, consider a horizontal cut at $A/L = 0.4$. Then, the control still works in more than 90 % of the cases.

Figure 3.9 further corroborates these results. A good approximation of the success rate f_c can be obtained if we assume that for successful control each node in the networks needs at least two incoming links which are subject to adaptation. One adapted link is not sufficient because it is not able to change due to the unity row-sum condition. Only if a second incoming link is present, the links can change in order to control the dynamics of the node because the effect of the adaptation of the first link on the row-sum can be counterbalanced by the second link. Figure 3.9 depicts f_c versus A/L as red circles for a fixed ratio of $P/L = 0.4, 0.6, 1$ in panels (a)–(c), respectively. The blue circles depict the fraction $p_{>1}$ of networks where all nodes have at least two incoming links. Obviously, $p_{>1}$ well approximates f_c although they are not identical indicating that cases exist where the network can be controlled though one node has less than two adapted incoming links, or where the control fails although each node has two incoming links. Note that an analytic expression for $p_{>1}$ can be derived, which yields the blue curves in Fig. 3.9. For details, see Sect. 10.8 in Ref. [74].

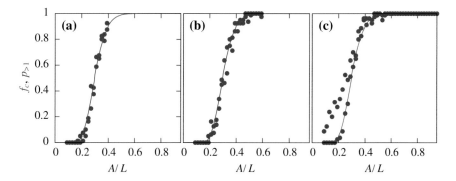

Fig. 3.9 Control of a subset of links for fixed ratio of **a** $P/L = 0.4$, **b** $P/L = 0.6$, and **c** $P/L = 1$. *Red circles* Success rate f_c. *Blue circles* Probability $p_{>1}$ that no node in the network has less than 2 incoming links which are adapted. *Blue line* Analytic calculation of $p_{>1}$ (cf. [74]). $K = 0.2$. 80 realizations for each value of A/L. Other parameters as in Fig. 3.7

3.4 Summary and Conclusions

We have applied a speed-gradient algorithm to adapt the topology of time-delay coupled oscillators to control cluster synchronization. The controller minimizes a goal function that is based on a generalized Kuramoto order parameter. The goal function is chosen according to the target cluster state, but independent of the ordering of the nodes. An additional term ensures amplitude synchronization. We find that this speed-gradient control scheme is very robust with respect to perturbations, different initial conditions, and coupling parameters. We have focused on the dependence on the coupling strength and delay time.

We have found that the distribution of link weights of the successfully controlled network is modulated by the coupling delay. A row-wise discrete Fourier transform of the coupling matrix gives insight into these delay modulations. Necessary conditions for the existence of a common radius and a common frequency give rise to restrictions affecting the first Fourier coefficients, while there is no restriction for the higher Fourier coefficients. We also found that the stability of the cluster states is only weakly affected by the higher Fourier coefficients. Thus, we conclude that the higher Fourier coefficients are mainly dependent on the random initial conditions and are therefore randomly distributed. On average, the network topology is therefore dominated by the first Fourier coefficients leading to the observed delay modulation.

Appropriate selection of the first Fourier coefficients leads to cluster states with a given common frequency. As an example, we have quenched the oscillations in a Stuart-Landau oscillator. This allows for construction of networks that exhibit a desired dynamical behavior.

In many real-world networks not all links are accessible to control. Therefore, we have considered random networks, where we have chosen a random subset of links to which we applied the adaptation algorithm. The other links remained fixed. We have found that the control is successful if the number of adapted links is equal or higher than approximately 30 % of all possible links, independently of the number of actual fixed links. For practical applications this opens up the possibility to apply the method more easily.

Since we have considered the paradigmatic Stuart-Landau oscillator as a generic model of the Hopf bifurcation, we expect broad applicability to control, for instance, synchronization of networks in medicine, chemistry or mechanical engineering or as self-organizing mechanisms in biological networks.

Acknowledgments This work was supported by Deutsche Forschungsgemeinschaft in the framework of SFB 910. JL acknowledges the support by the German-Russian Interdisciplinary Science Center (G-RISC) funded by the German Federal Foreign Office via the German Academic Exchange Service (DAAD). PH acknowledges support by the Federal Ministry of Education and Research (BMBF), Germany (grant no. 01GQ1001B).

References

1. S. Boccaletti, V. Latora, Y. Moreno, M. Chavez, D.U. Hwang, Phys. Rep. **424**(4–5), 175 (2006). doi:10.1016/j.physrep.2005.10.009
2. R. Albert, A.L. Barabasi, Rev. Mod. Phys. **74**(1), 47 (2002). doi:10.1103/revmodphys.74.47
3. M.E.J. Newman, SIAM Rev. **45**(2), 167 (2003). doi:10.1137/s0036144503
4. D.J. Watts, S.H. Strogatz, Nature **393**, 440 (1998)
5. A. Rapoport, Bull. Math. Biol. **19**, 257 (1957). doi:10.1007/bf02478417
6. P. Erdős, A. Rényi, Publ. Math. Debrecen **6**, 290 (1959)
7. M. Dhamala, V.K. Jirsa, M. Ding, Phys. Rev. Lett. **92**(7), 074104 (2004). doi:10.1103/physrevlett.92.074104
8. M. Zigzag, M. Butkovski, A. Englert, W. Kinzel, I. Kanter, Europhys. Lett. **85**(6), 60005 (2009)
9. C.U. Choe, T. Dahms, P. Hövel, E. Schöll, Phys. Rev. E **81**(2), 025205(R) (2010). doi:10.1103/physreve.81.025205
10. M. Chavez, D.U. Hwang, A. Amann, H.G.E. Hentschel, S. Boccaletti, Phys. Rev. Lett. **94**, 218701 (2005)
11. F. Sorrentino, E. Ott, Phys. Rev. E **76**(5), 056114 (2007). doi:10.1103/physreve.76.056114
12. R. Albert, H. Jeong, A.L. Barabasi, Nature **406**, 378 (2000)
13. W. Kinzel, A. Englert, G. Reents, M. Zigzag, I. Kanter, Phys. Rev. E **79**(5), 056207 (2009). doi:10.1103/physreve.79.056207
14. J. Lehnert, T. Dahms, P. Hövel, E. Schöll, Europhys. Lett. **96**, 60013 (2011). doi:10.1209/0295-5075/96/60013
15. A. Keane, T. Dahms, J. Lehnert, S.A. Suryanarayana, P. Hövel, E. Schöll, Eur. Phys. J. B **85**(12), 407 (2012). doi:10.1140/epjb/e2012-30810-x
16. E. Schöll, in *Advances in Analysis and Control of Time-Delayed Dynamical Systems*, ed. by J.-Q. Sun, Q. Ding (World Scientific, Singapore, 2013), Chap. 4, pp. 57–83
17. T. Gross, B. Blasius, J. R. Soc. Interface **5**(20), 259 (2008). doi:10.1098/rsif.2007.1229
18. T. Dahms, J. Lehnert, E. Schöll, Phys. Rev. E **86**(1), 016202 (2012). doi:10.1103/physreve.86.016202
19. I. Kanter, M. Zigzag, A. Englert, F. Geissler, W. Kinzel, Europhys. Lett. **93**(6), 60003 (2011)
20. I. Kanter, E. Kopelowitz, R. Vardi, M. Zigzag, W. Kinzel, M. Abeles, D. Cohen, Europhys. Lett. **93**(6), 66001 (2011)
21. M. Golubitsky, I. Stewart, *The Symmetry Perspective* (Birkhäuser, Basel, 2002)
22. K. Blaha, J. Lehnert, A. Keane, T. Dahms, P. Hövel, E. Schöll, J.L. Hudson, Phys. Rev. E **88**, 062915 (2013). doi:10.1103/physreve.88.062915
23. C.R.S. Williams, T.E. Murphy, R. Roy, F. Sorrentino, T. Dahms, E. Schöll, Phys. Rev. Lett. **110**(6), 064104 (2013). doi:10.1103/physrevlett.110.064104
24. E. Mosekilde, Y. Maistrenko, D. Postnov, *Chaotic Synchronization: Applications to Living Systems* (World Scientific, Singapore, 2002)
25. A. Ijspeert, Neural Netw. **21**(4), 642 (2008). doi:10.1016/j.neunet.2008.03.014
26. B. Blasius, A. Huppert, L. Stone, Nature (London) **399**, 354 (1999)
27. X. Lu, B. Qin, Phys. Lett. A **373**(40), 3650 (2009). doi:10.1016/j.physleta.2009.08.013
28. D. Hebb, *The Organization of Behavior: A Neuropsychological Theory*, new, edition edn. (Wiley, New York, 1949)
29. A.L. Fradkov, *Cybernetical Physics: From Control of Chaos to Quantum Control* (Springer, Heidelberg, Germany, 2007)
30. A.L. Fradkov, Physica D **128**(2), 159 (1999)
31. P. Erdős, A. Rényi, Publ. Math. Inst. Hung. Acad. Sci. **5**, 17 (1960)
32. W. Just, A. Pelster, M. Schanz, E. Schöll, Delayed complex systems: an overview, Theme Issue of Phil. Trans. R. Soc. A **368**, 301 (2010)
33. V. Flunkert, I. Fischer, E. Schöll, Dynamics, control and information in delay-coupled systems, Theme Issue of Phil. Trans. R. Soc. A **371**, 20120465 (2013)
34. K. Lüdge, *Nonlinear Laser Dynamics—From Quantum Dots to Cryptography* (Wiley-VCH, Weinheim, 2012)

35. C. Otto, *Dynamics of Quantum Dot Lasers—Effects of Optical Feedback and External Optical Injection*. Springer Theses (Springer, Heidelberg, 2014). doi:10.1007/978-3-319-03786-8
36. M.C. Soriano, J. García-Ojalvo, C.R. Mirasso, I. Fischer, Rev. Mod. Phys. **85**, 421 (2013)
37. C. Koch, *Biophysics of Computation: Information Processing in Single Neurons* (Oxford University Press, New York, 1999)
38. W. Kinzel, I. Kanter, in Handbook of Chaos Control, ed. by E. Schöll, H.G. Schuster (Wiley-VCH, Weinheim, 2008). Second completely revised and enlarged edition
39. J. Kestler, E. Kopelowitz, I. Kanter, W. Kinzel, Phys. Rev. E **77**(4), 046209 (2008). doi:10.1103/physreve.77.046209
40. A. Englert, W. Kinzel, Y. Aviad, M. Butkovski, I. Reidler, M. Zigzag, I. Kanter, M. Rosenbluh, Phys. Rev. Lett. **104**(11), 114102 (2010)
41. M. Zigzag, M. Butkovski, A. Englert, W. Kinzel, I. Kanter, Phys. Rev. E **81**, 036215 (2010). doi:10.1103/physreve.81.036215
42. V. Flunkert, S. Yanchuk, T. Dahms, E. Schöll, Phys. Rev. Lett. **105**, 254101 (2010). doi:10.1103/physrevlett.105.254101
43. A. Englert, S. Heiligenthal, W. Kinzel, I. Kanter, Phys. Rev. E **83**(4), 046222 (2011). doi:10.1103/physreve.83.046222
44. Y.N. Kyrychko, K.B. Blyuss, E. Schöll, Eur. Phys. J. B **84**, 307 (2011). doi:10.1140/epjb/e2011-20677-8
45. S. Heiligenthal, T. Dahms, S. Yanchuk, T. Jüngling, V. Flunkert, I. Kanter, E. Schöll, W. Kinzel, Phys. Rev. Lett. **107**, 234102 (2011). doi:10.1103/physrevlett.107.234102
46. V. Flunkert, S. Yanchuk, T. Dahms, E. Schöll, Contemp. Math. Fundam. Dir. **48**, 134 (2013). English version: J. Math. Sci. (Springer) (2014)
47. T. Dahms, Synchronization in delay-coupled laser networks. Ph.D. thesis, Technische Universität Berlin (2011)
48. O.V. Popovych, S. Yanchuk, P.A. Tass, Phys. Rev. Lett. **107**, 228102 (2011). doi:10.1103/physrevlett.107.228102
49. L. Lücken, J.P. Pade, K. Knauer, S. Yanchuk, EPL **103**, 10006 (2013). doi:10.1209/0295-5075/103/10006
50. W. Kinzel, Phil. Trans. R. Soc. A **371**, 20120461 (2013). doi:10.1098/rsta.2012.0461
51. Y.N. Kyrychko, K.B. Blyuss, E. Schöll, Phil. Trans. R. Soc. A **371**, 20120466 (2013). doi:10.1098/rsta.2012.0466
52. O. D'Huys, S. Zeeb, T. Jüngling, S. Heiligenthal, S. Yanchuk, W. Kinzel, EPL **103**, 10013 (2013). doi:10.1209/0295-5075/103/10013
53. M. Kantner, S. Yanchuk, Phil. Trans. R. Soc. A **371**, 20120470 (2013). doi:10.1098/rsta.2012.0470
54. Y.N. Kyrychko, K.B. Blyuss, E. Schöll, Chaos **24**, 043117 (2014). doi:10.1063/1.4898771
55. A. Gjurchinovski, A. Zakharova, E. Schöll, Phys. Rev. E **89**, 032915 (2014). doi:10.1103/physreve.89.032915
56. C. Wille, J. Lehnert, E. Schöll, Phys. Rev. E **90**, 032908 (2014). doi:10.1103/physreve.90.032908
57. O. D'Huys, T. Jüngling, W. Kinzel, Phys. Rev. E **90**, 032918 (2014)
58. K. Pyragas, Phys. Lett. A **170**, 421 (1992)
59. W. Just, T. Bernard, M. Ostheimer, E. Reibold, H. Benner, Phys. Rev. Lett. **78**, 203 (1997)
60. A. Ahlborn, U. Parlitz, Phys. Rev. Lett. **93**, 264101 (2004)
61. M.G. Rosenblum, A. Pikovsky, Phys. Rev. Lett. **92**, 114102 (2004)
62. P. Hövel, E. Schöll, Phys. Rev. E **72**, 046203 (2005)
63. E. Schöll, H.G. Schuster (eds.), Handbook of Chaos Control (Wiley-VCH, Weinheim, 2008). Second completely revised and enlarged edition
64. T. Dahms, P. Hövel, E. Schöll, Phys. Rev. E **76**(5), 056201 (2007). doi:10.1103/physreve.76.056201
65. C. Grebogi, Recent Progress in Controlling Chaos. Series on stability, vibration, and control of systems (World Scientific Publishing Company, Incorporated, 2010)

66. J. Lehnert, P. Hövel, V. Flunkert, P.Y. Guzenko, A.L. Fradkov, E. Schöll, Chaos **21**, 043111 (2011). doi:10.1063/1.3647320
67. A.A. Selivanov, J. Lehnert, T. Dahms, P. Hövel, A.L. Fradkov, E. Schöll, Phys. Rev. E **85**, 016201 (2012). doi:10.1103/physreve.85.016201
68. E. Schöll, A.A. Selivanov, J. Lehnert, T. Dahms, P. Hövel, A.L. Fradkov, Int. J. Mod. Phys. B **26**(25), 1246007 (2012). doi:10.1142/s0217979212460071
69. A. Selivanov, J. Lehnert, A.L. Fradkov, E. Schöll, Phys. Rev. E **91**, 012906 (2015). doi:10.1103/physreve.91.012906
70. Y. Kuramoto, *Chemical Oscillations, Waves and Turbulence* (Springer-Verlag, Berlin, 1984)
71. B. Andrievsky, A.L. Fradkov, Int. J. Bifurcation Chaos **9**(10), 2047 (1999)
72. A.L. Fradkov, A.Y. Pogromsky, IEEE Trans. Circuits Syst. I, Fundam. Theory Appl. **43**(11), 907 (1996)
73. J. Lehnert, P. Hövel, A.A. Selivanov, A.L. Fradkov, E. Schöll, Phys. Rev. E **90**(4), 042914 (2014). doi:10.1103/physreve.90.042914
74. J. Lehnert, Controlling synchronization patterns in complex networks. Springer Theses (Springer, Heidelberg, 2016)

Chapter 4
Controlling Oscillations in Nonlinear Systems with Delayed Output Feedback

Fatihcan M. Atay

Abstract We discuss the problem of controlling oscillations in weakly nonlinear systems by delayed feedback. In classical control theory, the objective of the control action is typically to drive the system to a stable equilibrium. Here we also study the possibility of driving the system to a stable limit cycle having a prescribed amplitude and frequency, as well as suppressing unwanted oscillations, using partial state information in the feedback. The presence of the delay in the output feedback turns out to play a crucial role in achieving these goals.

4.1 Introduction

Controlling the behavior of dynamical systems is a problem of practical importance in many applications. In classical control theory, the basic goal is usually stated as a *regulator problem*, namely, to obtain an asymptotically stable equilibrium solution which attracts all nearby initial conditions. A more sophisticated aim in oscillation control can be defined as obtaining a stable periodic solution with desired properties, such as oscillation at a given amplitude or frequency. We call this goal the *oscillator problem*. This chapter deals with the oscillator problem under delayed output feedback.

Feedback delays are an inevitable feature of many natural and man-made control mechanisms. While they are often seen as an undesired characteristic that can destabilize the system or complicate the analysis, positive uses of delays have also been studied. These go back to the 1950s [1], followed by other works in later years [2–5], where delays were used to enhance the system performance in various ways. Most of the analytical studies have so far focused on linear systems and stability. In the present chapter we consider feedback laws to control the amplitude and stability of oscillations in nonlinear systems. Moreover, we consider the problem from an *output*

F.M. Atay (✉)
Max Planck Institute for Mathematics in the Sciences, Inselstraße 22,
04103 Leipzig, Germany
e-mail: atay@member.ams.org

© Springer International Publishing Switzerland 2016
E. Schöll et al. (eds.), *Control of Self-Organizing Nonlinear Systems*,
Understanding Complex Systems, DOI 10.1007/978-3-319-28028-8_4

feedback point of view, where only partial information about the state of the system is available for the feedback control.

The main results we present can be rigorously derived in full generality for weak nonlinearities, such as for systems near a Hopf bifurcation, which is an important mechanism for generating oscillations in nonlinear systems. The analysis then starts by projecting the dynamics onto a center manifold and proceeds by investigating the resulting two-dimensional system. This general approach for oscillation control can be found in [6–8]. For the purpose of simplicity, here we will assume that such a reduction step has already been done and a two-dimensional system has been obtained. Therefore, we will study systems described by equations of the form

$$\ddot{x} + \omega^2 x + \varepsilon g(x, \dot{x}, \varepsilon) = \varepsilon f(x(t - \tau)). \tag{4.1}$$

Here $x \in \mathbb{R}$, and ω and $\varepsilon \ll 1$ are positive parameters. The left hand side of (4.1) describes the dynamics of the system after projection onto a two-dimensional center manifold corresponding to a pair of imaginary eigenvalues $\pm i\omega$, whereas the right hand side represents a feedback of position that is delayed by $\tau \geq 0$. The feedback is, at the moment, scaled by the parameter ε so that it has a comparable magnitude with the nonlinearity g; however, we will relax this assumption in Sect. 4.5 when we study frequency control.

The form of left hand side of (4.1) is quite general and includes several para-digmatic systems as special cases, for instance the van der Pol (with $g(x, \dot{x}, \varepsilon) = (x^2 - 1)\dot{x}$) and the Duffing oscillators (with $g(x, \dot{x}, \varepsilon) = \alpha x + \beta x^3 + \gamma \dot{x}$). Equations of the form (4.1) also come up in various biological and industrial settings, for example in the production of proteins [9, 10], orientation control in the fly [11, 12], neuromuscular regulation of movement and posture [10, 13, 14], acousto-optical bistability [15], metal cutting [16], vibration absorption [17], and control of the inverted pendulum [18]. Feedback loops with only partial state information is typical in many biological control mechanisms. Furthermore, the classical control-theoretic approach of using an observer to reconstruct the full state is not an option in natural systems. Hence, it is an interesting and challenging goal to discover the theoretical basis for control under partial and delayed information.

The regulator and oscillator problems under delayed feedback have been studied for nonlinear equations of type (4.1) in several previous works. Some of the most relevant ones for the purposes of this chapter using similar techniques can be listed as follows. Controlling the amplitude of oscillations was investigated in [19] for the van der Pol oscillator and later in [20] for more general oscillators (4.1). Controlling the frequency of oscillations is studied in [6]. Suppressing oscillations in networks has been treated in [21]. A general study for controlling systems near Hopf bifurcation using distributed delays is given in [7], and for networks of oscillators in [21].

In the following we will analyze (4.1) and show that the goal of the regulator problem (stabilizing the zero solution) can be achieved by a linear delayed feedback of the variable x, and the goal of the oscillator problem (obtaining a stable limit cycle at a given amplitude and/or modifying its frequency) can be achieved by a nonlinear feedback function. The conclusion holds for general nonlinearities g and using only the feedback of the position x. On the other hand, the presence of a positive delay in (4.1) turns out to be essential in attaining most of these goals.

4.2 Averaging Theory and Periodic Solutions

For notation, we will use $\|\cdot\|$ for the usual Euclidean norm and $D_i g$ for the partial derivative of the function g with respect to its ith argument. Without loss of generality, it will be assumed throughout that $\omega = 1$ in (4.1), which can always be achieved by a rescaling of the time $t \mapsto \omega t$. Furthermore, it will be assumed that the functions $f : \mathbb{R} \to \mathbb{R}$ and $g : \mathbb{R}^3 \to \mathbb{R}$ are C^2, $f(0) = 0$, and $g(0, 0, \varepsilon) = 0$ for all ε.

The main tool for the analysis of (4.1) will be averaging theory for delay differential equations. Consider amplitude-phase variables (r, θ) defined by the transformation

$$\begin{aligned} x(t) &= r(t)\cos(t + \theta(t)) \\ \dot{x}(t) &= -r(t)\sin(t + \theta(t)). \end{aligned} \tag{4.2}$$

In these new coordinates, (4.1) takes the form

$$\begin{aligned} \dot{r} &= \varepsilon \sin(t + \theta)(g - f) \\ \dot{\theta} &= \varepsilon \frac{1}{r} \cos(t + \theta)(g - f), \end{aligned} \tag{4.3}$$

where the arguments of f and g are expressed in terms of r and θ, i.e.,

$$\begin{aligned} g &= g\left(r(t)\cos(t + \theta(t)), -r(t)\sin(t + \theta(t)), \varepsilon\right) \\ f &= f\left(r(t - \tau)\cos(t - \tau + \theta(t - \tau))\right). \end{aligned} \tag{4.4}$$

When $\varepsilon = 0$, the solutions of (4.3) are constants, which correspond by (4.2) to the usual harmonic oscillations. Thus, (4.3) can be viewed as a time-dependent perturbation of a simple harmonic oscillator in the amplitude-phase variables, which can be analyzed by the method of averaging for small ε.

Letting $y = (r, \theta) \in \mathbb{R}^2$, the system (4.3) and (4.4) is a delay differential equation describing the relation between the instantaneous derivative $\dot{y}(t)$ and the present and past values of $y(t)$. A solution $y(t)$ of (4.3) describes a trajectory in the infinite-dimensional state space $\mathcal{C} := C([-\tau, 0], \mathbb{R}^2)$, namely, the Banach space of continuous functions mapping the interval $[-\tau, 0]$ to $\mathbb{R}^2$, equipped with the supremum norm, $\|f\| = \sup_{x \in [-\tau, 0]} f(x)$. A point y_t on a trajectory is a piece of the solution function over an interval of length τ, defined by $y_t(s) = y(t + s)$, $s \in [-\tau, 0]$. In this notation, (4.3) can be written as

$$\dot{y}(t) = \varepsilon h(t, y_t, \varepsilon) \tag{4.5}$$

where h is periodic in t with period $T = 2\pi$. The averaged equation corresponding to (4.5) is defined as

$$\dot{z}(t) = \varepsilon \bar{h}(z_t) \tag{4.6}$$

where

$$\bar{h}(\varphi) := \frac{1}{T} \int_0^T h(t, \varphi, 0) \, dt. \tag{4.7}$$

In (4.6) z_t is understood as a *constant* element of $\mathcal{C}$. One can intuitively understand this by noting that y is slowly changing by (4.5), so that $y_t(s) \equiv y(t) + \mathcal{O}(\varepsilon)$ for $s \in [-\tau, 0]$, i.e., y is almost constant over an interval of length τ. Thus, (4.6) is an ordinary differential equation. In this way, averaging reduces the infinite-dimensional system (4.5) to a finite dimensional one, (4.6). Furthermore, by the averaging theorem, hyperbolic equilibrium points of (4.6) correspond to hyperbolic periodic solutions of (4.5), with the same stability type [22].

We now return to our main equation (4.1) and its equivalent formulation (4.3) to apply averaging theory. We average the equation for r given in (4.3) in the sense of (4.7) to obtain

$$\dot{r} = \varepsilon \frac{1}{2\pi} \int_0^{2\pi} \sin(t + \theta) \, g(r \cos(t + \theta), -r \sin(t + \theta), 0) \, dt - \tag{4.8}$$

$$\varepsilon \frac{1}{2\pi} \int_0^{2\pi} \sin(t + \theta) f(r \cos(t - \tau + \theta)) \, dt.$$

Here, in accordance with (4.6), r and θ are treated as constants over one period. With the change of variables $u = -(t + \theta)$, and using the fact that the integrand is 2π-periodic in u, the first integral in (4.8) becomes

$$\varepsilon \frac{1}{2\pi} \int_{-\theta}^{-\theta - 2\pi} (\sin u) g(r \cos u, r \sin u, 0) \, du.$$

Similarly, with $u = t - \tau + \theta$, the second integral in (4.8) can be written as

$$-\varepsilon \frac{1}{2\pi} \int_{\theta - \tau}^{2\pi + \theta - \tau} \sin(u + \tau) f(r \cos u) \, du$$

$$= -\varepsilon \frac{\sin \tau}{2\pi} \int_0^{2\pi} f(r \cos u) \cos u \, du - \varepsilon \frac{\cos \tau}{2\pi} \int_0^{2\pi} f(r \cos u) \sin u \, du \tag{4.9}$$

where we have used the fact that the second integral in (4.9) is zero. Combining, we see that the averaged equation for r has the form

$$\dot{r} = -\varepsilon(F(r) + G(r)), \tag{4.10}$$

where

$$F(r) = \frac{\sin \tau}{2\pi} \int_0^{2\pi} f(r \cos t) \cos t \, dt, \tag{4.11}$$

$$G(r) = \frac{1}{2\pi} \int_0^{2\pi} g(r \cos t, r \sin t, 0) \sin t \, dt. \tag{4.12}$$

By the averaging theorem for delay differential equations [22] and the transformation (4.2), positive hyperbolic equilibria R of (4.10) yield hyperbolic periodic solutions of (4.1) of the form $x(t) \approx R \cos t$, with the same stability type. In other words, if $R > 0$ is such that $F(R) + G(R) = 0$ and $F'(R) + G'(R) \neq 0$, then (4.1) has a periodic solution which is orbitally asymptotically stable if $F'(R) + G'(R) > 0$, and unstable if $F'(R) + G'(R) < 0$, as long as $\varepsilon > 0$ is sufficiently small. In this way, studying nontrivial hyperbolic periodic solutions of (4.1) is reduced to investigating positive and hyperbolic equilibrium points of (4.10).

The stability argument extends to $R = 0$ and can be used to deduce the stability of the zero solution of (4.1). In fact, this can be done directly without resorting to averaging, but it is interesting to relate the conditions to the averaged quantities (4.11) and (4.12). Thus, linearization of (4.1) about the zero solution gives the characteristic equation

$$\Delta(\lambda, \varepsilon) := \lambda^2 + 1 + \varepsilon(D_1 g(0, 0, \varepsilon) + \lambda D_2 g(0, 0, \varepsilon)) - \varepsilon f'(0) e^{-\lambda \tau} = 0. \tag{4.13}$$

When $\varepsilon = 0$, there are two roots on the imaginary axis: $\lambda = \pm i$. By the implicit function theorem, the roots depend smoothly on ε in a neighborhood of $\varepsilon = 0$, and implicit differentiation of (4.45) gives

$$\mathrm{Re}[\lambda'(\varepsilon)|_{\varepsilon=0}] = -\frac{1}{2}(f'(0) \sin \tau + D_2 g(0, 0, 0)) \tag{4.14}$$

$$= -(F'(0) + G'(0)). \tag{4.15}$$

Hence, the roots λ move into the left (respectively, right) complex half-plane if $F'(0) + G'(0)$ is positive (resp., negative), and remain there for all sufficiently small $\varepsilon > 0$, indicating that the zero solution of (4.1) is asymptotically stable if $F'(0) + G'(0) > 0$ and unstable if $F'(0) + G'(0) < 0$. Thus, the stability of equilibrium solutions (regulator problem) and periodic orbits (oscillator problem) can be conveniently expressed within the same framework.

Remark 1 For calculations it is worthwhile to note that F and G defined in (4.11) and (4.12) are both odd functions of r; i.e.

$$F(-r) = -F(r) \quad \text{and} \quad G(-r) = -G(r) \quad \text{for all } r \in \mathbb{R}. \tag{4.16}$$

For details, see [20].

4.3 Linear Feedback

Classical control theory has been extensively developed for linear systems or their linearizations at suitable operating points. Hence, it is natural to first consider a linear feedback law, namely the case when f has the form

$$f(x) = k_1 x, \tag{4.17}$$

for some feedback gain $k_1 \in \mathbb{R}$. Then by (4.11),

$$F(r) = \frac{1}{2} r k_1 \sin \tau. \tag{4.18}$$

We first consider the regulator problem of stabilizing of the zero solution. From (4.14),

$$\text{Re}[\lambda'(\varepsilon)|_{\varepsilon=0}] = -\frac{1}{2}(k_1 \sin \tau + D_2 g(0, 0, 0)).$$

We thus immediately obtain that, for small $\varepsilon > 0$, the zero solution of (4.1) is asymptotically stable if $k_1 \sin \tau > -D_2 g(0, 0, 0)$, and unstable if $k_1 \sin \tau < -D_2 g(0, 0, 0)$.

For periodic solutions, we seek positive fixed points R of the averaged equation (4.10), i.e., of

$$\dot{r} = -\varepsilon \left(\frac{1}{2} r k_1 \sin \tau + G(r) \right), \tag{4.19}$$

which gives

$$k_1 \sin \tau = -2 \frac{G(R)}{R}. \tag{4.20}$$

We define the function

$$\bar{G}(r) := \frac{G(r)}{r} \tag{4.21}$$

and note that

$$\bar{G}'(r) = \frac{r G'(r) - G(r)}{r^2} = \frac{1}{r}(G'(r) - \bar{G}(r)). \tag{4.22}$$

Combining (4.18), (4.20), and (4.21), we have

$$F'(R) + G'(R) = R \bar{G}'(R).$$

Therefore, a positive solution R of (4.20) is a fixed point of the averaged equation (4.10) and its stability is determined only by the sign of $\bar{G}'(R)$. By the averaging theorem, such points correspond to periodic solutions of the original equation (4.1) with amplitude R, which are orbitally asymptotically stable if $\bar{G}'(R) > 0$, and unstable if $\bar{G}'(R) < 0$, for sufficiently small $\varepsilon > 0$.

Now the important observation is that, for any desired amplitude $R > 0$, it is possible to find a feedback gain k_1 such that (4.20) is satisfied, provided $\sin \tau \neq 0$. Hence, delayed linear feedback can be effective in modifying the amplitude of periodic solutions. The condition $\sin \tau \neq 0$ shows that a nonzero delay in the feedback is essential for this task. However, the *stability* of these periodic solutions depends only on the function $\bar{G}$, and hence on the nonlinearity g. So, linear feedback is helpful in solving the oscillator problem only to the extent allowed by the nonlinearity g (for an example see [19]). On the other hand, linear feedback is effective in the regulator problem since it can stabilize the zero solution. We illustrate with examples.

Example 2 Consider the celebrated van der Pol oscillator under delayed feedback

$$\ddot{x}(t) + \varepsilon(x^2 - 1)\dot{x} + 1 = \varepsilon k_1 x(t - \tau). \tag{4.23}$$

It is well known that the uncontrolled system ($k_1 = 0$) has an attracting limit cycle solution $x(t) \approx 2 \cos t$ for small ε whereas the origin is unstable. We will show that we can modify the amplitude of limit cycle oscillations or make the origin stable by an appropriate choice of feedback gain k_1. Now, (4.23) has the form (4.1) with $g(x, \dot{x}, \varepsilon) = (x^2 - 1)\dot{x}$ and $f(x) = k_1 x$. The averaged quantities are

$$G(r) = \frac{1}{2}r \left(\frac{r^2}{4} - 1 \right) \tag{4.24}$$

and F as in (4.18); so the averaged equation for (4.23) is

$$\dot{r} = -\varepsilon \frac{r}{2} \left(\frac{r^2}{4} - 1 + k_1 \sin \tau \right). \tag{4.25}$$

This equation has a fixed point at zero, and another one at $r = R = 2\sqrt{1 - k_1 \sin \tau}$ if $k_1 \sin \tau < 1$. We have $\bar{G}'(r) = r/4$, which is clearly positive for all $r > 0$; so the fixed point R is stable whenever it exists. Therefore, for $0 < \varepsilon \ll 1$, (4.23) can have a stable periodic solution with amplitude approximately $R = 2\sqrt{1 - k_1 \sin \tau}$. In the absence of feedback, i.e., when $k_1 = 0$, we recover the familiar periodic solution $x(t) \approx 2 \cos t$ of (4.23), but we also see that we can set the amplitude arbitrarily by changing k_1. Moreover, by choosing $k_1 \sin \tau > -D_2 g(0, 0, 0) = 1$, the limit cycle oscillations can be destroyed and the origin can be made stable. Both situations are depicted in Fig. 4.1.

Example 3 We make a small modification to the van der Pol oscillator of Example 2 and consider the nonlinearity g with reversed sign, i.e., $g = -(x^2 - 1)\dot{x}$, again with linear delayed feedback:

$$\ddot{x}(t) - \varepsilon(x^2 - 1)\dot{x} + 1 = \varepsilon k_1 x(t - \tau). \tag{4.26}$$

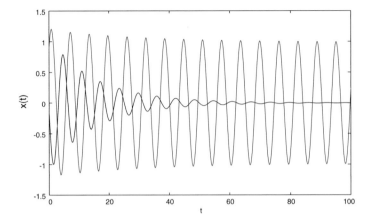

Fig. 4.1 Van der Pol oscillator under delayed feedback. Choosing a feedback gain of $k_1 = 0.75$ reduces the amplitude of limit cycle oscillations to 1 (*blue curve*), whereas increasing the gain to $k_1 = 2$ destroys the limit cycle and stabilizes the equilibrium point (*black curve*). *Parameter values* $\tau = \pi/2$ and $\varepsilon = 0.1$; random initial conditions

Now the averaged equation becomes

$$\dot{r} = -\varepsilon \frac{r}{2} \left(1 - \frac{r^2}{4} + k_1 \sin \tau \right), \tag{4.27}$$

which has a positive fixed point at $R = 2\sqrt{1 + k_1 \sin \tau}$ if $k_1 \sin \tau > -1$. As before, the amplitude of periodic solutions can be changed by appropriate choice of k_1 and τ. However, these solutions are all unstable because $\bar{G}'(R) = -R/4 < 0$. Thus, in this case the nonlinearity g does not allow the linear feedback to set up stable limit cycle oscillations at any amplitude R. (Note that the origin is locally stable as long as $k_1 \sin \tau > -D_2 g(0, 0, 0) = -1$.) In the next section we shall show how to overcome this limitation by adding a nonlinear term to the feedback function.

4.4 Nonlinear Feedback

As we have seen in Sect. 4.3, linear feedback is sufficient for the regulator problem but in general not for the oscillator problem. Therefore, we now turn to nonlinear feedback schemes. We show that, by adding a cubic term to the feedback function, the possibility of controlling oscillations through delayed feedback is greatly improved.

We consider a feedback function of the form

$$f(x) = k_1 x + k_3 x^3. \tag{4.28}$$

We will show that the coefficients k_i can be chosen so that the averaged equation (4.10) has a stable equilibrium point at a desired value R.

The averaged function F corresponding to (4.28) is also a cubic polynomial,

$$F(r) = q_1 r + q_3 r^3, \tag{4.29}$$

with

$$q_1 = \frac{1}{2} k_1 \sin \tau \quad \text{and} \quad q_3 = \frac{3}{8} k_3 \sin \tau. \tag{4.30}$$

Consequently,

$$F(r) + G(r) = r(q_1 + q_3 r^2 + \bar{G}(r)), \tag{4.31}$$

where the function $\bar{G}$ is defined in (4.21). Now let $R > 0$ be given. We will choose q_1 in a suitable manner, to be described shortly, and define q_3 in terms of q_1 as

$$q_3 = \frac{-q_1 - \bar{G}(R)}{R^2}. \tag{4.32}$$

With this choice of q_3, it follows from (4.31) that $F(R) + G(R) = 0$; so, R is an equilibrium point of the averaged equation (4.10). We will choose q_1 to ensure that R is a stable equilibrium, i.e., $F'(R) + G'(R) > 0$. From (4.31),

$$F'(R) + G'(R) = R(2Rq_3 + \bar{G}'(R)), \tag{4.33}$$

which is positive provided

$$q_3 > -\frac{\bar{G}'(R)}{2R}. \tag{4.34}$$

Using (4.34) in (4.32), the condition on q_1 is found as

$$q_1 < \frac{1}{2} R\bar{G}'(R) - \bar{G}(R). \tag{4.35}$$

From conditions (4.34) and (4.35), the feedback coefficients of (4.28) can then be calculated, using (4.30), as $k_1 = 2q_1(\sin \tau)^{-1}$ and $k_3 = 8q_3(3 \sin \tau)^{-1}$, whenever $\sin \tau \neq 0$. We thus have a procedure for feedback design to create a stable periodic solution with a prescribed amplitude R: First choose k_1 and/or τ so that

$$k_1 \sin \tau < R\bar{G}'(R) - 2\bar{G}(R). \tag{4.36}$$

Subsequently, calculate k_3 through the formula

$$k_3 = -\frac{8}{3R^2} \left(\frac{1}{2} k_1 + \frac{\bar{G}(R)}{\sin \tau} \right). \tag{4.37}$$

Then, for all sufficiently small $\varepsilon > 0$, the nonlinear feedback (4.28) ensures that the system (4.1) has an asymptotically orbitally stable periodic solution whose amplitude is $R + \mathcal{O}(\varepsilon)$. By a similar reasoning it can be seen that, by reversing the inequality in (4.36), one obtains an unstable periodic solution with amplitude $R + \mathcal{O}(\varepsilon)$.

Example 4 We consider the modified van der Pol equation of Example 3, this time with a nonlinear feedback:

$$\ddot{x}(t) - \varepsilon(x^2 - 1)\dot{x} + 1 = \varepsilon k_1 x(t - \tau) + \varepsilon k_3 x^3(t - \tau). \tag{4.38}$$

The averaged equation is

$$\dot{r} = -\varepsilon \frac{r}{2}\left(1 - \frac{r^2}{4} + k_1 \sin \tau + \frac{3r^2}{4}k_3 \sin \tau\right), \tag{4.39}$$

which has the positive fixed point

$$r = R = 2\sqrt{\frac{1 + k_1 \sin \tau}{1 - 3k_3 \sin \tau}} \tag{4.40}$$

whenever the radicand is positive. Furthermore,

$$R\bar{G}'(R) - 2\bar{G}(R) = -\frac{R^2}{4} - 2\left(\frac{1}{2} - \frac{R^2}{8}\right) = -1;$$

so, choosing $k_1 \sin \tau < -1$ satisfies (4.36) and ensures that the fixed point R is stable. Formula (4.37) then determines the remaining coefficient k_3. For instance, if it is desired to create stable oscillations at an amplitude of $R = 3$ with $\tau = \pi/2$, we can choose, e.g., $k_1 = -2$, and find $k_3 = 13/27$ from (4.40). Figure 4.2 shows the resulting limit cycle.

Remark 5 One may wonder why we have chosen to add a cubic term in (4.28). We note that if f is an even function, then (4.11) gives $F(-r) = F(r)$, so that, in view of (4.16), one has $F(r) \equiv 0$. Hence, F only depends on the odd part $f_o(x) = \frac{1}{2}(f(x) - f(-x))$ of f. Since for small ε the dynamics of (4.1) is determined by G and F, there is no loss of generality in assuming that f is an odd function. In this sense, (4.28) represents the simplest nonlinear feedback function (at least in the ring of polynomials). Together with the results of the previous section, it is seen that simple delayed feedback schemes can be quite powerful in oscillation control.

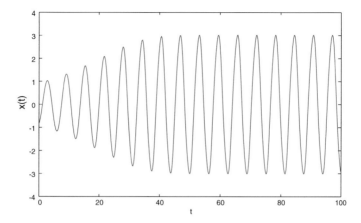

Fig. 4.2 The modified van der Pol oscillator (4.38) under nonlinear feedback exhibiting stable limit cycle oscillations at the prescribed amplitude 3. Parameter values are $\tau = \pi/2$, $\varepsilon = 0.1$, $k_1 = -2$, and $k_3 = 0.48$; random initial conditions

4.5 Controlling the Frequency of Oscillations

The results of the foregoing sections indicate that the control law (4.28) can be effective for controlling the stability of periodic solutions as well as specifying their amplitude. However, so far we have not discussed controlling the *frequency* of the oscillations. For this latter goal, it turns out that the feedback magnitude in (4.1) needs to be modified. Namely, we need to relax the assumption that the forcing term on the right hand side of (4.1) is of order ε. Therefore, we will now consider the slightly modified equation

$$\ddot{x} + x + \varepsilon g(x, \dot{x}, \varepsilon) = f(x(t - \tau)). \tag{4.41}$$

The reason for this change of right hand side can be understood as follows. The previous system (4.1) was viewed as an ε-perturbation of a simple harmonic oscillator. There, by using a suitable feedback function, we were able to create a stable limit cycle at a prescribed amplitude because the harmonic oscillator has periodic solutions of all amplitudes. However, all these solutions have the same frequency 1. Therefore, forcing the system with a feedback magnitude of order ε cannot change the frequency appreciably. In the case of (4.41), however, the unperturbed system

$$\ddot{x} + x = f(x(t - \tau)). \tag{4.42}$$

is no longer the simple harmonic oscillator; in fact, it is not a planar system anymore if $\tau \neq 0$. This may offer more possibilities for choosing a desired periodic solution at a certain amplitude *and* frequency. The price to be paid is that (4.42) is an infinite-dimensional system.

We can proceed in a similar way using averaging theory and the insight gained from the previous sections. From Sect. 4.4 we know that a cubic function of the form $f(x) = \varepsilon(k_1 x + k_3 x^3)$ can be used in (4.42) to control amplitude of oscillations, and we have observed that we would need a feedback term of higher magnitude if we are to have any hope of modifying frequencies significantly. We are therefore naturally led to trying the following feedback form

$$f(x) = kx + \varepsilon(k_1 x + k_3 x^3) \tag{4.43}$$

in (4.41).

With the choice (4.43) and small ε, (4.41) can be viewed as an perturbation of the linear system

$$\ddot{x} + x = kx(t - \tau). \tag{4.44}$$

Just like for the harmonic oscillator, we would like to know what variety of stable periodic solutions (4.44) has. For this purpose we seek purely imaginary solutions of the corresponding characteristic equation

$$\chi(\lambda) := \lambda^2 + 1 - ke^{-\lambda\tau} = 0. \tag{4.45}$$

The following result summarizes the frequency range about such solutions; for a proof see [6].

Lemma 6 ([6]) *Let $\Omega \in (\sqrt{2/5}, \sqrt{2})$ and $k = \Omega^2 - 1$. Let τ be an arbitrary nonnegative number if $\Omega = 1$, otherwise let $\tau = \pi/\Omega$. Then the characteristic equation (4.45) has precisely two roots $\lambda = \pm i\Omega$ on the imaginary axis and no roots with positive real parts.*

Thus, unlike the simple harmonic oscillator which has periodic solutions only with a single frequency, (4.44) has periodic solutions with a range of frequencies in the interval $(\sqrt{2/5}, \sqrt{2})$. Coming from a linear equation, these solutions can have arbitrary amplitudes since any multiple of a solution is also a solution. We are now in a familiar setting: after fixing one of these frequencies by choosing k as in the above Lemma, we can add $\mathcal{O}(\varepsilon)$ terms to the feedback to account for $\mathcal{O}(\varepsilon)$ nonlinearities in order to obtain stable limit cycles in (4.41) with a prescribed amplitude. In other words, we activate the coefficients k_1 and k_2 in the feedback law (4.43). We note, however, that the calculation of the averaged equations involves quite a different technique than the previous sections, namely the projection of the dynamics of (4.41) onto a center manifold corresponding to the roots $\lambda = \pm i\Omega$ of the characteristic equation (4.45). As the theory of center manifold reduction for delay differential equations is beyond our scope here, we refer the interested reader to [6] for details. The important thing to note is that the averaged quantities F and G are now given by

$$F(r) = \frac{1}{2\pi} \frac{4\Omega \sin(\Omega\tau) - 2\tau\left(1 - \Omega^2\right)\cos(\Omega\tau)}{\tau^2(1 - \Omega^2)^2 + 4\Omega^2} \int_0^{2\pi} \cos t \, f\left(r\cos t\right) dt \quad (4.46)$$

$$G(r) = \frac{1}{2\pi} \int_0^{2\pi} \frac{4\Omega \sin t + 2\tau\left(1 - \Omega^2\right)\cos t}{\tau^2(1 - \Omega^2)^2 + 4\Omega^2} \, g\left(r\cos t, \Omega r \sin t, 0\right) dt \quad (4.47)$$

instead of (4.11) and (4.12). With this change, the averaged equation still has the form (4.10) and the stability of its fixed points can be calculated as before.

More concretely, a nonlinear feedback function (4.43) can be constructed as follows: Given amplitude $R > 0$ and frequency $\Omega \in (\sqrt{2/5}, \sqrt{2})$, take $k = \Omega^2 - 1$, and choose $\tau = \pi/\Omega$ if $\Omega \neq 1$ (see Lemma 6) or take any τ such that $\sin \tau \neq 0$ if $\Omega = 1$. For $\mathcal{O}(\varepsilon)$ linear and cubic terms (4.28) in the feedback, (4.46) gives

$$F(r) = \frac{1}{2}\gamma k_1 r + \frac{3}{8}\gamma k_3 r^3, \quad (4.48)$$

where

$$\gamma = \frac{4\Omega \sin(\Omega\tau) - 2\tau\left(1 - \Omega^2\right)\cos(\Omega\tau)}{\tau^2(1 - \Omega^2)^2 + 4\Omega^2}, \quad (4.49)$$

which has a form similar to (4.29) with γ replacing $\sin \tau$. As in Sect. 4.4, choose the feedback coefficient k_1 satisfying (4.36) and determine k_3 through the formula (4.37), this time using (4.47) and (4.48) to calculate G and F. This determines all the feedback coefficients in (4.43). Then the averaging theorem yields that, for all sufficiently small $\varepsilon > 0$, the system (4.41) under the nonlinear feedback (4.43) has an asymptotically orbitally stable periodic solution of the form $x(t) \approx R \cos(\Omega t)$.

Remark 7 Recall that, by our standing assumption, time is rescaled in (4.41) so that the uncontrolled system ($f \equiv 0$) has frequency 1 in the rescaled time. Thus, the fact that the feedback term can set the frequency of the limit cycle to any $\Omega \in (\sqrt{2/5}, \sqrt{2})$ implies that it can reduce the frequency of the uncontrolled oscillator by as much as about 37 % or increase it by about 41 %.

Example 8 We return to the van der Pol oscillator used in Example 2, this time with the aim of changing *both* the frequency and amplitude of oscillations by delayed linear feedback. The controlled system is given by

$$\ddot{x}(t) + \varepsilon(x^2 - 1)\dot{x} + 1 = (k + \varepsilon k_1)x(t - \tau). \quad (4.50)$$

From $g(x, \dot{x}, \varepsilon) = (x^2 - 1)\dot{x}$ and (4.47) we calculate

$$G(r) = \frac{4\Omega^2}{\tau^2(1 - \Omega^2)^2 + 4\Omega^2} \times \frac{r}{2}\left(\frac{r^2}{4} - 1\right)$$

(compare with (4.24)), and from (4.48) and (4.49) we have $F(r) = \frac{1}{2}\gamma k_1 r$. If $\Omega \neq 1$ and τ is to be chosen according to Lemma 6 as $\tau = \pi/\Omega$, then (4.49) simplifies to

$$\gamma = \frac{2\pi\,\Omega\left(1-\Omega^2\right)}{\pi^2(1-\Omega^2)^2+4\Omega^4},$$

and the averaged equation (4.10) becomes

$$\dot{r} = -\varepsilon \frac{\Omega^4}{\pi^2(1-\Omega^2)^2+4\Omega^4} \times \frac{r}{2}\left(r^2 - 4 + \frac{2\pi\left(1-\Omega^2\right)}{\Omega^3}k_1\right).$$

There exists a positive fixed point

$$R = \sqrt{4 - 2\pi\left(1-\Omega^2\right)k_1/\Omega^3}\,, \qquad (4.51)$$

provided the radicand is positive. Note that $\bar{G}'(r) > 0$ for all $r > 0$, as in Example 2, so R is a stable fixed point. From (4.51) the value of k_1 can be determined as

$$k_1 = \frac{\Omega^3(4 - R^2)}{2\pi\left(1-\Omega^2\right)} \qquad (4.52)$$

for given values of R and Ω. For instance, to create a stable limit cycle at about 75 % of the frequency ($\Omega = 3/4$) and twice the amplitude ($R = 4$) of the uncontrolled van der Pol oscillator, we calculate $k = \Omega^2 - 1 = -7/16$ and $\tau = 4\pi/3$ from Lemma 6 and $k_1 = -81/14\pi$ from (4.52). Figure 4.3 shows the resulting limit cycle oscillations obtained for $\varepsilon = 0.01$ and random initial conditions.

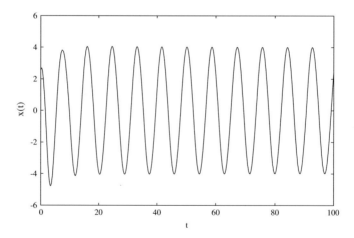

Fig. 4.3 Van der Pol oscillator of Example 8 exhibiting stable limit cycle oscillations at reduced frequency and increased amplitude

4.6   Conclusion

We have shown how delayed output feedback can be effectively used in the control of oscillatory behavior in weakly nonlinear systems. While the local stability of an equilibrium solution can be studied through a linear stability analysis, controlling periodic behavior in general requires nonlinear techniques. Here we have seen that linear feedback is capable of stabilizing the zero solution. Moreover, by adding nonlinear terms to the feedback function, it is possible to create stable limit cycle oscillations with any prescribed amplitude. In addition, delayed feedback can also modify the frequency of oscillations to a certain extent. In many cases these feats cannot be accomplished by undelayed feedback of position, exhibiting a *positive* use of delays in control.

References

1. G.H. Tallman, O.J.M. Smith, IRE Trans. Automat. Control **AC-3**, 14 (1958)
2. I.H. Suh, Z. Bien, IEEE Trans. Automat. Control **AC-24**, 370 (1979)
3. I.H. Suh, Z. Bien, IEEE Trans. Automat. Control **AC-25**, 600 (1980)
4. N. Shanmugathasan, R.D. Johnston, Int. J. Control **48**(3), 1137 (1988)
5. W.H. Kwon, G.W. Lee, S.W. Kim, Int. J. Control **52**(60), 1455 (1990)
6. F.M. Atay, in *Dynamics, Bifurcations and Control, Lecture Notes in Control and Information Sciences*, vol. 273, ed. by F. Colonius, L. Grüne (Springer-Verlag, Berlin, 2002), pp. 103–116. doi:10.1007/3-540-45606-6_7
7. F.M. Atay, Discrete Continuous Dyn. Syst. Series S **1**(2), 197 (2008). doi:10.1137/060673813
8. F.M. Atay, in *Complex Time-Delay Systems*, ed. by F.M. Atay, Understanding Complex Systems (Springer Berlin Heidelberg, 2010), pp. 45–62. doi:10.1007/978-3-642-02329-3_2
9. S. Hastings, J. Tyson, D. Webster, J. Differ. Equ. **25**, 39 (1977)
10. U. an der Heiden, J. Math. Biol. **8**, 345 (1979)
11. W. Reichardt, T. Poggio, Quart. Rev. Biophys. **3**, 311 (1976)
12. T. Poggio, W. Reichardt, Quart. Rev. Biophys. **9**, 377 (1976)
13. A. Beuter, J. Bélair, C. Labrie, Bull. Math. Biol. **55**, 525 (1993)
14. C.W. Eurich, J.G. Milton, Phys. Rev. E **54**, 6681 (1996)
15. R. Vallée, M. Dubois, M. Coté, C. Delisle, Phys. Rev. A **36**, 1327 (1987)
16. B.S. Berger, M. Rokni, I. Minis, Quart. Appl. Math. **51**, 601 (1993)
17. N. Olgac, B.T. Holm-Hansen, J. Sound Vib. **176**(1), 93 (1994)
18. F.M. Atay, Appl. Math. Lett. **12**(5), 51 (1999). doi:10.1016/S0893-9659(99)00056-7
19. F.M. Atay, J. Sound Vib. **218**(2), 333 (1998). doi:10.1006/jsvi.1998.1843
20. F.M. Atay, Int. J. Control **75**(5), 297 (2002). doi:10.1080/00207170110107265
21. F.M. Atay, J. Differ. Equ. **221**(1), 190 (2006). doi:10.1016/j.jde.2005.01.007
22. J.K. Hale, J. Differ. Equ. **2**, 57 (1966)

Chapter 5
Global Effects of Time-Delayed Feedback Control Applied to the Lorenz System

Anup S. Purewal, Bernd Krauskopf and Claire M. Postlethwaite

Abstract Time-delayed feedback control was introduced by Pyragas in 1992 as a general method for stabilizing an unstable periodic orbit of a given continuous-time dynamical system. The analysis of Pyragas control focused on its application to the normal form of a subcritical Hopf bifurcation, and it was initially concerned with stabilization near the Hopf bifurcation. A recent study considered this normal form delay differential equation model more globally in terms of its bifurcation structure for any values of system and control parameters. This revealed families of delay-induced Hopf bifurcations and secondary stability regions of periodic orbits. In this contribution we show that these results for the normal form are relevant in an application context. To this end, we present a case study of the well-known Lorenz system subject to Pyragas control to stabilize one of its two (symmetrically related) saddle periodic orbits. We find that, for a suitably chosen value of the 2π-periodic feedback phase, the controlled Hopf normal form describes qualitatively the bifurcation set and relevant stability regions that exist in the controlled Lorenz system down to the homoclinic bifurcation where the target saddle periodic orbit is born. In particular, there are secondary stability regions of periodic orbits in the Lorenz system. Finally, the normal form also describes correctly the effect of a delay mismatch.

5.1 Introduction

The control of unstable dynamics has been an area of significant research in recent years [1–4]. The control or stabilization of unstable periodic orbits, which occur naturally in many nonlinear dynamical systems, has been a particular focus [5–9], and a number of control schemes have been designed for this purpose. These include notably the method by Ott et al. [7] and time-delayed feedback control introduced by Pyragas [8, 10–12]—the subject of this investigation. The Pyragas control scheme adds a continuous time-delayed feedback term to the system to be stabilised, which

A.S. Purewal · B. Krauskopf (✉) · C.M. Postlethwaite
Department of Mathematics, University of Auckland, Auckland, New Zealand
e-mail: b.krauskopf@auckland.ac.nz

© Springer International Publishing Switzerland 2016   81
E. Schöll et al. (eds.), *Control of Self-Organizing Nonlinear Systems*,
Understanding Complex Systems, DOI 10.1007/978-3-319-28028-8_5

drives the dynamics to the target unstable periodic orbit. More formally, consider a
system of autonomous ordinary differential equations (ODEs)

$$\dot{y}(t) = g(y(t), \mu), \tag{5.1}$$

where $y \in \mathbb{R}^n$, $g : \mathbb{R}^n \times \mathbb{R}^m \mapsto \mathbb{R}^n$ is a smooth function and $\mu \in \mathbb{R}^m$ is a vector of
scalar parameters. Suppose (5.1) has an unstable periodic orbit Γ with the (parameter
dependent) period $T(\mu)$. Adding Pyragas control to (5.1) gives

$$\dot{y}(t) = g(y(t), \mu) + K[y(t - \tau) - y(t)],$$
$$\tau = T(\mu). \tag{5.2}$$

A control force is generated from the difference between the two signals $y(t)$ and
$y(t - \tau)$, where τ is set to the period $T(\mu)$ of the target periodic orbit Γ; the actual
feedback is determined by the $n \times n$ feedback gain matrix K. If the control is successful the feedback signal becomes smaller as Γ is approached and, at Γ, it becomes
zero. Hence, the Pyragas control scheme is non-invasive.

Pyragas control has been implemented successfully in a number of applications,
including laser [13, 14], electronic [15, 16] and engineering [17, 18] systems. The
control scheme has also been modified to include spatial feedback [19, 20] and to
control synchronization in coupled systems [21]. Recently, Postlethwaite et al. [22]
and Schneider and Bosewitz [23] have extended the standard Pyragas scheme to
systems with symmetries, which allows for the selection and stabilization of periodic solutions via their spatio-temporal pattern; this extension is particularly useful
for stabilizing periodic orbits of networks. For a comprehensive review of Pyragas
control and its various applications and extensions see [24] and the references therein.

It is important to realise that system (5.2) with Pyragas control is a delay differential equation (DDE) with a single fixed delay τ, rather than an ODE. For a DDE such
as (5.2), a continuous function $\phi(t)$ on the interval $[-\tau, 0]$ must be specified as an
initial condition [25], meaning that its phase space is the infinite dimensional space
$\mathbb{C}([-\tau, 0]; \mathbb{R}^n)$ of continuous functions from $[-\tau, 0]$ into $\mathbb{R}^n$. For DDEs with a finite
number of fixed delays any equilibrium has at most a finite number of unstable eigen-
values; similarly, any periodic solution has at most a finite number of Floquet multipliers outside of the unit circle in the complex plane [25, 26]. Therefore, standard
bifurcation theory (as known for ODEs) applies to DDEs such as (5.2); in particular,
one finds the usual codimension-one bifurcations: saddle-node, Hopf, saddle-node
of limit cycle (SNLC), period-doubling and torus (Neimark-Sacker) bifurcations.
On the other hand, it is quite difficult to study a DDE analytically, and a number of
numerical tools have been specifically designed for their bifurcation analysis [27].
Throughout this work, we use the package DDE-Biftool [28, 29], which is implemented in Matlab and allows for the numerical continuation of equilibria, periodic
orbits and their bifurcations of codimension one.

Most of the analysis of Pyragas control has concentrated on its application to the subcritical Hopf normal form

$$\dot{z} = (\lambda + i)z(t) + (1 + i\gamma)|z(t)|^2 z(t) + b_0 e^{i\beta}[z(t - \tau) - z(t)], \qquad (5.3)$$

introduced by Fiedler et al. [30]. Here $z = x + iy \in \mathbb{C}$ and $\lambda, \gamma \in \mathbb{R}$. The complex number $b_0 e^{i\beta}$ is the feedback gain, with feedback strength $b_0 \in \mathbb{R}$ and 2π-periodic phase β. System (5.3) has a Hopf bifurcation at $\lambda = 0$, denoted H_P, where an unstable periodic orbit Γ_P bifurcates from the origin (which is always an equilibrium). The periodic orbit Γ_P exists for $\lambda < 0$; it has amplitude $\sqrt{-\lambda}$ and period

$$T(\lambda) = \frac{2\pi}{1 - \gamma\lambda}, \qquad (5.4)$$

which is used in (5.3) by setting $\tau = T(\lambda)$. The analysis in [30] provided a counterexample to the so-called odd-number limitation [31], which states that the Pyragas scheme cannot stabilize periodic orbits with an odd number of Floquet multipliers. Subsequently, system (5.3) has become the standard example for the analysis of Pyragas control [23, 32–35]. Initially, the analysis was concerned with the stabilization of the target periodic orbit Γ_P near the Hopf bifurcation. Fiedler et al. [30] found that for small values of feedback gain b_0 the target periodic orbit Γ_P is stabilized in a transcritical bifurcation with a stable delay-induced periodic orbit. Just et al. [34] found that immediately above the threshold level of feedback strength

$$b_0^c = \frac{-1}{2\pi(\cos\beta + \gamma\sin\beta)}, \qquad (5.5)$$

the periodic orbit Γ_P bifurcates stably from H_P; the local bifurcation diagram near the point b_0^c on H_P was presented in [32].

In [36] we presented a detailed global bifurcation analysis of (5.3) throughout the parameter space, that is, well beyond a neighbourhood of the subcritical Hopf bifurcation. It revealed an overall structure with infinitely many delay-induced Hopf bifurcations and further bifurcations, including saddle-node of limit cycle and torus bifurcations. The overall domain of stability of the target periodic orbit was identified in parameter space, and we also found regions of stability of secondary periodic orbits. In other words, one must carefully choose parameter values to ensure that the control scheme is successful.

The question arises whether these global findings for the normal form system (5.3) have predictive power for any other system with a subcritical Hopf bifurcation. Indeed, normal form theory states that this will be the case, and Pyragas control will be successful, in the direct vicinity of the Hopf bifurcation. However, it is unclear a priori whether and how the overall bifurcation structure discussed in [36], away from the initial Hopf bifurcation, manifests itself in another context.

To address this question, we perform here a bifurcation analysis of the Lorenz system subject to Pyragas control to stabilize one of its saddle-periodic orbits. Postlethwaite and Silber [37] considered this system near its subcritical Hopf bifurcation and found that the mechanism of stabilization is the same as that for (5.3). Furthermore, Brown [38] showed that, close to the Hopf bifurcation, applying feedback directly to the Lorenz system is equivalent to applying feedback to the system after it has been reduced to the Hopf normal form. In other words, the dynamics locally near the Hopf bifurcation are the same for both ways of applying delayed feedback control.

These results form the starting point of our investigation into the global bifurcation structure of the Lorenz system with Pyragas control. We present one- and two-parameter bifurcation diagrams of the controlled Lorenz DDE and compare them directly with those of (5.3). We identify the overall domain of stability of the target periodic orbit in the controlled Lorenz system; it agrees well with that for the normal form when the feedback phase is chosen appropriately. Moreover, we find several domains of stability of further delay-induced periodic orbits; these exist between the subcritical Hopf bifurcation and the homoclinic bifurcation where the periodic orbit is created. Lastly, we also consider the effects of a delay mismatch in the Lorenz system subject to Pyragas control, where the delay τ is not exactly the period of the target periodic orbit. As for the controlled Hopf normal form, at least a linear approximation of the target period is required for Pyragas control to be successful. Overall, we conclude that the global bifurcation structure of the normal form (5.3) must be considered as relevant in an application context.

This contribution is organized as follows. Section 5.2 describes how Pyragas control is implemented in the Lorenz system. Section 5.3 presents the bifurcation analysis of the Lorenz system and compares it with that of the Hopf normal form. In Sect. 5.3.1 we recall from [36] the bifurcation set of (5.3) in the (λ, b_0)-plane. The bifurcation set of the controlled Lorenz system in the corresponding (ρ, b_0)-plane is presented in Sect. 5.3.2 for the standard values of the parameters; we illustrate the local nature of the control in Sect. 5.3.3 and investigate the influence of the feedback phase in Sect. 5.3.4. The effect of a delay mismatch is discussed briefly in Sect. 5.4. Some conclusions can be found in Sect. 5.5.

5.2 The Controlled Lorenz System

The Lorenz equations were derived by Lorenz [39] as a simplified model of thermal convection in the atmosphere; they take the form

$$
\begin{aligned}
\dot{x}(t) &= \sigma(y(t) - x(t)), \\
\dot{y}(t) &= \rho x(t) - y(t) - x(t)z(t), \\
\dot{z}(t) &= -\alpha z(t) + x(t)y(t).
\end{aligned}
\tag{5.6}
$$

Note that we use α, rather than the standard β, in the third equation of (5.6) to avoid confusion with the feedback phase β of (5.3). System (5.6) is perhaps the most famous example of a chaotic system. In particular, the well-known butterfly-shaped

Lorenz attractor can be found for the classical parameter values $\sigma = 10$, $\alpha = \frac{8}{3}$ and $\rho = 28$.

System (5.6) is most often studied for the parameter regime $\sigma = 10$, $\alpha = \frac{8}{3}$ and $\rho > 0$, that is, ρ is taken as the primary bifurcation parameter; this choice is also adopted here. The origin is always an equilibrium; it is stable for $\rho < 1$ and it loses stability in a supercritical pitchfork bifurcation at $\rho = 1$. For $\rho > 1$, there exist the two further equilibria

$$p^\pm = \left(\pm\sqrt{\alpha(\rho - 1)}, \pm\sqrt{\alpha(\rho - 1)}, \rho - 1 \right), \tag{5.7}$$

which are each other's counterparts under the symmetry $(x, y, z) \rightarrow (-x, -y, z)$ of the Lorenz system. The equilibria $p^\pm$ lose stability in a subcritical Hopf bifurcation at

$$\rho_H = \frac{\sigma(\sigma + \alpha + 3)}{(\sigma - \alpha - 1)} \approx 24.7368, \tag{5.8}$$

for $\sigma = 10$ and $\alpha = \frac{8}{3}$. For $\rho > \rho_H$, the equilibria $p^\pm$ are saddle points. From the Hopf bifurcation at ρ_H emanate two saddle periodic orbits $\Gamma^\pm$, which exist for $\rho_{\mathrm{hom}} < \rho < \rho_H$; at $\rho_{\mathrm{hom}} \approx 13.926$ the two periodic orbits disappear (for decreasing ρ) in a homoclinic bifurcation of the origin. This homoclinic bifurcation is also referred to as a homoclinic explosion point [40], and it is the source of the complicated dynamics exhibited in the Lorenz system. This type of dynamics is initially not attracting but of saddle type until, at $\rho = \rho_{\mathrm{het}} \approx 24.0579$, there is a heteroclinic connection between the origin and $\Gamma^\pm$. This heteroclinic bifurcation creates the chaotic attractor; for more information on this transition see, for example, [40–42] and references therein.

We apply Pyragas control to the Lorenz system in the manner first suggested by Postlethwaite and Silber [37] to stabilize the periodic orbit $\Gamma_P = \Gamma^+$ that bifurcates from the positive secondary equilibrium p^+ at the Hopf bifurcation given by (5.8) and is also denoted H_P. After a coordinate transformation that moves p^+ to the origin, the Lorenz system takes the form

$$\begin{aligned}
\dot{u}(t) &= \sigma(v(t) - u(t)), \\
\dot{v}(t) &= u(t) - v(t) - (\rho - 1)w(t) - (\rho - 1)u(t)w(t), \\
\dot{w}(t) &= \alpha(u(t) + v(t) - w(t) + u(t)v(t)).
\end{aligned} \tag{5.9}$$

When the Pyragas control term is added to (5.9) the system becomes

$$\begin{pmatrix} \dot{u}(t) \\ \dot{v}(t) \\ \dot{w}(t) \end{pmatrix} = J(\rho) \begin{pmatrix} u(t) \\ v(t) \\ w(t) \end{pmatrix} + \begin{pmatrix} 0 \\ -(\rho - 1)u(t)w(t) \\ \alpha u(t)v(t) \end{pmatrix} + \Pi \begin{pmatrix} u(t - \tau) - u(t) \\ v(t - \tau) - v(t) \\ w(t - \tau) - w(t) \end{pmatrix}, \tag{5.10}$$

where

$$J(\rho) = \begin{pmatrix} -\sigma & \sigma & 0 \\ 1 & -1 & -(\rho - 1) \\ \alpha & \alpha & -\alpha \end{pmatrix} \qquad (5.11)$$

and the feedback gain matrix Π is yet undetermined.

Close to the Hopf bifurcation point H_P, the target unstable periodic orbit Γ^+ lies on the two-dimensional centre manifold with an extra stable direction. One approach to applying Pyragas feedback would be to reduce the uncontrolled three-dimensional Lorenz system (5.9) to a two-dimensional system that governs the dynamics on the centre manifold. A normal form transformation (up to the cubic term) would then remove all nonlinear terms except the one that is proportional to $|z|^2 z$, which is the cubic term of the Hopf normal form. Thereby, the reduced system would be in Hopf normal form as in Eq. (5.3). Pyragas feedback control could then be applied to this system in exactly the same way as for the normal form case. However, it is unpractical from an application point of view to perform this reduction to normal form for every system to which one would like to apply Pyragas control—especially in an experimental setting, where the governing equations may even be unknown.

Here we consider the alternative approach of applying the Pyragas scheme, where feedback control is added to the original system but only applied in the respective unstable directions. This is achieved by defining the feedback gain matrix Π in (5.10) as

$$\Pi = EGE^{-1}, \qquad (5.12)$$

where

$$G = \begin{pmatrix} 0 & 0 & 0 \\ 0 & b_0 \cos \eta & -b_0 \sin \eta \\ 0 & b_0 \sin \eta & b_0 \cos \eta, \end{pmatrix} \qquad (5.13)$$

and E is the matrix of eigenvectors that puts $J(\rho_H)$ in Jordan normal form, that is,

$$E^{-1} J(\rho_H) E = \begin{pmatrix} \mu_H^S & 0 & 0 \\ 0 & 0 & -\omega_H \\ 0 & \omega_H & 0 \end{pmatrix}. \qquad (5.14)$$

The matrix G corresponds to the feedback gain of the Hopf normal form (applied to the last two coordinates), where $b_0 \in \mathbb{R}$ is the control amplitude with the convention that $b_0 \geq 0$. The parameter η is the 2π-periodic feedback phase. Note that, for ease of comparison later, we use the symbol η instead of the standard β as in the Hopf normal form case. Namely, due to the coordinate transformation (5.12), choosing a value of η in (5.13) is not the same as choosing the same value of β in (5.3). This is of course also true for the parameter b_0; however, since b_0 undergoes only a linear scaling, we use the same symbol for simplicity. Finally, conjugation with the matrix E ensures that the feedback is applied only in the centre eigenspace; a more detailed explanation for this choice of feedback gain can be found in [37].

A key ingredient of Pyragas control is a priori knowledge of a functional form of the period of the target periodic orbit in the underlying ODE, which needs to be set as the delay τ of the feedback term. As is the case for most nonlinear systems, and in contrast to the normal form, for the Lorenz system (5.6) there is no analytic expression for the period of the target period orbit Γ^+. Therefore, we proceed as in [37] by taking advantage of the fact that Γ^+ and its period can be continued numerically. More specifically, for fixed $\sigma = 10$ and $\alpha = \frac{8}{3}$ we continue Γ^+ in the parameter ρ in the interval $\rho_{\text{hom}} < \rho < \rho_H$ where Γ^+ exists; this single computation for the ODE (5.6) can be performed, for example, with the package AUTO [43]. The resulting data set forms the basis for defining $T(\rho)$ over the extended range $\rho_{\text{hom}} < \rho$. Very close to the homoclinic bifurcation at $\rho_{\text{hom}} \approx 13.926$, where numerical continuation becomes difficult, we use the approximation $T(\rho_{\text{hom}}) = -0.974 \log(\rho - \rho_{\text{hom}})$ to represent the period of Γ^+ going to infinity at the homoclinic bifurcation. Moreover, for $\rho_H < \rho$ we extrapolate the data set with the function

$$T(\rho) = \frac{T_H}{1 + 0.0528(\rho - \rho_H)}, \tag{5.15}$$

where $T_H \approx 0.6528$ is the period of Γ^+ at the Hopf bifurcation at $\rho_H \approx 24.7368$. Taking a spline through the overall data set ensures that the thus defined function $T(\rho)$ is continuous and has a continuous first derivative for all $\rho_{\text{hom}} < \rho$, and we set $\tau = T(\rho)$ in (5.10).

5.3   Global Bifurcation Analysis

We now compare the bifurcation set and stability regions of equilibria and periodic solutions of the controlled Lorenz system (5.10) in the (ρ, b_0)-plane with those of the controlled Hopf normal form (5.3) in the (λ, b_0)-plane.

5.3.1   Bifurcation Set of the Controlled Hopf Normal Form

Figure 5.1 illustrates how the initially unstable periodic orbit of the controlled Hopf normal form (5.3) is stabilized in a transcritical bifurcation when the feedback strength b_0 is chosen below b_0^c; here we consider the standard choice of $\gamma = -10$ and $\beta = \frac{\pi}{4}$ for the other parameters. Panel (a) shows the one-parameter bifurcation diagram in λ. The origin, that is, the equilibrium solution, appears along the bottom axis; it is initially stable before becoming unstable at the Hopf bifurcation point H_P. The unstable periodic orbit Γ_P bifurcates from the point H_P. The addition of feedback induces a further Hopf bifurcation H_L, from which a stable periodic orbit Γ_L bifurcates. The periodic orbits Γ_P and Γ_L exchange stability at the transcritical bifurcation TC. Thus, the target periodic orbit Γ_P is stabilized for λ-values below

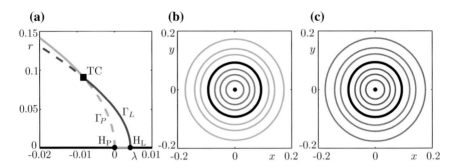

Fig. 5.1 Stabilization of the target periodic orbit Γ_P of (5.3) for $b_0 = 0.022$. Panel (**a**) shows the one-parameter bifurcation diagram in λ with the Hopf bifurcation points H_P (*black dot*) and H_L (*black dot*); the bifurcating periodic orbits Γ_P (*green*) and Γ_L (*red*) exchange stability at the transcritical bifurcation TC (*black square*); *solid* (*dashed*) *curves* represent stable (unstable) periodic orbits. Panels (**b**) and (**c**) show Γ_P and Γ_L, respectively, in projection onto the (x, y)-plane, in color when stable, grey when unstable, and black at TC. Here $\gamma = -10$ and $\beta = \frac{\pi}{4}$

TC. This exchange of stability is illustrated further in Fig. 5.1b, c with a selection of periodic orbits from the families Γ_P and Γ_L. The respective periodic orbit is unstable when shown in grey and stable when shown in colour; the origin from which they bifurcate is marked with a black dot. The thicker black periodic orbit is the one at the transcritical bifurcation TC, which is common to both families. Note that all periodic orbits of the contolled Hopf normal form (5.3) are exact circles.

Figure 5.2 shows the bifurcation set in the two-parameter (λ, b_0)-plane of (5.3), which will serve as a benchmark for comparison with the corresponding bifurcation set of the controlled Lorenz system. Figure 5.2a shows an overview of the (λ, b_0)-plane; its lower boundary is $b_0 = 0$ and its left-hand boundary is at $\lambda = \frac{1}{\gamma} = -0.1$, where the delay τ in (5.4) goes to infinity and becomes undefined. The Hopf bifurcation H_P is the vertical line at $\lambda = 0$. It is intersected at the point b_0^c by the Hopf bifurcation curve H_L, which actually forms a loop that intersects the curve H_P again at the double-Hopf bifurcation point HH_0; here the origin has two pairs of purely imaginary eigenvalues [44]. Both ends of H_L extend to the special point b_0^* on the left boundary of the (λ, b_0)-plane. In addition to H_L there exist further delay-induced Hopf bifurcation curves, which can be split into two distinct families, H_J^k and H_R^k. All of these curves of Hopf bifurcation emerge from the point b_0^*. Each curve in the family H_J^k has a J-shape with a vertical asymptote for a specific value of λ, while curves in H_R^k extend to infinity in both λ and b_0.

As is indicated by blue shading in Fig. 5.2, to the left of H_P the origin (the equilibrium) is stable in the region below the envelope formed by the curves H_L and H_J^k; to the right of H_P it is stable in the region bounded by H_L. Note that, since only some of the curves H_J^k are shown, this stability region is approximated near the left-hand boundary of the (λ, b_0)-plane by the line $b_0 = 0.35$.

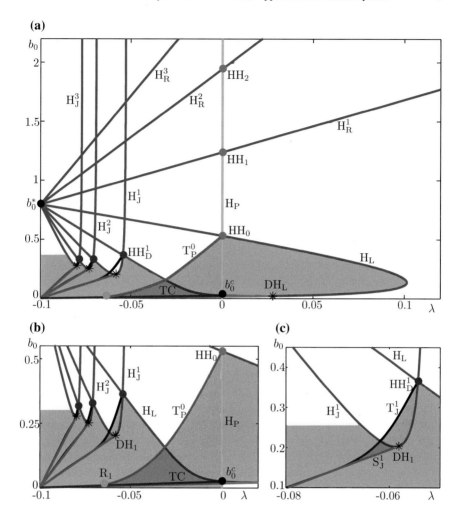

Fig. 5.2 Bifurcation set in the (λ, b_0)-plane of (5.3) for $\lambda > 1/\gamma$, showing the curves of Hopf bifurcation H_P (*green*), H_L, H_J^k and H_R^k (*red*), transcritical bifurcation TC (*purple*), torus bifurcation T_P^0 (*grey*) and T_J^1 (*black*), and SNLC bifurcation S_J^k (*blue*); also shown are points of double Hopf bifurcation HH_k (*dark green dots*) and HH_D^k (*violet dots*), of degenerate Hopf bifurcation DH_L and DH_k (*asterisks*), and of 1:1 resonance R_1 (*green dot*); notice also the point b_0^c (*black dot*) on the curve H_P. The stability region of the origin is indicated by *blue shading*, and regions where periodic orbits are stable are *shaded grey*. Panel (**a**) provides an overview of the (λ, b_0)-plane, and panels (**b**) and (**c**) are enlargements near the stability regions of Γ_P and Γ_J^1, respectively. Here $\gamma = -10$ and $\beta = \frac{\pi}{4}$

Above the point HH_0 the target periodic orbit Γ_P bifurcates unstably from H_P, because it has an additional complex conjugate pair of unstable Floquet multipliers. Between the points b_0^c and HH_0, on the other hand, Γ_P bifurcates stably from H_P for decreasing λ. As we have already seen in Fig. 5.1, the stability region of Γ_P

is bounded, for b_0 below b_0^c, by a transcritical bifurcation. In fact, the transcritical bifurcation curve TC emerges from the point b_0^c and ends at the left-hand boundary of the (λ, b_0)-plane at $b_0 = 0$. The left-hand boundary of the stability region is formed by a torus bifurcation curve T_P^0, which starts at HH_0 and ends on the transcritical bifurcation curve TC at the 1 : 1 resonance point R_1; this right-most part of the curve TC forms the lower boundary of the domain of stability of the target periodic orbit Γ_P. In other words, exactly in this region, which is shaded grey in Fig. 5.2 and enlarged in panel (b), the Pyragas control scheme achieves the stabilization of Γ_P. In the region between the curves TC, H_L and T_P^0 both the equilibrium and Γ_P are stable.

Each curve in the family H_R^k intersects H_P at further double-Hopf bifurcation point, above each of which Γ_P bifurcates with yet another complex conjugate pair of unstable Floquet multipliers. The curves in the family H_J^k also intersect in double-Hopf bifurcation points; the lowest of these give rise to curves T_J^k of torus bifurcation that extend to the left boundary at $b_0 = 0$; see Fig. 5.2a, b. Moreover, near the minimum of each curve in H_J^k there is a degenerate Hopf bifurcation, from which a curve S_J^k of SNLC bifurcation emanates that also connects to the left boundary at $b_0 = 0$. Between these two codimension-two points on the curve H_J^k a stable delay-induced periodic orbit Γ_J^k bifurcates. Its region of stability is bounded on the left by T_J^k and below by S_J^k. Figure 5.2c shows an enlargement of the stability region of Γ_J^1 (shaded grey); notice that there is a small region of bistability, between the curves H_J^1, T_J^1 and S_J^1, where both the equilibrium and Γ_J^1 are stable.

Overall, Fig. 5.2 shows a complicated bifurcation set in the (λ, b_0)-plane of (5.3) for the standard choice of fixed $\beta = \frac{\pi}{4}$ and $\gamma = -10$. It is organized by infinitely many curves of delay-induced Hopf bifurcations that extend to the left-hand boundary at $1/\gamma$ where the period $T(\lambda)$ goes to infinity. Double-Hopf and degenerate Hopf bifurcation points give rise to curves of torus and SNLC bifurcations that also extend to the left-hand boundary of the (λ, b_0)-plane. This allows us to characterize the stability region of the target periodic orbit Γ_P, as well as stability regions of delay-induced periodic orbits Γ_J^k; more details, including a study of the influence of β and γ on the bifurcation set, can be found in [36].

5.3.2 Bifurcation Set of the Controlled Lorenz System

Postlethwaite and Silber [37] studied the controlled Lorenz system (5.10) near its subcritical Hopf bifurcation H_P. They showed that the mechanism of stabilization of the target periodic orbit Γ_P is locally as that of the controlled Hopf normal form (5.3). More specifically, there also exists a critical level b_0^c of feedback amplitude b_0, immediately above which Γ_P bifurcates stably from H_P. Subsequently, Brown [38] performed a centre manifold reduction of (5.10) and derived the analytical expression

$$b_0^c = \frac{-\omega_0}{2\pi (\cos(\eta) + (-10.82 \sin(\eta)))}, \qquad (5.16)$$

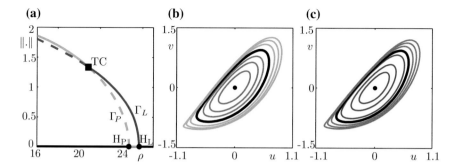

Fig. 5.3 Stabilization of the target periodic orbit Γ_P of (5.10) for $b_0 = 0.19$. Panel (**a**) shows the one-parameter bifurcation diagram in ρ with the periodic orbits Γ_P *(green)* and Γ_L *(red)* that bifurcate from H_P and H_L *(black dots)* and exchange stability at TC *(black square)*; *solid (dashed) curves* represent stable (unstable) periodic orbits. Panels (**b**) and (**c**) show Γ_P and Γ_L, respectively, in projection onto the (x, y)-plane, in color when stable, *grey* when unstable, and *black* at TC. Here $\sigma = 10$, $\alpha = \frac{8}{3}$ and $\eta = \frac{\pi}{4}$

where ω_0 is the linear frequency. Moreover, close to H_P, considering the controlled Lorenz system in the form (5.10) is equivalent to first computing the Hopf normal form of the Lorenz system on the centre manifold and then applying Pyragas feedback control. In particular, the local bifurcation structure in the (ρ, b_0)-plane of (5.10) near the point b_0^c on H_P is as found in (5.3); that is, it involves the delay-induced Hopf bifurcation H_L and the transcritical bifurcation curve TC. These results from [37, 38] are for $\sigma = 10$, $\alpha = \frac{8}{3}$ with the standard choice of $\eta = \frac{\pi}{4}$.

For the same choice of parameters, Fig. 5.3 shows the one-parameter bifurcation diagram of (5.10) in ρ for $b_0 = 0.19$, which is below $b_0^c \approx 0.2206$. The target periodic orbit Γ_P bifurcates unstably from H_P, while the delay-induced periodic orbit Γ_L bifurcates stably from H_L. Their amplitudes grow (as ρ is decreased) until Γ_P and Γ_L meet at the transcritical bifurcation TC, where they exchange stability. Hence, Γ_P is stabilized successfully for ρ-values below TC. A selection of periodic orbits from the families Γ_P and Γ_L along the respective branches is shown in projection onto the (u, v)-plane in Fig. 5.3b, c, respectively.

Direct comparison of Fig. 5.3 with Fig. 5.1 clearly illustrates that the mechnism of stabilization of the target periodic orbit Γ_P is indeed qualitatively exactly as predicted by the controlled Hopf normal form. The difference is that the periodic orbits Γ_P and Γ_L of (5.10) are not circular; notice, in particular, that the outer-most periodic orbits for $\rho = 16.0$ in Fig. 5.3b, c are already starting to deform characteristically as they approach the homoclinic bifurcation at $\rho = \rho_{\text{hom}}$.

We now present in Fig. 5.4 the bifurcation set and stability regions of the controlled Lorenz system (5.10) over a wide range of parameter values in the (ρ, b_0)-plane for the standard values of the other parameters. This figure is designed to allow for a direct, panel-by-panel comparison with the bifurcation set in the (λ, b_0)-plane of (5.3) in Fig. 5.2. The (ρ, b_0)-plane in Fig. 5.4 is bounded below by $b_0 = 0$.

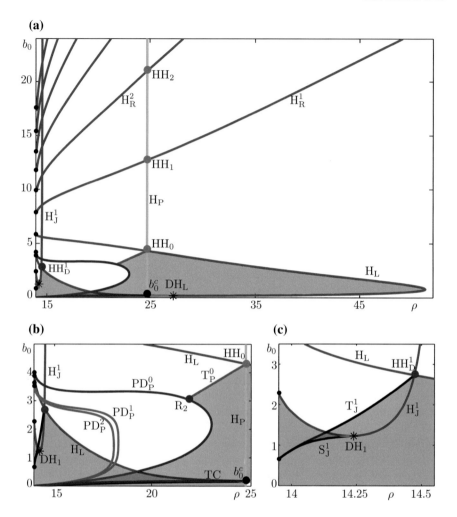

Fig. 5.4 Bifurcation set of (5.10) in the (ρ, b_0)-plane for $\rho > \rho_{\text{hom}}$, for $\eta = \frac{\pi}{4}$. Shown are curves of Hopf bifurcation H_P (*green*), H_L, H_J^1 and H_R^k (*red*), transcritical bifurcation TC (*purple*), torus bifurcation T_P^0 (*grey*) and T_J^1 (*black*), and SNLC bifurcation S_J^k (*blue*); also shown are points of double Hopf bifurcation HH_k (*dark green dots*) and HH_D^1 (*violet dot*), of degenerate Hopf bifurcation DH_L and DH_1 (*asterisks*), and of 1 : 1 resonance R_1 (*green dot*); notice also the point b_0^c (*black dot*) on the curve H_P. The stability region of the origin is *shaded blue*, and regions where periodic orbits are stable are *shaded grey*. Panel (**a**) provides an overview of the (λ, b_0)-plane, and panels (**b**) and (**c**) are enlargements near the stability regions of Γ_P and Γ_J^1, respectively. Here $\sigma = 10$ and $\alpha = \frac{8}{3}$

Its left-hand boundary is $\rho_{\text{hom}} \approx 13.926$, where Γ_P undergoes a homoclinic bifurcation and disppears and, hence, $\tau = T(\rho)$ goes to infinity as ρ_{hom} is approached from above; note that this is very similar to what happens at the left-hand boundary at $\lambda = \frac{1}{\gamma}$ in the (λ, b_0)-plane of (5.3) in Fig. 5.2.

As theory predicts [37, 38], in a strip around the vertical Hopf bifurcation curve H_P the bifurction set and stability regions in the (ρ, b_0)-plane in Fig. 5.4a agree with those in the (λ, b_0)-plane of in Fig. 5.2a. In particular, H_P is intersected twice by a curve H_L of delay-induced Hopf bifurcation, at the point b_0^c and at the double-Hopf bifurcation point HH_0. Along the interval between these points the target periodic orbit Γ_P bifurcates stably for decreasing values of ρ. The upper boundary of its stability region is the torus bifurcation curve T_P^0 emanating from HH_0, and the lower stability boundary is the curve TC emanating from the point b_0^c; see also Fig. 5.4b. Notice that Γ_P is unstable when it bifurcates from H_P above the point HH_0; as we found for the controlled Hopf normal form, Γ_P has an increasing number of complex-conjugate pairs of unstable Floquet multipliers for b_0 above the points HH_k, which are points of intersection of H_P with curves H_R^k.

There is actually considerable qualitative agreement between Figs. 5.4 and 5.2 beyond a small vertical strip around H_P. The Hopf bifurcation curves H_L and H_R^k and the curve H_J^1 in the (ρ, b_0)-plane of (5.10) are qualitatively as those in the (λ, b_0)-plane of (5.3), but with the difference that they have different end points (limiting values of b_0) on the left-hand boundary of the (ρ, b_0)-plane where $\rho = \rho_{\text{hom}}$. Note, in particular, that H_L forms a loop that intersects H_P twice. However, we only found the curve H_J^1 and no further curves of the family H_J^k. The stability region of p^+ in Fig. 5.4 is therefore bounded by only H_L and H_J^1; otherwise it is qualitatively as that in Fig. 5.2. Colloquially speaking, the structure of Hopf bifurcation curves in Fig. 5.4a is a 'chopped-off' version of that in Fig. 5.2a, with parts of curves to the left of $\lambda \approx -0.065$ missing. It seems reasonable to conjecture that this truncation effect is due to the presence of the homoclinic bifurcation at ρ_{hom}, where all periodic orbits of (5.10) disappear.

As is the case for the controlled Hopf normal form, the curve H_J^1 is associated with a region of stability of the delay-induced periodic orbit Γ_J^1; see Fig. 5.4c. More specifically, on H_J^1 there is also a point of double Hopf bifurcation HH_D^1 with an emerging curve T_J^1 of torus bifurcation, as well as a point of degenerate Hopf bifurcation DH_1 with an emerging curve S_J^1 of SNLC bifurcation. Together with the segment of H_J^1 between the points HH_D^1 and DH_1, the curves T_J^1 and S_J^1 form the boundary of the stability region of the bifurcating periodic orbit Γ_J^1; note also that T_J^1 and S_J^1 end at practically the same point on the left boundary of the (ρ, b_0)-plane. Comparison of Fig. 5.4c with Fig. 5.2c shows that the stability region of H_J^k is topologically the same for both controlled systems, meaning that it is bounded by the same configuration of bifurcation curves. Since we did not find further curves Γ_J^k for $k \geq 2$ of (5.10), we also did not find stability regions of associated delay-induced periodic orbits.

There is a notable difference between Figs. 5.4 and 5.2 when it comes to the stability region of the target periodic orbit Γ_P. As already discussed, the regions and its boundary curves T_P^0 and TC agree near H_P; however, the curve T_P^0 does not end at a $1:1$ resonance point on the curve TC, but rather on a $1:2$ resonance point R_2 on a curve PD_P^0 of period doubling bifurcation. The curve PD_P^0 was found in the local bifurcation analysis of (5.10) performed in [37]. This curve starts and ends

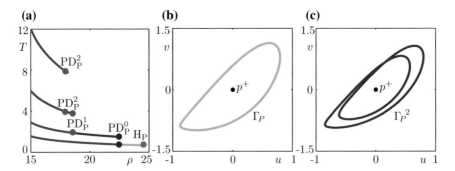

Fig. 5.5 Illustration of period-doubling of Γ_P in (5.10) for $b_0 = 1.5$. Panel (**a**) is the one-parameter bifurcation diagram in ρ, representing Γ_P and the period-doubled periodic orbits Γ_P^k by their periods. Panels (**b**) and (**c**) for $\rho = 22$ show Γ_P (*green*) and Γ_P^2 (*black*), respectively, in projection onto (u, v)-plane; also shown is p^+, which is a saddle point. Here $\sigma = 10$, $\alpha = \frac{8}{3}$ and $\eta = \frac{\pi}{4}$

at $\rho = \rho_{\text{hom}}$ and forms the left-hand boundary of the stability region of Γ_P in the (ρ, b_0)-plane, connecting to the point where $b_0 = 0$; see Fig. 5.4b. Also shown are curves PD_P^1 and PD_P^2 of further period-doubling bifurcations, which also start on the left-hand boundary of the (ρ, b_0)-plane, extend to the right and end at $(\rho_{\text{hom}}, 0)$.

The associated sequence of period-doublings of Γ_P is illustrated in Fig. 5.5. Panel (a) shows the one-parameter bifurcation diagram in ρ with Γ_P and the period-doubled orbits that bifurcate at PD_P^0 to PD_P^2. Panels (b) and (c) show the saddle periodic orbit Γ_P and the stable first period-doubled orbit Γ_P^2, which coexist immediately after the first period-doubling PD_P^0 (as ρ is decreased). We found evidence of further period-doubling bifurcations, suggesting that there may be a small chaotic attractor that contains the saddle periodic orbits Γ_P^k. This attractor is localised near p^+, is induced by the delay and should not be confused with the Lorenz attractor (which surrounds both p^+ and p^-); see also Sect. 5.3.3.

In spite of the differences discussed above, we can conclude from this comparison that the controlled Hopf normal form (5.3) has considerable predictive power well beyond the immediate vicinity of the Hopf bifurcation curve H_P where normal form theory applies. Indeed, the agreement between the bifurcation sets of (5.10) in the (ρ, b_0)-plane and of (5.3) in the (λ, b_0)-plane is quite remarkable for the standard parameter choice of $\beta = \eta = \frac{\pi}{4}$; the influence of the feedback phase η on this agreement will be investigated in Sect. 5.3.4.

5.3.3 Local Nature of the Control

As was explained in Sect. 5.2, Pyragas control is applied to the equilibrium p^+, which has been shifted to the origin in (5.10). In other words, the control acts only locally near p^+. Figure 5.6 shows with two phase portraits that the effect of Pyragas

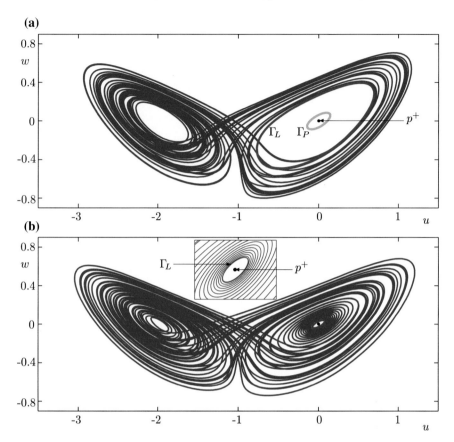

Fig. 5.6 Phase portraits of (5.10) in projection onto the (u, w)-plane for $b_0 = 0.3$ and $\rho = 24.5$ (**a**) and $\rho = 31$ (**b**). Shown are the equilibrium p^+ (*black dot*), the periodic orbits Γ_P (*green*) and Γ_L (*red*) and a trajectory (*blue*) on a chaotic attractor. Here $\sigma = 10$, $\alpha = \frac{8}{3}$ and $\eta = \frac{\pi}{4}$

control of Γ_P coexists with a chaotic attractor of (5.10). Panel (a) is for a parameter point from the region of stability of Γ_P in the (λ, b_0)-plane. The point p^+ is a saddle, as is the delay-induced periodic orbit Γ_L, whose stable manifold (not shown) bounds the points that converge to the stable orbit Γ_P. Figure 5.6b is for a parameter point from the region of stability of p^+ to the right of H_P. Now p^+ is an attractor, and the stable manifold (not shown) of the saddle periodic orbit Γ_L bounds the points that converge to p^+; see also the inset in panel (b).

The noteworthy feature of the two phase portraits of Fig. 5.6 is the fact that all points outside the stable manifold of Γ_L, which is a topological cylinder near Γ_L, converge to a chaotic attractor that switches irregularly between rotations around p^- and rotations around Γ_L; hence, this attractor has the same qualitative feature as the Lorenz attractor, and it can be regarded as its continuation or remainder in the controlled Lorenz system (5.10). The existence of this chaotic attractor is clear evidence that the addition of feedback only effects the dynamics of (5.10) close to p^+. In particular, the equilibrium p^- is still a saddle in the presence of feedback, because the control is not symmetric. Clearly, the controlled Hopf normal form (5.3) cannot be expected to describe this type of dynamics far from p^+. In fact, Fig. 5.6 shows that, when it is of codimension one, the stable manifold of Γ_L can be interpreted as the boundary of validity of the Hopf normal form.

5.3.4 Influence of the 2π-Periodic Feedback Phase η

So far we have kept the 2π-periodic phase η in (5.10) fixed at $\eta = \frac{\pi}{4}$, as is the convention [30, 34, 37] that is also used for fixing the feedback phase β in (5.3) to $\beta = \frac{\pi}{4}$. However, it is important to realise that, while η enters the feedback matrix G in a canonical way, G then undergoes the coordinate transformation (5.12) to enter (5.10) as the feedback gain matrix Π. Therefore, the actual phase of the feedback will not be η in general, and would need to be determined via a centre manifold and normal form reduction near the subcritical Hopf bifurcation H_P. Since we are interested in the overall dynamics, also away from H_P, we now consider η as an extra parameter. The idea is to check whether there is a value of η for which the associated bifurcation set of (5.10) in (ρ, b_0)-plane shows an even better qualitative agreement with the bifurcation set for $\beta = \frac{\pi}{4}$ of (5.3) in the (λ, b_0)-plane. For this purpose, we focus on the properties of the region of stability of the target periodic orbit Γ_P, for which we found a considerable difference between the two systems. A study of how the global bifurcation set of the controlled normal form (5.3) changes with β can be found in [36].

Figure 5.7 shows how the region of stability of Γ_P (shaded) changes in the (ρ, b_0)-plane as the feedback phase η is increased. The starting point is panel (a), which shows the domain of stability for $\eta = \frac{\pi}{4}$ from Fig. 5.4b where the stability region is bounded by the relevant parts of the curves H_P, TC, T_P^0 and PD_P^0. When η is increased, the curve PD_P^0 moves left in the (ρ, b_0)-plane, while the point R_2, where T_P^0 and PD_P^0 meet, moves along PD_P^0 towards the left-hand boundary of the (ρ, b_0)-plane; see Fig. 5.7b for $\eta = 2.50$. For about $\eta = 2.71$, as in Fig. 5.7c, the point R_2 has just reached the left-hand boundary at $\rho = \rho_{hom}$ and $b_0 = 0$; hence, the curve T_P^0 now extends all the way to this end point. As η is increased further, T_P^0 connects to the transcritical bifurcation curve TC at the $1:1$ resonance point R_1, which moves to the right; hence, the stability region of Γ_P is no longer bounded by a curve of period-doubling bifurcation. This situation is depicted in Fig. 5.7d for $\eta = 2.80$, which we identified as the value where the stability region of Γ_P of (5.10) agrees almost perfectly with that of the controlled Hopf normal form for $\beta = \frac{\pi}{4}$; compare with Fig. 5.4b. Notice, in particular, that the two stability regions and their boundary curves are topologically equivalent. We remark that, when η is increased even further, the stability region of Γ_P of (5.10) disappears at $\eta = \pi$, where the points b_0^c, HH_0 and R_1 all meet on H_P; this is the

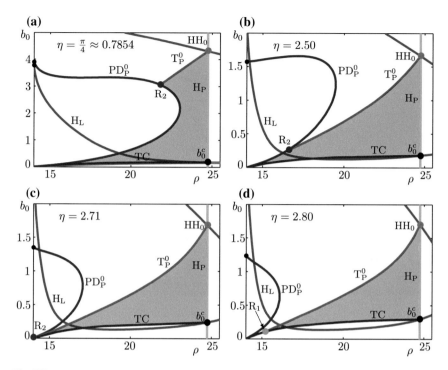

Fig. 5.7 The region of stability of Γ_P (*shaded*) in the (ρ, b_0)-plane of (5.10) for different values of increasing η as indicated in panels (**a**)–(**d**); shown are the curves H_P (*green*), H_L (*red*), T_P^0 (*grey*), TC (*purple*) and PD_P^0 (*dark blue*), and the points b_0^c (*black dot*), HH_0 (*dark green dot*), R_2 (*black dot*) and R_1 (*green dot*). Panel (**a**) is as Fig. 5.4b; here $\sigma = 10$ and $\alpha = \frac{8}{3}$

same mechanism that was identified in [36] for the controlled Hopf normal form, also for $\beta = \pi$.

We now consider the bifurcation set of (5.10) throughout the (ρ, b_0)-plane for the selected value $\eta = 2.80$. It is shown in Fig. 5.8 and should be compared with Fig. 5.2. Apart from the much better agreement between the respective regions of stability of Γ_P, enlarged in panel (b), the new feature with respect to Fig. 5.4 for $\eta = \frac{\pi}{4}$ is the existence of the second curve H_J^2 of the family H_J^k. However, this curve is extremely close to the left-hand boundary where $\rho = \rho_{\text{hom}}$, and we did not find a region of stability of the associated periodic orbit Γ_J^2. Note that the stability region of Γ_J^1 near the minimum of the curve H_J^1 remains qualitatively unchanged, except for the small difference that the curves T_J^1 and S_J^k now end at slightly different points on the left-hand boundary; compare Figs. 5.8c and 5.4c. Overall, we conclude that the controlled Hopf normal form (5.3) provides an excellent and global description of the controlled Lorenz system (5.10) over a very wide region of the (ρ, b_0)-plane when its feedback phase η is chosen as discussed.

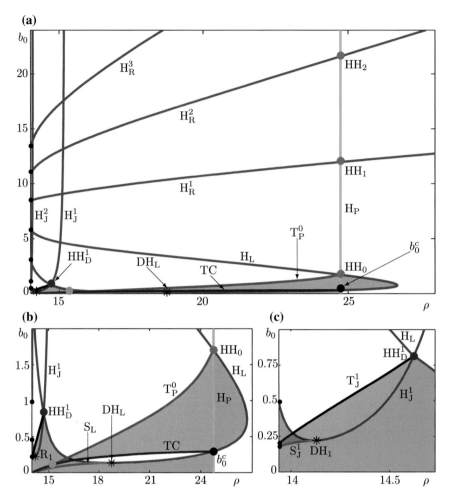

Fig. 5.8 Bifurcation set of (5.10) in the (ρ, b_0)-plane for $\rho > \rho_{\mathrm{hom}}$, where $\eta = 2.8$. Shown are curves of Hopf bifurcation H_P (*green*), H_L, H_J^1, H_J^2 and H_R^k (*red*), transcritical bifurcation TC (*purple*), torus bifurcation T_P^0 (*grey*) and T_J^1 (*black*), and SNLC bifurcation S_J^k (*blue*); also shown are points of double Hopf bifurcation HH_k (*dark green dots*) and HH_D^1 (*violet dot*), of degenerate Hopf bifurcation DH_L and DH_1 (*asterisks*), and of 1:1 resonance R_1 (*green dot*); notice also the point b_0^c (*black dot*) on the curve H_P. The stability region of the origin is *shaded blue*, and regions where periodic orbits are stable are *shaded grey*. Panel (**a**) provides an overview of the (λ, b_0)-plane, and panels (**b**) and (**c**) are enlargements near the stability regions of Γ_P and Γ_J^1, respectively. Here $\sigma = 10$ and $\alpha = \frac{8}{3}$

5.4   Delay Mismatch in the Controlled Lorenz System

When Pyragas control is implemented in practice, the functional form of the parameter-dependent period is often not known and needs to be approximated; see, for example, [13–15]. In [35] we studied the effect of the resulting delay mismatch on the overall stablity region of the target periodic orbit Γ_P of the controlled Hopf normal form (5.3). More specifically, we considered a constant and a linear approximation to the target period $T(\lambda)$ near the subcritical Hopf bifurcation H_P; the corresponding control scheme is considered successful in this context when the following three criteria are satisfied:

(i)   there exists a periodic orbit $\widehat{\Gamma}_P$ with amplitude and period close to Γ_P when it is stable;
(ii) any residual control force is sufficiently small;
(iii) in some suitably large vertical strip around the Hopf bifurcation H_P, the stability region of $\widehat{\Gamma}_P$ in the (λ, b_0)-plane is sufficiently close to that of Γ_P.

The main result of [35] is that the approximation of $T(\lambda)$ by its constant value $T_C(\lambda) = 2\pi$ at H_P does not result in a successful stabilization of the target periodic orbit according to the above criteria. However, for the linear approximation $\tau = T_L(\lambda) = 2\pi(1 + \gamma\lambda)$ all three stability criteria are satisfied and successful stabilization is achieved.

We now briefly consider the effect of delay mismatch in the controlled Lorenz system (5.10), where we again use $\eta = 2.80$. When the constant approximation of the period $T_C = T(\rho_H) = 0.6528$ is used for the delay τ of the Pyragas feedback then, as for the controlled Hopf normal form, stabilization is not successful. A linear approximation of $T(\rho)$ is given by

$$T_L(\rho) = -0.0345(\rho - \rho_H) + T(\rho_H) = -0.0345\rho + 1.5062 \qquad (5.17)$$

(again for $\sigma = 10$ and $\alpha = \frac{8}{3}$), where the slope -0.0345 of $T(\rho)$ at ρ_H is determined from the continuation data.

Figure 5.9 provides evidence that the controlled Lorenz system (5.10) with $\tau = T_L(\rho)$ does achieve successful stabilization of the periodic solution bifurcating from H_P; specifically, criteria (i)–(iii) are satisfied, where (iii) is now considered in the (ρ, b_0)-plane and with respect to the target period $T(\rho)$. We denote by $\widehat{\Gamma}_P$ the stabilized periodic orbit of (5.10) with $\tau = T_L(\rho)$ as given by (5.17); it needs to be compared with the target periodic orbit Γ_P of (5.10) with the exact period $T(\rho)$.

Figure 5.9a, b are one-parameter bifurcation diagrams in ρ, shown in terms of the norm and period, respectively. The branches of $\widehat{\Gamma}_P$ and Γ_P both bifurcate from H_P and agree very well in amplitude and period along the segment where they are stable, until T_P^0 is reached. Hence criterion (i) is satisfied, but notice that $\widehat{\Gamma}_P$ and Γ_P diverge considerably for $\rho \lesssim 22.5$. Panel (c) illustrates that, when $\widehat{\Gamma}_P$ is stable, the residual control forces $K_u[u(t - \tau) - u(t)]$, $K_v[v(t - \tau) - v(t)]$ and $K_w[w(t - \tau) - w(t)]$ are small compared to the amplitudes of the corresponding components u, v and

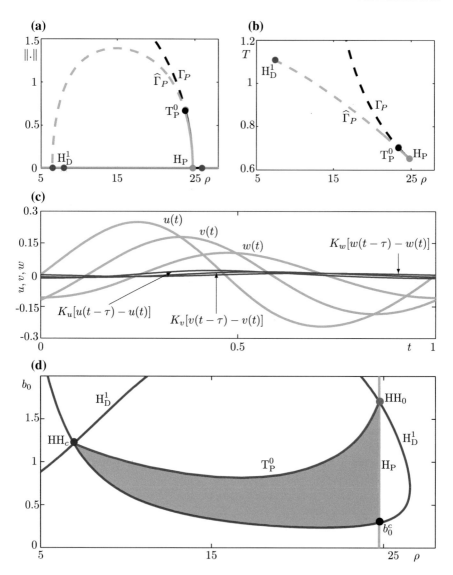

Fig. 5.9 System (5.10) with linear approximation of $T(\rho)$. Panels (**a**) and (**b**) show one-parameter bifurcation diagrams in ρ for $b_0 = 1.4$ with the branch of periodic orbits $\widehat{\Gamma}_P$ (*green*); also shown is the branch Γ_P (*black*) for the exact-period case; *solid (dashed) curves* indicate stable (unstable) periodic orbits. Panel (**c**) shows the u, v and w components of the solution profile of the stable periodic orbit $\widehat{\Gamma}_P$ (*green*) for $\rho = 24$ and $b_0 = 1.4$, together with the corresponding residual feedback components $K_u[u(t - \tau) - u(t)]$, $K_v[v(t - \tau) - v(t)]$ and $K_w[w(t - \tau) - w(t)]$ (*red*). Panel (**d**) shows the region of stability of $\widehat{\Gamma}_P$ (*shaded*) in the (ρ, b_0)-plane, which is bounded by the curves H_P (*green*), H_D^1 (*red*) and T_P^0 (*grey*). Here $\sigma = 10$, $\alpha = \frac{8}{3}$ and $\eta = 2.8$

w of this periodic solution. Here we define K_u as the sum of the terms in the first column of the matrix Π from (5.10), that is, of the feedback gain with respect to the variable u; the residuals K_v and K_w with respect to v and w are defined in the same way by summing over the second and third columns of Π, respectively. Specifically, the maximum amplitude of each residual control force is approximately 4 % of the amplitude of the corresponding solution; this is evidence that criterion (ii) is also satisfied. Finally, Fig. 5.9d shows the overall stablity region of $\widehat{\Gamma}_P$ in the (ρ, b_0)-plane. It is bounded by the segment of the curve H_P between b_0^c and HH_0, and it lies entirely to the left of H_P. Its upper boundary is the torus bifurcation curve T_P^0, which connects the double-Hopf points HH_0 on H_P and HH_c on H_D^1. The lower boundary of the stability region of $\widehat{\Gamma}_P$ is the curve H_D^1, which intersects H_P at b_0^c and has the point of self-intersection HH_c. A comparison reveals that for $18 \lesssim \rho \lesssim 26$ the stability region in the (ρ, b_0)-plane of $\widehat{\Gamma}_P$ in Fig. 5.9d is in close quantitative agreement with that of Γ_P in Fig. 5.8b. The main difference is that the lower boundary is formed by the Hopf bifurcation curve H_D^1, rather than the transcritical bifurcation curve TC. Nevertheless, we argue that criterion (iii) is satisfied as well.

Overall, we have shown that the controlled Hopf normal form (5.3) also correctly predicts the properties of the controlled Lorenz system (5.10) with regard to the effect of using a constant or linear approximation of the period $T(\rho)$ as the delay τ in the Pyragas control term.

5.5   Conclusions

The work presented here shows that the controlled Hopf normal form correctly predicts the observed dynamics of the controlled Lorenz system over a very large area of the relevant two-parameter plane—effectively, over the entire range where the target periodic orbit exists. Because the Pyragas control term is subject to a coordinate transformation when it is applied to the three-dimensional Lorenz system, its feedback phase needs to be adjusted to achieve the best overall agreement. Specifically, we confirmed the predicted existence of a further stable delay-induced periodic orbit in the controlled Lorenz system and showed that the effect of a delay mismatch is qualitatively the same for the two controlled systems. These results can be interpreted as a considerable extension of those of Brown [38], who showed that the controlled Hopf normal form accurately describes the dynamics in the vicinity of the subcritical Hopf bifurcation.

More generally, users of Pyragas control should be aware of the overall bifurcation set of the controlled Hopf normal form and the associated stability regions of different solutions. Depending on the system or experiment under consideration, there may well be some differences but, in principle, all global features of the controlled Hopf normal form should be expected in any system with Pyragas control where the target periodic orbit bifurcates from a subcritical Hopf bifurcation.

An interesting direction for future research is the study of Pyragas control applied to systems where an unstable periodic orbit bifurcates from a different bifurcation. Fiedler et al. [45] successfully stabilized rotating waves near a fold bifurcation, and it would be interesting to study the dynamics induced by Pyragas control near a fold bifurcation locally and globally.

References

1. B. Balachandran, T. Kalmár-Nagy, D.E. Gilsinn, *Delay Differential Equations: Recent Advances and New Directions* (Springer, New York, 2009)
2. A. Garfinkel, M.L. Spano, W.L. Ditto, J.N. Weiss, Controlling cardiac chaos. Science **257**, 1230–1235 (1992)
3. B. Peng, V. Petrov, K. Showalter, Controlling chemical chaos. J. Phys. Chem. **95**, 4957–4959 (1991)
4. S.J. Schiff, K. Jerger, D.H. Duong, T. Chang, M.L. Spano, W.L. Ditto, Controlling chaos in the brain. Nature **370**, 615–620 (1994)
5. S. Boccaletti, C. Grebogi, Y.C. Lai, H. Mancini, D. Maza, The control of chaos: theory and applications. Phys. Rep. **329**, 103–197 (2000)
6. A. Isidori, *Nonlinear Control Systems* (Springer, London, 1999)
7. E. Ott, C. Grebogi, J.A. Yorke, Controlling chaos. Phys. Rev. Lett. **64**, 2837–2837 (1990)
8. K. Pyragas, Continuous control of chaos by self-controlling feedback. Phys. Lett. A **170**, 421–428 (1992)
9. T. Shinbrot, C. Grebogi, E. Ott, J.A. Yorke, Using small perturbations to control chaos. Nature **363**, 411–417 (1993)
10. K. Pyragas, Control of chaos via extended delay feedback. Phys. Lett. A **206**, 323–330 (1995)
11. K. Pyragas, Delayed feedback control of chaos. Philos. Trans. Royal Soc. A: Math. Phys. Eng. Sci. **364**(1846), 2309–2334 (2006)
12. V. Pyragas, K. Pyragas, Adaptive modification of the delayed feedback control algorithm with a continuously varying time delay. Phys. Lett. A **375**(44), 3866–3871 (2011)
13. S. Bielawski, D. Derozier, P. Glorieux, Controlling unstable periodic orbits by a delayed continuous feedback. Phys. Rev. E **49**, 971–974 (1994)
14. S. Schikora, H.J. Wünsche, F. Henneberger, Odd-number theorem: optical feedback control at a subcritical Hopf bifurcation in a semiconductor laser. Phys. Rev. E **83**, 026203 (2011)
15. D.J. Gauthier, D.W. Sukow, H.M. Concannon, J.E.S. Socolar, Stabilizing unstable periodic orbits in a fast diode resonator using continuous time-delay autosynchronization. Phys. Rev. E **50**, 2343 (1994)
16. K. Pyragas, A. Tamaševičius, Experimental control of chaos by delayed self-controlling feedback. Phys. Lett. A **180**(1), 99–102 (1993)
17. A.L. Fradkov, R.J. Evans, B.R. Andrievsky, Control of chaos: methods and applications in mechanics. Philos. Trans. Royal Soc. A **364**, 2279–2307 (2006)
18. F.W. Schneider, R. Blittersdorf, A. Förster, T. Hauck, D. Lebender, J. Müller, Continuous control of chemical chaos by time delayed feedback. J. Phys. Chem. **97**, 12244–12248 (1993)
19. W. Lu, D. Yu, R.G. Harrison, Control of patterns in spatiotemporal chaos in optics. Phys. Rev. Lett **76**(18), 3316 (1996)
20. K.A. Montgomery, M. Silber, Feedback control of travelling wave solutions of the complex Ginzburg-Landau equation. Nonlinearity **17**(6), 2225 (2004)
21. E. Schöll, G. Hiller, P. Hövel, M.A. Dahlem, Time-delayed feedback in neurosystems. Philos. Trans. Royal Soc. A: Math. Phys. Eng. Sci. **367**(1891), 1079–1096 (2009)
22. C.M. Postlethwaite, G. Brown, M. Silber, Feedback control of unstable periodic orbits in equivariant Hopf bifurcation problems. Philos. Trans. Royal Soc. A **371**, (2013)

23. I. Schneider, M. Bosewitz, Eliminating restrictions of time-delayed feedback control using equivariance. Disc. Cont. Dyn. Syst. A **36**(1), 451–467 (2016)
24. E. Schöll, H.G. Schuster, *Handbook of Chaos Control*. (Wiley, 2008)
25. J.K. Hale, *Introduction to Functional Differential Equations* (Springer, New York, 1993)
26. R.D. Driver, *Ordinary and Delay Differential Equations* (Springer, New York, 1977)
27. D. Roose and R. Szalai. Continuation and bifurcation analysis of delay differential equations. In: H.M. Osinga B. Krauskopf, J. Galán-Vioque (eds.), *Numerical continuation methods for dynamical systems*, pp. 359–399. (Springer, Dordrecht, 2007)
28. K. Engelborghs, T. Luzyanina, G. Samaey, DDE-Biftool: a Matlab package for bifurcation analysis of delay differential equations. TW Report 305, (2000)
29. J. Sieber, Y. Kuznetsov, K. Engelborghs, Software package DDE-BIFTOOL, version 3.1; http://sourceforge.net/projects/ddebiftool/
30. B. Fiedler, V. Flunkert, M. Georgi, P. Hövel, E. Schöll, Refuting the odd-number limitation of time-delayed feedback control. Phys. Rev. Lett. **98**, 114101 (2007)
31. H. Nakajima, On analytical properties of delayed feedback control of chaos. Phys. Lett. A **232**(34), 207–210 (1997)
32. G. Brown, C.M. Postlethwaite, M. Silber, Time-delayed feedback control of unstable periodic orbits near a subcritical Hopf bifurcation. Phys. D: Nonlinear Phenom. **240**(910), 859–871 (2011)
33. H. Erzgräber, W. Just, Global view on a nonlinear oscillator subject to time-delayed feedback control. Phys. D: Nonlinear Phenom. **238**(16), 1680–1687 (2009)
34. W. Just, B. Fiedler, M. Georgi, V. Flunkert, P. Hövel, E. Schöll, Beyond the odd number limitation: a bifurcation analysis of time-delayed feedback control. Phys. Rev. E **76**, 026210 (2007)
35. A.S. Purewal, C.M. Postlethwaite, B. Krauskopf, Effect of delay mismatch in Pyragas feedback control. Phys. Rev. E **90**(5), 052905 (2014)
36. A.S. Purewal, C.M. Postlethwaite, B. Krauskopf, A global bifurcation analysis of the generic subcritical Hopf normal form subject to Pyragas time-delayed feedback control. SIAM J. Appl. Dynam. Syst. **13**(4), 1879–1915 (2014)
37. C.M. Postlethwaite, M. Silber, Stabilizing unstable periodic orbits in the Lorenz equations using time-delayed feedback control. Phys. Rev. E **76**(5), 056214 (2007)
38. G.C. Brown, An Analysis of Hopf Bifurcation Problems with Time-Delayed Feedback Control. Ph.D. thesis, Northwestern University (2011)
39. E.N. Lorenz, Deterministic nonperiodic flow. J. Atmos. Sci. **20**(2), 130–141 (1963)
40. C. Sparrow, *The Lorenz Equations: Bifurcations, Chaos, and Strange Attractors* (Springer, New York, 1982)
41. E.J. Doedel, B. Krauskopf, H.M. Osinga, Global bifurcations of the Lorenz manifold. Nonlinearity **19**(12), 2947 (2006)
42. E.J. Doedel, B. Krauskopf, H.M. Osinga, Global invariant manifolds in the transition to pre-turbulence in the Lorenz system. Indagationes Mathematicae **22**(3), 222–240 (2011)
43. E.J. Doedel, Auto- 07P: Continuation and bifurcation software for ordinary differential equations. In: A.R., Fairgrieve T.F., Kuznetsov Y.A., Oldeman B.E., Paffenroth, R.C., Sandstede, B., Wang, X.J., Zhang, C., (2007) with major contributions from Champneys. http://cmvl.cs.concordia.ca/auto/
44. J. Guckenheimer, P. Holmes, *Nonlinear Oscillations, Dynamical Systems, and Bifurcations of Vector Fields* (Springer, New York, 1983)
45. B. Fiedler, S. Yanchuk, V. Flunkert, P. Hövel, H.J. Wünsche, E. Schöll, Delay stabilization of rotating waves near fold bifurcation and application to all-optical control of a semiconductor laser. Phys. Rev. E **77**(6), 066207 (2008)

Chapter 6
Symmetry-Breaking Control of Rotating Waves

Isabelle Schneider and Bernold Fiedler

Abstract Our aim is the stabilization of time-periodic spatio-temporal synchronization patterns. Our primary examples are coupled networks of Stuart-Landau oscillators. We work in the spirit of Pyragas control by noninvasive delayed feedback. In addition we take advantage of symmetry aspects. For simplicity of presentation we first focus on a ring of coupled oscillators. We show how symmetry-breaking controls succeed in selecting and stabilizing unstable periodic orbits of rotating wave type. Standard Pyragas control at minimal period fails in this selection task. Instead, we use arbitrarily small noninvasive time-delays. As a consequence we succeed in stabilizing rotating waves—for arbitrary coupling strengths, and far from equilibrium.

6.1 Introduction

In their 1990 publication "Controlling Chaos" [1], Ott, Grebogi and Yorke presented a first control scheme to stabilize unstable periodic orbits in chaotic systems. Another particularly successful method for stabilizing periodic orbits was introduced by Kestutis Pyragas in 1992 using time-delayed feedback [2], as follows. Consider any autonomous system

$$\dot{z}(t) = F(z(t)), \tag{6.1}$$

say on a state space $z \in \mathbf{R}^N$ or $\mathbf{C}^N$. Suppose that there exists a solution $z^*(t)$ which is an unstable periodic orbit with minimal period $p > 0$. Pyragas suggests to stabilize the periodic orbit $z^*(t)$ by adding a delayed control term. It consists of the difference between the current state $z(t)$ and a delayed state $z(t - \tau)$ of the system (6.1). The resulting delayed feedback system takes the form

I. Schneider (✉) · B. Fiedler
Institut für Mathematik, Freie Universität Berlin, 14195 Berlin, Germany
e-mail: isabelle.schneider@fu-berlin.de

B. Fiedler
e-mail: fiedler@mi.fu-berlin.de

© Springer International Publishing Switzerland 2016
E. Schöll et al. (eds.), *Control of Self-Organizing Nonlinear Systems*,
Understanding Complex Systems, DOI 10.1007/978-3-319-28028-8_6

$$\dot{z}(t) = F(z(t)) + \boldsymbol{b}\big(z(t - mp) - z(t)\big). \qquad (6.2)$$

Here $\boldsymbol{b}$ is a scalar control parameter, or a control matrix. The time-delay $\tau = mp > 0$ is an integer multiple m of the minimal period p. By periodicity of $z(t)$ the control term in (6.2) vanishes on the periodic orbit and is therefore called *noninvasive*.

Many theoretical investigations and experimental implementations have shown the success of Pyragas control in stabilizing unstable periodic orbits. For an overview, see for example the survey paper by Pyragas [3]. Pyragas control can be applied without any explicit knowledge of the model (6.1), or its solutions $z(t)$. This is one of the main reasons for its widespread use in experiments.

Analytic results included the *odd-number-limitation*, which was formulated and proven by Nakajima in 1997 [4]. It states that in (generic) non-autonomous systems Pyragas control fails for periodic orbits with an odd number of unstable Floquet exponents, counting algebraic multiplicities of exponents with positive real part; see Just et al. [5]. Genericity basically requires the absence of a trivial zero Floquet exponent.

Autonomous systems do not depend on time, explicitly. Hence their nonstationary periodic solutions do possess a trivial Floquet exponent zero. For autonomous systems, the odd-number-limitation was in fact refuted by Fiedler et al. [6] in 2007. Subcritical Hopf bifurcation for the Stuart-Landau oscillator provided an analytically accessible counterexample.

Pyragas control for networks of coupled oscillators is a wide open subject of research. In fact the periodic solutions may exhibit various spatio-temporal symmetries, besides trivial complete synchrony. In diffusively coupled networks with positive coupling strength, stabilization by *standard* Pyragas control frequently turns out to fail, however, except for fully synchronized periodic orbits.

Symmetry-breaking control terms can overcome this limitation, and can target periodic orbits of prescribed spatio-temporal symmetry, separately. Our control terms still follow the main idea of Pyragas [2]—they are noninvasive and time-delayed. However the new control terms are able to select specific prescribed *spatio-temporal patterns* of the periodic orbits, as we will describe in more detail in Sect. 6.2.

A first step in the direction of using spatio-temporal symmetries was already proposed by Nakajima and Ueda in 1998 [7]. It was intended as a remedy to the odd-number-limitation [4], originally, which was believed to also hold for autonomous systems, at that time. Though the odd-number limitation has been refuted for autonomous systems, their approach remains a first paradigm for symmetric systems. For odd nonlinearities, $F(-z) = -F(z)$, p-periodic odd oscillations may arise which satisfy

$$z^*(t) = -z^*(t - p/2) \qquad (6.3)$$

at half period. The delayed feedback system is of the form

$$\dot{z}(t) = F(z(t)) + \boldsymbol{b}\big(z(t) + z(t - p/2)\big). \qquad (6.4)$$

Note how the control term (6.2) is noninvasive on odd oscillations (6.3), even though the time-delay $\tau = p/2$ has been reduced to *half* the minimal period p.

Using the network structure, another control term with half-period delay τ has been applied to a system of two diffusively coupled Stuart-Landau oscillators. See Fiedler et al. in 2010 [8]. The controlled system is of the following form:

$$\dot{z}_0(t) = f(z_0(t)) + a(z_1 - z_0) + \boldsymbol{b}(z_0(t) - z_1(t - p/2)) \tag{6.5}$$
$$\dot{z}_1(t) = f(z_1(t)) + a(z_0 - z_1) + \boldsymbol{b}(z_1(t) - z_0(t - p/2)). \tag{6.6}$$

Here the state vectors z_0 and z_1 denote the first and the second oscillator, respectively, for $z = (z_0, z_1)$. The parameter $a > 0$ denotes diffusive coupling. Note how this control scheme is noninvasive on p-periodic anti-phase oscillations

$$z_1^*(t) = z_0^*(t - p/2), \tag{6.7}$$

where the two oscillators are phase-locked at half-period p. In 2013, slightly more general control schemes were used by Bosewitz [9] and Bubolz [10] for the same system, overcoming certain limitations. We will review their contribution in Sect. 6.4.

Already for three equilaterally coupled Stuart-Landau oscillators, new challenges arise. Two different types of spatio-temporal symmetries arise, for p-periodic solutions (z_0, z_1, z_2) which are not fully synchronous. First we may encounter *discrete rotating waves*

$$z_1^*(t) = z_0^*(t - p/3), \qquad z_2^*(t) = z_0^*(t - 2p/3). \tag{6.8}$$

Their reflected counterpart

$$z_2^*(t) = z_0^*(t - p/3), \qquad z_1^*(t) = z_0^*(t - 2p/3), \tag{6.9}$$

rotates in the opposite direction. Another spatio-temporal symmetry type features double frequency oscillation of one node:

$$z_0^*(t) = z_0^*(t - p/2), \qquad z_2^*(t) = z_1^*(t - p/2). \tag{6.10}$$

Here the oscillators z_1 and z_2 of minimal period p are phase-locked at half-periods, whereas z_0 oscillates at double frequency. Again permutation of indices produces related solutions of analogous spatio-temporal symmetry type. See [11] and Fiedler [12] for an in-depth discussion in local and global bifurcation settings, respectively.

This simple example already demonstrates why the control term should be able to *select* periodic solutions with the desired symmetry type. We call a control term *pattern-selective* or *symmetry-breaking*, if it is noninvasive on exactly *one* periodic solution with a prescribed spatio-temporal pattern. First successful controls in this sense were presented by Schneider in 2011 [13]; see also [14] and closely related work by Postlethwaite et al. [15].

Choe et al. considered delay-coupled networks in [16, 17]. They were able to show that by tuning the coupling phase it is possible to control the stability of synchronous periodic orbits. Later they generalized their results by using arbitrary and distributed time-delays as well as nonlinear coupling terms [18].

Unifying all control terms, a general equivariant formulation has been presented, and its success has been proven near equivariant Hopf bifurcation in [14, 19] as well as in [15] in 2013:

$$\dot{z}(t) = F(z(t)) + \boldsymbol{b}\big(- z(t) + hz(t - \Theta(h)\,p)\big). \tag{6.11}$$

Here h and $\Theta(h)$ describe the spatio-temporal pattern of the periodic orbit such that the control term is again noninvasive on the periodic orbit. Again $\boldsymbol{b}$ denotes a suitably chosen control matrix. For details on the group theoretic formulation see Golubitsky, Stewart [11, 20], Fiedler [12], and Sect. 6.2.

For a ring of coupled oscillators, the most complete results on Pyragas stabilization, to date, have been presented in [21]. For rotationally symmetric oscillators, nonlinear controls have been constructed by Bosewitz [22]. Additionally, a sharp upper bound on the unstable Floquet multiplier allowing stabilization has been established in [22]. This upper bound depends on the time-delay τ; see also [23].

Our survey is organized as follows. We describe some general background on spatio-temporal patterns in Sect. 6.2. Sections 6.3 and 6.4 focus on a ring of n identical, diffusively coupled oscillators z_k, $k \bmod n$,

$$\dot{z}_k = f(z_k) + a(z_{k+1} - 2z_k + z_{k-1}), \tag{6.12}$$

where $a > 0$ is the diffusive coupling strength. The stabilization results for this model system will be discussed in Sect. 6.4. We emphasize the aspect of symmetry-breaking and compare different control schemes. We show that standard Pyragas control is indeed not able to stabilize any but the totally synchronous periodic orbit in our model system. In Sect. 6.5 we discuss results for general networks consisting of rotationally symmetric oscillators. In particular we establish that pattern-selective Pyragas control of rotating waves always succeeds, with sufficiently small delay τ and nonlinear control of order $1/\tau$. Section 6.6 summarizes our conclusions.

6.2   Spatio-Temporal Patterns: Theory

In this section we explain our concept of spatio-temporal symmetry patterns for time periodic solutions $z^*(t)$ of equivariant systems. We illustrate the abstract and rather general mathematical concept for the specific system (6.12) of a ring of diffusively coupled identical oscillators; see also Sect. 6.3. Our presentation follows [12].

For simplicity let us consider a linear action $z \mapsto gz$ of some group G of matrices g on $z \in \mathbf{R}^N$ or $\mathbf{C}^N$. We call the ODE system (6.1), $\dot{z} = F(z)$, *equivariant* under G, if $gz(t)$ is a solution of the ODE whenever $z(t)$ is, for any $g \in G$. Since the action of G is linear, this simply means that

$$F(gz) = g\,F(z) \tag{6.13}$$

for all elements $z \in \mathbf{R}^N$ or $\mathbf{C}^N$, and for all $g \in G$. We also call G an *equivariance group* of F in (6.1).

Consider our ring (6.12) of diffusively coupled oscillators, for example. Let D_n denote the dihedral group of n rotations and n reflections which leave the regular planar n-gon invariant. Then $G = D_n$ is an equivariance group of the oscillator ring (6.12). The n-gon is represented by the ring structure of the network. The vertices $k = 0, \ldots, n-1$ indicate the oscillators z_k, and the edges define symmetric diffusion coupling. More precisely $D_n = \langle \rho, \kappa \rangle$ is generated by the rotation ρ over $2\pi/n$, and the reflection κ through the bisector of some fixed n-gon vertex angle. The linear action of $g \in D_n$ on $z = (z_0, \ldots, z_{n-1})$ is given by index permutation:

$$(\rho z)_k = z_{k-1} \tag{6.14}$$
$$(\kappa z)_k = z_{-k} \tag{6.15}$$

for $k \bmod n$.

We now describe the spatio-temporal symmetry of any periodic solution $z^*(t)$ of any G-equivariant system $\dot{z} = F(z)$. Let $p > 0$ again denote the minimal period of $z^*(t)$, and let $\mathcal{O}^* = \{z^*(t) \mid 0 \le t < p\}$ denote the periodic orbit, as a set. We can then describe the *spatio-temporal symmetry* of $z^*(t)$ by a triplet (H, K, Θ), as illustrated in Fig. 6.1.

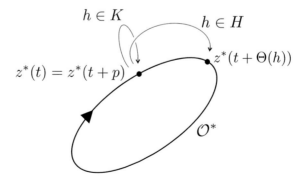

Fig. 6.1 Spatio-temporal symmetry $H^{\Theta} = \{(h, \Theta(h)) \in H \times S^1\}$ of any periodic orbit $z^*(t)$ with minimal period p. The group H fixes the periodic orbit as a set. The map $h \mapsto \Theta(h)$ indicates the (normalized) temporal phase shift on z^* effected by the spatial transformation $h \in H$. The kernel $K = \ker \Theta$ fixes any individual point on the periodic orbit, one by one

Let H denote the set of those group elements h in the equivariance group G which fix the orbit $\mathcal{O}^*$, as a set. In other words,

$$hz(t_0) = z(t_0 + \vartheta) \tag{6.16}$$

for some $t_0, \vartheta \in \mathbf{R}$. This definition makes sense because (6.16) holds for all $t_0 \in \mathbf{R}$, once it holds for any, by G-equivariance. Indeed, both $\tau \mapsto hz(t_0 + \tau)$ and $\tau \mapsto z(t_0 + \tau + \vartheta)$ are solutions of $\dot{z} = F(z)$ with the same initial condition at $\tau = 0$. Moreover the phase shift $\vartheta = \Theta(h)p$ is unique, for any given $h \in H$ and with normalized $\Theta(h)$ mod 1, by minimality of the period p of $z^*(t)$. This defines the *phase map*

$$\Theta : H \to S^1 = \mathbf{R}/\mathbf{Z}. \tag{6.17}$$

In fact G-equivariance implies that Θ is a group homomorphism:

$$\Theta(h_1 h_2) = \Theta(h_1) + \Theta(h_2) \text{ mod } 1, \tag{6.18}$$

for all $h_1, h_2 \in H$. Finally, let $K = \ker \Theta = \{h \in H \mid \Theta(h) = 0\}$ denote the kernel of the homomorphism Θ. Then $K \leq H$ is the set of group elements $h \in G$ which fix the periodic orbit $\mathcal{O}^*$, pointwise. In other words,

$$hz^*(t_0) = z(t_0) \tag{6.19}$$

for some (and hence for all) $t_0 \in \mathbf{R}$.

To summarize, the spatio-temporal symmetry of a periodic orbit $z^*(t)$ with minimal period p is characterized by a triplet (H, K, Θ). The phase map homomorphism $\Theta : H \to S^1$ describes the normalized time shifts

$$hz(t) = z\big(t + \Theta(h)p\big) \tag{6.20}$$

for all $t \in \mathbf{R}, h \in H$, and the normal subgroup $K := \ker \Theta$ of H describes the purely spatial symmetry of any periodic point $z^*(t_0)$. We sometimes abbreviate the triplet by the *twisted symmetry*

$$H^\Theta := \{(h, \Theta(h)) \mid h \in H\}. \tag{6.21}$$

The range of the phase map Θ is a subgroup of S^1, and range $\Theta \cong H/K$ by the homomorphism theorem. For compact (and in particular for finite) subgroups H, continuity of Θ implies compactness of range Θ. Therefore range $\Theta \leq S^1$ is either finite, or else coincides with S^1. We call $z^*(t)$ a *discrete wave*, in the former case, and a *rotating wave*, in the latter. Of course, finite equivariance G cannot lead to rotating waves.

By equivariance, $gz^*(t)$ is time periodic of minimal period p, whenever $z^*(t)$ itself is, for any fixed $g \in G$. The spatio-temporal symmetry H_g^Θ of $gz^*(z)$ is conjugate to the twisted symmetry H^Θ of $z^*(t)$ itself, i.e.

$$H_g^\Theta = (gHg^{-1})^\Theta, \quad \text{with} \quad \Theta(ghg^{-1}) := \Theta(h). \tag{6.22}$$

We also say that $z^*(t)$ and $gz^*(t)$ possess the same spatio-temporal *symmetry type*, differing only by conjugacy.

Let us now return to our example (6.12) of an oscillator ring with dihedral equivariance group $G = D_n$. The case $H = \langle \rho \rangle \cong \mathbf{Z}_n$ of the cyclic subgroup of rotations suggests $\Theta(\rho) := \pm 1/n \mod 1$ as a phase map with trivial kernel $K = \{id\}$. Golubitsky and Stewart coined the term "ponies-on-a-merry-go-round" for such *discrete rotating waves*. These are spatially discrete analogues of the above rotating waves where range $\Theta = S^1$. See (6.7), for the case $n = 2$, and (6.8), (6.9) for $n = 3$. Note how the cases $\Theta(\rho) = \pm 1/n$ are conjugate by the reflection $g = \kappa$; the right and left rotating discrete rotating waves belong to the same symmetry type.

More generally, we might observe discrete waves of the form

$$\Theta(\rho) = s/n \quad \mod 1, \tag{6.23}$$

for any integer $0 \leq s < n$, and we do! Then Θ possesses trivial kernel, if and only if $s \in \mathbf{Z}_n^*$ is a multiplicative unit. In general $K \cong \mathbf{Z}_{(s,n)}$, where (s, n) denotes the greatest common divisor of s and n. Again $\pm s$ belong to the same symmetry type, conjugated by the reflection κ.

Example (6.10) is of a different type: $H = \kappa \cong \mathbf{Z}_2$ and $\Theta(\kappa) = 1/2$. Of course the example generalizes to any n-ring, $n \geq 3$. We call $z^*(t)$ with this symmetry *standing waves*. For even n, only, the vertex $z_{n/2}(t)$ then oscillates at double frequency, as $z_0(t)$ always does. The reflection $\rho\kappa$ is nonconjugate to κ in $G = D_n$, for even n, and $H = \langle \rho\kappa \rangle \cong \mathbf{Z}_2$ with $\Theta(\rho\kappa) = 1/2$ does not feature any vertices with double frequency oscillations, in general.

Suppose next that $H = D_n$ arises in a spatio-temporal symmetry triplet (H, K, Θ) in the n-oscillator ring (6.12). Of course this is possible for $\Theta = 0$, i.e. for total synchrony $z_0 \equiv z_1 \equiv \cdots \equiv z_{n-1}$. We claim there is only one other possibility.

Indeed, $D_n/K = H/K \cong$ range $\Theta \leq S^1$ must be a nontrivial Abelian (in fact, cyclic) factor of D_n. Therefore K contains the commutator group $C(D_n) = [D_n, D_n]$ generated by all elements $g_1 g_2 g_1^{-1} g_2^{-2}$ with $g_1, g_2 \in D_n$. It is well-known that the commutator of D_n is generated by ρ^2. In particular the abelianization $D_n/C(D_n)$ of D_n is $\mathbf{Z}_2$, for odd n, and the Klein 4-group $\mathbf{Z}_2 \times \mathbf{Z}_2$ for even n. Because $D_n/K = H/K \leq S^1$ is cyclic this implies range $\Theta = \{0, 1/2\}$. Moreover $K = \mathbf{Z}_n$ for odd n. This already implies $z_0 \equiv z_1 \equiv \cdots \equiv z_{n-1}$ and triviality $\Theta = 0$. Next suppose n is even. Then K corresponds to one of the $\mathbf{Z}_2$ factors in $D_n/C(D_n)$, i.e. $K = \mathbf{Z}_n$ or $K = D_{n/2}$. We can discard total synchrony $K = \mathbf{Z}_n$, as before.

The only remaining option, n even and $K = D_{n/2}$, corresponds to clustering into the two clusters $z_0 = z_2 = \cdots = z_{n-2}$ and $z_1 = z_3 = \cdots = z_{n-1}$, each of size $n/2$.

By range $\Theta = \{0, 1/2\}$ the two clusters are half a period out of phase: $\Theta(\rho) = 1/2$. Note how our conclusions did not rely on any specific information about the underlying equations other than the symmetry aspect.

In Sects. 6.3 and 6.4 we focus on discrete right and left rotating waves of the type (6.23), as a target of Pyragas control. Rotating waves, i.e. range $\Theta = S^1$, are addressed in Sect. 6.5. We plan to treat standing waves elsewhere.

6.3 Spatio-Temporal Patterns: Application to Rings of Oscillators

We have seen how, both, standing waves and discrete rotating waves may appear, hand-in-hand and at the same parameters. Hopf bifurcation in oscillator rings (6.12) provides an example. To be specific, and for simplicity of presentation, let us consider identical Stuart-Landau oscillators,

$$f(z) = (\lambda + i + \gamma |z|^2) z, \tag{6.24}$$

for the individual dynamics in (6.12). Here λ is the real bifurcation parameter and the cubic coefficient γ is complex. In total, there are n types of discrete rotating waves and they all appear at equivariant Hopf bifurcations.

Proposition 1 (Equivariant Hopf bifurcations, [21]) *Consider the coupled oscillator ring (6.12), (6.24) of $n \geq 3$ identical, and identically diffusion coupled, Stuart-Landau oscillators. Hopf bifurcation occurs at the parameter values*

$$\lambda = \lambda_s = 2a(1 - \cos(2\pi s/n)), \quad s = 0, \ldots, n - 1. \tag{6.25}$$

The purely imaginary eigenvalues are normalized to $\pm i$, i.e. to unit frequency. They are of algebraic and geometric real multiplicity 4, for $s \notin \{0, n/2\}$. The associated eigenspace possesses complex dimension 2 and real dimension four. Both, standing waves and discrete rotating waves bifurcate, for each $s \notin \{0, n/2\}$. The discrete rotating waves are harmonic,

$$z_k(t) = r_s \exp\left(2\pi i \left(\frac{t}{p_s} + s\frac{k}{n}\right)\right), \tag{6.26}$$

for oscillators z_k, $k = 0, \ldots, n - 1$, respectively, and are phase shifted by $2\pi s/n$ between adjacent oscillators. See (6.23). Amplitudes r_s and minimal periods p_s are given explicitly by

$$r_s^2 = (\lambda_s - \lambda)/\text{Re}\,\gamma, \tag{6.27}$$
$$p_s = 2\pi/(1 + r_s^2 \,\text{Im}\,\gamma). \tag{6.28}$$

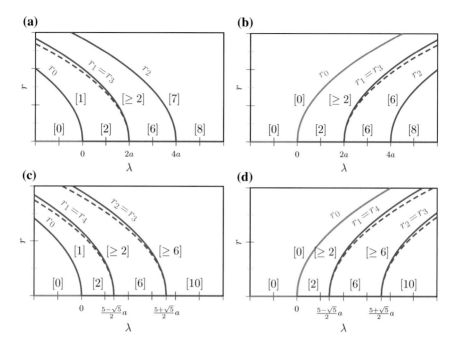

Fig. 6.2 Parameter-dependent stability of the equilibria and the bifurcating periodic orbits. The *upper row* shows the bifurcations for $n = 4$, while the *lower row* is for $n = 5$. In **a** and **c** subcritical bifurcations occur for Re $\gamma = 0.1$. In **b** and **d** supercritical bifurcations are plotted for Re $\gamma = -0.1$. The coupling parameter is $a = 0.2$. In brackets the number of unstable dimensions is denoted. Stable objects are colored in *green*, unstable ones in *red*. This figure, including its caption, has previously been published in [21]

In particular the discrete rotating waves bifurcate supercritically, *i.e. towards* $\lambda > \lambda_s$, *for* Re $\gamma < 0$, *and* subcritically, *i.e. towards* $\lambda < \lambda_s$, *for* Re $\gamma > 0$.

The minimal period p_s *grows with amplitude* (soft spring) *if* Im $\gamma < 0$, *and decreases with amplitude* (hard spring) *if* Im $\gamma > 0$.

Not surprisingly, the waves with index s and index $n - s$ bifurcate at the same point $\lambda_s = \lambda_{n-s}$. The resulting discrete left and right rotating waves are conjugate by reflection κ. See (6.22). See also Fig. 6.2 for $n = 4$ and $n = 5$ oscillators. Standing waves are known to also bifurcate at $\lambda_s = \lambda_{n-s}$, for $s \notin \{0, n/2\}$; see [12, 20]. They are not harmonic and their bifurcation direction and stability may differ from the discrete rotating waves. The Hopf bifurcation at $\lambda_0 = 0$ is a standard bifurcation with simple complex eigenvalues $\pm i$ and real two-dimensional eigenspace. This leads to the bifurcation of fully synchronous periodic solutions.

An elementary proof of Proposition 1 relies on the Ansatz (6.26). The harmonic character of the Ansatz (6.26) in time t is justified by the S^1-equivariance of the identical Stuart-Landau oscillators (6.24); see also Sect. 6.5. The discrete harmonic character of the Ansatz (6.26) in the oscillator nodes k is justified by the imposed discrete rotating wave symmetry $\Theta : \mathbf{Z}^n \to S^1 = \mathbf{R}/\mathbf{Z}$ in (6.12). In fact, both aspects can also be unified by studying the ansatz (6.26) as the definition of a rotating wave under the full equivariance group $G = D_n \times S^1$ of (6.12), (6.24). Insertion of (6.26) into (6.12) implies all other claims. For a more general analysis, in the framework of Sect. 6.2, see again [12, 20].

The fully synchronous periodic solution is the only one which may be stable in the uncontrolled system, locally at Hopf bifurcation.

Proposition 2 (Stability of the periodic orbits, [21]) *For $s \neq 0$, the bifurcating discrete rotating waves (6.26), enumerated by s, are unstable, both in the sub- and the supercritical case. For $s = 0$, i.e. the synchronous case, the periodic solution is unstable in the subcritical and stable in the supercritical case.*

For a complete proof of Proposition 2 see [21]. The essential step of the proof is to study the n dynamically invariant complex irreducible representation subspaces

$$X_s = \left\{ (z_0, \ldots, z_{n-1}) \mid z_k = e^{-2\pi i s/n} z_{k-1} \ \text{ for all } k \text{ mod } n \right\}. \tag{6.29}$$

On each subspace $x_s \in X_s$ the system (6.12), (6.24) reduces to one single complex-valued equation

$$\dot{x}_s = f(x_s) - 2a\left(1 - \cos(2\pi s/n)\right) x_s. \tag{6.30}$$

Here $x_s \in X_s$. Hopf bifurcation occurs at $\lambda = \lambda_s = 2a(1 - \cos(2\pi s/n))$, with the spatio-temporal symmetry of the bifurcating periodic solution inheriting the symmetry of the invariant subspace X_s.

6.4 Pyragas Stabilization for a Ring of Coupled Stuart-Landau Oscillators

In this section we aim to stabilize unstable discrete rotating waves with prescribed spatio-temporal symmetry pattern

$$z_k(t) = z_{k-1}(t - sp_s/n), \tag{6.31}$$

$k \bmod n$. In other words, an *index shift* ρ by 1 corresponds to a normalized *phase shift* $\Theta(\rho) = s/n$. The minimal period of the selected discrete rotating wave is denoted by p_s. The periodic orbit is uniquely determined by the choice of the parameter $s \in \{1, \ldots, n-1\} \bmod n$. We recall how the solutions for s and $-s$ are conjugate by reflection $\kappa : k \leftrightarrow -k$, mod n.

The stabilization problem has first been addressed for two coupled oscillators by Fiedler et al. [8], then for three coupled oscillators by Schneider [13, 14] and Postlethwaite et al. [15]. The general case of n oscillators has recently been discussed by Schneider and Bosewitz [21].

For an equivariant control term we choose

$$\boldsymbol{b}\big(-z(t) + h\,z(t-\tau)\big) \tag{6.32}$$

with suitable $\boldsymbol{b}$, h and $\tau = \Theta(h)p_s$. Note how the control term is noninvasive, by definition, on the periodic orbit (6.31); see also (6.20).

To achieve large stabilization regions, and preserve equivariance under $H = \mathbf{Z}_n$, it is suitable to employ complex circulant control matrices with constant diagonals, i.e.

$$\boldsymbol{b} = \begin{pmatrix} b_0 & b_1 & b_2 & \cdots & b_{n-1} \\ b_{n-1} & b_0 & b_1 & \cdots & b_{n-2} \\ b_{n-2} & b_{n-1} & b_0 & \cdots & b_{n-3} \\ \vdots & \vdots & \vdots & \ddots & \vdots \\ b_1 & b_2 & b_3 & \cdots & b_0 \end{pmatrix} \tag{6.33}$$

with constant diagonal coefficients $b_k \in \mathbf{C}$. Note the highly nonlocal character of (6.33) with all-to-all coupling. Equivalently to (6.33) we may employ n complex control parameters $\beta_0, \ldots, \beta_{n-1}$, one in each representation subspace X_s, $s = 0, \ldots, n-1$. The matrix $\boldsymbol{b}$ and the coefficients β are related via the invertible linear transformation

$$b_k := \sum_{s=0}^{n-1} \beta_s \exp\left(2\pi \mathrm{i}sk/n\right). \tag{6.34}$$

The control parameters β_s diagonalize the $\mathbf{Z}_n$-equivariant circulant matrix $\boldsymbol{b}$ via the representations of $\mathbf{Z}_n = \langle \rho \rangle$ on X_s, but represent inherently nonlocal coupling.

In the control term (6.32) we are still free to choose $h = \rho^m$. Noninvasivity on the discrete rotating wave (6.32) of type s is guaranteed by the delay

$$\tau = \Theta(h)p_s = \Theta(\rho^m)p_s = msp_s/n, \qquad \mathrm{mod}\ p_s, \tag{6.35}$$

and only by any such choice.

To be more precise suppose the control (6.32) with $h = \rho^m$, delay τ, and invertible circulant control matrix $\boldsymbol{b}$ is noninvasive on some solution $z^*(t)$ of the oscillator ring (6.12). Let ν denote the order of $h = \rho^m$ in $\mathbf{Z}_n$, i.e. $\nu > 0$ is minimal such that $\nu m \equiv 0$ mod n. Then noninvasivity implies

$$z^*(t) = h^\nu z(t) = z^*(t + \nu\tau). \tag{6.36}$$

Therefore z^* is time-periodic with minimal period p dividing

$$\nu\tau = s'p, \tag{6.37}$$

for some positive integer s'. Let (H, K, Θ) denote the spatio-temporal symmetry of z^*. Of course $hz(t) = z(t - \tau)$ only guarantees $H \geq \langle h \rangle$. To ensure $\langle h \rangle \geq \mathbf{Z}_n$ let us choose m and n co-prime. In particular the order ν of h becomes n. Let $m'm = 1$ mod n denote the multiplicative inverse of m. Then

$$s := n\Theta(\rho) = n\Theta(\rho^{mm'}) = n\Theta(h^{m'}) = nm'\Theta(h) = nm'\tau/p = \nu m'\tau/p = m's' \quad (6.38)$$

mod n. We have used $mm' = 1$, $\rho^m = h$, the definition of the phase map Θ, $\nu = n$, and (6.37), successively, in this line. This readily identifies s in (6.31) from the Pyragas data $h = \rho^m$ and τ, via $(m', m) = 1$ and $s' = n\tau/p$.

In principle, it is possible to use a time-delay $\tau > p_s$. However, in [21] it was found that larger time-delay diminishes the stabilization regions.

The following theorem tells us that the constructed control is indeed stabilizing.

Theorem 3 (Successful stabilization of discrete rotating waves [21]) *Consider the Hopf bifurcation of discrete rotating waves*

$$z_k^*(t) = z_{k-1}^*(t - sp_s/n), \quad (6.39)$$

of the Stuart-Landau ring

$$\dot{z}_k = (\lambda + \mathrm{i} + \gamma |z_k|^2) z_k + a(z_{k-1} - 2z_k + z_{k+1}), \quad (6.40)$$

k mod n, with $\lambda \in \mathbf{R}$, $a > 0$ and $\gamma \in \mathbf{C} \setminus \mathbf{R}_+$.

Then for every combination of s and m, with $s, m \in \{1, \ldots, n-1\}$ and m co-prime to n, there exists a positive constant $a_{m,s}$ such that the following conclusion holds for all real coupling constants $0 < a < a_{m,s}$, and sufficiently near the selected Hopf bifurcation at $\lambda_s = 2a(1 - \cos(2\pi s/n))$.

There exist open regions of complex control parameters $\beta_0, \ldots, \beta_{n-1}$ such that in the delayed feedback system

$$\dot{z} = f(z) + a(\varrho z - 2z + \varrho^{-1} z) + \mathbf{b} \left(-z(t) + \varrho^m z(t - \tau) \right) \quad (6.41)$$

with circulant control matrix $\mathbf{b} = (b_{kl})$, $b_{kl} = \sum_{j=1}^{n-1} \beta_j \exp \left(2\pi \mathrm{i} j (l - k)/n \right)$, the discrete rotating wave solution (6.39) is stabilized for a time-delay

$$\tau = msp_s/n \mod p_s, \quad (6.42)$$

$0 < \tau \leq p_s$. *The stabilization is noninvasive and pattern-selective, i.e. the control vanishes only on the discrete rotating wave (6.39).*

If m is not co-prime to n, then the control term is noninvasive on more than one discrete rotating wave. This can obstruct stabilization, as we will see below for standard Pyragas control, $m = n$, $\tau = p_s$.

Theorem 3 is proved in [21]. Some examples for stabilization regions in the complex parameters $\beta_0, \ldots, \beta_{n-1}$ are depicted in Fig. 6.3, for $n = 4$ oscillators, and in Fig. 6.4, for $n = 5$ oscillators.

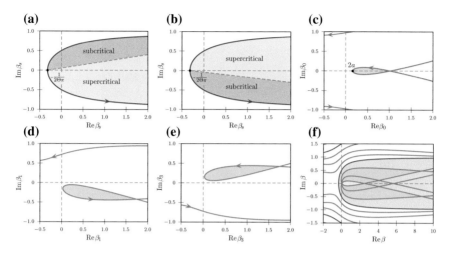

Fig. 6.3 Stabilization curves and regions for $n = 4$ coupled oscillators (6.12), (6.24) controlled as in (6.41) with index shift $m = 1$, for the third Hopf bifurcation (i.e. $s = 2$). The coupling constant was chosen as $a = 0.08$ in (**a–e**) and $a = 0.01$ in (**f**). The curve belonging to the parameter $\beta_s = \beta_2$ is drawn in *red*, while the curves corresponding to β_1, β_3 and β_4 are drawn in *green*. These figures have previously been published in [21]

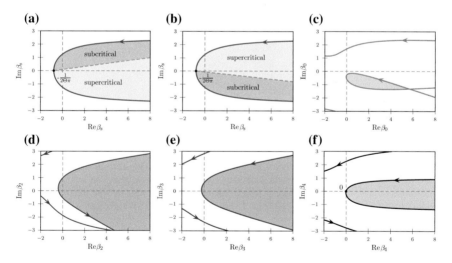

Fig. 6.4 Stabilization curves and regions for $n = 5$ coupled oscillators (6.12), (6.24) with coupling constant $a = 0.2$, controlled as in (6.41) with index shift $m = 1$, for $s = 1$. The curve belonging to the parameter $\beta_s = \beta_1$ is drawn in *red*, and the *green curve* corresponds to those parameters β_j for which $\lambda_j < \lambda_s = \lambda_1$ (see also Fig. 6.3). *Blue* is used for the curves β_2 and β_3, which occur for $\lambda_2 = \lambda_3 > \lambda_s = \lambda_1$. For the black curve in (**f**), corresponding to β_4, we find $\lambda_4 = \lambda_s = \lambda_1$. Note that we only find four different types of stabilization regions for the parameters β_j, depending on the relative positions of λ_s and λ_j. See also Fig. 6.2 (**c**) and (**d**). These figures have previously been published in [21]

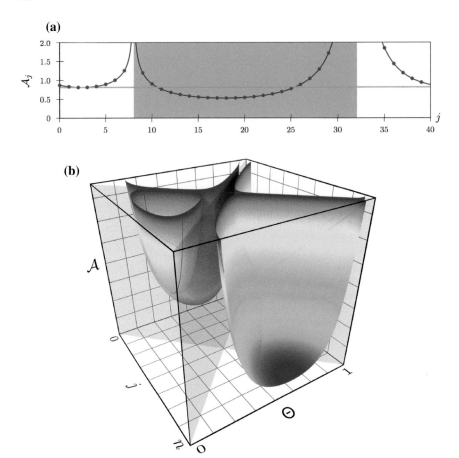

Fig. 6.5 Figure (**a**) shows the maximal coupling constant a for $n = 40, m = 1, s = 8$ and $\Theta = 1/5$. It is given by the minimum of $\mathcal{A}_j := A_j/|2\cos(2\pi j/n) - 2\cos(2\pi s/n)|$ for $j < 8$ and $j > 32$. The *gray area* given by $8 \leq j \leq 32$ is not relevant for the minimum. The maximal coupling constant a allowing stabilization of the system is marked as a *horizontal* (*green*) *line*. The *solid* (*red*) *curve* corresponds to the solutions for real $j \in [0, n]$. This *red curve* can also be seen in (**b**), where we see the general dependence of the threshold $\mathcal{A}_j$ on j as well as on Θ, for arbitrary n

The constants $a_{m,s}$ limit the maximal coupling strength for which a stabilization region exists. They depend on the spatio-temporal pattern, identified by s, and the chosen control term, identified by the index shift m. The following theorem describes this upper bound $a_{m,s}$ as a solution to a system of trigonometric equations. For a graphical depiction see Fig. 6.5.

Theorem 4 (Maximal coupling constant for a given control scheme [21]) *Under the above conditions, the maximal coupling constant $a_{m,s}$ is given implicitly by the minimum of all*

$$A_j/|2\cos(2\pi j/n) - 2\cos(2\pi s/n)|, \tag{6.43}$$

with either $0 \leq j < s$ *or* $n - s < j < n$. *Here* (A_j, ω_j), $j = 0, \ldots, n - 1$, *is the implicit solution of the system*

$$\sin \Omega_j \cos \Omega_j = -\omega_j \pi \Theta \tag{6.44}$$

$$\sin^2 \Omega_j = A_j \pi \Theta, \tag{6.45}$$

with $\Omega_j := \pi(mj/n - \Theta(1 + \omega_j))$, *and* $\Theta = ms/n \bmod 1, 0 < \Theta \leq 1$.

The upper bound $a_{m,s}$ for the coupling parameter a is strictly positive for the equivariant control type, and equal to zero for standard Pyragas control. For infinitesimal time-delay $\tau \searrow 0$ the threshold tends to infinity.

From Theorem 4, we can directly conclude that *standard* Pyragas control fails to stabilize any spatio-temporal pattern which is not completely synchronous. Indeed $h = \mathrm{id}$ and $\tau = p_s$ for standard Pyragas control. The system (6.44), (6.45) then simplifies to

$$\sin(\pi(1 + \omega_j)) \cos(\pi(1 + \omega_j)) = \omega_j \pi \tag{6.46}$$

$$\sin^2(\pi(1 + \omega_j)) = A_j \pi. \tag{6.47}$$

For any j, we only obtain the trivial solution $(A_j, \omega_j) = (0, 0)$. Hence we conclude $a_{0,s} = 0$ for all $s > 0$ and we obtain the following corollary.

Corollary 5 (Failure of standard Pyragas control [21]) *The discrete rotating wave*

$$z_k(t) = z_{k-1}(t - sp_s/n), \tag{6.48}$$

with $s \neq 0$ *cannot be stabilized by standard Pyragas control, i.e. by any delayed feedback system of the form*

$$\dot{z} = f(z) + a(\varrho z - 2z + \varrho^{-1}z) + b\left(-z(t) + z(t - p_s)\right) \tag{6.49}$$

which preserves at least $\mathbf{Z}_n$*-equivariance. In fact the solution* (6.48) *is unstable, sufficiently close to Hopf bifurcation, for any complex circulant* $n \times n$ *control matrix* b *as in* (6.33), *and any coupling constant* $a > 0$.

In the above control schemes, we have only used a single noninvasive control term. Linear combinations of such noninvasive control terms are an interesting extension. For example we may consider weighted sums of *all* noninvasive control terms,

$$\dot{z} = f(z) + a(\varrho z - 2z + \varrho^{-1}z) \tag{6.50}$$

$$+ \sum_{m=0}^{n-1} b_m\left(-z(t) + \varrho^m z(t - \tau_m)\right), \tag{6.51}$$

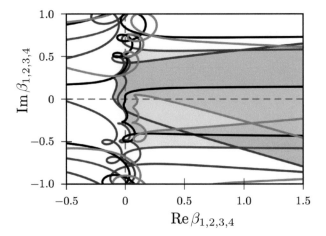

Fig. 6.6 Stabilization curves and regions for the sum of the delays as in with $n = 4$, $s = 1$, and $a = 0.2$ are shown. In accordance with the *color* coding of the other figures, β_0 is *green*, β_1 *red*, β_2 *blue* and β_3 *black*. This figure has previously been published in [21]

with appropriate time-delays $\tau_m = m s p_s / n \bmod p_s$ and control matrices $\boldsymbol{b}_m$. Such control terms also yield stabilization regions, see Fig. 6.6. However, the existence of non-empty control regions for not necessarily small enough coupling constants a remains an open problem.

6.5 Rotating Waves Under Free S^1-Actions

From an abstract view point we return to the general G-equivariant dynamics

$$\dot{z} = F(z) \tag{6.52}$$

of Sect. 6.2. We assume the group G to take the direct product form

$$G = \Gamma \times S^1 \tag{6.53}$$

$$g = (\gamma, \vartheta) \tag{6.54}$$

with elements $\gamma \in \Gamma$ and $\vartheta \in S^1 = \mathbf{R}/\mathbf{Z}$. We assume the linear action of the factor S^1 on z to be free; i.e. $\vartheta(z) = z$ only if $\vartheta = 0$ or $z = 0$. In other words, we can assume $z \in \mathbf{R}^{2n} \cong \mathbf{C}^n$ and

$$\vartheta(z) := \exp(2\pi i \vartheta) z. \tag{6.55}$$

For simplicity we will assume the other factor Γ to be a finite group.

Our task, in the present section, will be the Pyragas stabilization of a hyperbolic rotating wave solution

$$hz^*(t) = z^*(t + \Theta(h)p) \tag{6.56}$$

of (6.52), in the sense of Sect. 6.2. See in particular (6.16)–(6.21). For rotating waves z^* with spatio-temporal symmetry H^Θ, we recall that the homomorphism

$$\Theta : H \to S^1 \tag{6.57}$$

is assumed to be surjective. We will first observe that

$$z^*(t) = e^{i\omega t} z^*(0) \tag{6.58}$$

is necessarily *harmonic* with frequency $\omega = \pm 2\pi/p$. We then show how *nonlinear Pyragas stabilization* of the form

$$\dot{z}(t) = F(z(t)) + B(e^{i\omega\tau}z(t-\tau), z(t)) \tag{6.59}$$

can always succeed, in this setting, for any small enough $\tau > 0$ and suitably chosen complex B of order $1/\tau$. Here $B(z_1, z_2)$ is a vector-valued nonlinear and nonlocal control which vanishes on the diagonal

$$B(z, z) = 0. \tag{6.60}$$

We will only sketch the relevant arguments; for further mathematical details we refer to [21–23].

Any finite graph of linearly or nonlinearly coupled, neither necessarily identical nor identically coupled, Stuart-Landau oscillators $z_0, \ldots z_{n-1}$ provides an example for our setting. Here Γ is the automorphism group of the coupled oscillator system; the elements γ of Γ are those permutations of the vertex indices $0, \ldots, n-1$ which leave the system of coupled oscillators invariant. Recall $\Gamma = D_n$ for the symmetric diffusion coupling in a ring (6.12) of identical and identically coupled oscillators. Slightly more generally than the Stuart-Landau choice (6.24), we only require the nonlinearities $f_k(z_k)$ to satisfy the S^1-equivariance

$$f_k(e^{2\pi i\vartheta}\zeta) = e^{2\pi i\vartheta} f_k(\zeta) \tag{6.61}$$

for all $\zeta \in \mathbf{C}, \vartheta \in S^1$.

To illustrate our abstract approach, we first consider the elementary planar case of a single S^1-equivariant oscillator

$$\dot{z}(t) = f(z(t)) + b(e^{i\omega\tau}z(t-\tau) - z(t)), \tag{6.62}$$

$z \in \mathbf{C}$. For absent complex scalar control $b = 0$, any nonstationary periodic solution $z = z^*(t)$ with minimal period $p > 0$ must be harmonic,

$$z^*(t) = e^{i\omega t} z^*(0), \tag{6.63}$$

with amplitude $z^*(0) = r^* > 0$ and frequency $\omega = \pm 2\pi/p$. Note

$$\text{Re } f(r^*) = 0, \tag{6.64}$$
$$\text{Im } f(r^*) = \omega. \tag{6.65}$$

We do not assume small amplitude r^*, here or below. In co-rotating coordinates $\zeta(t) := \exp(-i\omega t)z(t)$ we obtain

$$\dot{\zeta}(t) = f(\zeta(t)) - i\omega\zeta(t) + b(\zeta(t - \tau) - \zeta(t)). \tag{6.66}$$

Note how (6.66) remains autonomous, by S^1-equivariance of (6.62). The circle of stationary solutions $\zeta^* = \exp(i\varphi)r^*$, $\varphi \in S^1$, testifies to noninvasivity of the Pyragas control scheme (6.62) on the harmonic rotating wave z^*, for all b.

Pyragas stabilization of z^* in (6.62) is equivalent to stabilization of the equilibrium $\zeta^* = r^*$ in (6.66), by some complex b. For small τ, we Taylor expand

$$\zeta(t - \tau) - \zeta(t) = -\tau\dot{\zeta}(t) + \cdots \tag{6.67}$$

with higher terms of order τ^2. In other words (6.66) reads

$$(1 + b\tau)\dot{\zeta} = f(\zeta) - i\omega\zeta + \cdots. \tag{6.68}$$

Fix $\beta := b\tau$ and consider $\tau \to 0$. Then a remarkably early result by Kurzweil in 1971 indeed justifies (6.68) as an ODE approximation, even in the nonlinear case; see [24]. In modern language the ODE reduction (6.68) corresponds to a center manifold reduction with rapid attraction in the infinitely many remaining directions. For the characteristic equation of the linearization of (6.66) at $\zeta = \zeta^* = r^*$ the justification is trivial, by an exponential Ansatz. Note how the trivial zero eigenvalue of the linearization at ζ^* remains unaffected by (6.68). The only other eigenvalue μ, alias the nontrivial Floquet exponents of z^*, can be rotated at will by a suitably fixed choice of $\beta = b\tau \in \mathbf{C}$. This choice stabilizes z^*, for sufficiently small $\tau > 0$, because all omitted terms are of order $b\tau^2 = \beta\tau = O(\tau)$ or higher. *This completes noninvasive Pyragas stabilization of large rotating waves, in the planar case.*

The above result is in marked contrast with controls based on $z(t)$ and $z(t - p)$, only. In fact suppose the controls are even allowed to be of the nonlinear form (6.59), (6.60). In the general S^1-equivariant planar case, the nontrivial Floquet exponent μ must then satisfy a constraint

$$p \,\text{Re}\, \mu < 9, \tag{6.69}$$

in order to enable noninvasive control; see [23]. In [22] the analogous bound $\vartheta p \operatorname{Re} \mu < 9$ was established for controls based on $z(t)$ and $z(t - \vartheta p)$, in the S^1-equivariant case.

Let us now return to the general case of $G = \Gamma \times S^1$-equivariant systems $\dot{z} = F(z)$. Our task is to stabilize a rotating wave z^* of spatio-temporal symmetry H^Θ. We pursue the nonlinear Pyragas scheme (6.59) with small delays $\vartheta > 0$ and bounded control ϑB.

The spatio-temporal symmetry H^Θ of the rotating wave z^* possesses the following general structure:

$$H = H_0 \times S^1, \qquad \text{and} \qquad (6.70)$$
$$\Theta(\gamma, \vartheta) = \Theta_0(\gamma) \pm \vartheta \qquad (6.71)$$

with a suitable finite subgroup $H_0 \leq \Gamma$, and a homomorphism $\Theta_0 : H_0 \to S^1$. Here we use that $G = \Gamma \times S^1$ with Γ finite, and the free action $z \mapsto e^{2\pi i \vartheta} z$ of S^1 on $z \in \mathbf{C}$. We omit mere mathematical details. Fixing $\gamma = \mathrm{id}$, $\Theta_0(\gamma) = 0$ in (6.71) and letting $t = \pm \vartheta p$, we obtain

$$z^*(t) = z^*(\pm \vartheta p) = z^*(0 + \Theta(\mathrm{id}, \vartheta)p) = (\mathrm{id}, \vartheta)z^*(0) \qquad (6.72)$$
$$= e^{2\pi i \vartheta} z^*(0) = e^{i\omega t} z^*(0)$$

with frequency $\omega = \pm 2\pi/p$. In particular $z^*(t)$ is harmonic, as claimed in (6.58), and the nonlinear Pyragas control scheme (6.59) is noninvasive on $z^*(t)$.

Noninvasive nonlinear Pyragas stabilization will be based on the pair

$$(z_1, z_2) = (hz(t - \Theta(h)p), z(t)) \qquad (6.73)$$

for some suitable $h = (\gamma, \pm \vartheta) \in H = H_0 \times S^1$, $\vartheta > 0$. By (6.70) we may pick $\gamma = \mathrm{id}$. Then the pair (6.73) becomes

$$\big((\pm \vartheta)z(t - \vartheta p), z(t)\big) = \big(e^{i\omega \tau} z(t - \tau), z(t)\big) \qquad (6.74)$$

with delay $\tau = \vartheta p > 0$. This shows that the noninvasive control scheme (6.59) is indeed based on the Pyragas difference (6.73).

To establish the success of (6.73) for small $\tau = \vartheta p > 0$ we proceed very much as in the planar case. In co-rotating coordinates $\zeta(t) := e^{-i\omega t} z(t)$, which freeze the harmonic rotation of $z^*(t)$, we obtain

$$\dot{\zeta}(t) = F(\zeta(t)) - i\omega \zeta(t) + B(\zeta(t - \tau), \zeta(t)). \qquad (6.75)$$

Here we have replaced the complex scalar b by the complex vector nonlinearity B. We have assumed that B commutes with the S^1-action on $z \in \mathbf{C}^N$. The justifiable approximation (6.68) then reads

$$(\mathrm{id} + \beta)\dot{\zeta} = F(\zeta) - \mathrm{i}\omega\zeta + \cdots . \tag{6.76}$$

with the complex nonlinearity $\beta := \tau \partial_1 B(\zeta, \zeta)$ chosen to be τ-independent.

Naively, at first, we might attempt to choose β to be a complex scalar multiple of an eigenprojection Q onto the strictly unstable eigenspace of the linearization $f'(\zeta^*) - \mathrm{i}\omega$ at $\zeta^* = z^*(0)$. Unfortunately the Jacobian $f'(\zeta^*)$ itself need not be *complex* linear, but will only be *real* linear in general. Indeed S^1-equivariance

$$f(\mathrm{e}^{\mathrm{i}\vartheta} z) = \mathrm{e}^{\mathrm{i}\vartheta} f(z), \tag{6.77}$$

for all $\vartheta \in \mathbf{R}/2\pi\mathbf{Z}$ and all $z \in \mathbf{C}^n$, only implies conjugacy

$$f'(\mathrm{e}^{\mathrm{i}\vartheta} z) = \mathrm{e}^{\mathrm{i}\vartheta} f'(z)\mathrm{e}^{-\mathrm{i}\vartheta} \tag{6.78}$$

at $z \neq 0$, but not complex linearity. In particular, the Floquet eigenprojection $Q = Q(z^*(t))$ depends on the footpoint $z^*(t)$ of the linearization and satisfies the same conjugacy

$$Q(\mathrm{e}^{\mathrm{i}\vartheta} \zeta^*) = \mathrm{e}^{\mathrm{i}\vartheta} Q(\zeta^*)\mathrm{e}^{-\mathrm{i}\vartheta}. \tag{6.79}$$

However, by hyperbolicity of ζ^* the unstable eigenspace is transverse to the group orbit $z^*(t) = \exp(\mathrm{i}\omega t)\zeta^*$ of ζ^*; the tangent $\dot{z}^*(0) = \mathrm{i}\omega\zeta^*$ to that group orbit provides the trivial Floquet exponent $\mu = 0$ which remains unaffected by Q, B, and β. Therefore we can define a *nonlinear* control $B = B(z_1, z_2)$, which vanishes on the diagonal $z_1 = z_2$, stays S^1-equivariant, and provides the appropriate linearization β in a full neighborhood of the group orbit $z^*(t)$, for $z_1 = \exp(\mathrm{i}\omega\tau)z^*(t - \tau) = z^*(t) = z_2$.

In conclusion, *noninvasive nonlinear Pyragas stabilization of large rotating waves also succeeds in the general $\Gamma \times S^1$-equivariant case with free S^1-action.* In the general case, our result requires a nonlinear vector-valued control term of the form (6.59), (6.60).

In [19], Schneider has studied the above problem with complex linear control matrices $B = b$ based on (6.59), (6.60), for rotating waves of small amplitudes near Hopf bifurcation from $z = 0$. In the supercritical case, control was successful in open regions of b. The subcritical case required an additional condition: the minimal period had to depend sufficiently strongly on amplitude ζ^*. This condition is satisfied, essentially, for a sufficiently nonlinear soft spring or hard spring case.

6.6 Summary

In this chapter, we have commented, and added to, recent results which show how equivariant Pyragas control succeeds in networks where standard Pyragas control fails. For a ring of n diffusively coupled Stuart-Landau oscillators, explicit complex linear control terms were constructed for each periodic orbit of discrete rotating wave type. These control terms are nonlocal. They use an interplay between *index shifts*

h and temporal *phase shifts* $\Theta(h)$ to select the spatio-temporal pattern, and they break the symmetry of the complete system. The control term is noninvasive only for exactly one type of discrete rotating waves. For full pattern-selectivity, the index shift m of $h = \rho^m$ must be chosen co-prime to the number n of oscillators coupled in the ring. For each discrete rotating wave, this provides several control terms for which a control is successful. In this case we have also provided an upper threshold on the maximally admissible coupling parameter, depending on the specific index shift.

For coupled Stuart-Landau oscillators, the additional S^1-equivariance of each oscillator allows additional conclusions. In this case, the existence of a noninvasive stabilizing control scheme on rotating waves can be guaranteed by choosing the time-delay small enough.

The control term will, however, become nonlocal and nonlinear in general. Only near Hopf bifurcation, results on stabilization by linear control have been obtained. The general results, on the other hand, are not limited to small amplitude. They do not require ring architecture. These results are based on equivariance under a free S^1-action as is provided by, but by no means limited to, the paradigm of coupled Stuart-Landau oscillators.

Acknowledgments This work originated at, and was supported by the SFB 910 "Control of Self-Organizing Nonlinear Systems: Theoretical Methods and Concepts of Application" of the Deutsche Forschungsgemeinschaft. The authors would like to thank Matthias Bosewitz for fruitful discussions and providing Fig. 6.5. Furthermore, we thank Eckehard Schöll, Ulrike Geiger, Xingjian Zhang, and all the members of the SFB for their many helpful suggestions.

References

1. E. Ott, C. Grebogi, J.A. Yorke, Phys. Rev. Lett. **64**(11), 1196 (1990)
2. K. Pyragas, Phys. Lett. A **170**(6), 421 (1992)
3. K. Pyragas, Phil. Trans. Roy. Soc. A **364**(1846), 2309 (2006)
4. H. Nakajima, Phys. Lett. A **232**(3), 207 (1997)
5. W. Just, T. Bernard, M. Ostheimer, E. Reibold, H. Benner, Phys. Rev. Lett. **78**(2), 203 (1997)
6. B. Fiedler, V. Flunkert, M. Georgi, P. Hövel, E. Schöll, Phys. Rev. Lett. **98**(11), 114101 (2007)
7. H. Nakajima, Y. Ueda, Phys. Rev. E **58**(2), 1757 (1998)
8. B. Fiedler, V. Flunkert, P. Hövel, E. Schöll, Phil. Trans. Roy. Soc. A **368**, 319 (2010)
9. M. Bosewitz, Stabilisierung gekoppelter Oszillatoren durch verzögerte Rückkopplungskontrolle. Bachelor Thesis, Freie Universität Berlin, 2013
10. K. Bubolz, Stabilisierung periodischer Orbits im System zweier gekoppelter Oszillatoren durch zeitverzögerte Rückkopplungskontrolle. Bachelor Thesis, Freie Universität Berlin, 2013
11. M. Golubitsky, I. Stewart, D. Schaeffer, *Singularities and Groups in Bifurcation Theory*, vol. 2. Applied Mathematical Sciences, vol. 69 (Springer, New York, 1988). doi:10.1007/978-1-4612-4574-2
12. B. Fiedler, *Global Bifurcation of Periodic Solutions with Symmetry*. Lecture Notes in Mathematics, vol. 1309 (Springer, Heidelberg, 1988). doi:10.1007/BFb0082943
13. I. Schneider, Stabilisierung von drei symmetrisch gekoppelten Oszillatoren durch zeitverzögerte Rückkopplungskontrolle. Bachelor Thesis, Freie Universität Berlin, 2011
14. I. Schneider, Phil. Trans. Roy. Soc. A **371**(1999), 20120472 (2013)

15. C.M. Postlethwaite, G. Brown, M. Silber, Phil. Trans. Roy. Soc. A **371**(1999), 20120467 (2013)
16. C.U. Choe, T. Dahms, P. Hövel, E. Schöll, Phys. Rev. E **81**(2), 025205 (2010)
17. C.U. Choe, H. Jang, V. Flunkert, T. Dahms, P. Hövel, E. Schöll, Dyn. Syst. **28**(1), 15 (2013)
18. C.U. Choe, R.S. Kim, H. Jang, P. Hövel, E. Schöll, Int. J. Dyn. Control **2**(1), 2 (2014)
19. I. Schneider, Equivariant Pyragas control. Master Thesis, Freie Universität Berlin, 2014
20. M. Golubitsky, I. Stewart, *The Symmetry Perspective: From Equilibrium to Chaos in Phase Space and Physical Space, Progress in Mathematics*, vol. 200 (Birkhäuser, Basel, 2003)
21. I. Schneider, M. Bosewitz, Disc. Cont. Dyn. Syst. A **36**, 451 (2016)
22. M. Bosewitz, Time-delayed feedback control of rotationally symmetric systems. Master Thesis, Freie Universität Berlin, 2014
23. B. Fiedler, in *Proceedings of 6th EUROMECH Nonlinear Dynamics Conference (ENOC-2008)*, 2008, ed. by A. Fradkov, B. Andrievsky
24. J. Kurzweil, in *Proceedings of the Symposium on Differential Equations and Dynamical Systems* (Springer, New York, 1971), pp. 47–49

Chapter 7
On the Interplay of Noise and Delay in Coupled Oscillators

Otti D'Huys, Thomas Jüngling and Wolfgang Kinzel

Abstract Coupling delays can play an important role in the dynamics of various networks, such as coupled semiconductor lasers, communication networks, genetic transcription circuits or the brain. A well established effect of a delay is to induce multistability: In oscillatory systems a delay gives rise to coexistent periodic orbits with different frequencies and oscillation patterns. Adding noise to the dynamics, the network switches stochastically between these delay-induced orbits. For phase oscillators, we compute analytically the distribution of frequencies, the robustness to noise and their dependence on system parameters as the coupling strength and coupling delay.

7.1 Introduction

In many areas of physics, biology and technology delay differential equations play an increasingly important role, as many systems can be modeled with a time-delay [1, 2]. For example between coupled semiconductor lasers [3], a time delay can arise due to the traveling time of light, while in gene regulatory networks a time-delay accounts for transcription and translation times [4]. Especially in networks a time delay can arise due to the traveling time of a signal from one node to another, as it is the case in neural, electronic, social, or communication networks [5–7].

One of the main effects of a delay is to induce multistability [8, 9]. In particular in oscillatory systems, a delay induces multiple periodic orbits. A well known example are the external cavity modes in a laser with delayed feedback [10], but the effect is universal for any dynamical system with a limit cycle [11, 12]. We highlight here the interaction of a coupling delay with network symmetry. The possible oscillation pat-

O. D'Huys (✉) · W. Kinzel
Universität Würzburg, Am Hubland, 97074 Würzburg, Germany
e-mail: otti.dhuys@physik.uni-wuerzburg.de

T. Jüngling
Universitat de les Illes Balears IFISC (CSIC-UIB), Campus UIB,
07122 Palma de Mallorca, Spain
e-mail: thomas@ifisc.uib-csic.es

© Springer International Publishing Switzerland 2016
E. Schöll et al. (eds.), *Control of Self-Organizing Nonlinear Systems*,
Understanding Complex Systems, DOI 10.1007/978-3-319-28028-8_7

terns, or spatio-temporal symmetries in the network dynamics, are determined by its topology [13]: In a ring topology, in-phase, anti-phase and out-of-phase oscillations are allowed. We show here how, in a delay system, these patterns can be mapped onto each other.

We discuss in Sect. 7.2 the oscillation patterns in coupled Kuramoto oscillators. The Kuramoto phase oscillator model has become a paradigmatic model to describe synchronization phenomena, as many limit cycle oscillators can be reduced to phase oscillators in the weak coupling regime [14–17]. Thanks to its simplicity, analytic calculations are possible, and we obtain explicit expressions for the phase relations and frequency of the oscillators, and their stability. The results for phase oscillators are shown to apply in a general context: we sketch the multistability mechanism and the interaction of a coupling delay and the ring topology in a general periodic system in Sect. 7.3.

In Sect. 7.4, we focus on the influence of noise, which is unavoidable in real networks. As in any multistable system, due to noise the oscillators can switch between different stable attractors. For phase oscillators it is possible to approximate the oscillators by a delay-free system; this allows to compute analytically the distribution of frequencies over the different periodic orbits and their average residence times. Especially in neural networks, coexistent patterns could be related to memory storage and temporal pattern recognition [18–20]; their robustness to noise is thus relevant in this context.

7.2 Periodic Orbits in a Ring of Delay-Coupled Phase Oscillators

In an oscillator network the possible oscillation patterns are determined by the network topology [13]. In a unidirectional ring, the topology allows for fully symmetric solutions, or in-phase oscillations, in which the oscillators behave all identically. Secondly, solutions with a spatio-temporal symmetry are possible; in this case the system is invariant under the combination of time-shift and a cyclic permutation of the elements, these are out-of-phase oscillations. In both cases the oscillators all have the same period and only differ by their relative phase.

Without delay, attractive coupling between oscillators typically leads to stable in-phase oscillations and repulsive coupling to stable anti-phase or out-of-phase oscillations. When the coupling is delayed, the coexistence of stable in-phase, anti-phase and out-of-phase oscillations has been reported analytically for Kuramoto and Stuart-Landau oscillators [21–23], numerically for, among/amongst others, neural and Stuart-Landau oscillators [18, 23–25] and experimentally in opto-electronic oscillators [26]. This multistability effect can be particularly easily illustrated in Kuramoto phase oscillators, which describe the oscillating dynamics by a single phase variable. We discuss here the periodic orbits in the most basic network, a single oscillator with feedback, the synchronized orbits of two delay-coupled oscil-

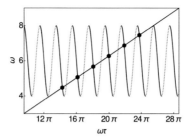

Fig. 7.1 Graphical determination of the different coexisting frequencies of a single oscillator with delayed feedback (Eq. 7.2). The intersections with the *thick* decreasing slopes of the sine function correspond to stable orbits, and are marked with a *circle*. Parameters are $\omega_0 = 6$, $\kappa = 2$ and $\tau = 10$

lators, and its generalization to a unidirectional ring. A single oscillator with delayed feedback is modeled by

$$\dot{\phi}(t) = \omega_0 + \kappa \sin(\phi(t - \tau) - \phi(t) + \theta). \tag{7.1}$$

The oscillator has a natural frequency ω_0, the other parameters are the coupling delay τ, the coupling strength $\kappa > 0$ and the coupling phase θ. As the dynamics is invariant under a transformation $\phi(t) \to \phi(t) + \tilde{\omega}t$, $\omega_0 \to \omega_0 + \tilde{\omega}$, $\theta \to \theta - \tilde{\omega}\tau$, the coupling phase θ can be omitted without loss of generality.

We seek solutions with a constant frequency of the form $\phi(t) = \omega_k t$. Substituting such a solution in the model (7.1), we find the locking frequencies ω_k by solving a transcendental equation

$$\omega_k = \omega_0 - \kappa \sin(\omega_k \tau). \tag{7.2}$$

The stability of an orbit with a locking frequency ω_k is determined by the characteristic equation for the growth rate λ of a small perturbation,

$$\lambda = -\kappa \cos(\omega_k \tau)(1 - e^{-\lambda \tau}). \tag{7.3}$$

Thus, frequencies for which $\kappa \tau \cos \omega_k \tau > 1$ holds, are stable. A graphical determination of the frequencies ω_k is shown in Fig. 7.1. Clearly, the number of coexisting orbits increases with coupling strength κ, as the frequencies range between $\omega_{\max} = \omega_0 + \kappa$ and $\omega_{\min} = \omega_0 - \kappa$, and with the delay time τ. In the long delay limit $\kappa \tau \gg 1$, the stable frequencies ω_k are approximated by $2n\pi/\tau$.

Coupling two identical oscillators, two spatio-temporal symmetric periodic patterns are possible: in the fully symmetric state the oscillators are in-phase with each other $\phi_1(t) = \phi_2(t)$. In the anti-phase state we have $\phi_1(t) = \phi_2(t + T/2) = \phi_2(t) + \pi$, where $T = 2\pi/\omega_k$ denotes the period of oscillation. The anti-phase state is invariant under the spatio-temporal symmetry $\phi_1 \to \phi_2, t \to t + T/2$. Two mutually coupled phase oscillators with delay are modeled as

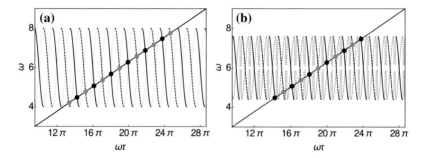

Fig. 7.2 Graphical determination of the locking frequencies of two **a** identical and **b** nonidentical delay-coupled phase oscillators. Intersections with the thick full (*dashed*) *line*, marked by a *black* (*gray*) *dot*, correspond to stable in-phase (anti-phase) orbits. In panel (**b**) a stable orbit is labeled as in-phase if $-\pi/2 < \delta < \pi/2$. Parameters are $\omega_0 = 6$, $\kappa = 2$, $\tau = 10$ and (**b**) $\Delta = 0.8$

$$\dot{\phi}_1(t) = \omega_0 + \kappa \sin(\phi_2(t - \tau) - \phi_1(t))$$
$$\dot{\phi}_2(t) = \omega_0 + \kappa \sin(\phi_1(t - \tau) - \phi_2(t)). \tag{7.4}$$

The in-phase periodic states are identical to the solutions of a single oscillator with feedback, for the anti-phase oscillations we have $\omega_k = \omega_0 + \kappa \sin(\omega_k \tau)$. As shown in Fig. 7.2a, the frequencies of stable in-phase and anti-phase orbits alternate each other over the whole frequency range. For long delays the stable frequencies are hence approximated as $\omega_k = n\pi/\tau$.

It is straightforward to extend the results of two identical coupled phase oscillators to a unidirectional ring of N oscillators. Such a system is then modeled as

$$\dot{\phi}_n(t) = \omega_0 + \kappa \sin(\phi_{n+1}(t - \tau) - \phi_n(t)), \tag{7.5}$$

with $N + 1 \equiv 1$. The coupling topology allows for in-phase oscillations $\phi_n(t) = \phi_{n+1}(t) = \omega_k t$ and several out-of-phase oscillation patterns $\phi_n(t) = \phi_{n+1}(t - mT/N) = \omega_k t + n\Delta\phi$, with $\Delta\phi = 2m\pi/N$. The corresponding frequencies are given by $\omega_k = \omega_0 - \kappa \sin(\omega_k \tau + \Delta\phi)$. For strong coupling and long delay, the frequencies are approximated by $\omega_k \approx 2m\pi/(N\tau)$. Comparing the different oscillation patterns, we find that all solutions have the same range $[\omega_0 - \kappa, \omega_0 + \kappa]$, and different oscillation patterns alternate each other. Moreover, solutions with a different spatio-temporal symmetry can be mapped onto each other by adjusting the coupling phase $\theta \rightarrow \theta + \Delta\phi$ [24].

The stability of the different solutions is similar in a single feedback system and in a unidirectional ring of any number of elements [21, 22, 27]. By applying a master stability function approach [28], and decomposing the system into the eigenmodes of a unidirectional ring, we find a characteristic equation for the growth rate λ of a small perturbation

$$\lambda = -\kappa \cos(\omega_k \tau - \Delta\phi)(1 - \gamma_m e^{-\lambda\tau}), \tag{7.6}$$

where $\gamma = e^{2m\pi i/N}$ denotes the eigenvalue of the adjacency matrix in the mth direction. In the long delay limit, only the magnitude of this factor γ plays a role [29, 30]. Hence, for long enough delay, a solution is stable if $\cos(\omega_k \tau - \Delta\phi) > 0$, irrespective of the number of oscillators in the ring or oscillation pattern of the specific solution.

When the symmetry between the oscillators is broken, the oscillators can still lock to a common frequency if the coupling is strong enough [31]. Their phase differences are however no longer equal to $2\pi/N$. We sketch here the solution for two nonidentical delay-coupled phase oscillators, modeled as

$$\dot{\phi}_n(t) = \omega_{0n} + \kappa \sin(\phi_{n+1}(t - \tau) - \phi_n(t)), \tag{7.7}$$

with $\omega_{01,02} = \omega_0 \pm \Delta/2$, and Δ being the detuning between the oscillators. The oscillators can then lock to a solution $\phi_1(t) = \phi_2(t) + \delta = \omega_k t$, with the phase difference δ and the locking frequency ω_k given by

$$\omega_k = \omega_0 - \kappa \sin(\omega_k \tau) \cos \delta$$
$$\sin \delta = \frac{\Delta}{2\kappa \cos(\omega_k \tau)}. \tag{7.8}$$

The graphical solution of Eq. (7.8) is shown in Fig. 7.2b: The locking frequencies and their stability properties are similar as for identical oscillators, the frequency range is smaller due to detuning. We find the same locking condition $\Delta < 2\kappa$ as for instantaneous coupling; for vanishing detuning we recover the in-phase and anti-phase states for identical oscillators. However, in the delayed case locked and desynchronized solutions can coexist [32].

7.3   Delay, Multistability and Oscillation Patterns

For phase oscillators, a delay induces multistability: as discussed in the previous section, the number of coexistent stable solutions scales with the frequency range and thus the coupling strength, the delay time, and in a unidirectional ring, with the number of elements. Moreover, in-phase and out-of-phase solutions share the same stability properties. We show here that these properties extend to periodic orbits in delay-coupled systems in general. Following the approach of Yanchuk and Perlikowski [11], we assume a general oscillator with delayed feedback

$$\dot{x} = f(x(t), x(t - \tau)), \tag{7.9}$$

with $x \in \mathbb{R}^m$. If the system has a periodical solution $x(t) = x(t + T)$ for a given delay $\tau = \tau_0$, by construction, this is also a solution when the delay is equal to $\tau_n = \tau_0 + nT$. This mechanism leads to a structure in the (ω, τ)-plane of repeating branches of solutions. With varying delay, the frequency of the original periodic

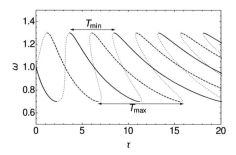

Fig. 7.3 Alternating branches of in-phase (*full line*) and anti-phase (*dashed line*) periodic orbits of two identical coupled phase oscillators in the (ω, τ)-plane. The minimal and maximal period, which are the upper and lower distance between the branches, are also indicated. Parameters are $\omega_0 = 1$ and $\kappa = 0.3$

orbit varies between a maximal value $\omega_{\max} = 2\pi/T_{\min}$, for $\tau = \tau_{01}$ and a minimal frequency $\omega_{\min} = 2\pi/T_{\max}$ occurring at $\tau = \tau_{02}$, forming a branch stretching over a delay interval $[\tau_{01}, \tau_{02}]$. The nth branch is then stretched over a larger delay interval $[\tau_{01} + nT_{\min}, \tau_{02} + nT_{\max}]$. The number of coexisting orbits at a particular (large) delay τ can then be estimated in the following way: Assuming $\tau \approx lT_{\min} \approx mT_{\max}$, the number of orbits approximated as $l - m \approx (\omega_{\max} - \omega_{\min})\frac{\tau}{2\pi}$. Consequently, the difference between the frequencies can be approximated as $2\pi/\tau$.

Extending this study to unidirectional rings of delay coupled oscillators, we consider a set of N identical nonlinear systems coupled with delay

$$\dot{x}_n = f(x_n(t), x_{n-1}(t - \tau)), \tag{7.10}$$

where $0 \equiv N$. Suppose that this network allows for an in-phase periodic solution $x_n(t) = x_{n-1}(t) = x_n(t + T)$ for a coupling delay $\tau = \tau_0$. This implies that there exists an out-of-phase periodic solution $x_n(t) = x_{n-1}(t - T/N) = x_n(t + T)$ at a delay $\tau_1 = \tau_0 + T/n$. Similarly, we find out-of-phase branches associated to all the other out-of-phase patterns that are allowed in the ring: a pattern corresponding to $x_n(t) = x_{n-1}(t - kT/N) = x_n(t + T)$ can be found at a delay $\tau_k = \tau_0 + kT/N$. Consequently, in-phase and out-of-phase branches alternate each other in the (ω, τ)-plane. For two mutually delay-coupled Kuramoto oscillators, these characteristic branches are shown in Fig. 7.3.

These in-phase and out-of-phase branches share the same stability properties. Linearizing Eq. (7.10) around an in-phase periodic solution at $\tau = \tau_0$, we obtain a set of equations for a perturbation $\delta(t)$

$$\dot{\delta}(t) = D_1 f(x(t))\delta(t) + \gamma_m D_2 f(x(t - \tau_0))\delta(t - \tau_0), \tag{7.11}$$

where $D_1 f(x(t))$ and $D_2 f(x(t - \tau_0))$ are the derivatives of $f(x(t), x(t - \tau))$ with respect to the first and second variable respectively, evaluated along the periodic orbit $x(t)$; These matrices are time-dependent and have a period equal to the oscillation period T. The factor $\gamma_m = e^{2mi\pi/N}$ arises from the master stability function [28], as we evaluate the perturbation along the eigenvectors of the adjacency matrix. It does not play a role for the stability in the long delay limit, as only its magnitude matters [29]. To solve such equation (7.11), we assume a perturbation of the form $\delta(t) = p(t)e^{\lambda t}$, with $p(t)$ being T-periodic. We find a stability equation

$$\dot{p}(t) = (D_1 f(x(t)) - \lambda 1)\, p(t) + \gamma_m e^{-\lambda \tau_0} D_2 f(x(t - t_0)) p(t - t_0), \qquad (7.12)$$

where we used the periodicity of the $x(t)$ and $p(t)$ to replace τ_0 by $t_0 = \tau_0 \ \mathrm{mod}\ T$ when possible. The delay hence only appears in the exponential term, as the value of t_0 depends only on the frequency of the orbit (i.e. on the position on the the branch) and not on the number of the branch.

At a delay $\tau_k = \tau_0 + kT/N$, the same solution $x(t)$ reappears as an out-of-phase state $x_n(t) = x_{n-1}(t - kT/N) \equiv x(t)$. Applying a perturbation $\delta(t) = (p(t), \gamma_m p(t + kT/N), \ldots)e^{\lambda t}$, we find a stability equation

$$\begin{aligned}\dot{p}(t) &= (D_1 f(x(t)) - \lambda 1)\, p(t) + \gamma_m e^{-\lambda \tau_k} D_2 f(x(t + kT/N - \tau_k)) p(t + kT/N - \tau_k) \\ &= (D_1 f(x(t)) - \lambda 1)\, p(t) + \gamma_m e^{-\lambda \tau_k} D_2 f(x(t - t_0)) p(t - t_0).\end{aligned} \qquad (7.13)$$

Upon a factor $e^{\lambda kT/N}$, this stability equation for out-of-phase oscillations is the same as for the in-phase oscillations. The stability equations for solutions with the same frequency only differ by the value of the delay, irrespective of the oscillation pattern. As the solutions of Eq. (7.11) scale as $\lambda = i\sigma + \gamma(\sigma)/\tau$ in the long delay limit [11], the stability of a solution solely depends on its frequency and neither on the branch on which a solution is located nor on the oscillation pattern. Thus, for long enough delay, if in-phase oscillations are stable (over a finite delay interval), and their period is sufficiently shorter than the delay time $T \ll \tau$, this implies that other oscillation patterns, with a similar waveform and a frequency difference of $2k\pi/N\tau$, are stable as well, and vice versa.

One can apply these ideas as well to the unstable periodic orbits (UPOs) that lie densely in a chaotic attractor: if an attractor of a chaotic delay system contains an orbit with a period $T \ll \tau$, one will find orbits with slightly different stability properties at frequencies $\omega = 2\pi/T \pm k\Delta\omega$, with $\Delta\omega = 2\pi/\tau_0$. The chaotic attractor for two delay-coupled units contains then at least twice as many UPOs as the single system: it contains all the in-phase orbits that exist in the single attractor, and these in-phase orbits are alternated with anti-phase orbits having very similar (in)stability properties. A ring of N elements has a chaotic attractor containing minimally N times as many periodic orbits as the single system.

7.4 Interaction of Noise and Delay

Multistability is an inherent property of delayed oscillatory systems, as shown in the previous sections. Consequently, due to the presence of noise in the system, the oscillator can switch between coexistent orbits. Such switching behavior has been observed experimentally between the external cavity modes of a semiconductor laser subject to delayed external feedback [33], or numerically, between different chaotic attractors in the Lang-Kobayashi model [8] or between periodic orbits in a ring of neurons [19]. In neuroscience these multistable orbits have been linked to memory storage: the average residence times of the orbits would then relate to the length of memory.

We describe these mode hopping statistics in coupled phase oscillators. Since the periodic orbits are explicitly known, it is possible to construct a potential and compute the distribution of frequencies and their respective lifetimes [34]. We add Gaussian white noise $\xi_n(t)$, with a variance $2D$, to each phase oscillator described as in Eq. (7.5)

$$\dot{\phi}_n = \omega_{0n} + \kappa \sin(\phi_{n+1}(t - \tau) - \phi_n(t)) + \xi_n(t). \tag{7.14}$$

7.4.1 Mode Hopping in a Single Oscillator with Feedback

We first discuss a single oscillator with delayed feedback. A typical timetrace of the phase $\phi(t)$, with several hoppings between the deterministic frequencies ω_k, is shown in Fig. 7.4a. Numerically the mode hoppings are determined by evaluating $\omega(t) \equiv (\phi(t) - \phi(t - \tau))/\tau$ (shown in Fig. 7.4b), as this frequency measure distinguishes clearly between the different ω_k, and appears as a driving term of the dynamics.

To describe the mode hopping dynamics, the delay system is approximated by an undelayed system. It is then possible to define a Langevin equation and to compute the frequency distributions and average residence times of the different periodic orbits analytically. We follow the method of Mørk et al. [35] for the description of mode hopping between external cavity modes in a laser with delayed feedback. Assuming that the oscillator resides in one of the periodic orbits during a delay interval, we

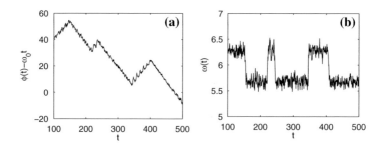

Fig. 7.4 a Phase evolution $\phi(t) - \omega_0 t$ of a Kuramoto oscillator with delayed feedback and white noise. We subtracted the natural frequency $\omega_0 t$ for better visibility of the mode hoppings. **b** The frequency measure $\omega(t) = (\phi(t) - \phi(t-\tau))/\tau$ is a good indicator for the mode hoppings. Parameters are $\omega_0 = 6, \kappa = 2, \tau = 10$ and $D = 0.5$; the oscillator has six stable periodic orbits, with respective frequencies $\omega_1 \approx 4.48$, $\omega_2 \approx 5.07$, $\omega_3 \approx 5.67$, $\omega_4 \approx 6.27$, $\omega_5 \approx 6.87$ and $\omega_6 \approx 7.46$, shown in Fig. 7.1

Fig. 7.5 Potential for a
Kuramoto oscillator with
feedback (Eq. 7.16).
Parameters are $\omega_0 = 6$,
$\kappa = 2, \tau = 10$

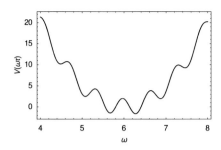

approximate the instantaneous frequency $\dot{\phi}(t - \tau)$ by the frequency averaged over
the future delay interval plus its noise source

$$\dot{\phi}(t - \tau) \approx \frac{\phi(t) - \phi(t - \tau)}{\tau} + \xi(t - \tau). \qquad (7.15)$$

Using this approximation, one obtains a closed equation for the phase difference
$x(t) = \phi(t) - \phi(t - \tau)$, that can be written in terms of a potential,

$$\dot{x}(t) = -\frac{dV(x)}{dx} + \tilde{\xi}(t) \text{ with}$$

$$V(x) = \frac{1}{2\tau}(x - x_0)^2 - \kappa \cos x, \qquad (7.16)$$

with $x_0 = \omega_0 \tau$ and $\tilde{\xi}(t) = \xi(t) - \xi(t - \tau)$. As the noise sources $\xi(t)$ and $\xi(t - \tau)$
are independent, the simplified oscillator is effectively subject to a magnified noise
strength of $\langle \tilde{\xi}^2(t) \rangle = 4D$. The potential $V(x)$, which has a parabolic shape modulated
by a cosine function, is shown in Fig. 7.5. The local minima $x_k = \omega_k \tau$ correspond
to the deterministic frequencies, while the local maxima x_m correspond to unstable
solutions: By calculating the potential extrema, Eq. (7.2) is recovered.

The distribution of frequencies $p(\omega)$ is then given by

$$p(\omega) \propto e^{-\frac{V(\omega\tau)}{2D}} = e^{-\frac{\tau}{4D}(\omega-\omega_0)^2} e^{\frac{\kappa \cos \omega\tau}{2D}}. \qquad (7.17)$$

This distribution $p(\omega)$ has a Gaussian envelope with mean ω_0 and variance $\sigma^2 = 2D/\tau$, corresponding to the probability distribution of a Wiener process. The feedback term, which only appears in the second factor of Eq. (7.17), determines the
location and the shape of the different peaks: With increasing feedback strength the
extrema in the distribution become more pronounced, for vanishing feedback the
Wiener process distribution is recovered. In contrast to the deterministic system, the
number of attended orbits grows as $\sqrt{D\tau}$, and is independent of the coupling strength
κ. In Fig. 7.6a the analytical result for the simplified system (Eq. 7.17) is compared
with numerical simulations of the original delay system (Eq. 7.15); the theoretical
results provide an excellent approximation for the distribution of frequencies.

Fig. 7.6 The frequency
distribution (*gray*) for an
oscillator with feedback,
together with its analytical
approximation (Eq. 7.17) in
black. The *dash-dotted line*
shows the Gaussian
envelope. The parameters are
$\kappa = 2$, $\tau = 10$, $\omega_0 = 6$,
$D = 0.5$. Reprinted figure
with permission from [34],
Copyright (2014) by the
American Physical Society

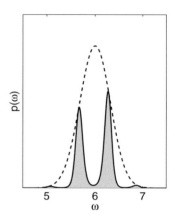

Also the average residence times can be estimated by the potential model (7.16): In the limit of low noise, the escape rates from an orbit with frequency ω_k are given by the Kramers rate [36, 37]

$$r(\omega_k) = \frac{\sqrt{-V''(\omega_k\tau)V''(x_m)}}{2\pi}e^{-\frac{\Delta V}{2D}}.$$

The average residence time $T_0(\omega_k)$ reads then

$$T_0(\omega_k) \approx \frac{1}{r_+(\omega_k) + r_-(\omega_k)},$$

where the suffix denotes whether the oscillator hops to a mode with a higher or a lower frequency. For strong coupling and large feedback delay $\kappa\tau \gg 1$, using the approximations $\omega_k\tau \approx 2n\pi$ and $x_m = (2n + 1)\pi$, the average residence time is further approximated as

$$T_0(\omega_k) \approx \frac{\pi}{\kappa}\frac{e^{\frac{\kappa}{D}+\frac{\pi^2}{4\tau D}}}{\cosh\left(\frac{\pi(\omega_k-\omega_0)}{2D}\right)}. \tag{7.18}$$

Thus, the average residence time $T_0(\omega_k)$, as to be expected, increases exponentially with the feedback strength κ, which determines the depth of the potential wells, and decreases with the noise strength D. The feedback delay τ has little influence on the residence times (for long delays the average residence time no longer depends on τ). Moreover, the range of frequencies with a long lifetime scales with the noise strength D, the number of relatively robust orbits thus scales as $D\tau$. A comparison between the theoretical and the numerical average residence times (Eq. 7.18) is shown in Fig. 7.7, and the approximation gives good results. However, upon the typical exponential decay predicted by theory, there are maxima at multiples of the delay time, which

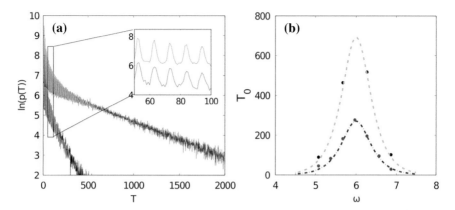

Fig. 7.7 **a** Logarithm of the residence time distribution $\ln(p(T))$, for a Kuramoto oscillator with delayed feedback, for the orbits with frequencies ω_2 and ω_3. **b** Mean residence time of the orbits $\omega_{2,3,4,5}$ versus their frequency for a single oscillator (*upper black dots*) together with the theoretical approximation (Eq. 7.18) (*upper dashed gray curve*). The *lower gray dots* and the *lower dashed curve* represent the mean average residence times of the orbits and their theoretical approximation (Eq. 7.22) respectively for two identical coupled systems. The parameters are $\kappa = 3$, $\tau = 10$, $\omega_0 = 6$ and $D = 0.5$. Reprinted figure with permission from [34], Copyright (2014) by the American Physical Society

are not captured by the delay-free approximation (shown in the inset). These peaks result from a known stochastic resonance effect in delay systems [8, 38, 39]: A mode hopping causes a perturbation, which repeats itself a delay time later, thereby increasing the probability for another mode hopping.

7.4.2 Mode Hopping in Two Mutually Delay-Coupled Oscillators

As a next step we consider two mutually delay-coupled phase oscillators. The phase evolution of two identical delay-coupled oscillators is shown in Fig. 7.8. In the coupled system, a frequency transition takes place in two steps: first one oscillator, the leader, changes its frequency, a delay time later the other oscillator, the laggard, follows. Such leader-laggard behavior is a typical signature of a stochastic phenomenon in delay-coupled systems [40]. Looking at the evolution of the driving terms $\phi_{1,2}(t) - \phi_{2,1}(t - \tau)$, shown in Fig. 7.8b, we find that during a mode hopping the driving term of the leader increases or decreases with approximately 2π, while the laggard changes its frequency without a phase jump. To quantify the frequency of the coupled oscillators we use the mean frequency of the two nodes averaged over the past delay interval $\omega(t) = (\phi_1(t) + \phi_2(t) - \phi_1(t - \tau) - \phi_2(t - \tau))/(2\tau)$; we thus capture the frequency transition of the leading oscillator. If the oscillators are identical, leader and laggard changes role randomly. This is still the case for detuned

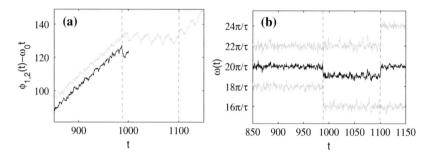

Fig. 7.8 **a** The phase evolution $\phi_1 - \omega_0 t$ (*gray*) and $\phi_2 - \omega_0 t$ (*black*) of two identical noisy Kuramoto oscillators coupled with delay. We subtracted the natural frequency ω_0 for better visibility of the mode hoppings. **b** The time evolution of the frequency $\omega(t) = (\phi_1(t) + \phi_2(t) - \phi_1(t - \tau) - \phi_2(t - \tau))/(2\tau)$ for two coupled oscillators (*black*), together with the phase differences $x_1(t)/\tau = (\phi_1(t) - \phi_2(t - \tau))/\tau$ (*upper gray curve*) and $x_2(t)/\tau = (\phi_2(t) - \phi_1(t - \tau))/\tau$ (*lower gray curve*). The *dashed lines* indicate the mode hoppings. Parameters are $\omega_0 = 6, \kappa = 3, \tau = 10$ and $D = 0.5$

oscillators, however, the oscillator with the higher natural frequency more often leads the transition to a faster frequency, while the slower oscillator plays more often the role of leader, if the system switches to a lower frequency.

To define a delay-free Langevin formalism, we rewrite the system as a function of the driving terms $x_1(t)$ and $x_2(t)$, defined as $x_{1,2}(t) = (\phi_{1,2}(t) - \phi_{2,1}(t - \tau))$. We assume that the oscillators are locked to the same fixed frequency over the delay interval, and thus, $\dot{\phi}_1(t - \tau)$ and $\dot{\phi}_2(t - \tau)$ only differ in the contribution of the noise. Using the main reduction

$$\dot{\phi}_{1,2}(t - \tau) \approx (x_1(t) + x_2(t))/(2\tau) + \xi_{1,2}(t - \tau), \tag{7.19}$$

the system can be rewritten as a function of a two-dimensional potential:

$$\dot{x}_{1,2}(t) = -\frac{\partial V_{1,2}}{\partial x_{1,2}} + \tilde{\xi}_{1,2}(t) \text{ with}$$

$$V(x_1, x_2) = \frac{1}{4\tau}(2x_0 - x_1 - x_2)^2 + \frac{\Delta}{2}(x_1 - x_2)$$
$$- \kappa (\cos x_1 + \cos x_2), \tag{7.20}$$

with $x_0 = \omega_0 \tau$ and $\tilde{\xi}_{1,2}(t) = \xi_{1,2}(t) - \xi_{2,1}(t - \tau)$. The potential is shown in Fig. 7.9. The wells are located at $(x_1, x_2) = (\omega_k \tau + 2n\pi - \delta, \omega_k \tau - 2n\pi + \delta)$, with δ being determined by (7.8). The phase difference between the two oscillators, and hence the delay phase differences x_1 and x_2, are only determined upon multiples of 2π, leading to a 4π-periodic potential with respect to $x_1 - x_2 = x_A$. The frequency of the system is given by the average frequency $\omega = (x_1 + x_2)/(2\tau)$, and a single frequency ω_k corresponds to multiple potential minima (x_1, x_2). There are thus two pathways for a transition $\omega_k \to \omega_{k\pm 1}$: x_1 changes with almost 2π, while x_2 remains

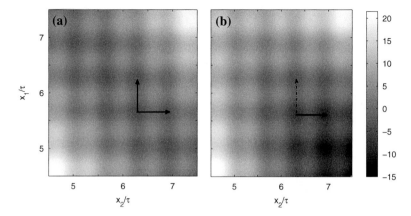

Fig. 7.9 Two-dimensional potential for two coupled Kuramoto oscillators, without (**a**) and with (**b**) detuning. The *arrows* indicate the two pathways for a transition between two frequencies, *thicker arrows* correspond to more probable pathways. Parameters are $\omega_0 = 6$, $\kappa = 2$, $\tau = 10$ and (**b**) $\Delta = 0.8$

almost constant, and $\phi_1(t)$ is the leader in the transition, and vice versa. For identical oscillators these pathways are equally probable and transitions always take place between orbits with a different oscillation pattern, as these have a minimal potential barrier between them.

If the oscillators are identical, we obtain the frequency distribution $p(\omega)$ by integrating over the double phase difference x_A. We find

$$p(\omega) \propto e^{-\frac{\tau}{2D}(\omega-\omega_0)^2} I_0(\kappa \cos \omega\tau/D), \qquad (7.21)$$

with $I_0(y)$ being the modified Bessel function of the first kind, $I_0(y) = \sum \frac{y^{2n}}{2^{2n}(n!)^2}$.

Like for the single oscillator, the frequency distribution can be written as the product of a Gaussian envelope, which depends on the delay time and noise strength, and a factor which involves the coupling strength, and determines the shape and location of the separate peaks. However, the variance of the envelope is only half of the variance of the single feedback system. Moreover, the Bessel function $I_0(y)$ is symmetric: peaks corresponding to in-phase and anti-phase orbits alternate each other, their height only depends on their respective frequencies ω_k and not on the oscillation pattern. Numerical and theoretical results for the frequency distribution are compared in Fig. 7.10a. The agreement is excellent.

To calculate the average residence times for identical mutually coupled oscillators, we use that, in the low noise limit, all the transitions take place via the two optimal pathways. We obtain

$$T_0(\omega_k) = \frac{1}{2r_+(k) + 2r_-(k)} \approx \frac{\pi}{2\kappa} \frac{e^{\frac{\kappa}{D} + \frac{\pi^2}{8\tau D}}}{\cosh\left(\frac{\pi(\omega_k - \omega_0)}{2D}\right)}. \qquad (7.22)$$

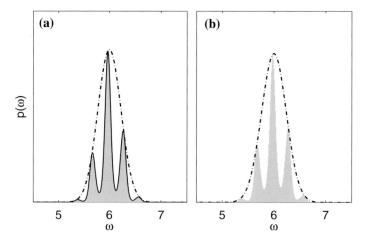

Fig. 7.10 The frequency distributions (*gray*) **a** two coupled identical identical oscillators and **b** two detuned oscillators. The analytical approximation Eq. (7.21) is plotted in *black*, the *dash-dotted lines* show the respective Gaussian envelopes. The parameters are $\kappa = 2, \tau = 10, \omega_0 = 6, D = 0.5$, and **c** $\Delta = 0.8$. Reprinted figure with permission from [34], Copyright (2014) by the American Physical Society

This corresponds to half of the lifetime of the orbits of a single oscillator with a feedback delay 2τ; As we have two oscillators, the transition probability doubles. The numerical and theoretical results are compared in Fig. 7.7b.

For nonzero detuning $\Delta > 0$, the potential (Eq. 7.20) is tilted, as is shown in Fig. 7.9b. As a result, the phase difference $x_A(t)$ between the oscillators preferentially increases during a mode hopping. The most probable, and the least probable transition pathway between two frequencies are also sketched on Fig. 7.9b. If the detuning is large enough we assume that for all the transitions to a higher frequency the faster oscillator x_2 is the leader, and for the transitions to a lower frequency the slow oscillator x_1 leads the dynamics. For $\kappa\tau$ sufficiently large, the envelope can then be approximated by assuming detailed balance

$$p(\omega_k)r_+(\omega_k) = p(\omega_{k+1})r_-(\omega_{k+1}) \Leftrightarrow$$
$$\frac{p(\omega_{k+1})}{p(\omega_k)} \approx e^{-\frac{\Delta V(\omega_{k+1}\to\omega_k)-\Delta V(\omega_k\to\omega_{k+1})}{2D}}$$
$$\approx e^{-\frac{\tau}{2D}\left((\omega_{k+1}-\omega_0)^2-(\omega_k-\omega_0)^2+\frac{2\delta}{\tau}(2\omega_0-\omega_{k+1}-\omega_k)\right)}$$
$$\approx e^{-\frac{\tau}{2D}\left(1-\frac{2\delta}{\pi}\right)\left((\omega_{k+1}-\omega_0)^2-(\omega_k-\omega_0)^2\right)}. \tag{7.23}$$

This corresponds to a Gaussian envelope of the frequency distribution with mean ω_0 and variance $\sigma^2 = D/(\tau(1-\epsilon))$, with $\epsilon = 2\arcsin(\Delta/2\kappa)/\pi > 0$. The distribution of frequencies broadens due to the detuning. In Fig. 7.10b we show the approximated Gaussian envelope together with the simulated distribution of frequencies.

7.4.3   Generalization to a Unidirectional Ring

This formalism can easily be generalized to a unidirectional ring of identical oscillators. Defining $x_n = \phi_n(t) - \phi_{n+1}(t - \tau)$ as the respective driving terms and approximating the instantaneous frequencies of the oscillators by the mean frequency averaged over the delay interval and their noise source,

$$\dot{\phi}_n(t - \tau) \approx \frac{1}{N\tau} \sum_{l=1}^{N} x_l + \xi_n(t - \tau),$$

one finds an N-dimensional potential

$$V(x_1, \ldots, x_N) = \frac{N}{2\tau}(x_0 - x_S)^2 + \kappa \sum_{l=1}^{N} \cos x_l, \tag{7.24}$$

with $x_S = \frac{1}{N} \sum_{l=1}^{N} x_l$. The frequency of the system is then measured by $\omega(t) = x_S(t)/\tau$. Although it is not possible to calculate the frequency distribution explicitly, it is straightforward to show that the parabolic term in Eq. (7.24) results in a Gaussian envelope. Its variance is given by $\sigma^2 = 2D/(N\tau)$, and thus scales inversely with the total delay in the ring $N\tau$. As the frequency difference between the orbits is approximated by $2\pi/(N\tau)$, the number of attended orbits scales as $\sqrt{DN\tau}$. Moreover, the potential is symmetric with respect to the different oscillation patterns, so that each pattern is equally often visited in the long delay limit.

For low noise and large $\kappa\tau$, the average residence times scale inversely with the number of transition pathways, and thus with the number of oscillators in the ring, and they depend weakly on the total roundtrip delay. They are approximated by

$$T_0(\omega_k) = \frac{1}{Nr_+(\omega_k) + Nr_-(\omega_k)} \approx \frac{\pi}{N\kappa} \frac{e^{\frac{\kappa}{D} + \frac{\pi^2}{4N\tau D}}}{\cosh\left(\frac{\pi(\omega_k - \omega_0)}{2D}\right)}. \tag{7.25}$$

7.4.4   Mode Hopping in Coupled FitzHugh-Nagumo Oscillators

The Kuramoto model is a weak coupling approximation for general limit cycle oscillators. In particular, the Kuramoto approximation applies when the coupling mainly influences the oscillation phase, while the waveform or oscillation amplitude is hardly affected. To confirm whether the analytic results obtained for Kuramoto oscillators apply in such a general context, we compared the analytic results for phase oscillators

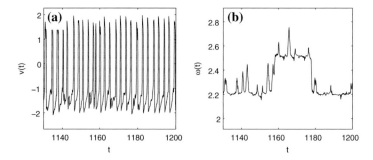

Fig. 7.11 a Time trace and **b** frequency evolution of a FitzHugh-Nagumo oscillator with feedback. Parameters are $\epsilon = 0.01$, $a = 0.9$, $k = 0.2$, $\tau = 20$ and $\tilde{D} = 0.0145$

to a delay-coupled FitzHugh-Nagumo system. This oscillator is frequently used to model neural dynamics, and has a spiking rather than a sinusoidal waveform. A ring of FitzHugh-Nagumo oscillators is modeled by

$$\epsilon \dot{v}_n(t) = v_n(t) - \frac{v_n^3(t)}{3} - w_n(t) + k(v_{n+1}(t - \tau) - v_n(t))$$
$$\dot{w}_n(t) = v_n(t) + a + \xi_n(t), \tag{7.26}$$

with $(v_{N+1}, u_{N+1}) \equiv (v_1, u_1)$, and $\xi_n(t)$ being Gaussian white noise with a variance $2\tilde{D}$. In Fig. 7.11 we show a timetrace, which shows two frequency transitions, together with the frequency evolution.

The mode hopping can be analyzed in a similar way as for phase oscillators. Defining the phase of the oscillators by the Hilbert-transform of the fast variable, $\phi_n(t) = \arg(\mathcal{H}(v_n(t)))$, the frequency is measured by $\omega(t) = \sum(\phi_n(t) - \phi_n(t - \tau))/(N\tau)$. Looking at the distribution of frequencies, and their respective residence times (Fig. 7.12), a similar picture as for phase oscillators, with multiple peaks, separated by a difference of $2\pi/N$ is recovered. The analytic results provide an accurate description for the mode hopping statistics: the scaling properties with delay time and number of elements are reproduced, both for the frequency distribution and the average residence times; the single and coupled system can be fitted by a single set of parameters for a Kuramoto system.

7.5 Final Remarks

For stochastic delay systems there is no corresponding theory yet to the well-known Fokker-Planck formalism. Although the reduction technique used here is far from precise, the correspondence between the analytic delay-free results and the numerical results of the delay system is excellent. The underlying mechanism for this agreement remains to a large extent an open question. However, it is related to the fact that on

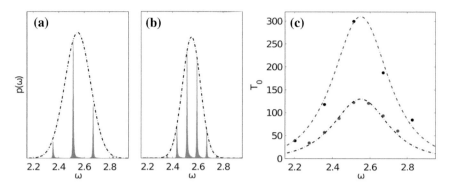

Fig. 7.12 Frequency distribution $p(\omega)$ for one (**a**) and two (**b**) coupled FitzHugh-Nagumo oscillators with delay, with $\epsilon = 0.01$, $a = 0.9$, $k = 0.2$, $\tau = 40$ and $\tilde{D} = 0.0145$. The *dashed curves* show the Gaussian envelope for the corresponding Kuramoto system with $\omega_0 = 2.55$, $D = 0.2$ and $\tau = 40$. In panel (**c**) the corresponding average residence times are shown for one (*upper black dots*) and two (*lower gray dots*) oscillators, the *dashed curves* represent the Kuramoto approximation for one (Eq. 7.18, *upper curve*) and two (7.22, *lower curve*) elements for $\omega_0 = 2.55$, $D = 0.2$, $\tau = 40$ and $\kappa = 0.815$. Reprinted figure with permission from [34], Copyright (2014) by the American Physical Society

timescales much shorter than the delay $t_0 \ll \tau$, the influence of the feedback is not visible: the two point distribution of $\phi(t) - \phi(t - t_0)$ is identical to the one of a random walk; the instantaneous frequency can thus be approximated as such. Only on timescales equal or larger than the delay, the dynamics (i.e. the timetrace) of an oscillator with delayed feedback differs significantly from a random walk. A similar phenomenon is observed in other stochastic systems driven by a nonlinear delay term [41].

The comparison between the deterministic and the stochastic system is noteworthy: The deterministic range of frequencies depends on the coupling strength. In contrast, in the stochastic system it is determined by the noise level and delay time; the frequency range is the same as for a random walk. The robustness of the periodic orbits against noise, measured by the average residence time, depends mainly on the coupling strength and only weakly on the delay time. Moreover, we found that the number of attended orbits scales as $\sqrt{DN\tau}$, whereas the number of orbits with a reasonably long residence time scales as $DN\tau$.

There are similarities between delay-coupled stochastic and chaotic systems as well, in the sense that they show similar correlation functions. Coupled in a unidirectional ring, the oscillators are almost always synchronized, in the sense that they have the same frequency ω_k. However, they spend, for long enough coupling delays, as much time in in-phase as in out-of-phase orbits. Consequently the oscillators are not correlated at zero lag, but the cross-correlation shows maxima at multiples of the coupling delay corresponding to the traveling time between the nodes. Also in a chaotic attractor a delay induces multiple periodic orbits, and thus, it is not surprising that we find the same correlation pattern [42, 43]. However, the analogy between chaotic and stochastic systems does not go any further, as they show different scaling behavior with the delay time and the number of elements in the ring.

We discussed here only unidirectional rings: these are the only networks, for which a potential can be constructed. As soon as one of the oscillators receives multiple inputs, the formalism fails. The unidirectional is also the only network, where the delay introduces a mapping between different oscillation patterns.

For more complicated networks, such as bidirectional rings or globally coupled oscillators, or even two nodes with feedback, the dynamics becomes much richer. One of the main effects observed in globally coupled oscillators is the strong preference for in-phase oscillations; they are more often attended and more robust to noise than out-of-phase oscillations. Besides in-phase and out-of-phase oscillations, other dynamical patterns can be observed.

References

1. T. Erneux, *Applied Delay Differential Equations* (Springer, New York, 2009)
2. V. Flunkert, I. Fischer, E. Schöll (eds.), *Theme Issue "Dynamics, control and information in delay-coupled systems"* (Phil. Trans. Roy. Soc. A, 2013), vol. 371 (1999), p. 20120465
3. M.C. Soriano, J. Garcia-Ojalvo, C.R. Mirasso, I. Fischer, Rev. Mod. Phys. **85**, 421 (2013)
4. L. Chen, K. Aihara, IEEE Trans. Circuits Syst. I **49**, 602 (2002)
5. G. Stepan (ed.), *Theme Issue "Delay effects in brain dynamics"* (Phil. Trans. Roy. Soc. A, 2009), vol. 367 (1891), pp. 1059–1212
6. W. Wang, J. Slotine, IEEE Trans. Automat. Control **51**, 712717 (2006)
7. W. Lu, F.M. Atay, J. Jost, Netw. Heterogen. Media **6**, 329 (2011)
8. C. Masoller, Phys. Rev. Lett. **88**, 034102 (2002)
9. S. Kim, S.H. Park, C.S. Ryu, Phys. Rev. Lett. **79** (1997)
10. R. Lang, K. Kobayashi, IEEE J. Quantum Electron. **16**, 347 (1980)
11. S. Yanchuk, P. Perlikowski, Phys. Rev. E **79**, 046221 (2009)
12. J. Sieber, M. Wolfrum, M. Lichtner, S. Yanchuk, Discrete Continuous Dyn. Syst. **33**, 3109 (2013)
13. M. Golubitsky, I. Stewart, *The Symmetry Perspective* (Birkhauser, 2003)
14. Y. Kuramoto, *Chemical Oscillations, Waves and Turbulence* (Springer, Berlin, 1984)
15. Y. Kuramoto, Int. J. Bifurcation Chaos Appl. Sci. Eng. **7**, 789 (1997)
16. H. Daido, Int. J. Bifurcation Chaos Appl. Sci. Eng. **7**, 807 (1997)
17. J. Acebrón, L. Bonilla, C.P. Vicente, F. Ritort, R. Spigler, Rev. Mod. Phys. **77**, 137 (2005)
18. J. Foss, A. Longtin, B. Mensour, J. Milton, Phys. Rev. Lett. **76**, 708 (1996)
19. J. Foss, F. Moss, J. Milton, Phys. Rev. E **55**, 4536 (1997)
20. M. Rabinovich, P. Varona, A. Selverston, H. Abarbanel, Rev. Mod. Phys. **78**, 1213 (2006)
21. O. D'Huys, R. Vicente, T. Erneux, J. Danckaert, I. Fischer, Chaos **18**, 037116 (2008)
22. O. D'Huys, I. Fischer, J. Danckaert, R. Vicente, Chaos **20**, 043127 (2010)
23. P. Perlikowski, S. Yanchuk, O.V. Popovych, P.A. Tass, Phys. Rev. E **82**, 036208 (2010)
24. C.U. Choe, T. Dahms, P. Hövel, E. Schöll, Phys. Rev. E **81**, 025205(R) (2010)
25. A. Vüllings, E. Schöll, B. Lindner, Eur. Phys. J. B **87**, 31 (2014)
26. C.R.S. Williams, F. Sorrentino, T.E. Murphy, R. Roy, Chaos **23**, 043117 (2013)
27. M. Earl, S. Strogatz, Phys. Rev. E **67**, 036204 (2003)
28. L.M. Pecora, T.L. Carroll, Phys. Rev. Lett. **80**, 2108 (1998)
29. V. Flunkert, S. Yanchuk, T. Dahms, E. Schöll, Phys. Rev. Lett. **105**(25), 254101 (2010). doi:10.1103/PhysRevLett.105.254101
30. O. D'Huys, I. Fischer, J. Danckaert, R. Vicente, Phys. Rev. E **83**, 046223 (2011)

31. H.G. Schuster, P. Wagner, Progress Theoret. Phys. **81**, 939 (1989)
32. B. Lysyansky, Y. Maestrenko, P. Tass, Int. J. Bifurcat. Chaos **18**, 1791 (2008)
33. J. Mørk, B. Tromborg, J. Mark, IEEE J. Quantum Electron. **28**, 93 (1992)
34. O. D'Huys, T. Jüngling, W. Kinzel, Phys. Rev. E **90**, 032918 (2014)
35. J. Mørk, B. Semkow, B. Tromborg, Electron. Lett. **26**, 609 (1990)
36. H. Kramers, Physica **7**, 284 (1940)
37. B. McNamara, K. Wiesenfeld, Phys. Rev. A **39**, 4854 (1989)
38. T. Ohira, Y. Sato, Phys. Rev. Lett. **82**, 2811 (1999)
39. C. Masoller, Physica D **168–169**, 171 (2002)
40. K. Hicke, X. Porte, I. Fischer, Phys. Rev. E **88**, 052904 (2013)
41. T. Jüngling, O. D'Huys, W. Kinzel, Phys. Rev. E **91**, 062918 (2015)
42. T. Heil, I. Fischer, W. Elsäßer, J. Mulet, C. Mirasso, Phys. Rev. Lett. **86**, 795 (2001)
43. G. Van der Sande, M. Soriano, I. Fischer, C. Mirasso, Phys. Rev. E **77**, 055202 (2008)

Chapter 8
Noisy Dynamical Systems with Time Delay: Some Basic Analytical Perturbation Schemes with Applications

Wolfram Just, Paul M. Geffert, Anna Zakharova and Eckehard Schöll

Abstract Systems with time delay, a rather prominent branch in applied dynamical systems theory, constitute a special case of functional differential equations for which the general mathematical theory is fairly well developed and largely parallels the theory of ordinary differential equations. Hence analytic concepts like bifurcation theory, adiabatic elimination, global attractors, invariant manifolds and others can be used to study dynamical behaviour of systems with time delay if some care is applied to take special features of infinite dimensional phase spaces into account. Simple analytic perturbation schemes, frequently used to gain insight for ordinary differential equations, can be applied to time delay dynamics as well. However, such approaches seem to be used infrequently within the physics community, probably because of a lack of easily accessible expositions. Here we review some elementary and well established concepts for the analytical treatment of time delay dynamics, even when subjected to noise. We cover normal form reduction and adiabatic elimination, stochastic linearisation of time delay dynamics with noise, and some elements of weakly nonlinear- and bifurcation analysis. These tools will be illustrated with applications in control problems, time delay autosynchronisation, coherence resonance, and the computation and structure of power spectra in noisy time delay systems.

8.1 Introduction

At about half a century ago the notion of chaos and the importance of nonlinear properties has become relevant in quite diverse branches of science, with the effect that the well developed specialised mathematical discipline of dynamical systems theory became one of the major subjects in applied sciences. Since then, dynamical systems

W. Just (✉) · P.M. Geffert
School of Mathematical Sciences, Queen Mary University of London,
Mile End Road, London E1 4NS, UK
e-mail: w.just@qmul.ac.uk

W. Just · P.M. Geffert · A. Zakharova · E. Schöll
Institut für Theoretische Physik, Technische Universität Berlin,
Hardenbergstr. 36, 10623 Berlin, Germany

© Springer International Publishing Switzerland 2016
E. Schöll et al. (eds.), *Control of Self-Organizing Nonlinear Systems*,
Understanding Complex Systems, DOI 10.1007/978-3-319-28028-8_8

theory has been enriched and diversified by the incorporation of other, important aspects. Noise and imperfections are certainly one of the major issues appearing in applications. The emphasis of this aspect can be traced back for about a century [1–3]. The constructive role of noise becomes apparent when nonlinear dynamical behaviour is considered (see e.g. [4, 5]), and this aspect plays a major role for the theory of systems far from equilibrium which is still under development (see e.g. [6]). The various facets of rigorous approaches and a functional analytic setting are rather well developed, with even quite accessible expositions being available [7, 8] at the expense of some rigour. In addition, stochastic dynamics has been cast into the language of modern dynamical systems theory, using tools like invariant manifolds and bifurcations [9].

The inclusion of time-delayed interactions and the investigation of the corresponding non-Markovian dynamics has as well a fairly long tradition, mainly triggered by topics in engineering (see e.g. [10] and references therein) or by linear response theories in solid state physics [11, 12]. The proper inclusion of nonlinear dynamical aspects has added additional stimulus to this branch of dynamical systems theory [13–17], in particular, for applications in fast optical systems and communication networks where the speed of light becomes a dynamically relevant quantity [18]. Rigorous mathematical approaches for time delay systems are fairly well developed within a suitable functional theoretical setting [19], underpinning the numerical approaches which largely parallel developments in finite dimensional systems, like, e.g., numerical continuation for uncovering bifurcation structures in time delay dynamics [20].

Here we want to consider analytical approaches for dynamical systems which may have both of the previously mentioned facets in conjunction, i.e., nonlinear stochastic time delay systems. We will focus on simple analytical perturbation schemes and therefore sacrifice to a large extent any mathematical rigour. Instead we focus on formal expansion schemes, which have been well developed and which are in fact quite well known in the context of ordinary differential equations. What we present here is not completely new and can be found, at least indirectly, in the existing specialised literature. However, we have the impression that such approaches are probably not very well known within an applied dynamical systems community. As a consequence of our strategy we will largely skip recent topical developments in time delay dynamics, as these topics normally still defy any systematic analytical treatment. For instance, the general theory of systems with distributed and time dependent delays is well developed and can be dealt with by numerical tools, but analytical results in closed form are rare. Most challenging are systems with state dependent delays were even the fundamental theory is still not completely developed. As for synchronisation in coupled time delay systems master stability function techniques [21] can be applied in the case of uniform time delay and nice results, e.g., in the limit of large delay times can be deduced analytically [22, 23], emphasising the role of strong and weak instabilities. However, the power of this approach to disentangle the topology of the coupling from the properties of the single site dynamics disappears when more complicated delay structures are considered. For these advanced cases one still relies largely on numerical simulations with little input from systematic analytical expansions. Similar caveats apply if one intends to generalise Fokker-Planck techniques to noisy time delay dynamics. While it is still possible to derive an equation of motion for the distribution function, the approach suffers from the shortcoming that no closed simple equation of motion can be obtained. Nevertheless, such ideas can be used to study linear dynamical systems [24] and the limit of small delay and thus can quantitatively capture corrections to the Markovian limit [25].

The aim of our contribution is modest. We want to summarise some elementary formally consistent analytic perturbation tools for the analysis of delay dynamics with and without noise, which are valid beyond the limit of small time delay. To some extent our approach can be illustrated by the intuitive averaging technique. Averaging techniques, multiple scaling expansions, and weakly nonlinear analysis all describe similar approaches to deal with oscillatory systems which are in some sense close to a linear case. These concepts have the appeal of an immediate intuitive interpretation, they are quite flexible, they have appeared in various contexts, such as nonlinear oscillators (see, e.g., [26] for an introduction) or coherent pattern formation [27], and can be easily generalised to capture time delay dynamics as well (see, e.g., [28, 29] for some examples in time-delayed feedback control). Multiple scaling techniques are closely related and often largely identical to adiabatic elimination schemes [30], which at the rigorous level can be formalised in terms of centre manifold and normal form reductions. For deterministic time delay dynamics the entire setup is very well established, but not easy to apply by non specialists as one has to master a fairly heavy technical framework which is involved in a rigours description [19]. Some of the more traditional and less general literature is probably easier to access for non-mathematicians [10, 31] and it is this kind of approach we intend to pursue here.

The analytic expansion schemes which are at focus of our interest will largely be based on a thorough understanding of linear time delay dynamics. Even though linear systems are straightforward to solve, either in the deterministic or in the stochastic case, we will summarise the essential features in Sect. 8.2. For those readers familiar with expansion schemes in dynamical systems theory it will not come as a surprise that the proper understanding of the adjoint equation will turn out to be the key for analytic perturbation schemes. As one by-product of our discussion we will describe as well the essential structure of an adiabatic elimination scheme, i.e., the centre manifold reduction in weakly nonlinear noisy time delay systems. We will use two particular examples to illustrate a few aspects of the analytic expansion schemes. Section 8.3 is devoted to time-delayed feedback control, a simple setup for the study of deterministic time delay dynamics. We will use this case to discuss in some detail the analytic investigation of instabilities in time delay systems. The results will show a few, probably general, features shared by time-delayed feedback control systems. As an example for stochastic time delay dynamics we will analyse in Sect. 8.4 a simple model for coherence resonance subjected to time-delayed feedback. We will use this case to illustrate centre manifold reduction and adiabatic elimination in a noisy time delay system, to finally compute the stationary distribution. Correlation functions will be dealt with by stochastic linearisation, to demonstrate the benefit of mean-field methods in a weakly nonlinear setting.

8.2   Some Basic Features of Linear and Weakly Nonlinear Systems

Analytic perturbation schemes for dynamical systems rely to a large extent on the detailed analysis of linear systems. Let us recall some of those ideas in the context of time delay dynamics. All these issues can be found in the literature with various degrees of rigour, and here we largely follow the approach used in [32]. We will focus on an elementary formally consistent approach which does not require any sophisticated mathematical background.

8.2.1   Linear Equations and Eigenvalue Decompositions

Consider a linear inhomogeneous delay differential equation with a single time delay τ

$$\dot{x} = \underline{\underline{A}}\,\underline{x} + \underline{\underline{B}}\,\underline{x}(t - \tau) + \underline{h}(t) \tag{8.1}$$

where we allow for a vector valued variable $\underline{x} \in \mathbb{R}^n$ with the coefficients being given by square matrices $\underline{\underline{A}}$ and $\underline{\underline{B}}$. The inhomogeneity $\underline{h}(t)$ can be either a deterministic drive or a stochastic forcing. In fact, with a slight abuse of notation we mean by linear equation a linear equation with constant coefficients. The case of time dependent coefficients would require a completely different approach and would hardly allow for explicit solutions in closed form.

As usual, the corresponding homogeneous system can be reduced to a nonlinear eigenvalue problem

$$\Lambda \underline{u}_\Lambda = \left(\underline{\underline{A}} + \exp(-\Lambda\tau)\underline{\underline{B}} \right) \underline{u}_\Lambda \tag{8.2}$$

when using a particular solutions of exponential type, $\underline{x}(t) = \exp(\Lambda t)\underline{u}_\Lambda$. The eigenvalues are determined by the quasipolynomial

$$\det \left(\underline{\underline{A}} - \Lambda\underline{\underline{I}} + \exp(-\Lambda\tau)\underline{\underline{B}} \right) = 0 \tag{8.3}$$

where $\underline{\underline{I}}$ denotes the identity matrix. Such a transcendental characteristic equation is not straightforward to solve (see, e.g., [10, 19, 33] for some approaches) as it normally has an infinite set of complex valued solutions. It is however fairly straightforward to see that the real parts of the eigenvalues cannot become arbitrarily large. If that would be the case then the exponential contribution $\exp(-\Lambda\tau)$ in Eq. (8.3) tends to zero and the eigenvalues would need to tend towards the finite eigenvalues of $\underline{\underline{A}}$. A similar argument shows that eigenvalues with large imaginary part necessarily have their real parts tending to minus infinity. Hence, even though there are infinitely many eigenmodes almost all are exponentially damped, meaning that the delay dynamics becomes essentially finite dimensional. In Eq. (8.1) we have considered delays appearing in the state variable $\underline{x}(t)$ only, the so called "retarded" case. Let us add a remark on time delays appearing in the derivative. If we replace the second term in Eq. (8.1), say, by $\underline{\underline{B}}\,\dot{\underline{x}}(t - \tau)$ then the corresponding quasipolynomial (8.3) contains a term $\Lambda \exp(-\Lambda\tau)\underline{\underline{B}}$. If we now apply the previous reasoning then it is easily possible to have eigenvalues with large imaginary parts and finite negative real parts by keeping $\Lambda \exp(-\Lambda\tau)$ finite. Hence, for such types of equations we have a huge number of modes which are weakly damped, a case which reminds us of properties of Hamiltonian systems. That is one of the reasons why systems with the delay appearing in the highest derivative behave differently and have been somehow misleadingly termed as "neutral" delay systems. Here we entirely focus on the retarded case.

For solving the original problem, Eq. (8.1), i.e., to perform a decomposition in eigenmodes the adjoint equation plays a central role. The adjoint dynamics is in fact a dynamical system running backwards in time, and it seems to be quite challenging to find elementary expositions for time dependent cases in the literature [31]. Here we will just need the adjoint eigenvalue equation which reads

$$\underline{v}_\Lambda^\dagger \Lambda = \underline{v}_\Lambda^\dagger \left(\underline{\underline{A}} + \exp(-\Lambda\tau)\underline{\underline{B}} \right). \tag{8.4}$$

The eigenvectors obey a kind of "pseudo orthogonality" in the sense that for $\Lambda \neq \Lambda'$

$$\underline{v}_{\Lambda'}^\dagger \underline{u}_\Lambda + \int_{-\tau}^0 \exp(-\Lambda'(\theta + \tau)) \exp(\Lambda\theta) \underline{v}_{\Lambda'}^\dagger \underline{\underline{B}}\, \underline{u}_\Lambda d\theta$$

$$= \underline{v}_{\Lambda'}^\dagger \frac{\Lambda - \Lambda' + \left(\exp(-\Lambda'\tau) - \exp(-\Lambda\tau) \right) \underline{\underline{B}}}{\Lambda - \Lambda'} \underline{u}_\Lambda = 0. \tag{8.5}$$

For the last step we used the eigenvalue equations (8.2) and (8.4). We are now able to decompose the inhomogeneous system into eigenmodes if we introduce an appropriate bilinear form which is inspired by Eq. (8.5)

$$(V_\Lambda | X_t) = \underline{v}_\Lambda^\dagger \underline{x}(t) + \int_{-\tau}^0 \exp(-\Lambda(\theta + \tau)) \underline{v}_\Lambda^\dagger \underline{\underline{B}}\, \underline{x}(t + \theta) d\theta. \tag{8.6}$$

One can consider Eq. (8.6) to be just a useful abbreviation. Naively, it could be viewed as a kind of scalar product, even though one has to keep in mind that the underlying phase space is not a Hilbert space. By straightforward differentiation of Eq. (8.6) one easily verifies that Eq. (8.1) reduces to

$$\frac{d}{dt} (V_\Lambda | X_t) = \Lambda (V_\Lambda | X_t) + \underline{v}_\Lambda^\dagger \underline{h}(t) \tag{8.7}$$

i.e., we have decomposed the original problem into a set of decoupled scalar equations. It is possible to reconstruct the solution $\underline{x}(t)$ given the expressions (8.6).

However, in the linear case it is often simpler to use alternative methods to compute solutions, e.g., by Laplace or Fourier transforms.

8.2.2 Adiabatic Elimination and Centre Manifold Reduction

Dynamical systems with slow modes allow for the adiabatic elimination of fast degrees of freedom. Such a case occurs for instance in the neighbourhood of time independent states when almost all eigenvalues of the linear part are negative and just a few eigenvalues have small positive or vanishing real part. Within the context of time delay dynamics let us assume that by an appropriate coordinate transformation the time independent state is the trivial solution $\underline{x} = 0$, and that the linear part of the equations of motion is given by Eq. (8.1), i.e.,

$$\dot{\underline{x}}(t) = \underline{\underline{A}}\,\underline{x}(t) + \underline{\underline{B}}\,\underline{x}(t - \tau) + \underline{f}(\underline{x}(t), \underline{x}(t - \tau)) + \underline{\underline{g}}(\underline{x}(t), \underline{x}(t - \tau))\underline{\xi}(t). \quad (8.8)$$

Here $\underline{f}$ denotes the higher order nonlinear terms, and we allow as well for the inclusion of a Gaussian white noise with correlation function $\langle \underline{\xi}(t)\underline{\xi}^T(t') \rangle = \delta(t - t')\underline{\underline{I}}$. We assume that all eigenvalues of the linear equation have negative real part apart from a complex conjugate purely imaginary pair $\pm i\Omega$. The corresponding two eigenfunctions constitute the slow modes, and we are able to reduce the original system (8.8) effectively to the dynamics on this two-dimensional manifold. In order to keep the notation as simple as possible and to avoid the functional analytic description of the phase space let us just recall that the integration of a delay differential equation requires the knowledge of the solution at times $t + \theta$ with $\theta \in [-\tau, 0)$ denoting the history. Hence, on our slow manifold we express the solution in the form

$$\underline{x}(t + \theta) = C(t)\exp(i\Omega\theta)\underline{u}_{i\Omega} + \bar{C}(t)\exp(-i\Omega\theta)\underline{\bar{u}}_{i\Omega} + \underline{R}(C(t), \theta) \quad (8.9)$$

where $\exp(i\Omega\theta)\underline{u}_{i\Omega}$ is the relevant slow eigenmode, $C(t) \in \mathbb{C}$ denotes the coordinate on the slow manifold, $\bar{C}(t)$ its complex conjugate, and $\underline{R}$ abbreviates the terms of higher than first order in C. In geometric terms the higher order contributions take care of the deviation of the centre manifold from the linear eigenspace, i.e., $\underline{R}$ takes the curvature of the centre manifold into account. The splitting given by Eq. (8.9) has some ambiguity and we have some flexibility to define the higher order terms. We use this freedom to require that the higher order terms are "orthogonal" to the linear eigenspace in the sense of the bilinear form, Eq. (8.6). Hence, the condition $(V_{i\Omega}|R) = 0$ results in

$$(V_{i\Omega}|X_t) = C(t)\left(\underline{v}_{i\Omega}^{\dagger}\underline{u}_{i\Omega} + \int_{-\tau}^{0}\exp(i\Omega\theta)\exp(-i\Omega(\theta+\tau))\underline{v}_{i\Omega}^{\dagger}\underline{\underline{B}}\,\underline{u}_{i\Omega}d\theta\right)$$

$$+\bar{C}(t)\left(\underline{v}_{i\Omega}^{\dagger}\underline{\bar{u}}_{i\Omega} + \int_{-\tau}^{0}\exp(-i\Omega\theta)\exp(-i\Omega(\theta+\tau))\underline{v}_{i\Omega}^{\dagger}\underline{\underline{B}}\,\underline{\bar{u}}_{i\Omega}d\theta\right)$$

$$= C(t)\left(\underline{v}_{i\Omega}^{\dagger}\underline{u}_{i\Omega} + \tau\exp(-i\Omega\tau)\underline{v}_{i\Omega}^{\dagger}\underline{\underline{B}}\,\underline{u}_{i\Omega}\right). \tag{8.10}$$

The contribution containing the complex conjugate amplitude $\bar{C}(t)$ vanishes because of the orthogonality (8.5). Since the abbreviation, Eq. (8.6) or (8.10), obeys the equation of motion (8.7) when we identify the inhomogeneous part $\underline{h}$ with the nonlinear and stochastic contributions in Eq. (8.8), we arrive at

$$\dot{C}(t) = i\Omega C(t) + \frac{\underline{v}_{i\Omega}^{\dagger}\underline{f}(\underline{x}(t),\underline{x}(t-\tau)) + \underline{v}_{i\Omega}^{\dagger}\underline{\underline{g}}(\underline{x}(t),\underline{x}(t-\tau))\underline{\xi}(t)}{\underline{v}_{i\Omega}^{\dagger}\underline{u}_{i\Omega} + \tau\exp(-i\Omega\tau)\underline{v}_{i\Omega}^{\dagger}\underline{\underline{B}}\,\underline{u}_{i\Omega}}. \tag{8.11}$$

If we take into account that we consider solutions on our slow manifold, i.e., $\underline{x}$ obeys Eq. (8.9), we obtain a closed ordinary stochastic differential equation for the amplitude C. The impact of the time delay is contained in the coefficients via the spectrum and the eigenvectors. If our original system has only cubic nonlinearities, a case which for simplicity is often considered in applications, see e.g. [34, 35], then the lowest order linear approximation in Eq. (8.9) is sufficient. Otherwise, one would need to compute the nonlinear corrections $\underline{R}$ to the invariant manifold using dynamical invariance. The details are essentially identical to the procedure used for ordinary differential equations. In addition, some greater care is needed when one deals with stochastic dynamics. In fact, the underlying mathematical considerations may become quite involved (see, e.g., [9] for an introduction in the case of stochastic differential equations).

The goal to arrive at an effective equation of motion can be achieved by different, largely equivalent, methods. Multiple scaling techniques (e.g. [26, 36]) are rather efficient to derive equations of motion on slow time scales, even though one needs to identify first a small expansion parameter. To some extent these methods inherently involve a time scale argument which eliminates as well nonresonant terms from equations of motion and thus perform both, a centre manifold and a normal form reduction in one step. Above all, the approaches sketched here rely on a detailed understanding of linear equations of motion, which to a large extent is the backbone of most analytical perturbation schemes.

8.2.3 Linear Stochastic Equations and Power Spectral Density

An important example of a linear driven system is a stochastic system where the driving force is given by white noise

$$\underline{h}(t) = \underline{\underline{g}}\,\underline{\xi}(t), \qquad \langle \underline{\xi}(t)\underline{\xi}^T(t')\rangle = \delta(t - t')\underline{\underline{I}} \tag{8.12}$$

with $2\underline{\underline{D}} = \underline{\underline{g}}\,\underline{\underline{g}}^T$ denoting the corresponding diffusion matrix. If we require in addition the noise to be Gaussian, then the two-point correlations entirely determine the stationary behaviour of the system. In fact, Eqs. (8.1) and (8.12) define a stochastic process (see, e.g., [37] for a detailed rigorous account). Here we just concentrate on the formal computation of correlation functions and power spectral densities. The simplest approach is along the lines of signal processing and uses a Fourier representation of the noise

$$\underline{\xi}(t) = \frac{1}{\sqrt{2\pi}} \int_{-\infty}^{\infty} \underline{\xi}_{\omega}\exp(i\omega t)d\omega. \tag{8.13}$$

The expansion coefficients are assumed to be uncorrelated Gaussian random variables $\langle \underline{\xi}_{\omega}\underline{\xi}_{\omega'}^T\rangle = \delta(\omega - \omega')\underline{I}$ to ensure Gaussian white noise. The condition $\underline{\xi}_{\omega} = \bar{\underline{\xi}}_{-\omega}$ keeps the noise real-valued. The formal Fourier transform of Eq. (8.1) immediately results in

$$\underline{x}_{\omega} = \left(i\omega\underline{\underline{I}} - \underline{\underline{A}} - \underline{\underline{B}}\exp(-i\omega\tau)\right)^{-1}\underline{\underline{g}}\,\underline{\xi}_{\omega} \tag{8.14}$$

where we implicitly assume that all the eigenvalues of the homogeneous equation have negative real part and that we just focus on the stationary properties of the process. Using the properties of the Fourier coefficients $\underline{\xi}_{\omega}$ the correlation matrix is then easily evaluated as

$$\langle \underline{x}(t)\underline{x}^T(t')\rangle = \frac{1}{2\pi} \int_{-\infty}^{\infty} \int_{-\infty}^{\infty} \langle \underline{x}_{\omega}\underline{x}_{\omega'}^T\rangle \exp(i\omega t + i\omega't')d\omega d\omega'$$

$$= \frac{1}{2\pi} \int_{-\infty}^{\infty} \underline{\underline{\Sigma}}_{\omega}\exp(i\omega(t - t'))d\omega \tag{8.15}$$

where we have introduced the resolvent matrix

$$\underline{\underline{\Sigma}}_{\omega} = \left(i\omega\underline{\underline{I}} - \underline{\underline{A}} - \underline{\underline{B}}\exp(-i\omega\tau)\right)^{-1} 2\underline{\underline{D}}\left(-i\omega\underline{\underline{I}} - \underline{\underline{A}}^T - \underline{\underline{B}}^T\exp(i\omega\tau)\right)^{-1}. \tag{8.16}$$

The diagonal elements of the matrix (8.16) are just the power spectral densities of the individual components, whereas the off-diagonal elements are the Fourier transform of the corresponding cross correlation function. It is of course obvious that the time delay enters in the same way as in the eigenvalue problem (8.2), and that the eigenvalues determine the complex poles of the resolvent matrix.

8.3 Weakly Nonlinear Analysis of Time-Delayed Feedback Control

Time-delayed feedback control [38, 39] is a convenient setup to control the stability of periodic solutions. As a paradigmatic example we will use a Duffing-Van der Pol oscillator to illustrate various aspects of the analytic perturbation schemes. The equations of motion are given by

$$
\begin{aligned}
\dot{x}_1(t) &= \omega x_2(t) \\
\dot{x}_2(t) &= -\omega x_1(t) + \varepsilon(\sigma x_1(t) - \nu x_1^3(t) - \mu x_2(t)(1 - x_1^2(t))) \\
&\quad -\kappa(x_1(t) - x_1(t - \tau)).
\end{aligned}
\tag{8.17}
$$

Here ω denotes the linear frequency, ν governs the nonlinear part of the potential, and μ the nonlinear damping. Time-delayed feedback with control amplitude κ and delay time τ is applied to stabilise periodic orbits with period $T = \tau$, where the equality between period T and time delay τ ensures the control method to be non-invasive. For the purpose of a perturbation expansion ε denotes a small expansion parameter. In the leading order, $\varepsilon = 0$, and without time-delayed feedback, $\kappa = 0$, the system admits a continuum of harmonic solutions, $x_1(t) = |a| \cos(\omega t + \varphi)$, $x_2(t) = -|a| \sin(\omega t + \varphi)$ for any $a \in \mathbb{R}$. Including the nonlinear terms only a few of those, if at all, will result in periodic orbits. Typically the period T of those orbits will depend on nonlinear contributions as well, and multiple scaling techniques or Lindsted type expansions are required to cope with these issues. It is technically simpler, in particular in the presence of time delay, if we include a small detuning σ in the perturbation and require that the resulting orbit, at least at first order, still has period $T = 2\pi/\omega$.

We can perform the perturbation expansion even in a more general setup if we use vector notation. Then Eq. (8.17) reads

$$
\underline{\dot{x}}(t) = \underline{\underline{A}}\,\underline{x}(t) + \varepsilon \underline{f}(\underline{x}(t)) - \underline{\underline{K}}(\underline{x}(t) - \underline{x}(t - \tau)), \qquad \tau = 2\pi/\omega
\tag{8.18}
$$

where the linear part admits a pair of imaginary eigenvalues with right- and left-eigenvectors $\underline{u}$ and $\underline{v}^\dagger$, respectively

$$
\underline{\underline{A}}\,\underline{u} = i\omega \underline{u}, \quad \underline{v}^\dagger \underline{\underline{A}} = i\omega \underline{v}^\dagger.
\tag{8.19}
$$

8.3.1 Computation of Periodic Orbits

For $\varepsilon = 0$ Eq. (8.18) clearly admits harmonic solutions $a \exp(i\omega t)\underline{u} + c.c..$ Hence, for the nonlinear system we first seek to compute periodic solutions by a simple series expansion

$$\underline{x}_*(t) = \underline{x}^{(0)}(t) + \varepsilon \underline{x}^{(1)}(t) + \cdots$$
$$\underline{x}^{(0)}(t) = a \exp(i\omega t)\underline{u} + \bar{a}\exp(-i\omega t)\underline{\bar{u}}$$
$$\underline{x}^{(1)}(t) = \underline{x}^{(1)}(t+T). \tag{8.20}$$

Equation (8.18) in first order gives the inhomogeneous system

$$\underline{\dot{x}}^{(1)}(t) = \underline{\underline{A}}\,\underline{x}^{(1)}(t) + \underline{f}(\underline{x}^{(0)}(t)). \tag{8.21}$$

In the case considered here, $T = \tau$, the delay term drops from this existence condition because the control term is finally non-invasive. Since the solution of Eq. (8.21) has to be periodic, see Eq. (8.20), a solvability or Fredholm condition applies. The condition can be derived straightforwardly if we multiply Eq. (8.21) with $\exp(-i\omega t)\underline{v}^\dagger$, the solution of the adjoint problem. Then the time derivative and the linear contribution combine to result in the total derivative of $\exp(-i\omega t)\underline{v}^\dagger \underline{x}^{(1)}(t)$ and integration over the period $T = \tau = 2\pi/\omega$ yields

$$0 = \int_0^\tau \frac{d}{dt}\left(\exp(-i\omega t)\underline{v}^\dagger \underline{x}^{(1)}(t)\right)dt = \int_0^\tau \exp(-i\omega t)\underline{v}^\dagger \underline{f}(\underline{x}^{(0)}(t))dt. \tag{8.22}$$

The condition, Eq. (8.22), can be expressed in terms of an effective drift

$$0 = \frac{1}{\tau \underline{v}^\dagger \underline{u}}\int_0^\tau \exp(-i\omega t)\underline{v}^\dagger \underline{f}(\underline{x}^{(0)}(t))dt = aF(|a|^2). \tag{8.23}$$

A simple phase argument shows that the integral in Eq. (8.23) depends only on the modulus $|a|$ apart from a single amplitude factor. If we denote by F_R and F_I the real and the imaginary part of F, respectively, then the condition for the periodic orbit results in

$$F_R(|a|^2) = 0, \quad F_I(|a|^2) = 0. \tag{8.24}$$

The first of these conditions determines the value of the amplitude of the periodic orbit, Eq. (8.20), whereas the second equation states the constraint that the period of this orbit is not renormalised by the nonlinear contributions. Such a constraint can be, for instance, satisfied by including a small linear part $\varepsilon \underline{\underline{A}}^{(1)}\underline{x}$ in the perturbation, which properly renormalises the linear frequency. In addition, there exists of course the trivial solution $a = 0$ resulting in a small amplitude periodic orbit according to Eq. (8.20). For the system without delayed feedback, $\underline{\underline{K}} = 0$, the expression F_R acts as the derivative of an effective potential and the stability of the orbit is directly related to the derivative of F_R.

For our model system, Eq. (8.17), the eigenvectors in Eq. (8.19) are given by $\underline{u} = (1, i)^T$ and $\underline{v}^\dagger = (1, -i)$. From the definition (8.23) and the perturbation according to Eq. (8.17) we obtain

$$F(|a|^2) = F_R(|a|^2) + i F_I(|a|^2) = -\mu(1 - |a|^2)/2 + i(3\nu|a|^2 - \sigma)/2. \quad (8.25)$$

Hence the finite amplitude periodic orbit has amplitude $|a| = 1$. It will turn out to be an unstable orbit for $\mu > 0$ as one expects to be the case for a weakly nonlinear Van der Pol oscillator. The second condition in Eq. (8.24) on the frequency renormalisation requires $\sigma = 3\nu$. In fact, the nonlinear potential part of the Duffing oscillator introduces a chirp (anisochronicity), a dependence of the period on the amplitude of the oscillation, which is then compensated by the appropriate detuning σ.

8.3.2 Linear Stability and Strongly Stable Domain

So far the time delay has not played any essential role for the existence of periodic orbits, because of the noninvasive character of the feedback. Of course, the situation is different when stability considerations become relevant. Assume the periodic orbit, Eq. (8.20), is known. Linear stability is governed by the variational equation, which in turn using an exponential ansatz, $\delta \underline{x}(t) = \exp(\Lambda t)\underline{w}(t)$, can be converted into a Floquet eigenvalue problem

$$\dot{\underline{w}}(t) + \Lambda \underline{w}(t) = (\underline{\underline{A}} + \varepsilon \underline{Df}(\underline{x}_*(t)) - \underline{\underline{K}}(1 - \exp(-\Lambda \tau)))\underline{w}(t), \quad \underline{w}(t) = \underline{w}(t + \tau).$$
$$(8.26)$$

The symbol Df denotes the Jacobian matrix. If we consider Eq. (8.26) in lowest order, i.e., for $\varepsilon = 0$, then the time dependence drops and the equation reduces to an ordinary (nonlinear) eigenvalue problem. The exponents in this order are determined by the usual quasipolynomial (see Eq. (8.3))

$$\det\left(\underline{\underline{A}} - \Lambda \underline{\underline{I}} - \underline{\underline{K}}(1 - \exp(-\Lambda \tau))\right) = 0. \quad (8.27)$$

The condition that all the solutions of Eq. (8.27) have nonpositive real part is a necessary constraint for the stability of the orbit. If we use $\Lambda = i\Omega$ in Eq. (8.27) we are able to determine the stability boundaries in the parameter space. Our stability condition so far is determined by the control matrix and the dominant linear part of the dynamics, and thus essentially reflects the stability of the control loop. In the leading order, $\varepsilon = 0$, neither the nonlinear part of the dynamics nor the actual shape of the orbit has entered the analysis.

For our example, Eq. (8.17), the characteristic equation (8.27) in leading order reads

$$0 = \det \begin{pmatrix} -\Lambda & \omega \\ -\omega - \kappa(1 - \exp(-\Lambda\tau)) & -\Lambda \end{pmatrix} = \Lambda^2 + \omega^2 + \kappa\omega(1 - \exp(-\Lambda\tau)).$$
(8.28)

If we choose $\Lambda = i\Omega$ to determine the boundaries of the control domain, it is evident that $\Omega\tau = \pi + 2\pi n$, $n \in \mathbb{Z}$. Hence we obtain from Eq. (8.28) the thresholds $\kappa = \omega(n + 3/2)(n - 1/2)/2$. The stable interval where Eq. (8.28) has no solution with positive real part is given by

$$\kappa \in (-3\omega/8, 5\omega/8).$$
(8.29)

Outside this interval there exists at least one Floquet exponent with positive real part of order one. As our stability condition has been derived in leading order we call Eq. (8.29) the strongly stable domain.

8.3.3 Perturbation Expansion of the Eigenvalue Problem and Weak Instabilities

Floquet exponents which in leading order already have a nonvanishing real part do not change the stability properties if the small perturbation is taken into account. Hence the strongly stable domain, Eq. (8.29), is a necessary constraint for stability. However, within this interval and in order $\varepsilon = 0$ there still occur two neutral modes with leading Floquet exponent zero. It is easy to verify that

$$\underline{w}^{(0)}(t) = \alpha \exp(i\omega t)\underline{u} + \bar{\alpha} \exp(-i\omega t)\underline{\bar{u}}$$
(8.30)

solves Eq. (8.26) for $\varepsilon = 0$ and $\Lambda = 0$. Equation (8.30) determines a two-dimensional subspace, parametrised by α and $\bar{\alpha}$, which contains the Goldstone mode of the periodic orbit. That means no matter what kind of perturbation we apply one of the Floquet exponents remains zero. However, the second exponent may become nonzero and thus results in an additional stability condition. We aim for computing such a Floquet exponent using the straightforward series expansions

$$\underline{w}(t) = \underline{w}^{(0)}(t) + \varepsilon\underline{w}^{(1)}(t) + \cdots, \quad \Lambda = 0 + \varepsilon\Lambda^{(1)} + \cdots.$$
(8.31)

Then Eq. (8.26) in first order reads

$$\dot{\underline{w}}^{(1)}(t) = \underline{\underline{A}}\,\underline{w}^{(1)}(t) + \underline{h}(t), \quad \underline{w}^{(1)}(t) = \underline{w}^{(1)}(t + \tau)$$
(8.32)

with inhomogeneous part

$$\underline{h}(t) = \left(D\underline{f}(\underline{x}^{(0)}(t)) - \Lambda^{(1)} - \Lambda^{(1)}\tau\,\underline{\underline{K}} \right) \underline{w}^{(0)}(t). \tag{8.33}$$

As in the previous section the existence of a periodic solution of Eq. (8.32) puts secular constraints on the inhomogeneous part, which can be easily derived if we use the neutral modes of the adjoint problem, $\exp(-i\omega t)\underline{v}^\dagger$ and $\exp(i\omega t)\underline{\bar{v}}^\dagger$. If we multiply Eq. (8.32) with one of these, the time derivative and the linear term combine to give a total derivative and integration over one period finally yields the secular conditions

$$\int_0^\tau \exp(-i\omega t)\underline{v}^\dagger \underline{h}(t)dt = 0, \quad \int_0^\tau \exp(i\omega t)\underline{\bar{v}}^\dagger \underline{h}(t)dt = 0. \tag{8.34}$$

In view of Eqs. (8.30) and (8.33) these conditions constitute a two-dimensional linear system for α and $\bar{\alpha}$. The vanishing of the determinant yields the condition for the eigenvalue $\Lambda^{(1)}$ at first order. The integrals occurring in Eq. (8.34) can, in fact, be written in terms of the previously introduced effective drift, Eq. (8.23). If we take in Eq. (8.23) derivatives with respect to a or $\bar{a}$ we obtain

$$a^2 F'(|a|^2) = \int_0^\tau \exp(-i\omega t)\underline{v}^\dagger D\underline{f}(\underline{x}^{(0)}(t))\underline{\bar{u}}\exp(-i\omega t)dt/(\tau\underline{v}^\dagger\underline{u})$$

$$F(|a|^2) + |a|^2 F'(|a|^2) = \int_0^\tau \exp(-i\omega t)\underline{v}^\dagger D\underline{f}(\underline{x}^{(0)}(t))\underline{u}\exp(i\omega t)dt/(\tau\underline{v}^\dagger\underline{u}). \tag{8.35}$$

Employing the property (8.24) the two secular conditions, Eq. (8.34), can be written as the homogeneous system

$$\begin{pmatrix} |a|^2 F'(|a|^2) - \Lambda^{(1)} - \gamma\Lambda^{(1)} & a^2 F'(|a|^2) \\ \bar{a}^2 \bar{F}'(|a|^2) & |a|^2 \bar{F}'(|a|^2) - \Lambda^{(1)} - \bar{\gamma}\Lambda^{(1)} \end{pmatrix} \begin{pmatrix} \alpha \\ \bar{\alpha} \end{pmatrix} = 0, \tag{8.36}$$

where the abbreviation

$$\gamma = \tau\underline{v}^\dagger \underline{\underline{K}}\,\underline{u}/(\underline{v}^\dagger\underline{u}) \tag{8.37}$$

takes the effect of the control loop into account. Equation (8.36) finally results in the characteristic polynomial

$$0 = \left(\Lambda^{(1)}\right)^2 |1 + \gamma|^2 - 2\Lambda^{(1)}\mathrm{Re}\left((1 + \bar{\gamma})|a|^2 F'(|a|^2)\right) \tag{8.38}$$

which determines the two small Floquet exponents in first order. Clearly one of the solutions $\Lambda^{(1)} = 0$ corresponds to the Goldstone mode, while the other may take nontrivial values and may induce a weak instability of the orbit. Without control $\gamma = 0$ we have $\Lambda^{(1)} = 2|a|^2 F'_R(|a|^2)$ which, as already mentioned above, determines the stability properties of the periodic orbit without delayed feedback. In cases when the coefficient of the leading term in Eq. (8.38) vanishes the results have to be considered with some care.

For our model, Eq. (8.17), the effective drift, Eq. (8.25), gives $F'(|a|^2) = (\mu + 3i\nu)/2$ so that the orbit with amplitude $|a| = 1$ and without delayed feedback $\kappa = 0$

is unstable for $\mu > 0$, as already mentioned. With the eigenvectors stated above we obtain from Eq. (8.37) $\gamma = -i\tau\kappa/2$, and the polynomial (8.38) results in the nontrivial eigenvalue $\Lambda^{(1)}|1 + i\tau\kappa/2|^2 = \mu - 3\tau\kappa\nu/2$. Thus the weak stability condition induced by this branch reads

$$\kappa\nu > \mu\omega/(3\pi). \tag{8.39}$$

In this particular example a nonzero value for the nonlinear potential, i.e., a chirp is required for stabilisation. The two conditions, Eqs. (8.29) and (8.39), give a fairly simple shape for the control domain in the parameter space. The perturbative results are in quite good agreement with numerical results obtained for finite but small values of the expansion parameter, see Fig. 8.1. Overall the perturbative results can provide some insight which type of feedback could be used to achieve stabilisation, even though the analytic expressions are limited to the weakly nonlinear regime

The weakly nonlinear analysis of the model (8.17) has very much in common with the discussion of the corresponding Hopf normal form (Stuart Landau oscillator) subjected to time delayed feedback [40–42]. The analysis provided here is able to link the control domain with the actual parameters of the original equation of motion. In fact, the procedure applied to our example can be almost verbatim transferred to the discussion of the general weakly nonlinear system

$$\underline{\dot{x}}(t) = \underline{A}\,\underline{x}(t) + \underline{B}\,\underline{x}(t-\tau) + \varepsilon\underline{f}(\underline{x}(t),\underline{x}(t-\tau)) \tag{8.40}$$

if we assume that the linear part, i.e., the equation for $\varepsilon = 0$ supports periodic solutions.

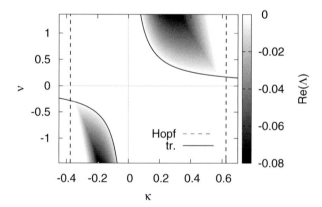

Fig. 8.1 Control domain for the periodic orbit of the Duffing-Van der Pol oscillator, Eq. (8.17), for $\mu = 1.0$, $\omega = 1.0$, $\sigma = 3\nu$, and delay τ coinciding with the period of the orbit. *Dashed lines*: Analytic result of the strong stability boundary, Eq. (8.29). *Solid lines*: Analytic result for the weak stability boundary, Eq. (8.39). *Shading* indicates numerical results for the real part of the leading Floquet exponent for $\varepsilon = 0.05$

8.4 Coherence Resonance Modulated by Time-Delayed Feedback

The counterintiuitive phenomenon of coherence resonance [4, 43] was originally discovered for excitable systems. It implies that noise-induced oscillations become most regular for an optimum non-zero value of the noise intensity. It has been shown that coherence resonance can be enhanced or suppressed by applying time-delayed feedback in systems with type-I [44] and type-II [45] excitability. Recently, coherence resonance has also been found in non-excitable systems with a subcritical Hopf bifurcation [46–48] and its modulation by time-delayed feedback has been demonstrated theoretically for a subcritical Hopf normal form [29], and confirmed in experiments with an electronic circuit for a generalised Van der Pol system [49]. It is important to note that the pure coherence resonance effect for non-excitable systems is observed for a subcritical Hopf bifurcation and not for the supercritical case. The standard Van der Pol model close to a supercritical Hopf bifurcation has been investigated in the presence of delay and noise [45, 50, 51], but here no coherence resonance, i.e., no non-monotonic dependence of the correlation time upon the noise intensity, is found. Coherence resonance has also been observed in real-world systems, for instance, in microwave dynamical systems such as a five-cavity delayed-feedback klystron oscillator at the self-excitation threshold [52], in lasers with saturable absorber [53], with optical feedback [54–56], with optical injection [57], or in semiconductor superlattices [58, 59].

Here we investigate the impact of time-delayed feedback in a non-excitable model of coherence resonance. We will apply perturbation techniques to a model which can be considered as the normal form of a subcritical Hopf bifurcation, see [29]. The calculation of the stationary amplitude probability distribution and of correlation functions will be at the centre of interest. Consider a two-variable quintic normal form of a nonlinear oscillator

$$\dot{x}_1(t) = (\lambda + r^2(t) - r^4(t))x_1(t) - 2\pi x_2(t) - K(x_1(t) - x_1(t - \tau)) + \sqrt{2D}\xi_1(t)$$
$$\dot{x}_2(t) = (\lambda + r^2(t) - r^4(t))x_2(t) + 2\pi x_1(t) - K(x_2(t) - x_2(t - \tau)) + \sqrt{2D}\xi_2(t)$$
$$r^2(t) = x_1^2(t) + x_2^2(t) \tag{8.41}$$

subjected to time-delayed feedback with control amplitude K and to isotropic Gaussian white noise of strength D, with vanishing mean $\langle\xi_k(t)\rangle = 0$ and correlation function $\langle\xi_k(t)\xi_\ell(t')\rangle = \delta(t - t')\delta_{k\ell}$. The timescale has been adjusted such that the linear frequency is given by $\omega = 2\pi$, and λ denotes the bifurcation parameter. For simplicity of the perturbative treatment we will only allow integer values of the delay τ. The system mimics a subcritical Hopf bifurcation, including the saddle node bifurcation of the unstable limit cycle at $\lambda = -1/4$, which is the main cause for the coherence resonance phenomenon.

8.4.1 Adiabatic Elimination and Stationary Probability Distribution

To derive an effective simple stochastic differential equation let us first analyse the bifurcations of the trivial stationary state $x_1 = x_2 = 0$ of the deterministic part of the model (8.41). The variational equation results in an eigenvalue problem of the form Eq. (8.2) and the characteristic equation (8.3) reads

$$
0 = \det \begin{pmatrix} \lambda - \Lambda - K(1 - \exp(-\Lambda\tau)) & -2\pi \\ 2\pi & \lambda - \Lambda - K(1 - \exp(-\Lambda\tau)) \end{pmatrix}
$$
$$
= (\lambda - \Lambda - 2\pi i - K(1 - \exp(-\Lambda\tau)))(\lambda - \Lambda + 2\pi i - K(1 - \exp(-\Lambda\tau))). \quad (8.42)
$$

Obviously, for integer values for the delay τ and $\lambda = 0$, $\Lambda = \pm i\Omega = \pm 2\pi i$ is a purely imaginary pair of eigenvalues, giving rise to a Hopf bifurcation at $\lambda = 0$. It is in fact possible to show that all the other eigenvalues have negative real part if K is sufficiently small, for instance, the condition $|K\tau| \leq 1$ is sufficient. But the corresponding techniques are fairly nontrivial and a discussion of the related algebraic and numerical concepts can be found, e.g., in [10, 19, 60]. We can apply the techniques to derive the effective stochastic differential equation, Eq. (8.11), when rewriting the model (8.41) using the notation of Eq. (8.8)

$$
\underline{\underline{A}} = \begin{pmatrix} -K & -2\pi \\ 2\pi & -K \end{pmatrix}, \quad \underline{\underline{B}} = \begin{pmatrix} K & 0 \\ 0 & K \end{pmatrix}
$$
$$
\underline{f}(\underline{x}(t), \underline{x}(t-\tau)) = (\lambda + r^2(t) - r^4(t)) \begin{pmatrix} x_1(t) \\ x_2(t) \end{pmatrix}
$$
$$
\underline{\underline{g}}(\underline{x}(t), \underline{x}(t-\tau)) = \sqrt{2D} \begin{pmatrix} 1 & 0 \\ 0 & 1 \end{pmatrix}. \quad (8.43)
$$

For $\Lambda = i\Omega = 2\pi i$ the eigenvectors of the linear part (see Eqs. (8.2) and (8.4)) are easily computed as $\underline{u}_{i\Omega} = (1, -i)^T$ and $\underline{v}^\dagger_{i\Omega} = (1, i)$. Equation (8.9) then tells us that on the slow manifold the phase space variables can be expressed in terms of the amplitude $C(t)$

$$
x_1(t+\theta) = C(t)\exp(i2\pi\theta) + \bar{C}(t)\exp(-i2\pi\theta) + R_1(C(t), \theta)
$$
$$
x_2(t+\theta) = -iC(t)\exp(i2\pi\theta) + i\bar{C}(t)\exp(-i2\pi\theta) + R_2(C(t), \theta). \quad (8.44)
$$

In particular, neglecting the nonlinear contributions in Eq. (8.44) we have $x_1(t) = 2\text{Re}(C(t)) + \cdots$, $x_2(t) = 2\text{Im}(C(t)) + \cdots$ and $r^2(t) = 4|C(t)|^2 + \cdots$. Thus, the amplitude C has a a direct meaning in terms of the original phase space values. Finally, using the higher order terms given in Eq. (8.43) the effective stochastic differential equation (8.11) reads

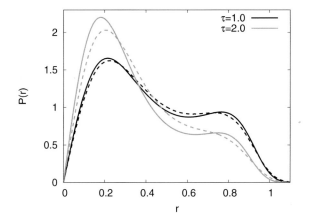

Fig. 8.2 Stationary probability distribution of the radial variable $r = (x_1^2 + x_2^2)^{1/2}$ for the stochastic delay system, Eq. (8.41), with parameters $\lambda = -0.26$, $K = 0.5$, $D = 0.015$, and different values of the delay τ. *Solid lines*: Distribution according to the analytic perturbation expansion, Eq. (8.46). *Dashed lines*: Results obtained from numerical simulations

$$\dot{C}(t) = 2\pi i C(t) + \frac{(\lambda + |2C(t)|^2 - |2C(t)|^4)2C(t) + \sqrt{2D}(\xi_1(t) + i\xi_2(t))}{2(1 + K\tau)}$$
(8.45)

where we have evaluated the nonlinear terms using the lowest order linear approximation on the slow manifold, Eq. (8.44). The time-delayed feedback results in an effective rescaling of the time scale, which in turn can be converted to a rescaling of the noise intensity D. As the system (8.45) has a phase symmetry, the stationary distribution is spherically symmetric as well and can be computed by standard methods quite easily [8]. If we use the radial variable $r = 2|C|$ the stationary distribution reads

$$P(r) = Nr \exp\left(\frac{r^2(\lambda/2 + r^2/4 - r^4/6)}{D/(1 + K\tau)}\right)$$
(8.46)

where N denotes the normalisation factor. The analytic approximation, which has been obtained as an expansion at the bifurcation point, gives in fact a rather accurate description even if parameter values deviate substantially from the bifurcation point (see Fig. 8.2).

The formal derivation of the effective stochastic model (8.45) has not been entirely systematic, as we were including higher order nonlinear terms to keep the system globally stable. Such kind of heuristic approach is often used in the physics literature, and can in principle be dealt with by computing the higher order nonlinear corrections to the slow manifold (8.44).

8.4.2 Dynamical Correlations and Power Spectral Density

Correlation functions and power spectral densities are at the heart of detecting coherence resonance phenomena. The evaluation of dynamical correlation functions in nonlinear stochastic systems is an inherently difficult task. If time delay is added then one cannot even resort to spectral theories, e.g., using Fokker-Planck operators. On the contrary, the evaluation of correlations in linear systems is almost trivial by comparison. Hence, one often aims at approximating the original nonlinear dynamics by a suitable linear system [61], even at the expense of a systematic approximation scheme. Following ideas developed in the context of Fokker-Planck systems with nonlinear drift [62] one tries to approximate nonlinear terms in the equations of motion by linear ones, using some kind of optimisation approach. To illustrate the main idea let us consider the stochastic time delay equation

$$\underline{\dot{x}}(t) = \underline{f}(\underline{x}(t)) + \underline{\underline{B}}\,\underline{x}(t - \tau) + \underline{\underline{g}}\,\underline{\xi}(t) \tag{8.47}$$

which covers as well the model of interest, Eq. (8.41) when we choose the nonlinear term $\underline{f}(\underline{x})$ appropriately. We intend to replace this nonlinear function by a linear contribution $\underline{\underline{A}}\,\underline{x}$ by making the "best choice" for the coefficient matrix in the sense that we minimise the mean square deviation $\langle (\underline{f}(\underline{x}) - \underline{\underline{A}}\,\underline{x})^T (\underline{f}(\underline{x}) - \underline{\underline{A}}\,\underline{x}) \rangle$. The minimisation yields

$$\underline{\underline{A}} = \langle \underline{f}(\underline{x})\underline{x}^T \rangle \langle \underline{x}\,\underline{x}^T \rangle^{-1} \tag{8.48}$$

with Eqs. (8.1) and (8.12) being the "best" linear approximation of Eq. (8.47). The evaluation of Eq. (8.48) requires the computation of static expectation values, a feature which is quite common in dynamical theories of this kind. One could deduce such values either from simulations or from alternative theories, like, e.g., those developed in the previous section. The impact of the nonlinearity on the dynamics has been effectively condensed in the few parameters, Eq. (8.48). The computation of dynamical correlation functions now becomes a trivial task as we can resort to the exact result available for linear systems, see Eqs. (8.15) and (8.16). There are certainly alternatives to arrive at a suitable linear model, i.e., the scheme outlined here is by no means a unique way to solve the task. In addition, the accuracy of the method is difficult to predict a priori. Like most mean-field schemes the approach does not rely on a small expansion parameter. But the scheme has the potential to capture at least qualitatively the main features of the correlations as the impact of the time delay has been fully taken into account.

For our model (8.41) we have, using the notation of Eq. (8.47)

$$\underline{\underline{B}} = \begin{pmatrix} K & 0 \\ 0 & K \end{pmatrix}, \quad \underline{g} = \sqrt{2D} \begin{pmatrix} 1 & 0 \\ 0 & 1 \end{pmatrix} \tag{8.49}$$

while the components of the nonlinear part are given by $f_1 = (\lambda + r^2 - r^4 - K) x_1 - 2\pi x_2$ and $f_2 = (\lambda + r^2 - r^4 - K)x_2 + 2\pi x_1$. Since our model is

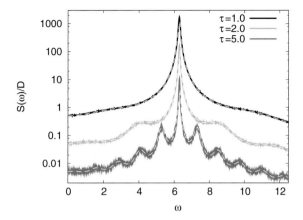

Fig. 8.3 Normalised power spectral density $S(\omega)/D$ of the model, Eq. (8.41), for $\lambda = -0.26$, $K = 0.5$, $D = 0.015$ and different values of the delay. *Solid lines*: Numerical simulations of the stochastic delay differential equation. *Dashed lines* (*white*): Analytic approximation according to Eq. (8.52). For better visibility the *top* and the *bottom* spectrum is shifted by ± 10 dB, respectively

spherically symmetric all static cross correlations vanish, i.e., $\langle x_k x_\ell \rangle = 0$ and $\langle (r^2 - r^4)x_k x_\ell \rangle = 0$ for $k \neq \ell$. The diagonal components coincide and expectation values can be written in terms of the radial variable, $\langle x_k x_k \rangle = \langle r^2 \rangle / 2$ and $\langle (r^2 - r^4)x_k x_k \rangle = \langle (r^2 - r^4)r^2 \rangle / 2$. Thus, Eq. (8.48) results in

$$\underline{A} = \begin{pmatrix} \lambda_{eff} - K & -2\pi \\ 2\pi & \lambda_{eff} - K \end{pmatrix}, \quad \lambda_{eff} = \lambda + \langle (r^4 - r^6) \rangle / \langle r^2 \rangle. \qquad (8.50)$$

Within our approximation the entire impact of the nonlinear contributions is a static renormalisation of the bifurcation parameter. The required expectation values can be either obtained from the approximation, Eq. (8.46), or from simulations. One can also aim at a self-consistent computation using the Gaussian stationary solution of the effective model, Eqs. (8.1) and (8.12), even though the actual stationary density is by no means well approximated by a normal distribution, see Fig. 8.2. Finally, the correlation matrix is easily evaluated from Eq. (8.16). The result is conveniently stated if we introduce the abbreviation

$$z(\omega) = i\omega - \lambda_{eff} + K(1 - \exp(-i\omega\tau)). \qquad (8.51)$$

The diagonal elements coincide, $(\Sigma_\omega)_{11} = (\Sigma_\omega)_{22} = S(\omega)$, and represent the power spectral density of each component

$$S(\omega) = 2D\frac{|z(\omega)|^2 + (2\pi)^2}{|z(\omega) - 2\pi i|^2 |z(\omega) + 2\pi i|^2} = D(|z(\omega) - 2\pi i|^{-2} + |z(\omega) + 2\pi i|^{-2}).$$
$$(8.52)$$

For the cross correlations we obtain

$$
\begin{aligned}
(\Sigma_\omega)_{12} = -(\Sigma_\omega)_{21} &= 2D\frac{2\pi\,(z(\omega) - z(-\omega))}{|z(\omega) - 2\pi i|^2 |z(\omega) + 2\pi i|^2} \\
&= 2D\frac{4\pi i\,(\omega + K\,\sin(\omega\tau))}{|z(\omega) - 2\pi i|^2 |z(\omega) + 2\pi i|^2}.
\end{aligned}
\tag{8.53}
$$

The results of this phenomenological approximation scheme are quite encouraging as the main features of the power spectral density are captured quantitatively to a fairly high degree of precision, see Fig. 8.3.

8.5 Concluding Remarks

Formal analytic expansion schemes, even if they just cover very simple setups, play their role for the investigation of dynamical systems, in particular for models including noise and time delay. Initially, the methods have been largely developed within an engineering context for ordinary differential equations. The tools turn out to be useful for the understanding of basic mechanisms in dynamical systems and provide an analytic overview of structures in phase- and parameter space. Even for contemporary research questions, such as the relevance of coupling topologies in networks with noise or time delay, or global dynamical aspects such as manifolds determining basins of attraction, simple analytic perturbation schemes could give viable input. For instance, Melnikov's method has been empirically generalised to cope with noise and time delay, but systematic studies are still lacking.

Our selection of perturbation expansion schemes was of course incomplete, and there are plenty of related approaches available in the literature. Essentially all of these have limitations from a mathematical point of view. Unlike the impression which is sometimes given in the physics literature formal expansions do not provide rigorous proofs. Caveats apply as well when applications are considered. Most real world problems do not come in a setting which allows for the application of analytic expansion schemes. Hence, numerical approaches, say either direct simulations or sophisticated continuation techniques to track bifurcations, are certainly often the method of choice to determine the behaviour of a particular dynamical systems. Having said that, simple analytic formal expansion schemes such as those sketched above, combined with rigorous methods or numerical approaches can provide useful insight for complex dynamical behaviour of systems with noise and time delay.

Acknowledgments This work was supported by DFG in the framework of SFB 910.

References

1. A. Einstein, Ann. Phys. **322**, 549 (1905)
2. H.A. Kramers, Physica **7**, 284 (1940)
3. P. Hänggi, P. Talkner, M. Borkovec, Rev. Mod. Phys. **62**, 251 (1990)
4. A. Pikovsky, J. Kurths, Phys. Rev. Lett. **78**, 775 (1997)
5. P. Gammaitoni, P. Hänggi, P. Jung, F. Marchesoni, Rev. Mod. Phys. **70**(1), 223 (1998)
6. R. Klages, W. Just, C. Jarzynski (eds.), *Nonequilibrium Statistical Physics of Small Systems: Fluctuation Relations and Beyond* (Wiley-VCH, Weinheim, 2013)
7. W. Horsthemke, R. Lefever, *Noise-Induced Transitions* (Springer, Berlin, 1984)
8. H. Risken, *The Fokker-Planck Equation: Methods of Solution and Applications* (Springer, Berlin, 1989)
9. L. Arnold, *Random Dynamical Systems* (Springer, Berlin, 2002)
10. R. Bellmann, K.L. Cooke, *Differential-Difference Equations* (Acad. Press, New York, 1963)
11. E. Fick, G. Sauermann, *The Quantum Statistics of Dynamic Processes* (Springer, Berlin, 1990)
12. R. Kubo, M. Toda, N. Hashitsume, *Statistical Physics 2 (Nonequilibrium statistical mechanics)* (Springer, Berlin, 1991)
13. T. Erneux, *Applied Delay Differential Equations* (Springer, Berlin, 2009)
14. F.M. Atay (ed.), *Complex Time-Delay Systems* (Springer, Berlin, Heidelberg, 2010)
15. W. Just, A. Pelster, M. Schanz, E. Schöll (eds.), *Delayed Complex Systems*, Theme Issue of Phil. Trans. R. Soc. A **368**, 301–513 (2010)
16. V. Flunkert, I. Fischer, E. Schöll (eds.), *Dynamics, control and information in delay-coupled systems*, Theme Issue of Phil. Trans. R. Soc. A **371**, 20120465 (2013)
17. J.Q. Sun, G. Ding (eds.), *Advances in Analysis and Control of Time-Delayed Dynamical Systems* (World Scientific, Singapore, 2013)
18. M. Soriano, J. Garca-Ojalvo, C. Mirasso, I. Fischer, Rev. Mod. Phys. **85**, 421 (2013)
19. J.K. Hale, S.M. Verduyn Lunel, *Introduction to Functional Differential Equations* (Springer, New York, 1993)
20. K. Engelborghs, *DDE-BIFTOOL: a Mathlab package for bifurcation analysis of delay differential equations*. http://www.cs.kuleuven.ac.be/koen/delay/ddebiftool.shtml
21. L. Pecora, T. Carroll, Phys. Rev. Lett. **80**, 2109 (1998)
22. S. Yanchuk, M. Wolfrum, P. Hövel, E. Schöll, Phys. Rev. E **74**, 026201 (2006)
23. M. Wolfrum, S. Yanchuk, P. Hövel, E. Schöll, Eur. Phys. J. Special Topics **191**, 91 (2010)
24. T.D. Frank, P.J. Beck, R. Friedrich, Phys. Rev. E **68**, 021912 (2003)
25. T. Frank, Phys. Rev. E **72**, 011112 (2005)
26. A. Nayfeh, *Perturbation Methods* (Wiley-VCH, Weinheim, 2000)
27. M.C. Cross, P.C. Hohenberg, Rev. Mod. Phys. **65**(3), 851 (1993)
28. W. Just, H. Benner, C. v. Loewenich. Physica D **199**, 33 (2004)
29. P.M. Geffert, A. Zakharova, A. Vüllings, W. Just, E. Schöll, Eur. Phys. J. B **87**, 291 (2014)
30. H. Haken, *Synergetics: Introduction and Advanced Topics* (Springer, Berlin, 2004)
31. A. Halanay, *Differential Equations: Stability, Oscillations, Time Lags* (Acad. Press, New York, 1966)
32. A. Amann, E. Schöll, W. Just, Physica A **373**, 191 (2007)
33. R.M. Corless, G.H. Gonnet, D.E.G. Hare, D.J. Jeffrey, D.E. Knuth, Adv. Comp. Math. **5**, 329 (1996)
34. F. Giannakopoulos, A. Zapp, Physica D **159**, 215 (2001)
35. B.F. Redmond, V.G. LeBlanc, A. Longtin, Physica D **166**, 131 (2002)
36. N.G. van Kampen, Phys. Rep. **124**(2), 69 (1985)
37. U. Küchler, B. Mensch, Stoch. Stoch. Rep. **40**, 23 (1992)
38. K. Pyragas, Phys. Lett. A **170**, 421 (1992)
39. E. Schöll, H.G. Schuster (eds.), *Handbook of Chaos Control* (Wiley-VCH, Weinheim, 2008)
40. B. Fiedler, V. Flunkert, M. Georgi, P. Hövel, E. Schöll, Phys. Rev. Lett. **98**, 114101 (2007)
41. W. Just, B. Fiedler, M. Georgi, V. Flunkert, P. Hövel, E. Schöll, Phys. Rev. E **76**, 026210 (2007)

42. G. Brown, C.M. Postlethwaite, M. Silber, Physica D **240**, 859 (2011)
43. G. Hu, T. Ditzinger, C.Z. Ning, H. Haken, Phys. Rev. Lett. **71**, 807 (1993)
44. R. Aust, P. Hövel, J. Hizanidis, E. Schöll, Eur. Phys. J. Spec. Top. **187**, 77 (2010)
45. N.B. Janson, A.G. Balanov, E. Schöll, Phys. Rev. Lett. **93**, 010601 (2004)
46. O.V. Ushakov, H.J. Wünsche, F. Henneberger, I.A. Khovanov, L. Schimansky-Geier, M.A. Zaks, Phys. Rev. Lett. **95**, 123903 (2005)
47. A. Zakharova, T. Vadivasova, V. Anishchenko, A. Koseska, J. Kurths, Phys. Rev. E **81**, 011106 (2010)
48. A. Zakharova, A. Feoktistov, T. Vadivasova, E. Schöll, Eur. Phys. J. Spec. Top. **222**, 2481 (2013)
49. V. Semenov, A. Feoktistov, T. Vadivasova, E. Schöll, A. Zakharova, Chaos **25**, 033111 (2015)
50. E. Schöll, A.G. Balanov, N.B. Janson, A.B. Neiman, Stoch. Dyn. **5**, 281 (2005)
51. J. Pomplun, A. Amann, E. Schöll, Europhys. Lett. **71**, 366 (2005)
52. B. Dmitriev, Y. Zharkov, S. Sadovnikov, V. Skorokhodov, A. Stepanov, Tech. Phys. Lett. **37**, 1082 (2011)
53. J.L.A. Dubbeldam, B. Krauskopf, D. Lenstra, Phys. Rev. E **60**, 6580 (1999)
54. G. Giacomelli, M. Giudici, S. Balle, J.R. Tredicce, Phys. Rev. Lett. **84**, 3298 (2000)
55. J.F.M. Avila, H.L.D. de S. Cavalcante, J.R.R. Leite. Phys. Rev. Lett. **93**, 144101 (2004)
56. C. Otto, B. Lingnau, E. Schöll, K. Lüdge, Opt. Express **22**, 13288 (2014)
57. D. Ziemann, R. Aust, B. Lingnau, E. Schöll, K. Lüdge, Europhys. Lett. **103**, 14002 (2013)
58. J. Hizanidis, E. Schöll, Phys. Rev. E **78**, 066205 (2008)
59. Y. Huang, H. Qin, W. Li, S. Lu, J. Dong, H.T. Grahn, Y. Zhang, Europhys. Lett. **105**, 47005 (2014)
60. M.E. Bleich, J.E.S. Socolar, Phys. Lett. A **210**, 87 (1996)
61. A. Vüllings, E. Schöll, B. Lindner, Eur. Phys. J. B **87**, 31 (2014)
62. J. Kottalam, K. Lindenberg, B.J. West, J. Stat. Phys. **42**, 979 (1986)

Chapter 9
Study on Critical Conditions and Transient Behavior in Noise-Induced Bifurcations

Zigang Li, Kongming Guo, Jun Jiang and Ling Hong

Abstract In this work, the stochastic sensitivity function method, which can describe the probabilistic distribution range of a stochastic attractor, is extended to the non-autonomous dynamical systems by constructing a $1/N$-period stroboscopic map to discretize a continuous cycle into a discrete one. With confidence ranges of a stochastic attractor and the global structure of the deterministic nonlinear system, like chaotic saddle in basin of attraction and/or saddle on basin boundary as well as its stable and unstable manifolds, the critical noise intensity for the occurrence of transition behavior due to noise-induced bifurcations may be estimated. Furthermore, to efficiently capture the stochastic transient behaviors after the critical conditions, an idea of evolving probabilistic vector (EPV) is introduced into the Generalized Cell Mapping method (GCM) in order to enhance the computation efficiency of the numerical method. A Mathieu-Duffing oscillator under external and parametric excitation as well as additive noise is studied as an example of application to show the validity of the proposed methods and the interesting phenomena in noise-induced explosive and dangerous bifurcations of the oscillator that are characterized respectively by an abrupt enlargement and a sudden fast jump of the response probability distribution are demonstrated. The insight into the roles of deterministic global structure and noise as well as their interplay is gained.

9.1 Introduction

Noise is ubiquitous in nature and engineering systems that are all inherently nonlinear. Uncertain disturbances or noise on nonlinear dynamical systems often evoke some unexpected and even coherent responses. Various noise-induced behaviors have been found, such as noise-induced chaos [1, 2], stochastic bifurcation [3, 4], noise-induced

Z. Li · J. Jiang (✉) · L. Hong
State Key Laboratory for Strength and Vibration, Xi'an Jiaotong University,
Xi'an 710049, China
e-mail: jun.jiang@mail.xjtu.edu.cn

K. Guo
School of Electromechanical Engineering, Xidian University, Xi'an 710071, China

© Springer International Publishing Switzerland 2016   169
E. Schöll et al. (eds.), *Control of Self-Organizing Nonlinear Systems*,
Understanding Complex Systems, DOI 10.1007/978-3-319-28028-8_9

intermittency [5, 6], noise-induced hopping [7, 8] and so on. It is well accepted now that the interplay between the global structure of a deterministic nonlinear system and the noise determines the final forms of noise-induced responses. The global structure of the deterministic nonlinear system includes attractors and unstable invariant sets as well as their stable and unstable manifolds, and usually sets a basic frame for noise-induced responses. Depending upon the noise intensity, the noise-induced responses may completely follow the frame of the structure or alter it by inducing connection or collision within its sub-structures, which can induce bifurcations quite similar to its deterministic counterparts.

An interesting classification on the bifurcations in deterministic (dissipative) non-linear dynamical systems is presented in [9], where three categories are defined as safe, explosive and dangerous bifurcations. The explosive bifurcations are defined as catastrophic global bifurcations with an abrupt enlargement of the attracting set but with no jump to a remote disconnected attractor. The dangerous bifurcations are catastrophic bifurcations with blue-sky disappearance of the attractor with a sudden fast jump to a distant unrelated attractor. Undoubtedly, these two kinds of bifurcations have very important engineering meaning since they imply an abrupt and large shift of the operation state in a machine or system with the continuous variation of a parameter that may even induce possible damage or destruction on it.

Since uncertain disturbance is generally unavoidable in the real engineering environment, it is thus of great interest to exploit if noise-induced bifurcations, like their deterministic counterparts, may occur, possibly even at the system parameters that are far from the bifurcation points of the corresponding deterministic nonlinear dynamical system. It is also crucial to predict when such bifurcations, if they exist, may take place when the noise intensity works as the bifurcation parameter, and to determine how the transient responses of the noise-induced large transitions evolve when such bifurcations occur. The present work will tackle the problems.

As known when a dynamical system is under excitation of Gaussian white noise, the probabilistic description of the stochastic responses is governed by the Fokker-Planck-Kolmogorov (FPK) equation. It is also known that it is quite difficult to solve FPK equation analytically even for quite simple stochastic systems. For the case of weak noise, the probability density can be well approximated through quasipotential [10]. On basis of this method, the stochastic sensitivity function (SSF) was proposed in [11] and successfully applied to analyze the sensitivity of stationary points and periodic cycles in differential dynamical systems as well as that of fixed points and periodic solutions in discrete dynamical systems [12]. Compared with other FPK equation-based methods, SSF is useful and easier for obtaining an approximate analytical description of the probability distribution. However, when the sensitivity of periodic attractors in non-autonomous dynamical systems under external and/or parametric excitations is analyzed, a boundary value problem of matrix differential equations must be solved. In this work a method that discretizes a periodic attractor of a non-autonomous nonlinear dynamical system into a discrete map by $1/N$-period stroboscopic map is proposed, and so only the matrix algebra equations need to be solved in order to get the stochastic sensitivity function of the periodic attractor. With the stochastic sensitivity functions, confidence ellipses along every sections of the periodic attractor at each $1/N$-period interval can be constructed for a given fiducial probability and used to depict the distribution ranges of the stochastic attractor in the state space. Furthermore, by combining the confidence ellipses and the deterministic global structure of the system, the critical conditions for the occurrence of noise-induced transition phenomena under a given fiducial probability can be estimated, for instance, when a confidence ellipse starts to touch the manifolds of a certain saddle-typed invariant set [13].

For the case with large noise intensity, Monte-Carlo simulation (MCS) is a well-known method to get the distribution of a stochastic attractor, and can be applied in a straightforward manner and to different kinds of stochastic systems. But it is too expensive in computations to carry out a systematic investigation. Thus, several approximated numerical methods on solving FPK equation have been developed, including Finite Element Method [14], path integral method [15, 16], and Generalized Cell Mapping method (GCM) [17, 18]. Like many other numerical methods for stochastic analysis, computation efficiency is still a crucial problem that GCM needs to be solved with effort.

In this study, a method is proposed in order to enhance the computation efficiency of GCM when the problem that the noise-induced response with the initial probability distribution focusing around a given deterministic attracting set is dealt with. From a practical point of view, the given deterministic attracting set may represent the designed operating state of an engineering system and the working state will be adjusted with effort just near it in the real engineering environment. So the evolution of a stochastic response that starts from an initial probability distribution occupying only a very small region in the state space and gradually converges to a stochastic steady-state occupying a certain region in the state space within a given fiducial probability is of interest. Thus, the traditional GCM that deals with a priori defined sufficiently large chosen region in the state space is not quite efficient for the problem. Therefore, an idea of evolving probabilistic vector (EPV) is introduced in this work. By using EPV, only the one-step transition probability of the cells in the chosen region, whose probabilities are within a given fiducial probability, will be calculated, instead of all the cells within the chosen region in the state space. In this way, the dimension of the probabilistic vector in the present GCM method (GCM with EPV), which varies with the evolution of the stochastic response, is greatly reduced and usually much smaller than that of the corresponding fix-sized probabilistic vector.

This work is organized as follows: In Sect. 9.2, we first briefly introduce the concept of stochastic sensitivity function (SSF) and then present the algorithm to obtain SSF of periodic attractors in non-autonomous nonlinear system by constructing a $1/N$-period stroboscopic map. In Sect. 9.3, we briefly introduce the GCM method with short-time Gaussian and then describe the idea of evolving probabilistic vector and devise the corresponding algorithm. In Sect. 9.4, we apply the proposed methods to a Mathieu-Duffing oscillator under parametric and external excitation as well as additive noise to study the transient behavior of the noise-induced explosive and dangerous bifurcations. Finally, the conclusions are drawn in Sect. 9.5.

9.2 An Analytical Method for Stochastic Responses to Weak Noise

9.2.1 Concept of Stochastic Sensitivity Function

Consider a continuous noise-disturbed dynamical system

$$\dot{\vec{x}} = \vec{f}(\vec{x}) + \varepsilon \vec{\sigma}(\vec{x})\vec{\xi}(t) \tag{9.1}$$

where $\vec{\xi}$ is n-dimensional Gaussian white noise, $\vec{\sigma}$ is $n \times n$ matrix which defines the relation between the noise and the system state, ε is the noise intensity.

The stationary probability density function ρ of a stochastic attractor for system (Eq. 9.1) obeys the steady-state FPK equation

$$\frac{\varepsilon^2}{2} \sum_{i,j=1}^{n} \frac{\partial^2 [\vec{\sigma}\vec{\sigma}^T]_{ij} \rho}{\partial x_i \partial x_j} - \sum_{i=1}^{n} \frac{\partial f_i \rho}{\partial x_i} = 0 \tag{9.2}$$

when the limit $\varepsilon \to 0$ is taken, the Wentzel-Kramers-Brillouin approximation [19] of ρ can be made

$$\rho \approx K exp\left(-\frac{\Phi(\vec{x})}{\varepsilon^2}\right) \tag{9.3}$$

where $\Phi(\vec{x})$ is the so-called quasi-potential [20] and K is the normalizing factor. In a small neighborhood of a stationary point or cycle, the first approximation of quasi-potential is made and ρ takes the form as a Gaussian distribution [10]

$$\rho \approx K exp\left(-\frac{\frac{1}{2}\Delta(\vec{x})^T \vec{W}^+(\vec{x})\Delta(\vec{x})}{\varepsilon^2}\right) \tag{9.4}$$

with covariance matrix $\varepsilon^2 \vec{W}$, while $\Delta(\vec{x})$ is the deviation of point x from the deterministic attractor. So matrix $\vec{W}$ characterizes the spatial arrangement and size of the stochastic attractor around the corresponding deterministic attractor and is defined as stochastic sensitivity function (SSF) of the deterministic attractor, while $\vec{W}^+$ is its pseudo-inverse matrix (If $\vec{W}$ is full-rank, using its inverse matrix $\vec{W}^-$ instead). Through eigenvalue or singular value decomposition of $\vec{W}$, the most and least sensitive directions and the degree of sensitiveness in these directions of every point in the attractor can be obtained.

Similarly, SSF can also be utilized to analyze the sensitivity of attractors in discrete stochastic dynamical systems [12], which will help to get SSF in non-autonomous dynamical system in next section.

9.2.2 Stochastic Sensitivity Function of Periodic Attractors in the Non-autonomous Systems

Consider a continuous non-autonomous dynamical system with the form

$$\dot{\vec{x}} = \vec{f}(\vec{x}, t) \tag{9.5}$$

where $\vec{f}(\vec{x}, t)$ is n-dimensional non-autonomous vector field depending on both state $\vec{x}$ and time t.

When there is a periodic attractor with period T in non-autonomous system (Eq. 9.5), a stroboscopic map is often used to investigate the character of the attractor (a cycle in the state space). Let $\varphi_t(\vec{x})$ be the flow generated by the vector field of Eq. (9.5), and denote the periodic attractor by $\varphi_t(\vec{x}_0)$ where $\vec{x}_0$ is any point passed by the cycle when $t = t_0$. So, $\varphi_{nT}(\vec{x}_0) = \vec{x}_0$. Taking a snapshot of the flow at discrete times $t = t_0 + k\Delta t$ (k is positive integer), a continuous time trajectory $\vec{x}(t)$ is divided into a discrete time trajectory $\vec{x}_k$. Point $\vec{x}_{k+1}$ is determined only by $\vec{x}_k$ through integrating (Eq. 9.5) forward with time Δt from $\vec{x}_k$. So a map can be defined as

$$\vec{x}_{k+1} = \varphi_{\Delta t}(\vec{x}_k) \tag{9.6}$$

which is called stroboscopic map. And sections

$$\Sigma_k = \{(\vec{x}, t) \in R^n \times R \mid t = t_0 + k\Delta t\} \tag{9.7}$$

are stroboscopic sections.

Traditionally, the sampling time interval Δt will be chosen to coincide with period of $\varphi_t(\vec{x}_0)$, say, $\Delta t = T$, and recall that $\varphi_{nT}(\vec{x}_0) = \vec{x}_0$, it can be seen that a periodic attractor in Eq. (9.5) forms a fixed point in map Eq. (9.6) and all the stroboscopic sections Σ_k share the same character. So it is regarded as only one section and the subscript k can be omitted. In the following text this kind of stroboscopic map is called 1-period stroboscopic map. The sensitivity analysis of any point $\vec{x}_0$ in the cycle of a periodic attractor can be used to calculate SSF of the fixed point in 1-period stroboscopic map, whose stroboscopic section contains $\vec{x}_0$. However, though the algorithm to get SSF of fixed point of maps is raised in [12], for most of the nonlinear dynamical systems, the explicit expression of 1-period stroboscopic map cannot be obtained.

Note that, if $\Delta t \to 0$, the linear approximation of map Eq. (9.6) can be taken in the interval $[t_0 + k\Delta t, t_0 + (k + 1)\Delta t]$

$$\vec{x}_{k+1} = exp(\vec{J}_k \Delta t)\vec{x}_k \tag{9.8}$$

where $\vec{J}_k = \frac{\partial \vec{f}}{\partial \vec{x}}\big|_{x=x_k, t=t_0+k\Delta t}$ is Jacobian matrix of Eq. (9.5) at point $\vec{x}_k$ and time $t_0 + k\Delta t$.

So, the sampling time interval Δt of stroboscopic map can be set to

$$\Delta t = T/N, \quad N \gg 1 \tag{9.9}$$

and a new stroboscopic map can be written in the form Eq. (9.8). This new map is named a $1/N$-period stroboscopic map. Because of the period-N character, let $\Sigma_k = \Sigma_{k+N}$, there are only N stroboscopic sections to be considered. Through this new map, the original periodic attractor Γ in Eq. (9.5) is discretized into a period-N cycle $\Gamma^* = \{\vec{x}_1, ..., \vec{x}_N\}$ by N stroboscopic sections $\{\Sigma_1, ..., \Sigma_N\}$.

Now consider system Eq. (9.5) subject to stochastic disturbance

$$\dot{\vec{x}} = \vec{f}(\vec{x}, t) + \varepsilon \vec{\sigma}(\vec{x})\vec{\xi}(t) \tag{9.10}$$

The $1/N$-period stroboscopic map of system Eq. (9.10) is written as

$$\vec{x}_{k+1} = exp(\vec{J}_k \Delta t)\vec{x}_k + \varepsilon \vec{\sigma}(\vec{x}_k)\Delta \vec{w} \tag{9.11}$$

where

$$\Delta \vec{w} = \sqrt{\Delta t}\vec{\xi}$$

is an increment of Wiener process during time interval $[t_0 + k\Delta t, t_0 + (k+1)\Delta t]$. So Eq. (9.10) is discretized into a map disturbed by noise with intensity $\varepsilon\sqrt{\Delta t}$.

According to [12], if the deterministic period-N cycle in Eq. (9.8) is an attractor, it is always exponentially stable. The SSF of period point $\vec{x}_k$ satisfies:

$$\vec{W}_{k+1} = \vec{F}_k \vec{W}_k \vec{F}_k^T + \vec{Q} \tag{9.12}$$

where

$$\vec{F}_k = exp(\vec{J}_k \Delta t), \vec{Q} = \vec{\sigma}\vec{\sigma}^T \tag{9.13}$$

For period-N cycle, it is obvious that:

$$\vec{W}_k = \vec{W}_{k+N} \tag{9.14}$$

Without loss of generality, let $k = 1$, based on Eq. (9.12), it can be deduced that:

$$\vec{W}_{1+N} = \vec{B}\vec{W}_1\vec{B}^T + \overline{\vec{Q}} \tag{9.15}$$

where

$$\begin{cases} \vec{B} = \vec{F}_N \vec{F}_{N-1}...\vec{F}_2 \vec{F}_1 \\ \overline{\vec{Q}} = \vec{Q} + \vec{F}_N \vec{Q}\vec{F}_N^T + ... + \vec{F}_N...\vec{F}_2 \vec{Q}\vec{F}_2^T...\vec{F}_N^T \end{cases} \tag{9.16}$$

According to Eq. (9.14), $\vec{W}_1$ can be calculated using matrix equation:

$$\vec{W}_1 = \vec{B}\vec{W}_1\vec{B}^T + \overline{\vec{Q}} \tag{9.17}$$

and $\vec{W}_2, \vec{W}_3, \dots, \vec{W}_N$ can then be calculated by the recurrence relation described by Eq. (9.12).

After $\vec{W}_k$ is calculated, a confidence ellipse that represents the spatial distribution of stochastic states concentrated near point $\vec{x}_k$ in stroboscopic section Σ_k can be obtained using the following equation:

$$(\vec{x} - \vec{x}_k)^T \vec{W}_k^{-1}(\vec{x} - \vec{x}_k) = \varepsilon^2 \Delta t \chi_P^2 \tag{9.18}$$

where χ_P^2 computes the inverse of n-dimensional Chi-square cumulative distribution function with fiducial probability density P_f, with which the points in the stochastic attractor are contained in the ellipse. Specially, when $n = 2$, namely, for two-dimensional system, $\chi_P^2 = -2\ln(1 - P_f)$. In Eq. (9.18) the SSF, $\vec{W}_k$, also relies on time interval Δt. When $\Delta t \to 0$, the confidence ellipse described by Eq. (9.18) will converges [21].

9.2.3 Estimation of Critical Conditions for Noise-Induced Transition

As it is known, under the (small) stochastic disturbances or noise, random trajectories will leave deterministic attractors and form probabilistic distribution around them. That is, the system is expected to spend most of its time in the vicinity of one of the stable states. However, when the intensity of noise exceeds some critical values, the large noise-induced transition phenomena will be detected and the stochastic response probability will no longer distribute around the corresponding deterministic attractors. It is found that the collision between the stochastic response around a deterministic attractor and the unstable manifolds of a chaotic saddle in the basin of attraction of the attractor will cause an abrupt increase of probabilistic distribution of the stochastic response and form the so-call noise-induced chaos with positive Lyapunov exponents [22]. It is also well known that the system will escape from a basin of attraction following an optimal escape path during the activation process [23]. For the purpose of the present work, the noise-induced transition phenomena with a given fiducial probability near 1 are of interest. So the critical noise intensity under a given fiducial probability is estimated by using the knowledge from the confidence ellipses and the global structure of the deterministic system.

To judge if a stochastic attractor around a deterministic attractor will evolve into a noise-induced chaos by colliding with a chaotic saddle in its basin of attraction, a critical noise intensity is estimated, from Eq. (9.18), as

$$\varepsilon_c = \sqrt{\min \frac{(\vec{x}_{cs} - \overline{\vec{x}}_k)^T \vec{W}_k^{-1}(\vec{x}_{cs} - \overline{\vec{x}}_k)}{2\Delta t \ln(1 - P_f)}} \tag{9.19}$$

where $\vec{x}_{cs}$ denotes the points on the chaotic saddle, min stands for minimization.

It is relatively difficult to estimate the critical noise intensity for a given fiducial probability by which system will escape from the stable states across the boundary of their attraction basin. It is found, also see Figs. 9.6 and 9.7 below, that the optimal escape path with dominant probability is usually around the unstable manifolds of the saddle on the boundary of the attraction basin. Thus, it is important to understand why the activation process can start, during which trajectories will leave the stable states and escape to the saddle on the boundary of attraction basin along the direction of its unstable manifold under noise. We believe that the global structure of the corresponding deterministic nonlinear system might provide some useful information, besides a boundary problem needs to be solved with challenge to get the activation energy [23].

9.3   An Efficient Method for Transient Stochastic Responses

9.3.1   Generalized Cell Mapping with Short-Time Gaussian Approximation

The response of a N-dimensional nonlinear system subjected to additive and/or multiplicative Gaussian white noise excitations is well known to be a diffusion Markov process. Based on the Generalized Cell Mapping method (GCM), the continuous state spaces R^N is discretized into a cell space with a countably infinite number of hyper-cubes that are called *cells*. The cells used to cover a pre-defined chosen region in the state space are finite and called *global cells*, which will be indexed by integers from 1 to N. The probability evolution of the stochastic system is described by a homogeneous Markov chain in the cell space as

$$\boldsymbol{P} \cdot \boldsymbol{p}(n) = \boldsymbol{p}(n+1) \tag{9.20}$$

where $\boldsymbol{p}(n)$ denotes the probabilistic vector describing the probability of each cell at nth step, and $\boldsymbol{P}$ the one-step transition probability matrix of the stochastic system. Let $p_i(n)$ be the ith element of $\boldsymbol{p}(n)$ that indicates the probability of the response in ith cell at n-step mappings. Let P_{ij} be the (i, j)th element of $\boldsymbol{P}$ that describes the probability in ith cell after *one*-step mapping from jth cell. P_{ij} and $p_i(n)$ can be determined by the following formulae

$$P_{ij} = \int_{C_i} p(\boldsymbol{x}, t | \boldsymbol{x}_j, t_0) d\boldsymbol{x} = \int_{C_i} p(\boldsymbol{x}, \tau | \boldsymbol{x}_j, t_0) d\boldsymbol{x}, \qquad p_i(n) = \int_{C_i} p(\boldsymbol{x}, n\tau) d\boldsymbol{x} \tag{9.21}$$

where $\tau = t - t_0$ denotes a mapping time step; C_i is the domain occupied by ith cell in R^N, and $p(\boldsymbol{x}, \tau | \boldsymbol{x}_j, t_0)$ and $p(\boldsymbol{x}, n\tau)$ represent the one-step transition probability and the probability under n-steps mapping in R^N, respectively.

Λ Gauss-Legendre quadrature is applicable to estimate the above integral in domain C_i because the domain is finite and the weight function is one. This means that probabilities in ith cell are discretely expressed by that at Gauss quadrature points in the cell. Therefore, based on this rule

$$P_{ij} = \sum_{k=1}^{S_i} A_k p(x^k, \tau | x_j, t_0), \qquad p_i(n) = \sum_{k=1}^{S_i} A_k p(x^k, n\tau) \qquad (9.22)$$

where x_j is the geometrical center of jth cell; x^k is the kth Gauss quadrature point, namely, the zeros of the Legendre polynomials of chosen order; S_i is the number of Gauss quadrature points in ith cell, and A_k is the quadrature factor. The advantages of this method are that appropriate accuracy and high efficiency can be achieved since only Gauss points, which are of unequal division characteristics, are chosen, instead of large number of random points within each cell, to determine the one-step transition probability matrix.

To release the difficulty of huge time-consumption in solving nonlinear stochastic equations based on sampling methods, like straightforward MCS to estimate the one-step transition probability matrix P_{ij}, a short-time Gaussian approximation approach proposed in [24] is adopted. This method considers the fact that $p(x, \tau | x_j, t_0)$ is approximately Gaussian when an additive and/or a multiplicative noise excitation is applied as Gauss white noises and the mapping time-step τ is sufficiently small. The distribution can be approximately specified by the mean and the variance. The moment evolution equations can be derived by applying the Itô calculus and the Gaussian closure or higher order cumulant-neglect closure methods for an appropriate accuracy [25, 26]. The first and second order moments of $x(t)$ need to be evaluated by integrating moment equations from $t = 0$ to $t = \tau$.

In the simulation, only the geometrical center of each cell is taken as the initial mapping point, instead of all inner points in a cell. For an N-dimensional stochastic system, P_{ij} is expressed as

$$P_{ij} = \sum_{k=1}^{S_i} \frac{1}{(2\pi)^{N/2} |B(\tau)|^{1/2}} exp \left\{ -\frac{1}{2} \left[x^k - m(\tau) \right]^T B(\tau)^{-1} \left[x^k - m(\tau) \right] \right\}$$

$$(9.23)$$

where $m(\tau)$ represents the short-time mean value vector and $B(\tau)$ the short-time covariance matrix at time τ, which can be solved from the moment evolution equations.

9.3.2 Generalized Cell Mapping with Evolving Probabilistic Vector

For many practical problems with noise-induced transitions phenomena, more attentions are focused on probability evolution initially localized around a given deterministic attracting set. In this case, most of the response realizations are concentrated within a finite local region in the state space, especially by considering a prescribed fiducial probability P_f. Thus, the cells within the chosen region in the state space need not be treated equally. Borrowing the idea from Point Mapping under Cell Reference method [27, 28], the cells in the chosen region will be classified into active cells and inactive cells. An *active cell* represents the cell whose probability density function (PDF) is within the prescribed fiducial probability, and an *inactive cell* is the cell whose PDF is outside the prescribed fiducial probability, as shown in Fig. 9.1. The cells outside the chosen region are defined as a *sink cell*, which is assumed to have a zero probability.

Traditionally, P in GCM is a matrix of dimension $N \times N$ with (i, j)th element of P_{ij} being the transition probability from jth cell to ith cell. According to the Markov chain formulation Eq. (9.20), if the probability $p_r(n)$ of rth cell in the probabilistic vector $p(n)$ is too small (whose probability is outside prescribed fiducial probability), the probability of its image cells, say ith cell denoted by $p_i(n + 1)$, will be almost close to zero. So the corresponding one-step transition probability P_{ir} in P need not to be computed and stored. The cells, like rth cell, are the inactive cells and can be neglected in the computation of the short-time mapping, that is, $P_{ir} p_r = 0$ when rth cell is an inactive cell (see Fig. 9.1b).

So the probabilistic vector $p(n)$ in the present work is no longer a vector with a fixed length N as in the traditional GCM, rather its length will vary and equal the number of active cells at nth-step mapping evolving with the system response from the initial probability distribution. Let r denotes the inactive cell whose probability is outside the fuducial probability, then the evolving probabilistic vector is governed by

$$\begin{cases} P_{ij} p_j(n) = 0 & j = r \\ P_{ij} p_j(n) = p_i(n + 1) & j \neq r \end{cases} \quad j = 1, 2, 3, ..., N \qquad (9.24)$$

The fiducial probability in each mapping is expressed by

$$P_f = \sum_i \sum_{j \neq r} P_{ij} p_j(n) \qquad (9.25)$$

In this way, a Generalized Cell Mapping method with an evolving probabilistic vector, or GCM with EPV in short, is set up, by which only the active cells, whose number is much less than that of the global cells N, are involved in the computation of one-step transition probability matrix. So the computation and the storage of a very large fixed-size one-step transition probability matrix are avoided, and both

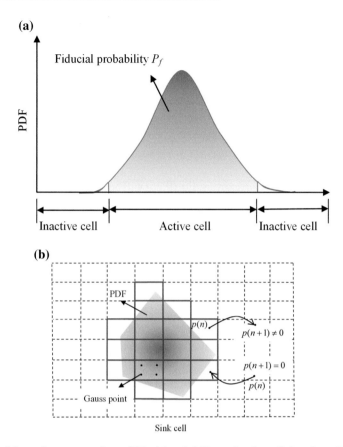

Fig. 9.1 Schematic representations of fiducial probability and active cell, inactive cell, sink cell: **a** fiducial probability; **b** active cells (*red*), inactive cells and sink cell

computation consumption and memory storage are much more reduced to make the method even more suitable for investigation of high-dimensional systems. Furthermore, the computational efficiency is seldom influenced by extending of chosen region while using the same cell size, which is very useful for detecting the transient-state responses and large transition problems in stochastic systems as shown by the example below.

9.4 Noise-Induced Bifurcations: An Example of Application

To demonstrate the capability of the proposed method in efficiently capturing transient responses in noise-induced bifurcations, a Mathieu-Duffing oscillator under excitations of additive noises, which has the following form, is studied:

$$\frac{d^2x}{dt^2} + 25x^3 + 0.173\frac{dx}{dt} + [2.62 - 0.456S(1 - \cos 2t)]x = 0.92S(1 - \cos 2t) + \varepsilon w(t)$$
$$(9.26)$$

where S is a bifurcation parameter, ε is the noise intensity and $w(t)$ is Gaussian with its mean and correlation function satisfying

$$\begin{cases} E[w(t)] = 0 \\ E[w(t)w(t+\tau)] = \delta(\tau) \end{cases} \qquad (9.27)$$

For the purpose of stochastic analysis, Eq. (9.26) will be first converted to a set of stochastic differential equation in the Itô sense, following rules of Itô and Stratonovich stochastic calculus,

$$\begin{cases} \frac{dx_1}{dt} = x_2 \\ \frac{dx_2}{dt} = -25x_1^3 - 0.173x_2 - 2.62x_1 + 0.456S(1 - \cos 2t)x_1 \\ \quad + 0.92S(1 - \cos 2t) + \varepsilon w(t) \end{cases} \qquad (9.28)$$

9.4.1 Dynamics of the Deterministic System

When $\varepsilon = 0$, Eq. (9.28) becomes a deterministic nonlinear system under both forced and parameter excitations. The bifurcation diagram of Eq. (9.28) with the variation of S is drawn in Fig. 9.2. As can be seen, there are two response branches in the parameter range $S \in [3.3, 4.4]$. The upper response branch in Fig. 9.2 consists of a period-doubling cascade to chaos, a period-3 window, another period-doubling cascade to chaos and ends with a boundary crisis at about $S = 4.26$. The lower response branch is a period-1 motion starting from about $S = 3.46$ though saddle-node bifurcation. According to the classification in [9], dangerous bifurcation occurs at $S = 4.26$ when increasing S on the upper branch, and at $S = 3.46$ when decreasing S following the lower branch.

To see the global structure within the parameter range with the coexistence of period-3 and period-1 attractor, the two-scaled global analysis method [27, 28] is employed to determine the attractors, saddles and their manifolds. Figure 9.3 shows the global structure at $S = 3.80$.

9.4.2 Transient Responses in Noise-Induced Bifurcations

Below the transient responses in noise-induced bifurcations are studied by the proposed method, GCM with EPV, in case that $\varepsilon \neq 0$.

Let $m_{pq} = E[x^p \dot{x}^q]$, based on the short-time Gaussian approximation, the moment evolution equations for Eq. (9.28) are derived by applying Gaussian closure as

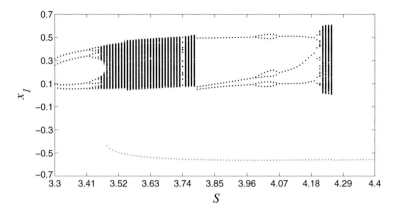

Fig. 9.2 Bifurcation diagram of the deterministic Mathieu-Duffing oscillator with the variation of S

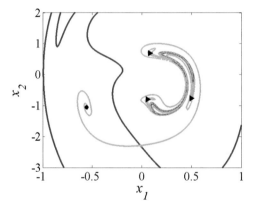

Fig. 9.3 Global structure of deterministic Mathieu-Duffing oscillator when $S = 3.80$, where a *black dot* stands for period-1 (P-1) attractor, *three black triangles* for period-3 (P-3) attractor, *dark blue points* for chaotic saddle, and *green dot* for a saddle point on basin boundary with its stable and unstable manifolds being the *red* and *light blue solid curves* respectively

$$
\begin{cases}
\dot{m}_{10} = m_{01} \\
\dot{m}_{01} = -25(3m_{20}m_{10} - 2m_{10}^3) - 0.173m_{01} - [2.62 - 0.456S(1 - \cos 2t)]m_{10} \\
\qquad + 0.92S(1 - \cos 2t) \\
\dot{m}_{20} = 2m_{11} \\
\dot{m}_{11} = m_{02} - 25(3m_{20}^2 - 2m_{10}^4) - 0.173m_{11} - [2.62 - 0.456S(1 - \cos 2t)]m_{20} \\
\qquad + 0.92S(1 - \cos 2t)m_{10} \\
\dot{m}_{02} = -50(3m_{20}m_{11} - 2m_{10}^3m_{01}) - 0.346m_{02} - [5.24 - 0.912S(1 - \cos 2t)]m_{11} \\
\qquad + 1.84S(1 - \cos 2t)m_{01} + \varepsilon^2
\end{cases}
\tag{9.29}
$$

By using the proposed GCM with EPV, the chosen region of x and $\dot{x}$ in the state space is taken to be $[-1.0, 1.0] \times [-3.0, 2.0]$, which will be covered by 500×500

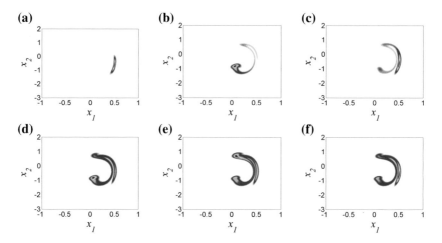

Fig. 9.4 Noise-induced explosive bifurcation: pseudo-color images of the evolving PDFs of system Eq. (9.28) when $S = 3.80$ and $\varepsilon = 0.05$ from initial PDF around $(0.5, -1)$: **a** $t = T$; **b** $t = 4T$; **c** $t = 6T$; **d** $t = 28T$; **e** $t = 29T$; **f** $t = 30T$

cells with the resolution of 0.004×0.012. The fiducial probability $P_f = 0.95$ is used in all of the computations below. The noise intensity is taken as the bifurcation parameter to see when the noise-induced bifurcations occur, what kind of noise-induced bifurcation takes place and how the transient responses of the system evolve.

It is not hard to imagine that the stochastic responses will mainly concentrate around the deterministic attractors when the noise intensity is sufficient small. For instance, when $\varepsilon = 0.02$, the stationary PDFs of Eq. (9.28) at $S = 3.80$ will mainly concentrate within a small vicinity around attractor P-1 or P-3, depending upon in which basin of attraction the initial PDF locates. In this case no noise-induced bifurcation occurs and the stochastic responses can be well estimated by the stochastic sensitivity function technique introduced in the above section.

Noise-induced explosive bifurcation: As the noise intensity increases, say to $\varepsilon = 0.05$, while no qualitative change takes place on the stochastic responses around P-1 attractor, the stochastic responses around P-3 attractor have a significant qualitative change, by which the most of response realizations start to escape from the vicinity of P-3 attractor and evolve along the chaotic saddle and then go back to P-3 attractor repeatedly. Figure 9.4 shows a transient evolution of the response PDF from the initial PDF around point $(0.5, -1)$ that locates in the basin of attraction of P-3 attractor (see Fig. 9.3). It is found that the PDF of the responses in this case gradually goes out of the region of the initial PDF and evolves on a region covering both P-3 attractor and chaotic saddle only after 6 time periods (see Fig. 9.4c). The stationary PDF of the stochastic responses undertakes a recurrent transition in a periodic way in period three as shown by Fig. 9.4d–f. Since the stochastic response increases the response region abruptly through the connection of P-3 attractor and the chaotic saddle when the noise intensity exceeds certain critical value, we will call the phenomenon a

Fig. 9.5 Stationary PDFs when $\varepsilon = 0.10$: the initial PDF lays around $(-0.5, -1)$ for P-1 attractor, and around $(0.5, -1)$ for P-3 attractor

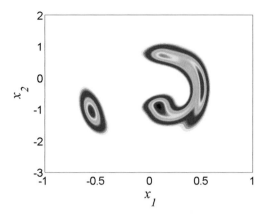

noise-induced explosive bifurcation in accordance to the deterministic definition in [9].

If the noise intensity is further increased to $\varepsilon = 0.10$, the stationary PDF of stochastic responses starting from an initial PDF in the basin of attraction of P-3 attractor will lose the periodic recurrent feature and become a noise-induced chaos as shown by Fig. 9.5. Meanwhile, the PDF of other stochastic responses around the coexisting period-1 attractor is unchanged when the initial PDF is put in its basin of attraction as shown by Fig. 9.5.

Noise-induced dangerous bifurcation: As the noise intensity further increases, for instance, $\varepsilon = 0.15$ there will be only one steady-state PDF of the stochastic responses in system (9.25) when $S = 3.80$. Especially, when the initial PDF locates in the basin of attraction of P-1 in Fig. 9.3, the stochastic response is no longer in the region around the P-1 attractor but jumps to the region around the distant unrelated stochastic attractor around P-3 and chaotic saddle. Figure 9.6 shows the transient evolution of the PDF at the parameters. This suggests that a noise-induced saddle-node bifurcation occurs due to the collision between the stochastic P-1 attractor with a stochastic saddle on the basin boundary and causes the blue-sky disappearance of PDF for the stochastic responses around P-1 attractor. In accordance to the definition in [9], we will call it a *noise-induced dangerous bifurcation*.

We will show another type of noise-induced dangerous bifurcation, namely, noise-induced boundary crisis at the same noise intensities as above, that is, $\varepsilon = 0.15$. The only difference lies on that $S = 4.18$. This means that we shift the deterministic parameter a little closer to the bifurcation point of the deterministic boundary crisis that however still locates in the period-3 window (see Fig. 9.2). In this case the stochastic chaos will collide the stochastic saddle on the basin boundary and disappear abruptly. Figure 9.7 shows the transient evolution of the PDF at the parameters, by which the PDF evolves from the initial one in the basin of P-3 attractor with chaotic saddle as a noise-induced chaos response, and almost surely escapes from the region and goes into the region around P-1 attractor. In both of above cases, it is seen that the noise-induced transitions go across basin boundary through the stochastic saddle on

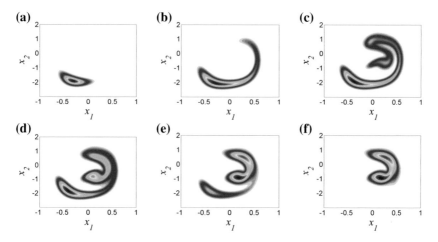

Fig. 9.6 Noise-induced dangerous bifurcation: pseudo-color images of the evolving PDFs of system Eq. (9.28) when $S = 3.80$ and $\varepsilon = 0.15$ from initial PDF around $(-0.5, -1)$: **a** $t = T$; **b** $t = 3T$; **c** $t = 5T$; **d** $t = 20T$; **e** $t = 35T$; **f** $t = 50T$

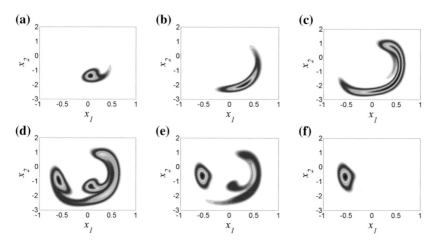

Fig. 9.7 Noise-induced dangerous bifurcation: pseudo-color images of the evolving PDFs of system Eq. (9.28) when $S = 4.18$ and $\varepsilon = 0.15$ from initial PDF around $(0.5, -1)$: **a** $t = T$; **b** $t = 3T$; **c** $t = 4T$; **d** $t = 10T$; **e** $t = 18T$; **f** $t = 50T$

it, which serves as a *bridge*. The noise-induced dangerous bifurcation has profound engineering meaning because it indicates that even though the system is designed far from the deterministic dangerous bifurcation points, the bifurcation can still be induced by noise.

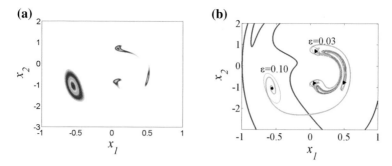

Fig. 9.8 **a** Stationary PDFs around P-1 attractor when $\varepsilon = 0.10$ with the initial PDF lays near $(-0.5, -1)$, and around P-3 attractor when $\varepsilon = 0.03$ with the initial PDF lays near $(0.5, -1)$. **b** Corresponding confidence ellipses (*dark red circles*) for the two attractors, with a fiducial probability $P_f = 0.95$

9.4.3 Estimation of the Critical Noise Intensity

As mentioned above that the sensitivity character of the periodic attractors under weak noise can be well determined through the method introduced in Sect. 9.2.2 (see also [21]). Figure 9.8a shows the stochastic attractors around period-1 attractor when $\varepsilon = 0.10$ and around period-3 attractor when $\varepsilon = 0.03$ calculated by GCM with EPV for a fiducial probability $P_f = 0.95$. Actually, the two noise intensities for the two different attractors give the approximate critical noise intensities for the occurrance of the noise-induced dangerous bifurcation around period-1 attactor and the noise-induced explosive bifurcation around period-3 attractor respectively. Figure 9.8b draws the corresponding confidence ellipses, with a fiducial probability $P_f = 0.95$, on the global structure of the deterministic system. It can be seen that the confidence ellipses for period-3 attractor at $\varepsilon = 0.03$ almost touch with the chaotic saddle in its basin of attraction. When $\varepsilon = 0.04$, the noise-induced explosive bifurcation is detectable by our numerical method. Depending upon the points depicting the chaotic saddle, formula (9.16) indicates the critical noise intensity should be in between 0.03 and 0.04. For the confidence ellipse around period-1 attractor with $\varepsilon = 0.10$, it is found that the ellipse just grows outside the spiral loop of the unstable manifold of the saddle around the attractor. By considering that the period-1 attractor possesses stronger capability of attraction within the spiral part of the unstable manifold, it is not hard to imagine that the stochastic trajectories must well get out of the restriction with the help of stronger noise before they can escape. When $\varepsilon = 0.11$, the noise-induced dangerous bifurcation is detectable with a fiducial probability $P_f = 0.95$ by our numerical method. So intuitively, the critical noise intensity for the occurrence of the noise-induced dangerous bifurcation is estimated inbetween 0.10 and 0.11. Certainly, there are still a lot of works to do in order to more accurately predict the critical noise intensity for the escape in a more rigorous way.

9.5 Conclusions

In practical engineering problems, the transient probability density function (PDF) of stochastic responses under noise excitation of an engineering system, starting from an initial distribution localizes near a given deterministic attracting set that represents its operating state, is of great interest. The sensitivity of the periodic attractors is analyzed by discretizing the non-autonomous system into a discrete $1/N$-period stroboscopic map in this work. In this way, boundary value problems of matrix differential equations were avoided by solving only matrix algebra equations. The method can provide analytical description on the sensitivity character and the probabilistic distribution of a stochastic attractor in the case of weak noise. In order to obtain the critical noise intensity of noise-induced transition phenomena, SSF is used to judge if the corresponding confidence ellipse is in touch with the manifolds of certain saddle-typed invariant sets or their manifolds.

In order to efficiently capture the transient behaviors in noise-induced bifurcations in the case of large noise, an idea of evolving probabilistic vector (EPV) is introduced into the Generalized Cell Mapping method (GCM) in this paper in the form of a Generalized Cell Mapping method with evolving probabilistic vector, or GCM with EPV. So much smaller evolving probabilistic vector and one-step probability transition matrix are used in the proposed method in comparison with the fix-sized one in the traditional GCM. Thus, the efficiency in both computation and storage is largely enhanced making the method applicable for high-dimensional systems.

A Mathieu-Duffing oscillator under excitation of additive noise is studied to show the noise-induced explosive and dangerous bifurcations of the oscillator. The transient and stationary PDFs are captured by the GCM with EPV either with an abrupt enlargement or a sudden fast jump of the probability distribution of the stochastic responses. The profound engineering meaning of the noise-induced dangerous bifurcation is emphasized because it indicates that even though the system is designed far from the deterministic dangerous bifurcation points, the bifurcation can still be induced by noise.

References

1. W.M. Zhang, O. Tabata, T. Tsuchiya et al., Phys. Lett. A **375**(32), 2903–2910 (2011)
2. T. Tél, Y.C. Lai, Phys. Rev. E **81**(056208), 1–8 (2010)
3. K. Malick, P. Marcq, Eur. Phys. J. D **36**, 119–128 (2003)
4. W. Xu, Q. He, T. Fang et al., Int. J. Bifurcat. Chaos **13**(10), 3115–3123 (2003)
5. K. Suso, F. Ulrike, Phys. D **181**, 222–234 (2003)
6. I. Bashkirtseva, L. Ryashko, Phys. A **392**, 295–306 (2013)
7. F.T. Arecchi, R. Badii, A. Politi, Phys. Rev. A **32**, 402–408 (1985)
8. K. Suso, F. Ulrike, Phys. Rev. E **66**(015207(R)), 1–4 (2002)
9. J.M.T. Thompson, H.B. Stewart, Y. Udea, Phys. Rev. E **49**(2), 1019–1027 (1994)
10. M. Freidlin, A.D. Wentzel, *Random Perturbations in Dynamical Systems* (Springer, New York, 1984)
11. I.A. Bashkirtseva, L.B. Ryashko, Dynam. Syst. Appl. **11**(2), 293–309 (2002)

12. I.A. Bashkirtseva, L.B. Ryashko, I.N. Tsvetkov, Dynamics of continuous. Discrete Impulsive Syst. Series A: Math. Anal. **17**, 501–515 (2010)
13. K.M. Guo, J. Jiang, Acta Phys. Sin. **63**(19), 190503 (2014) (in Chinese)
14. B.F. Spencer, L.A. Bergman, Nonlinear Dynam. **4**(4), 357–372 (1993)
15. M.F. Wehner, W.G. Wolfer, Phys. Rev. A **27**(5), 2663–2670 (1983)
16. M. Di Paola, R. Santoro, Probab. Eng. Mech. **24**(3), 300–311 (2009)
17. C.S. Hsu, J. Appl. Mech. **48**, 634–642 (1981)
18. J.Q. Sun, C.S. Hsu, J. Appl. Mech. **55**, 694–701 (1987)
19. R.V. Roy, J. Appl. Mech. **62**(2), 496–504 (1995)
20. R. Graham, T. Tél, Phys. Rev L **52**(1), 9–12 (1984)
21. K.M. Guo, J. Jiang, Phys. Lett. A **378**(34), 2518–2523 (2014)
22. T. Tél, Y.C. Lai, Int. J. Bifurcat. Chaos **18**(2), 509–520 (2008)
23. S. Beri, R. Mannella, D.G. Luchinsky et al., Phys. Rev. E **77**, 036131 (2005)
24. J.Q. Sun, C.S. Hsu, J. Appl. Mech. **57**, 1018–1025 (1990)
25. J.Q. Sun, C.S. Hsu, J. Appl. Mech. **54**, 649–655 (1987)
26. S.F. Wojtkiewicz, B.F. Spencer, L.A. Bergman, Int. J. Nonlin. Mech. **31**(5), 657–684 (1996)
27. J. Jiang, Theor. Appl. Mech. Lett. **1**, 063001 (2011)
28. J. Jiang, Chinese Phys. Lett. **29**(5), 050503 (2012)

Chapter 10
Analytical, Optimal, and Sparse Optimal Control of Traveling Wave Solutions to Reaction-Diffusion Systems

Christopher Ryll, Jakob Löber, Steffen Martens, Harald Engel and Fredi Tröltzsch

Abstract This work deals with the position control of selected patterns in reaction-diffusion systems. Exemplarily, the Schlögl and FitzHugh-Nagumo model are discussed using three different approaches. First, an analytical solution is proposed. Second, the standard optimal control procedure is applied. The third approach extends standard optimal control to so-called sparse optimal control that results in very localized control signals and allows the analysis of second order optimality conditions.

10.1 Introduction

Beside the well-known Turing patterns, reaction-diffusion (RD) systems possess a rich variety of self-organized spatio-temporal wave patterns including propagating fronts, solitary excitation pulses, and periodic pulse trains in one-dimensional media. These patterns are "building blocks" of wave patterns like target patterns, wave segments, and spiral waves in two as well as scroll waves in three spatial dimensions, respectively. Another important class of RD patterns are stationary, breathing, and moving localized spots [1–7].

Several control strategies have been developed for purposeful manipulation of wave dynamics as the application of closed-loop or feedback-mediated control loops with and without delays [8–11] and open-loop control that includes external spatio-temporal forcing [10, 12–14], optimal control [15–17], and control by imposed geometric constraints and heterogeneities on the medium [18, 19]. While feedback-mediated control relies on a continuous monitoring of the system's state, open-loop control is based on a detailed knowledge of the system's dynamics and its parameters.

Experimentally, feedback control loops have been developed for the photosensitive Belousov-Zhabotinsky (BZ) reaction. The feedback signals are obtained from wave activity measured at one or several detector points, along detector lines, or in

C. Ryll · F. Tröltzsch
Institut für Theoretische Physik, Technische Universität Berlin, 10623 Berlin, Germany

J. Löber · S. Martens (✉) · H. Engel
Institut für Theoretische Physik, Technische Universität Berlin, 10623 Berlin, Germany
e-mail: steffenmartens@win.tu-berlin.de

© Springer International Publishing Switzerland 2016   189
E. Schöll et al. (eds.), *Control of Self-Organizing Nonlinear Systems*,
Understanding Complex Systems, DOI 10.1007/978-3-319-28028-8_10

a spatially extended control domain including global feedback control [8, 9, 20]. Varying the excitability of the light-sensitive BZ medium by changing the globally applied light intensity forces a spiral wave tip to describe a wide range of hypocycloidal and epicycloidal trajectories [21, 22]. Moreover, feedback-mediated control loops have been applied successfully in order to stabilize unstable patterns in experiments such as unstable traveling wave segments and spots [11]. Two feedback loops were used to guide unstable wave segments in the BZ reaction along pre-given trajectories [23]. An open loop control was successfully deployed in dragging traveling chemical pulses of adsorbed CO during heterogeneous catalysis on platinum single crystal surfaces [24]. In these experiments, the pulse velocity was controlled by a laser beam creating a movable localized temperature heterogeneity on an addressable catalyst surface, resulting in a V-shaped pattern [25]. Dragging a one-dimensional chemical front or phase interface to a new position by anchoring it to a movable parameter heterogeneity was studied theoretically in [26, 27].

Recently, an open-loop control for controlling the position of traveling waves over time according to a prescribed *protocol of motion* $\vec{\phi}(t)$ was proposed that preserves simultaneously the wave profile [28]. Although position control is realized by external spatio-temporal forcing, i.e., it is an open-loop control, no detailed knowledge about the reaction dynamics as well as the system parameters is needed. We have already demonstrated the ability of position control to accelerate or decelerate traveling fronts and pulses in one spatial dimension for a variety of RD models [29, 30]. In particular, we found that the analytically derived control function is close to a numerically obtained optimal control solution. A similar approach allows to control the two-dimensional shape of traveling wave solutions. Control signals that realize a desired wave shape are determined analytically from nonlinear evolution equations for isoconcentration lines as the perturbed nonlinear phase diffusion equation or the perturbed linear eikonal equation [31]. In the work at hand, we compare our analytic approach for position control with optimal trajectory tracking of RD patterns in more detail. In particular, we quantify the difference between an analytical solution and a numerically obtained result to optimal control. Thereby, we determine the conditions under which the numerical result approaches the analytical result. This establishes a basis for using analytical solutions to speed up numerical computations of optimal control and serves as a consistency check for numerical algorithms.

We consider the following controlled RD system

$$\partial_t \vec{u}(\vec{x}, t) - \mathcal{D}\Delta\vec{u}(\vec{x}, t) + \vec{R}(\vec{u}(\vec{x}, t)) = \mathcal{B}\vec{f}(\vec{x}, t). \tag{10.1}$$

Here, $\vec{u}(\vec{x}, t) = (u_1(\vec{x}, t), \ldots, u_n(\vec{x}, t))^T$ is a vector of n state components in a bounded or unbounded spatial domain $\Omega \subset \mathbb{R}^N$ of dimension $N \in \{1, 2, 3\}$. $\mathcal{D}$ is an $n \times n$ matrix of diffusion coefficients which is assumed to be diagonal, $\mathcal{D} = \text{diag}(D_1, \ldots, D_n)$, because the medium is presumed to be isotropic. Δ represents the N-dimensional Laplace operator and $\vec{R}$ denotes the vector of n reaction kinetics which, in general, are nonlinear functions of the state. The vector of control signals $\vec{f}(\vec{x}, t) = (f_1(\vec{x}, t), \ldots, f_m(\vec{x}, t))^T$ acts at all times and everywhere within the spatial domain Ω. The latter assumption is rarely justified in experiments, where the application of control signals is often restricted to subsets of Ω. However, notable exceptions as, e.g., the already mentioned photosensitive BZ reaction, exist. Here, the light intensity is deployed as the control signal such that the control acts everywhere within a two-dimensional domain.

Equation (10.1) must be supplemented with an initial condition $\vec{u}(\vec{x}, t_0) = \vec{u}_0(\vec{x})$ and appropriate boundary conditions. A common choice are no-flux boundary conditions at the boundary $\Sigma = \partial\Omega \times (0, T)$, $\partial_n \vec{u}(\vec{x}, t) = \vec{0}$, where $\partial_n \vec{u}$ denotes the component-wise spatial derivative in the direction normal to the boundary $\Gamma = \partial\Omega$ of the spatial domain.

Typically, the number m of independent control signals in Eq. (10.1) is smaller than the number n of state components. We call such a system an *underactuated* system. The $n \times m$ matrix $\mathcal{B}$ determines which state components are directly affected by the control signals. If $m = n$ and the matrix $\mathcal{B}$ is regular, it is called a *fully actuated* system.

Our main goal is to identify a control $\vec{f}$ such that the state $\vec{u}$ follows a desired spatio-temporal trajectory $\vec{u}_d$, also called a desired distribution, as closely as possible everywhere in space Ω and for all times $0 \le t \le T$. We can measure the distance between the actual solution $\vec{u}$ of the controlled RD system Eq. (10.1) and the desired trajectory $\vec{u}_d$ up to the terminal time T with the non-negative functional

$$J(\vec{u}) = \|\vec{u} - \vec{u}_d\|^2_{L^2(Q)}, \tag{10.2}$$

where $\|\cdot\|^2_{L^2(Q)}$ is the L^2-norm defined by

$$\|\vec{h}\|^2_{L^2(Q)} = \int\limits_0^T\!\!\int\limits_\Omega d\vec{x}\, dt\, \left\{ \vec{h}(\vec{x}, t)^2 \right\}, \tag{10.3}$$

in the space-time-cylinder $Q := \Omega \times (0, T)$. The functional Eq. (10.2) reaches its smallest possible value, $J = 0$ if and only if the controlled state $\vec{u}$ equals the desired trajectory almost everywhere in time and space.

In many cases, the desired trajectory $\vec{u}_d$ cannot be realized exactly by the control, cf. Ref. [32], for examples. However, one might be able to find a control which enforces the state $\vec{u}$ to follow $\vec{u}_d$ as closely as possible as measured by J. A control $\vec{f} = \vec{f}$ is *optimal* if it realizes a state $\vec{u}$ which minimizes J. The method of optimal control views J as a constrained functional subject to $\vec{u}$ satisfying the controlled RD system Eq. (10.1).

Often, the minimum of the objective functional J, Eq. (10.2), does not exist within appropriate function spaces. Consider, for example, the assumption that the controlled state, obtained as a solution to the optimization problem, is continuous in time and space. Despite that a discontinuous state $\vec{u}$, leading to a smaller value for $J(\vec{u})$ than any continuous function, might exist, this state is not regarded as a solution to the optimization problem. Furthermore, a control enforcing a discontinuous state may diverge at exactly the points of discontinuity; examples in the context of dynamical systems are discussed in Ref. [32]. For that reason, the *unregularized* optimization problem, Eq. (10.2), is also called a singular optimal control problem. To ensure the existence of a minimum of J within appropriate function spaces and bounded control signals, additional (inequality) constraints such as bounds for the control signal can be introduced, cf. Ref. [33]. Alternatively, it is possible to add a so-called Tikhonov-regularization term to the functional Eq. (10.2) which is quadratic in the control,

$$J(\vec{u}, \vec{f}) = \|\vec{u} - \vec{u}_d\|^2_{L^2(Q)} + \nu \|\vec{f}\|^2_{L^2(Q)}. \tag{10.4}$$

The L^2-norm of the control $\vec{f}$ is weighted by a small coefficient $\nu > 0$. This term might be interpreted as a control cost to achieve a certain state $\vec{u}$. Since the control $\vec{f}$ does not come for free, there is a "price" to pay. In numerical computations, $\nu > 0$ serves as a regularization parameter that stabilizes the algorithm. For the numerical results shown in later sections, we typically choose ν in the range $10^{-8} \leq \nu \leq 10^{-3}$. While $\nu > 0$ guarantees the existence of an optimal control $\vec{f}$ in one and two spatial dimensions even in the absence of bounds on the control signal [33], it is not known whether Tikhonov-regularization alone also works in spatial dimensions N larger than two. Here, we restrict our investigations to one and two spatial dimensions. The presence of the regularization term causes the states to be further away from the desired trajectories than in the case of $\nu = 0$. Thus, the case $\nu = 0$ is of special interest. Naturally, the solution $\vec{u}$ for $\nu = 0$ is the closest (in the $L^2(Q)$-sense) to the desired trajectory $\vec{u}_d$ among all optimal solutions associated with any $\nu \geq 0$. Therefore, it can be seen as the limit of realizability of a certain desired trajectory $\vec{u}_d$.

In addition to the weighted L^2-norm of the control, other terms can be added to the functional Eq. (10.4). An interesting choice is the weighted L^1-norm such that the functional reads

$$J(\vec{u}, \vec{f}) = \|\vec{u} - \vec{u}_d\|^2_{L^2(Q)} + \nu \|\vec{f}\|^2_{L^2(Q)} + \kappa \|\vec{f}\|_{L^1(Q)}. \tag{10.5}$$

For appropriate values of $\kappa > 0$, the corresponding optimal control becomes sparse, i.e., it only acts in some localized regions of the space-time-cylinder, while it vanishes identically everywhere else. Therefore, it is also called *sparse control* or *sparse optimal control* in the literature, see Refs. [34–37]. In some sense, we can interpret the areas with non-vanishing sparse optimal control signals as the most sensitive areas of the RD patterns with respect to the desired control goal. A manipulation of the RD pattern in these areas is most efficient, while control signals applied in other regions have only weak impact. Furthermore, the weighted L^1-norm enables the analysis of solutions with a Tikhonov-regularization parameter ν tending to zero. This allows to draw conclusions about the approximation of solutions to unregularized problems by regularized ones.

In Sect. 10.2, we present an analytical approach for the control of the position of RD patterns in fully actuated systems. These analytical expressions are solutions to the unregularized ($\nu = 0$) optimization problem, Eq. (10.2), and might provide an appropriate initial guess for numerical optimal control algorithms. Notably, neither the controlled state nor the control signal suffering from the problems are usually associated with unregularized optimal control; both expressions yield continuous and bounded solutions under certain assumptions postulated in Sect. 10.2. In Sect. 10.3, we explicitly state the optimal control problem for traveling wave solutions to the Schlögl [1, 38] and the FitzHugh-Nagumo model [39, 40]. Both are well-known models to describe traveling fronts and pulses in one spatial dimension, solitary spots and spiral waves in two spatial dimensions, and scroll waves in three spatial dimensions [4, 10, 41, 42]. We compare the analytical solutions from Sect. 10.2 with a numerically obtained regularized optimal control solution for the position control of a traveling front solution in the one-dimensional Schlögl model in Sect. 10.3.3. In particular, we demonstrate the convergence of the numerical result to the analytical solution for decreasing values v. The agreement becomes perfect within numerical accuracy if v is chosen sufficiently small. Section 10.4 discusses sparse optimal control in detail and presents numerical examples obtained for the FitzHugh-Nagumo system. Finally, we conclude our findings in Sect. 10.5.

10.2   Analytical Approach

Below, we sketch the idea of analytical position control of RD patterns proposed previously in Refs. [28, 29]. For simplicity, we consider a single-component RD system of the form

$$\partial_t u(x,t) - D\partial_x^2 u(x,t) + R(u(x,t)) = f(x,t), \qquad (10.6)$$

in a one-dimensional infinitely extended spatial domain $\Omega = \mathbb{R}$. The state u as well as the control signal f are scalar functions and the system Eq. (10.6) is fully actuated. Usually, Eq. (10.6) is viewed as a differential equation for the state u with the control signal f acting as an inhomogeneity. Alternatively, Eq. (10.6) can also be seen as an expression for the control signal. Exploiting this relation, one simply inserts the desired trajectory u_d for u in Eq. (10.6) and obtains for the control

$$f(x,t) = \partial_t u_d(x,t) - D\partial_x^2 u_d(x,t) + R(u_d(x,t)). \qquad (10.7)$$

In the following, we assume that the desired trajectory u_d is sufficiently smooth everywhere in the space-time-cylinder Q such that the evaluation of the derivatives $\partial_t u_d$ and $\partial_x^2 u_d$ yields continuous expressions. We call a desired trajectory u_d *exactly realizable* if the controlled state u equals u_d everywhere in Q, i.e., $u(x,t) = u_d(x,t)$. For the control signal given by Eq. (10.7), this can only be true if two more conditions are satisfied. First, the initial condition for the controlled state must coincide with the initial state of the desired trajectory, i.e., $u(x,t_0) = u_d(x,t_0)$. Second, all boundary conditions obeyed by u have to be obeyed by the desired trajectory u_d as well. Because of $u(x,t) = u_d(x,t)$, the corresponding unregularized functional J,

Eq. (10.2), vanishes identically. Thus, the control f is certainly a control which minimizes the unregularized functional J and, in particular, it is optimal.

In conclusion, we found a solution to the unregularized optimization problem Eq. (10.2). The solution for the controlled state is simply $u(x, t) = u_d(x, t)$, while the solution for the control signal is given by Eq. (10.7). Even though we are dealing with an unregularized optimization problem, the control signal as well as the controlled state are continuous and bounded functions, provided that the desired trajectory u_d is sufficiently smooth in space and time.

Generalizing the procedure to multi-component RD systems in multiple spatial dimensions, the expression for the control reads

$$\vec{f}(x, t) = \mathcal{B}^{-1}(\partial_t \vec{u}_d(x, t) - \vec{\mathcal{D}} \Delta \vec{u}_d(x, t) + \vec{R}(\vec{u}_d(x, t))). \qquad (10.8)$$

Once more, the initial and boundary conditions for the desired trajectory $\vec{u}_d$ have to comply with the initial and boundary conditions of the state $\vec{u}$. Clearly, the inverse of $\vec{\mathcal{B}}$ exists if and only if $\vec{\mathcal{B}}$ is a regular square matrix, i.e., the system must be fully actuated. We emphasize the generality of the result. Apart from mild conditions on the smoothness of the desired distributions $\vec{u}_d$, Eq. (10.8) yields a simple expression for the control signal for arbitrary $\vec{u}_d$.

Next, we exemplarily consider the position control of traveling waves (TW) in one spatial dimension. Traveling waves are solutions to the uncontrolled RD system, i.e., Eq. (10.1) with $\vec{f} = \vec{0}$. They are characterized by a wave profile $\vec{u}(x, t) = \vec{U}_c(x - ct)$ which is stationary in a frame of reference $\xi = x - ct$ co-moving with velocity c. The wave profile $\vec{U}_c$ satisfies the following ordinary differential equation (ODE),

$$\vec{\mathcal{D}} \vec{U}_c''(\xi) + c \vec{U}_c'(\xi) - \vec{R}(\vec{U}_c(\xi)) = \vec{0}, \quad \xi \in \Omega \subset \mathbb{R}. \qquad (10.9)$$

The prime denotes differentiation with respect to ξ. Note that stationary solutions with a vanishing propagation velocity $c = 0$ are also considered as traveling waves. The ODE for the wave profile, Eq. (10.9), can exhibit one or more homogeneous steady states. Typically, the wave profile $\vec{U}_c$ approaches either two different steady states or the same steady state for $\xi \to \pm\infty$. This fact can be used to classify traveling wave profiles. Front profiles connect different steady states for $\xi \to \pm\infty$ and are found to be heteroclinic orbits of Eq. (10.9). Pulse profiles join the same steady state and are found to be homoclinic orbits [43]. Furthermore, all TW solutions are localized in the sense that their spatial derivatives of any order $m \geq 1$ decay to zero, $\lim_{\xi \to \pm\infty} \partial_\xi^m \vec{U}_c(\xi) = 0$.

We propose a spatio-temporal control signal $\vec{f}(x, t)$ which shifts the traveling wave according to a prescribed protocol of motion $\phi(t)$ while simultaneously preserving the uncontrolled wave profile $\vec{U}_c$. Correspondingly, the desired trajectory reads

$$\vec{u}_d(x, t) = \vec{U}_c(x - \phi(t)). \qquad (10.10)$$

Note that the desired trajectory is localized for all values of $\phi(t)$ because the TW profile $\vec{U}_c$ is localized. The initial condition for the state is $\vec{u}(x, t_0) = \vec{U}_c(x - x_0)$ which fixes the initial value of the protocol of motion as $\phi(t_0) = x_0$. Then, the solution Eq. (10.8) for the control signal becomes

$$\vec{f}(x, t) = \mathcal{B}^{-1}(-\dot{\phi}(t)\vec{U}'_c(x - \phi(t)) - \mathcal{D}\vec{U}''_c(x - \phi(t)) + \vec{R}(\vec{U}_c(x - \phi(t))))$$
$$(10.11)$$

with $\dot{\phi}(t)$ denoting the derivative of $\phi(t)$ with respect to time t. Using Eq. (10.9) to eliminate the non-linear reaction kinetics $\vec{R}$, we finally obtain the following analytical expression for the control signal

$$\vec{f}(x, t) = (c - \dot{\phi}(t))\mathcal{B}^{-1}\vec{U}'_c(x - \phi(t)) =: \vec{f}_{an}.$$
$$(10.12)$$

Remarkably, any reference to the reaction function $\vec{R}$ drops out from the expression for the control. This is of great advantage for applications without or with only incomplete knowledge of the underlying reaction kinetics $\vec{R}$. The method is applicable as long as the propagation velocity c is known and the uncontrolled wave profile $\vec{U}_c$ can be measured with sufficient accuracy to calculate the derivative $\vec{U}'_c$.

Being an open loop control, a general problem of the proposed position control is its possible inherent instability against perturbations of the initial conditions as well as other data uncertainty. However, assuming protocol velocities $\dot{\phi}(t)$ close to the uncontrolled velocity c, $\dot{\phi} \sim c$, the control signal Eq. (10.12) is small in amplitude and enforces a wave which is relatively close to the uncontrolled TW. Since the uncontrolled TW is presumed to be stable, the controlled TW might benefit from that and a stable open loop control is expected. This expectation is confirmed numerically for a variety of controlled RD systems [28] and also analytically in Ref. [29].

Despite the advantages of our analytical solution stated above, there are limits for it as well. The restriction to fully actuated systems, i.e., systems for which $\mathcal{B}^{-1}$ exists, is not always practical. In experiments with RD systems, the number of state components is usually much larger than one, while the number of control signals is often restricted to one or two. Thus, the question arises if the approach can be extended to underactuated systems with a number of independent control signals smaller than the number of state components. This is indeed the case but entails additional assumptions about the desired trajectory. In the context of position control of TWs, it leads to a control which is not able to preserve the TW profile for all state components, see Ref. [28]. The general case is discussed in the thesis [32] and is not part of this paper.

Moreover, in applications, it is often necessary to impose inequality constraints in form of upper and lower bounds on the control. For example, the intensity of a heat source deployed as control is bounded by technical reasons. Even worse, if the control is the temperature itself, it is impossible to attain negative values. Since the control signal $\vec{f}_{an}$ for position control is proportional to the slope of the controlled wave profile $\vec{U}'_c$, the magnitude of the applied control may locally attain non-realizable values. In our analytic approach no bounds for the control signal are imposed. The control signal $\vec{f}$ as given by Eq. (10.8) is optimal only in case of a vanishing Tikhonov-regularization parameter $\nu = 0$, cf. Eq. (10.4). Moreover, desired trajectories $\vec{u}_d$ which do not comply with initial as well as boundary conditions or are non-smooth might be requested. Lastly, the control signal $\vec{f}$ cannot be used in systems where only a restricted region of the spatial domain Ω is accessible by control. While all these cases cannot be treated within the analytical approach proposed here, optimal control can deal with many of these complications.

10.3 Optimal Control

In the following, we recall the optimal control problem and sketch the most important analytical results to provide the optimality system.

10.3.1 The Control Problem

For simplicity, we explicitly state the optimal control problem for the FitzHugh-Nagumo system [39, 40]. The FitzHugh-Nagumo system is a two-component model $\vec{u} = (u, v)^T$ for an activator u and an inhibitor v,

$$\begin{aligned}
\partial_t u(\vec{x}, t) - \Delta u(\vec{x}, t) + R(u(\vec{x}, t)) + \alpha \, v(\vec{x}, t) &= f(\vec{x}, t), \\
\partial_t v(\vec{x}, t) + \beta \, v(\vec{x}, t) - \gamma \, u(\vec{x}, t) + \delta &= 0,
\end{aligned} \tag{10.13}$$

in a bounded Lipschitz-domain $\Omega \subset \mathbb{R}^N$ of dimension $1 \leq N \leq 3$. Since the single-component control f appears solely on the right-hand side of the first equation, this system is underactuated. Allowing a control in the second equation is fairly analogous. The kinetic parameters α, β, γ, and δ are real numbers with $\beta \geq 0$. Moreover, the reaction kinetics are given by the nonlinear function $R(u) = u(u - a)(u - 1)$ for $0 \leq a \leq 1$. Note that the equation for the activator u decouples from the equation for the inhibitor v for $\alpha = 0$, cf. Eq. (10.13), resulting in the Schlögl model [1, 38], sometimes also called the Nagumo model. We assume homogeneous Neumann-boundary conditions for the activator u and $u(\vec{x}, 0) = u_0(\vec{x})$, $v(\vec{x}, 0) = v_0(\vec{x})$ are given initial states belonging to $L^\infty(\Omega)$, i.e., they are bounded.

The aim of our control problem is the tracking of desired trajectories $\vec{u}_d = (u_d, v_d)^T$ in the space-time cylinder Q and to reach desired terminal states $\vec{u}_T = (u_T, v_T)^T$ at the final time T. In contrast to the analytic approach from Sect. 10.2, these desired trajectories are neither assumed to be smooth nor compatible with the given initial data or boundary conditions. For simplicity, we assume their boundedness, i.e., $(u_d, v_d)^T \in (L^\infty(Q))^2$ and $(u_T, v_T)^T \in (L^\infty(\Omega))^2$. The goal of reaching the desired states is expressed as the minimization of the objective functional

$$
\begin{aligned}
J(u, v, f) = \frac{1}{2} & \left(c_T^U \|u(\cdot, T) - u_T\|_{L^2(\Omega)}^2 + c_T^V \|v(\cdot, T) - v_T\|_{L^2(\Omega)}^2 \right) \\
+ \frac{1}{2} & \left(c_d^U \|u - u_d\|_{L^2(Q)}^2 + c_d^V \|v - v_d\|_{L^2(Q)}^2 \right) + \frac{\nu}{2} \|f\|_{L^2(Q)}^2 .
\end{aligned}
\tag{10.14}
$$

This functional is slightly more general than the one given by Eq. (10.2) because it also takes into account the terminal states. We emphasize that the given non-negative coefficients c_d^U, c_d^V, c_T^U, and c_T^V can also be chosen as functions depending on space and time. In some applications, this turns out to be very useful [44]. The control signals can be taken out of the set of admissible controls

$$
\mathcal{F}_{ad} = \{ f \in L^\infty(Q) : f_a \leq f(\vec{x}, t) \leq f_b, \text{ for } (\vec{x}, t) \in Q \}.
\tag{10.15}
$$

The bounds $-\infty < f_a < f_b < \infty$ model the technical capacities for generating controls.

Under the previous assumptions, the controlled RD equations (10.13) have a unique weak solution denoted by $(u_f, v_f)^T$ for a given control $f \in \mathcal{F}_{ad}$. This solution is bounded, i.e., u_f, $v_f \in L^\infty(Q)$, cf. [44]. If the initial data $(u_0, v_0)^T$ are continuous, then u_f and v_f are continuous on $\bar{\Omega} \times [0, T]$ with $\bar{\Omega} = \Omega \cup \partial\Omega$ as well. Moreover, the control-to-state mapping $G := f \mapsto (u_f, v_f)^T$ is twice continuously (Fréchet-) differentiable. A proof can be found in Ref. [44, Theorem 2.1, Corollary 2.1, and Theorem 2.2]. Expressed in terms of the solution $(u_f, v_f)^T$, the value of the objective functional depends only on f, $J(u, v, f) = J(u_f, v_f, f) =: F(f)$, and the optimal control problem can be formulated in a condensed form as

$$
\text{(P)} \quad \text{Min } F(f), \quad f \in \mathcal{F}_{ad}.
\tag{10.16}
$$

Referring to [44, Theorem 3.1], we know that the control problem (P) has at least one (optimal) solution $\bar{f}$ for all $\nu \geq 0$. To determine this solution numerically, we need the first and second-order derivatives of the objective functional F. Since the mapping $f \mapsto (u, v)^T$ is twice continuously differentiable, so is $F : L^p(Q) \longrightarrow \mathbb{R}$. Its first derivative $F'(f)$ in the direction $h \in L^p(Q)$ can be computed as follows:

$$
F'(f)h = \int_0^T \!\!\! \int_\Omega d\vec{x} \, dt \, \{ (\varphi_f + \nu f) h \},
\tag{10.17}
$$

where φ_f denotes the first component of the so-called adjoint state (φ_f, ψ_f). It solves a linearized FitzHugh-Nagumo system, backwards in time,

$$
\begin{aligned}
-\partial_t \varphi_f - \Delta \varphi_f + R'(u_f)\varphi_f - \gamma \, \psi_f &= c_d^U (u_f - u_d), \\
-\partial_t \psi_f + \beta \, \psi_f + \alpha \, \varphi_f &= c_d^V (v_f - v_d)
\end{aligned}
\tag{10.18}
$$

with homogeneous Neumann-boundary and terminal conditions $\varphi_f(\vec{x}, T) = c_T^U$ $(u_f(\vec{x}, T) - u_T(\vec{x}))$ and $\psi_f(\vec{x}, T) = c_T^V(v_f(\vec{x}, T) - v_T(\vec{x}))$ in Ω.

This first derivative is used in numerical methods of gradient type. Higher order methods of Newton type need also the second derivative $F''(f)$. It reads

$$F''(f)h^2 = \int_\Omega d\vec{x} \left\{ c_T^U \eta_h(\vec{x}, T)^2 + c_T^V \zeta_h(\vec{x}, T)^2 \right\}$$

$$+ \int_0^T\!\!\!\int_\Omega d\vec{x}\, dt \left\{ [c_d^U - R''(u_f)\varphi_f]\eta_h^2 + c_d^V \zeta_h^2 \right\} + v \int_0^T\!\!\!\int_\Omega d\vec{x}\, dt \left\{ h^2 \right\}$$

$$\tag{10.19}$$

in a single direction $h \in L^p(Q)$. In this expression, the state $(\eta_h, \zeta_h) := G'(f)h$ denotes the solution of a linearized FitzHugh-Nagumo system similar to Eq. (10.18), see Ref. [44, Theorem 2.2] for more information.

10.3.2   First-Order Optimality Conditions

We emphasize that the control problem (P) is not necessarily convex. Although the objective functional $J(u, v, f)$ is convex, in general, the nonlinearity of the mapping $f \mapsto (u_f, v_f)^T$ will lead to a non-convex functional, F. Therefore, (P) is a problem of non-convex optimization, possibly leading to several local minima instead of a single global minimum.

As in standard calculus, we invoke first-order necessary optimality conditions to find a (locally) optimal control f, denoted by $\bar{f}$. In the case of unconstrained control, i.e., $\mathcal{F}_{ad} := L^p(Q)$, the first derivative of F must be zero, $F'(\bar{f}) = 0$. Computationally, this condition is better expressed in the weak formulation

$$F'(\bar{f})f = \int_0^T\!\!\!\int_\Omega d\vec{x}\, dt \left\{ (\bar{\varphi} + v\bar{f})f \right\} = 0 \quad \forall f \in L^p(Q) \tag{10.20}$$

where $\bar{\varphi}$ denotes the first component of the adjoint state associated with $\bar{f}$. If $\bar{f}$ is not locally optimal, one finds a descent direction d such that $F'(\bar{f})d < 0$. This is used for methods of gradient type.

If the restrictions $\mathcal{F}_{ad}$ are given by Eq. (10.15), then Eq. (10.20) does not hold true in general. Instead, the variational inequality

$$F'(\bar{f})(f - \bar{f}) = \int_0^T\!\!\!\int_\Omega d\vec{x}\, dt \left\{ (\bar{\varphi} + v\bar{f})(f - \bar{f}) \right\} \geq 0 \quad \forall f \in \mathcal{F}_{ad} \cap U(\bar{f})$$

$$\tag{10.21}$$

must be fulfilled, cf. [45]. Here, $U(\bar{f}) \subset L^p(Q)$ denotes a neighborhood of $\bar{f}$. Roughly speaking, it says that in a local minimum we cannot find an admissible direction of descent. A gradient method would stop in such a point. A pointwise discussion of Eq. (10.21) leads to the following identity:

$$\bar{f}(\vec{x}, t) = \text{Proj}_{[f_a, f_b]}\left(-\frac{1}{\nu}[\bar{\varphi}(\vec{x}, t)]\right) \quad \text{for } \nu > 0. \tag{10.22}$$

Here, $\text{Proj}_{[f_a, f_b]}(x) = \min\{\max\{f_a, x\}, f_b\}$ denotes the projection to the interval $[f_a, f_b]$ such that $\bar{f}(\vec{x}, t)$ belongs to the set of admissible controls $\mathcal{F}_{\text{ad}}$ defined in Eq. (10.15). According to Eq. (10.22), as long as $\bar{\varphi}$ does not vanish, a decreasing value $\nu \geq 0$ yields an optimal control growing in amplitude until it attains its bounds f_a or f_b, respectively. Thus, the variational inequality Eq. (10.21) leads to so-called bang-bang-controls [45] for $\nu = 0$ and $\bar{\varphi} \neq 0$. These are control signals which attain its maximal or minimal possible values for all times and everywhere in the spatial domain Ω. A notable exception is the case of exactly realizable desired trajectories and $\nu = 0$, already discussed in Sect. 10.2. In this case, it can be shown that $\bar{\varphi}$ vanishes [32] and Eq. (10.22) cannot be used to determine the control signal $\bar{f}$.

Numerically, solutions to optimal control are obtained by solving the controlled RD system Eq. (10.13) and the adjoint system, Eq. (10.18), such that the last identity, Eq. (10.22), is fulfilled. In numerical computations with very large or even missing bounds f_a, f_b, Eq. (10.22) becomes ill-conditioned if ν is close to zero. This might lead to large roundoff errors in the computation of the control signal and can affect the stability of numerical optimal control algorithms.

10.3.3 Example 1: Analytical and Optimal Position Control

In 1972, Schlögl discussed the auto-catalytic trimolecular RD scheme [1, 38] as a prototype of a non-equilibrium first order phase transition. The reaction kinetics R for the chemical with concentration $u(x, t)$ is cubic and can be casted into the dimensional form $R(u) = u(u - a)(u - 1)$. The associated controlled RD equation reads

$$\partial_t u - \partial_x^2 u + u\,(u - a)\,(u - 1) = f(x, t), \quad 0 < a < 1,$$

in one spatial dimension, $\Omega = \mathbb{R}$. A linear stability analysis of the uncontrolled system reveals that $u = 0$ and $u = 1$ are spatially homogeneous stable steady states (HSS), while $u = a$ is an unstable homogeneous steady state. In an infinite one-dimensional domain, the Schlögl model possesses a stable traveling front solution whose profile is given by

$$U_c(\xi) = 1/\left(1 + \exp\left(\xi/\sqrt{2}\right)\right) \tag{10.23}$$

in the frame of reference $\xi = x - ct$ co-moving with front velocity c. This front solution establishes a heteroclinic connection between the two stable HSS for $\xi \to \pm\infty$ and travels with a velocity $c = (1 - 2a)/\sqrt{2}$ from the left to the right.

As an example, we aim to accelerate a traveling front according to the following protocol of motion

$$\phi(t) = -10 + ct + \frac{10 - 1/\sqrt{2}}{200} t^2, \qquad (10.24)$$

while keeping the front profile as close as possible to the uncontrolled one. In other words, our desired trajectory reads $u_d(x, t) = U_c(x - \phi(t))$ and, consequently, the initial conditions of both the controlled and the desired trajectory are $u_0(x) = u_d(x, 0) = U_c(x + 10)$. In our numerical simulations, we set $T = 20$ for the terminal time T, $\Omega = (-25, 25)$ for the spatial domain, and the threshold parameter is kept fixed at $a = 9/20$. Additionally, we choose the terminal state to be equal to the desired trajectory, $u_T(x) = u_d(x, T)$, and set the remaining weighting coefficients to unity, $c_d^U = c_T^U = 1$, in the optimal control problem. The space-time plot of the desired trajectory u_d is presented in Fig. 10.1a for the protocol of motion $\phi(t)$ given by Eq. (10.24).

Below, we compare the numerically obtained solution to the optimal control problem (P) with the analytical solution from Sect. 10.2 for the Schlögl model. The Schlögl model arises from Eq. (10.12) by setting $\alpha = 0$ and ignoring the inhibitor variable v. Consequently, all weighting coefficients associated with the inhibitor trajectory are set to zero, $c_d^V = c_T^V = 0$, in the functional J, Eq. (10.14).

Figure 10.1b depicts the solution for the analytical position control f_{an} which is valid for a vanishing Tikhonov regularization parameter $\nu = 0$. The numerically obtained optimal control $\bar{f}$ for $\nu = 10^{-5}$, shown in Fig. 10.1c, does not differ visually from the analytic one. Both are located at the front position where the slope is maxi-

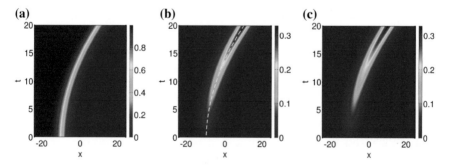

Fig. 10.1 **a** Space-time plot of the desired trajectory $u_d(x, t) = U_c(x - \phi(t))$ with the protocol of motion $\phi(t)$ as given in Eq. (10.24), **b** analytic position control signal $f_{an}(x, t)$, Eq. (10.12), and **c** numerically obtained optimal control $\bar{f}$ for Tikhonov regularization parameter $\nu = 10^{-5}$ are presented. The magnitude of the control signal is color-coded. In the *center panel* (**b**), the *dashed line* represents $\phi(t)$. The remaining parameter values are $a = 9/20$, $T = 20$, and $c_d^U = c_T^U = 1$

Table 10.1 The distance $\|\bar{f} - f_{\mathrm{an}}\|_2$ between the analytical control signal f_{an}, valid for $\nu = 0$, and the optimal control $\bar{f}$ obtained numerically for finite $\nu > 0$ decreases with decreasing values of ν (top row). Similarly, the optimally controlled state trajectory u_f approaches the desired trajectory u_d, measured by $\|u_f - u_d\|_2$, for smaller values of ν (bottom row)

ν	1	E-1	E-2	E-3	E-4	E-5	E-6
$\|\bar{f} - f_{\mathrm{an}}\|_2$	4.57E-4	1.14E-4	2.50E-5	1.01E-5	8.40E-6	8.30E-6	8.29E-6
$\|u_f - u_d\|_2$	4.77E-4	7.49E-5	8.34E-6	8.52E-7	8.55E-8	8.56E-9	8.56E-10

mal, $\vec{u}_d = 0.5$ (dashed line in Fig. 10.1b), and their magnitudes grow proportional to $\dot{\phi}(t)$. For a quantitative comparison, we compute the distance between analytical and optimal control signal $\|\bar{f} - f_{\mathrm{an}}\|_2$ in the sense of $L^2(Q)$, Eq. (10.3), and normalize it by the size of the space-time-cylinder $|Q| = T\,|\Omega|$,

$$\|h\|_2 := \frac{1}{|Q|}\|h\|_{L^2(Q)}. \tag{10.25}$$

The top row of Table 10.1 displays the distance $\|\bar{f} - f_{\mathrm{an}}\|_2$ as a function of the regularization parameter ν. Even for a large value $\nu = 1$, the distance is less than 5×10^{-4}. Decreasing the value of ν results in a shrinking distance $\|\bar{f} - f_{\mathrm{an}}\|_2$ until it saturates at $\simeq 8 \times 10^{-6}$. The saturation is due to numerical and systematic errors. Numerical computations are affected by errors arising in the discretization of the spatio-temporal domain and the amplification of roundoff errors by the ill-conditioned expression for the control, Eq. (10.22). A systematic error arises because the optimal control is computed for a bounded interval $\Omega = (-25, 25)$ with homogeneous Neumann-boundary conditions while the analytical result is valid only for an infinite domain.

Another interesting question is how close the controlled state u_f approaches the desired trajectory u_d. The bottom row of Table 10.1 shows the distance between the optimally controlled state trajectory u_f and the desired trajectory for different values ν. Similarly as for the control signal, the difference lessens with decreasing values ν. Note that the value does not saturate and becomes much smaller than the corresponding value for the difference between control signals. Here, no discretization errors arise because a discretized version of the desired trajectory is used as the target distribution. Nevertheless, systematic errors arise because neither the initial and final desired state nor the desired trajectory obey Neumann-boundary conditions. This results in an optimal control signal exhibiting bumps close to the domain boundaries. However, the violation of boundary conditions can be reduced by specifically designed protocols of motion. The further the protocol of motion keeps the controlled front away from any domain boundary the smaller is the violation of homogeneous Neumann-boundary conditions since the derivatives of traveling front solution Eq. (10.23) decay exponentially for large $|x|$. An alternative way of rigorously avoiding artifacts due to the violation of boundary conditions is the introduction of additional control terms acting on the domain boundaries, see Ref. [31].

For the example discussed above, the numerically obtained optimal control $\bar{f}$ for $\nu > 0$ is computed with a Newton-Raphson-type root finding algorithm. This iterative algorithm relies on an initial guess for the control signal which is often chosen to be random or uniform in space. The closer the initial guess is to the final solution, the fewer steps are necessary for the Newton-Raphson method to converge to the final solution. The similarity of the numerically obtained and analytical control solution, see Fig. 10.1 and Table 10.1, motivates the utilization of the analytical result f_{an} as an initial guess in numerical algorithms. Even for a simple single component RD system defined on a relatively small spatio-temporal domain Q as discussed in this section, the computational speedup is substantial. The algorithm requires only 2/3 of the computation time compared to random or uniform starting values for the control. In particular, we expect even larger speedups for simulations with larger domain sizes.

10.4 Sparse Optimal Control

In applications, it might be desirable to have localized controls acting only in some sub-areas of the domain. So-called sparse optimal controls provide such solutions without any a priori knowledge of these sub-areas. They result in a natural way because the control has the most efficient impact in these regions to minimize the objective functional.

For inverse problems, it has been observed that the use of an L^1-term in addition to the L^2-regularization leads to sparsity [46–48]. The idea to use the L^1-term goes back to Ref. [49].

To our knowledge, sparse optimal controls were first discussed in the context of optimal control in Ref. [34]. In that paper, an elliptic linear model was discussed. Several publications followed, investigating semi-linear elliptic equations, parabolic linear, and parabolic semi-linear equations; we refer, for instance, to Refs. [35–37] among others.

In this section, we follow the lines of Refs. [44, 50] and recall the most important results for the sparse optimal control of the Schlögl-model and the FitzHugh-Nagumo equation.

10.4.1 The Control Problem

In optimal control, sparsity is obtained by extending the objective functional J by a multiple of $j(f) := \|f\|_{L^1(Q)}$, the L^1-norm of the control f. Therefore, recalling that $J(u_f, v_f, f) =: \mathcal{F}(f)$, we consider the problem

$$(\text{P}_{sp}) \qquad \text{Min } \mathcal{F}(f) + \kappa\, j(f), \quad f \in \mathcal{F}_{ad}$$

for $\kappa > 0$. The first part F of the objective functional is differentiable, while the L^1-part is not.

Our goal is not only to derive first-order optimality conditions as in the previous section but also to observe the behavior of the optimal solutions for increasing κ and ν is tending to zero. For that task, we also need to introduce second-order optimality conditions.

As before, there exists at least one locally optimal solution f to the problem $(\mathrm{P_{sp}})$, denoted by $\bar{f}$. We refer to Ref. [44, Theorem 3.1] for more details. While $\mathcal{F}$ is twice continuously differentiable, the second part $j(f)$ is only Lipschitz convex but not differentiable. For that reason, we need the so-called subdifferential of $j(f)$. By subdifferential calculus and using directional derivatives of $j(f)$, we are able to derive necessary optimality conditions.

10.4.2 First-Order Optimality Conditions

We recall some results from Refs. [44, 50]. Due to the presence of $j(f)$ in the objective functional, there exists a $\bar{\lambda} \in \partial j(\bar{f})$ such that the variational inequality Eq. (10.21) changes to

$$\int\limits_0^T\!\!\int\limits_\Omega d\vec{x}\, dt \,\left\{ (\bar{\varphi} + \nu\bar{f} + \kappa\bar{\lambda})(f - \bar{f}) \right\} \geq 0 \quad \forall f \in \mathcal{F}_{\mathrm{ad}} \cap U(\bar{f}). \tag{10.26}$$

For the problem $(\mathrm{P_{sp}})$, a detailed and extensive discussion of the first-order necessary optimality condition leads to very interesting conclusions, namely

$$\bar{f}(\vec{x}, t) = 0, \ \text{if and only if } |\bar{\varphi}(\vec{x}, t)| \leq \kappa, \tag{10.27}$$

$$\bar{f}(\vec{x}, t) = \mathrm{Proj}_{[f_a, f_b]}\left(-\frac{1}{\nu}[\bar{\varphi}(\vec{x}, t) + \kappa\bar{\lambda}(\vec{x}, t)] \right), \tag{10.28}$$

$$\bar{\lambda}(\vec{x}, t) = \mathrm{Proj}_{[-1, +1]}\left(-\frac{1}{\kappa}\bar{\varphi}(\vec{x}, t) \right) \tag{10.29}$$

if $\nu > 0$. We refer to Refs. [36, Corollary 3.2] and [51, Theorem 3.1] in which the case $\nu = 0$ is discussed as well.

The relation in Eq. (10.27) leads to the sparsity of the (locally) optimal solution $\bar{f}$, depending on the sparsity parameter κ. In particular, the larger the choice of κ is, the smaller does the support of $\bar{f}$ become. To be more precise, there exists a value $\kappa_0 > \infty$ such that for every $\kappa \geq \kappa_0$ the only local minimum $\bar{f}$ is equal to zero. Obviously, this case is ridiculous and thus, one needs some intuition to find a suitable value κ. We emphasize that $\bar{\lambda}$ is unique, see Eq. (10.29), which is important for numerical calculations.

10.4.3 Example 2: Optimal and Sparse Optimal Position Control

For the numerical computations, we follow the lines of Ref. [44] and use a non-linear conjugate gradient method. The advantage of using a (conjugate) gradient method lies in the simplicity in its implementation and in the robustness of the method to errors in the solution process. Moreover, it allows to solve the systems Eq. (10.13) and the adjoint system separately. The disadvantage is clearly the fact that it might cause a huge amount of iterations to converge, cf. Ref. [44, Sect. 4].

Hence, we modify our approach by the use of *Model Predictive Control* [52, 53]. The idea is quite simple: Instead of optimizing the whole time-horizon, we only take a very small number of time steps, formulate a sub-problem, and solve it. Then, the first computed time-step of the solution $\bar{f}$ of this smaller problem is accepted on $[0, t_1]$ and is fixed. A new sub-problem is defined by going one time-step further and so on. Although the control gained in this way is only sub-optimal, it leads to a much better convergence-behavior in many computations.

Next, we revisit the task to extinguish a spiral wave by controlling its tip dynamics such that the whole pattern moves out of the spatial domain towards the Neumann boundaries [9, 21, 54]. To this goal, following Ex. 6 from Ref. [44, Sect. 4], we set the protocol of motion to $\vec{\phi}(t) = (0, \min\{120, 1/16\,t\})^T$ and $u_d(\vec{x}, t) := u_{\text{nat}}(\vec{x} - \vec{\phi}(t), t)$, where u_{nat} denotes the naturally developed spiral wave solution of the activator u to Eq. (10.13) for $f = 0$. In our numerical simulation, we take only 4 time-steps in each sub-problem of the receding horizon. More-over, we set the kinetic parameters in the FHN model, Eq. (10.13), to $a = 0.005$, $\alpha = 1, \beta = 0.01, \gamma = 0.0075$, and $\delta = 0$. Further, we fix the simulation domain $\Omega = (-120, 120) \times (-120, 120)$, the terminal time $T = 2000$, $\nu = 10^{-6}$ as Tikhonov parameter, and $f_a = -5$ and $f_b = 5$ as bounds for the control, respectively. As initial states $(u_0, v_0)^T$ a naturally developed spiral wave whose core is located at $(0, 0)$ is used; u_0 is presented in Fig. 10.2a.

In addition, an observation-function $c_d^U \in L^\infty(Q)$ instead of the constant factor $c_d^U \in \mathbb{R}$ is used with a support restricted to the area close to the desired spiral-tip. To be more precise, $c_d^U(\vec{x}, t) = 1$ holds only in the area defined by all $(\vec{x}, t) \in Q$ such that $|\vec{x} - \vec{\bar{x}}(t)| \le 20$ and vanishes identically otherwise. The other coefficients c_d^V, c_T^U, and c_T^V are set equal to zero.

The reason for the choice of such an observation-function is clear: a most intriguing property of spiral waves is that, despite being propagating waves affecting all accessible space, they behave as effectively localized particles-like objects [55]. The particle-like behavior of spirals corresponds to an effective localization of so called *response functions* [56, 57]. The asymptotic theory of the spiral wave drift [58] is based on the idea of summation of elementary responses of the spiral wave core position and rotation phase to elementary perturbations of different modalities and at different times and places. This is mathematically expressed in terms of the response functions. They decay quickly with distance from the spiral wave core and are almost equal to zero in the region where the spiral wave is insensitive to small perturbations.

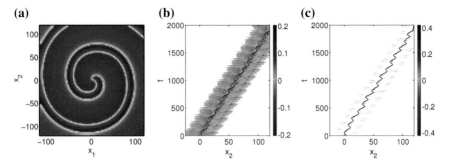

Fig. 10.2   a Spiral wave solution of the activator u to Eq. (10.13) with $\vec{f} = 0$. The latter is used as initial state $u_0(\vec{x}) = u_{\mathrm{nat}}(\vec{x} - \vec{\phi}(0), 0)$ in the problem (P$_{\mathrm{sp}}$). **b** Numerically obtained sub-optimal control ($\kappa = 0$) and **c** sparse sub-optimal control solution ($\kappa = 1$), both shown in the x_2–t–plane for $x_1 = 0$ with associated spiral-tip trajectory (*black line*). The magnitude of the control signal is color-coded. The remaining system parameters are $a = 0.005$, $\alpha = 1$, $\beta = 0.01$, $\gamma = 0.0075$, and $\delta = 0$. In the optimal control algorithms, we set $\nu = 10^{-6}$, $f_a = -5$, and $f_b = 5$

The numerical results for the sup-optimal control ($\kappa = 0$) and for the sparse sub-optimal control ($\kappa = 1$) are depicted in Fig. 10.2b, c, respectively. One notices that the prescribed spiral tip trajectory is realized for both choices for the sparsity parameter κ, viz., $\kappa = 0$ and $\kappa = 1$. The traces of the spiral tip is indicated by the solid lines in both panels. Since the spiral tip rotates rigidly around the spiral core which moves itself on a straight line according to $\vec{\phi}(t)$, one observes a periodic motion of the tip in the x_2–t–plane. In addition, the area of non-zero control (colored areas) is obviously much smaller for non-zero sparsity parameter κ compared to the case $\kappa = 0$, cf. Fig. 10.2b, c. However, in this example we observed that the amplitude of the sparse control is twice as large compared to optimal control ($\kappa = 0$).

10.4.4   Second-Order Optimality Conditions and Numerical Stability

To avoid this subsection to become too technical, we only state the main results from Ref. [50]. We know for an unconstrained problem with differentiable objective-functional that it is sufficient to show $F'(\bar{f}) = 0$ and $F''(\bar{f}) > 0$ to derive that $\bar{f}$ is a local minimizer of F if F is a real-valued function of one real variable. More details about the importance of second order optimality conditions in the context of PDE control can be found in Ref. [59].

In our setting, considering all directions $h \neq 0$ out of a certain so-called critical cone C_f, the condition for $\nu > 0$ reads

$$F''(\bar{f})h^2 > 0.$$

Then, $\bar{f}$ is a locally optimal solution of (P$_{sp}$). The detailed structure of C_f is described in Ref. [50]; also the much more complicated case $\nu = 0$ is discussed there.

The second-order sufficient optimality conditions are the basis for interesting questions, e.g., the stability of solutions for perturbed desired trajectories and desired states [50]. Moreover, we study the limiting case of Tikhonov parameter ν tending to zero.

10.4.5   Tikhonov Parameter Tending to Zero

In this section, we investigate the behavior of a sequence of optimal controls and the corresponding states as solutions of the problem (P$_{sp}$) as $\nu \to 0$. For this reason, we denote our control problem (P$_\nu$), the associated optimal control with $\bar{f}_\nu$, and its associated states with $(\bar{u}_\nu, \bar{v}_\nu)$ for a fixed $\nu \geq 0$. Since $\mathcal{F}_{ad}$ is bounded in $L^\infty(Q)$, any sequence of solutions $\{\bar{f}_\nu\}_{\nu>0}$ of (P$_\nu$) has subsequences converging weakly* in $L^\infty(Q)$. For a direct numerical approach, this is useless, but we can deduce interesting consequences of this convergence using second order sufficient optimality conditions.

Assume that the second order sufficient optimality conditions of Ref. [50, Theorem 4.7] are satisfied. Then, we derive a Hölder rate of convergence for the states

$$\lim_{\nu \to 0} \frac{1}{\sqrt{\nu}} \left\{ \|\bar{u}_\nu - \bar{u}_0\|_{L^2(Q)} + \|\bar{v}_\nu - \bar{v}_0\|_{L^2(Q)} \right\} = 0 \qquad (10.30)$$

with $(\bar{u}_\nu, \bar{v}_\nu) = G(\bar{f}_\nu)$ and $(\bar{u}_0, \bar{v}_0) = G(\bar{f}_0)$. We should mention that this estimate is fairly pessimistic. All of our numerical tests show that the convergence rate is of order ν, i.e., we observe a Lipschitz rather than a Hölder estimate [50]. As mentioned in Ref. [50], it should also be possible to prove Lipschitz stability and hence, to confirm the linear rate of convergence for $\nu \to 0$ with a remarkable amount of effort.

10.4.6   Example 3: Sparse Optimal Control with Tikhonov
Parameter Tending to Zero

Finally, we consider a traveling pulse solution in the FitzHugh-Nagumo system in one spatial dimension $N = 1$. Here, the limiting case of vanishing Tikhonov parameter, $\nu = 0$, is of our special interest. We observe that Newton-type methods yield very high accuracy even for very small values of $\nu > 0$. This allows us to study the convergence behavior of solutions for ν tending to zero as well.

Following Ref. [50] and in contrast to the last example in Sect. 10.4.3, we solve the full forward-backward-system of optimality. We stress that this is numerically possible solely for non-vanishing value of ν. However, we constructed examples where an exact solution of the optimality system for $\nu = 0$ is accessible as shown in

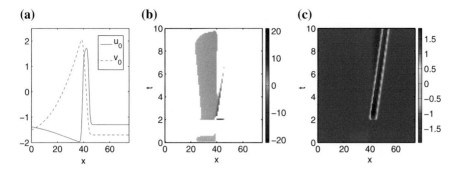

Fig. 10.3 a Segment of a traveling pulse solution $(u_0, v_0)^T$ in the uncontrolled FitzHugh-Nagumo system, Eq. (10.13) with $f = 0$. **b** Numerically obtained sparse optimal control solution $\bar{f}_\nu$ for almost vanishing Tikhonov parameter, $\nu = 10^{-10}$, and **c** the associated optimal state $\bar{u}_\nu$. The amplitude of the control signal is color-coded. The kinetic parameter values are set to $\alpha = 1$, $\beta = 0$, $\gamma = 0.33$, $\delta = -0.429$, and $R(u) = u(u - \sqrt{3})(u + \sqrt{3})$

Ref. [50, Sect. 5.3]. In this sequel, our reference-solution, denoted by $\bar{u}_{\text{ref}}$, will be the solution of (P_ν) for $\nu := \nu_{\text{ref}} = 10^{-10}$. For smaller values, the numerical errors do not allow to observe a further convergence. The distance $\|\bar{u}_\nu - \bar{u}_{\text{ref}}\|_{L^2(Q)}$ stagnates between $\nu = 10^{-10}$ and $\nu_{\text{ref}} < 10^{-10}$.

Next, we treat the well-studied problem of pulse nucleation [60, 61] by sparse optimal control. We aim to start and to stay in the lower HSS for the first two time-units, i.e., $u_d(\vec{x}, t) = -1.3$ for $t \in (0, 2)$. Then, the activator state shall coincide instantaneously with the traveling pulse solution u_{nat}, i.e., $u_d(\vec{x}, t) = u_{\text{nat}}(\vec{x}, t - 2)$. To get the activator profile u_{nat}, we solve Eq. (10.13) for $f = 0$ and its profile is shown in Fig. 10.3a.

In our optimal control algorithms, we set the parameters to $\Omega = (0, 75)$, $T = 10$, $\alpha = 1$, $\beta = 0$, $\gamma = 0.33$, and $\delta = -0.429$. Moreover, here we use a slightly different nonlinear reaction kinetics $R(u) = u(u - \sqrt{3})(u + \sqrt{3})$ in Eq. (10.13) but this does not change the analytical results. The upper and lower bounds for the control f are set to very large values, viz., $f_a = -100$ and $f_b = 100$. In addition, the coefficients in Eq. (10.14) are kept fixed, viz., $c_d^U = 1$ and $c_T^U = c_d^V = c_T^V = 0$.

Our numerical results obtained for a sparse optimal control $\bar{f}_\nu$ acting solely on the activator u, cf. Eq. (10.13), are presented in Fig. 10.3b, c. In order to create a traveling pulse solution from the HSS $u_d = -1.3$, the optimal control resembles a step-like excitation with high amplitude at $x \simeq 40$. Since the Tikhonov parameter is set to $\nu = 10^{-10}$, large control amplitudes are to be expected and indicate that in the unregularized case, even a delta distribution might appear. Because this excitation is supercritical, a new pulse will nucleate. In order to inhibit the propagation of this nucleated pulse to the left, the control must act at the back of the pulse as well. Thus, we observes a negative control amplitude acting in the back of the traveling pulse. We emphasize that the desired shape of the pulse is achieved qualitatively. The realization of the exact desired profile can not be expected due to a non-vanishing sparse parameter $\kappa = 0.01$. Even for this respectively small value, the sparsity of the optimal control shows.

Table 10.2 The comparison of the distance $\|\bar{u}_\nu - \bar{u}_{\mathrm{ref}}\|_{L^2(Q)}$ between the numerically obtained states $\bar{u}_\nu$ and the numerically obtained reference-solution $\bar{u}_{\mathrm{ref}}$, computed for $\nu = 10^{-10}$, for decreasing values of $\nu > 0$

ν	1E-3	1E-4	1E-5	1E-6	1E-7	1E-8	1E-9
$\|\bar{u}_\nu - \bar{u}_{\mathrm{ref}}\|_{L^2(Q)}$	1.58E-1	1.16E-2	1.33E-3	1.35E-4	1.35E-5	1.34E-6	1.31E-7

Since the displayed control and state are computed for an almost vanishing value $\nu = 10^{-10}$, we take the associated state as reference-state $\bar{u}_{\mathrm{ref}}$ in order to study the dependence of the distance $\|\bar{u}_\nu - \bar{u}_{\mathrm{ref}}\|_{L^2(Q)}$ on $\nu > 0$. From Table 10.2, one notices the already mentioned Lipschitz-rate of convergence for decreasing values $\nu > 0$, $\|\bar{u}_\nu - \bar{u}_{\mathrm{ref}}\|_{L^2(Q)} \propto \nu$. This observation is consistent with results from [50] for various other examples.

10.5 Conclusion

Optimal control of traveling wave patterns in RD systems according to a prescribed desired distribution is important for many applications.

Analytical solutions to an unregularized optimal control problem can be obtained with ease from the approach presented in Sect. 10.2. In particular, the control signal can be obtained without full knowledge about the underlying nonlinear reaction kinetics in case of position control. Moreover, they are a good initial guess for the numerical solution of regularized optimal control problems with small regularization parameter $\nu > 0$, thereby achieving a substantial computational speedup as discussed in Sect. 10.3.3. Generally, the analytical expressions may serve as consistency checks for numerical optimal control algorithms.

For the position control of fronts, pulses, and spiral waves, the control signal is spatially localized. By applying sparse optimal position control to reaction-diffusion systems, as discussed in Sect. 10.4, the size of the domains with non-vanishing control signals can be further decreased. Importantly, the method determines sparse controls without any a priori knowledge about restrictions to certain subdomains. Additionally, sparse control allows to study second order optimality conditions that are not only interesting from the theoretical perspective but also for numerical Newton-type algorithms.

References

1. F. Schlögl, Z. Phys. **253**, 147 (1972)
2. A. Winfree, Science **175**, 634 (1972)
3. J.J. Tyson, J.P. Keener, Phys. D **32**, 327 (1988)
4. R. Kapral, K. Showalter (eds.), *Chemical Waves and Patterns* (Kluwer, Dordrecht, 1995)

5. Y. Kuramoto, *Chemical Oscillations, Waves, and Turbulence* (Courier Dover Publications, New York, 2003)
6. J. Murray, *Mathematical Biology* (Springer-Verlag, Berlin, 2003)
7. A. Liehr, *Dissipative Solitons in Reaction Diffusion Systems: Mechanisms, Dynamics, Interaction*, vol. 70 (Springer Science & Business Media, 2013)
8. V.S. Zykov, G. Bordiougov, H. Brandtstädter, I. Gerdes, H. Engel, Phys. Rev. Lett. **92**, 018304 (2004)
9. V.S. Zykov, H. Engel, Phys. D **199**, 243 (2004)
10. A. Mikhailov, K. Showalter, Phys. Rep. **425**, 79 (2006)
11. J. Schlesner, V.S. Zykov, H. Engel, E. Schöll, Phys. Rev. E **74**, 046215 (2006)
12. M. Kim, M. Bertram, M. Pollmann, A. von Oertzen, A.S. Mikhailov, H.H. Rotermund, G. Ertl, Science **292**, 1357 (2001)
13. V.S. Zykov, G. Bordiougov, H. Brandtstädter, I. Gerdes, H. Engel, Phys. Rev. E **68**, 016214 (2003)
14. J.X. Chen, H. Zhang, Y.Q. Li, J. Chem. Phys. **130**, 124510 (2009)
15. H.W. Engl, T. Langthaler, P. Mansellio, in *Optimal Control of Partial Differential Equations*, ed. by K.H. Hoffmann, W. Krabs (Birkhäuser Verlag, Basel, 1987), pp. 67–90
16. W. Barthel, C. John, F. Tröltzsch, Z. Angew, Math. und Mech. **90**, 966 (2010)
17. R. Buchholz, H. Engel, E. Kammann, F. Tröltzsch, Comput. Optim. Appl. **56**, 153 (2013)
18. G. Haas, M. Bär, I.G. Kevrekidis, P.B. Rasmussen, H.H. Rotermund, G. Ertl, Phys. Rev. Lett. **75**, 3560 (1995)
19. S. Martens, J. Löber, H. Engel, Phys. Rev. E **91**, 022902 (2015)
20. V.S. Zykov, H. Brandtstädter, G. Bordiougov, H. Engel, Phys. Rev. E **72(R)**, 065201 (2005)
21. O. Steinbock, V.S. Zykov, S.C. Müller, Nature **366**, 322 (1993)
22. A. Schrader, M. Braune, H. Engel, Phys. Rev. E **52**, 98 (1995)
23. T. Sakurai, E. Mihaliuk, F. Chirila, K. Showalter, Science **296**, 2009 (2002)
24. J. Wolff, A.G. Papathanasiou, H.H. Rotermund, G. Ertl, X. Li, I.G. Kevrekidis, Phys. Rev. Lett. **90**, 018302 (2003)
25. J. Wolff, A.G. Papathanasiou, I.G. Kevrekidis, H.H. Rotermund, G. Ertl, Science **294**, 134 (2001)
26. B.A. Malomed, D.J. Frantzeskakis, H.E. Nistazakis, A.N. Yannacopoulos, P.G. Kevrekidis, Phys. Lett. A **295**, 267 (2002)
27. P.G. Kevrekidis, I.G. Kevrekidis, B.A. Malomed, H.E. Nistazakis, D.J. Frantzeskakis, Phys. Scr. **69**, 451 (2004)
28. J. Löber, H. Engel, Phys. Rev. Lett. **112**, 148305 (2014)
29. J. Löber, Phys. Rev. E **89**, 062904 (2014)
30. J. Löber, R. Coles, J. Siebert, H. Engel, E. Schöll, in *Engineering of Chemical Complexity II*, ed. by A. Mikhailov, G. Ertl (World Scientific, Singapore, 2015)
31. J. Löber, S. Martens, H. Engel, Phys. Rev. E **90**, 062911 (2014)
32. J. Löber, Optimal trajectory tracking. Ph.D. thesis, TU Berlin (2015)
33. K.H. Hoffmann, G. Leugering, F. Tröltzsch (eds.), *Optimal Control of Partial Differential Equations*, *ISNM*, vol. 133 (Birkhäuser Verlag, 1998)
34. G. Stadler, Comput. Optim. Appl. **44**, 159 (2009)
35. G. Wachsmuth, D. Wachsmuth, ESAIM Control Optim. Calc. Var. **17**, 858 (2011)
36. E. Casas, R. Herzog, G. Wachsmuth, SIAM J. Optim. **22**, 795 (2012)
37. E. Casas, F. Tröltzsch, SIAM J. Control Optim. **52**, 1010 (2014)
38. Y. Zeldovich, D. Frank-Kamenetsky, Dokl. Akad. Nauk SSSR **19**, 693 (1938)
39. J. Nagumo, Proc. IRE **50**, 2061 (1962)
40. R. FitzHugh, Biophys. J. **1**, 445 (1961)
41. A. Azhand, J.F. Totz, H. Engel, Eur. Phys. Lett. **108**, 10004 (2014)
42. J.F. Totz, H. Engel, O. Steinbock, New J. Phys. **17**, 093043 (2015)
43. J. Guckenheimer, P. Holmes, *Nonlinear Oscillations, Dynamical Systems, and Bifurcations of Vector Fields*, vol. 42 (Springer Science & Business Media, 1983)
44. E. Casas, C. Ryll, F. Tröltzsch, Comp. Meth. Appl. Math. **13**, 415 (2013)

45. F. Tröltzsch, *Optimal Control of Partial Differential Equations. Theory, Methods and Applications*, vol. 112 (American Math. Society, Providence, 2010)
46. I. Daubechies, M. Defrise, C. De Mol, Commun. Pure Appl. Math. **57**, 1413 (2004)
47. C.R. Vogel, *Computational Methods for Inverse Problems*, vol. 23 (Siam, 2002)
48. T.F. Chan, X.C. Tai, SIAM, J. Sci. Comput. **25**, 881 (2003)
49. L.I. Rudin, S. Osher, E. Fatemi, Phys. D **60**, 259 (1992)
50. E. Casas, C. Ryll, F. Tröltzsch, SIAM J. Control Optim. **53**, 2168 (2015)
51. E. Casas, SIAM J. Control Optim. **50**, 2355 (2012)
52. A. Propoi, Avtomat. i Telemekh **24**, 912 (1963)
53. E.F. Camacho, C. Bordons, *Model Predictive Control* (Springer-Verlag, London Limited, 1999)
54. J. Schlesner, V.S. Zykov, H. Brandtstädter, I. Gerdes, H. Engel, New J. Phys. **10**, 015003 (2008)
55. I.V. Biktasheva, V.N. Biktashev, Phys. Rev. E **67**, 026221 (2003)
56. H. Henry, V. Hakim, Phys. Rev. E **65**, 046235 (2002)
57. I.V. Biktasheva, D. Barkley, V.N. Biktashev, A.J. Foulkes, Phys. Rev. E **81**, 066202 (2010)
58. J.P. Keener, Phys. D **31**, 269 (1988)
59. E. Casas, F. Tröltzsch, Jahresbericht der Deutschen Mathematiker-Vereinigung (2014)
60. A. Mikhailov, L. Schimansky-Geier, W. Ebeling, Phys. Lett. A **96**, 453 (1983)
61. I. Idris, V.N. Biktashev, Phys. Rev. Lett. **101**, 244101 (2008)

Chapter 11
Recent Advances in Reaction-Diffusion Equations with Non-ideal Relays

Mark Curran, Pavel Gurevich and Sergey Tikhomirov

Abstract We survey recent results on reaction-diffusion equations with discontinuous hysteretic nonlinearities. We connect these equations with free boundary problems and introduce a related notion of spatial transversality for initial data and solutions. We assert that the equation with transverse initial data possesses a unique solution, which remains transverse for some time, and also describe its regularity. At a moment when the solution becomes nontransverse, we discretize the spatial variable and analyze the resulting lattice dynamical system with hysteresis. In particular, we discuss a new pattern formation mechanism—*rattling*, which indicates how one should reset the continuous model to make it well posed.

M. Curran (✉)
Institute of Mathematics I, Free University of Berlin, Arnimallee 7,
14195 Berlin, Germany
e-mail: mark.curran88@gmail.com

P. Gurevich
Institute of Mathematics I, Free University of Berlin, Arnimallee 3,
14195 Berlin, Germany
e-mail: gurevich@math.fu-berlin.de

P. Gurevich
Peoples' Friendship University of Russia, Miklukho-Maklaya Str. 6,
117198 Moscow, Russia

S. Tikhomirov
Max Planck Institute for Mathematics in the Sciences, Inselstraße 22,
04103 Leipzig, Germany
e-mail: sergey.tikhomirov@gmail.com

S. Tikhomirov
Chebyshev Laboratory, St. Petersburg State University, 14th Line, 29b,
199178 Saint Petersburg, Russia

© Springer International Publishing Switzerland 2016
E. Schöll et al. (eds.), *Control of Self-Organizing Nonlinear Systems*,
Understanding Complex Systems, DOI 10.1007/978-3-319-28028-8_11

211

11.1 Introduction

11.1.1 Motivation

In this chapter we will survey recent results on reaction-diffusion equations with a hysteretic discontinuity defined at every spatial point. We also refer to [1–3] and the more recent surveys by Visintin [4, 5] for other types of partial differential equations with hysteresis.

The equations we are dealing with in the present chapter were introduced in [6, 7] to describe growth patterns in colonies of bacteria (Salmonella typhirmurium). In these experiments, bacteria (non-difussing) are fixed to the surface of a petri dish, and their growth rate responds to changes in the relative concentrations of available nutrient and a growth-inhibiting by-product. The model asserts that at a location where there is a sufficiently high amount of nutrient relative to by-product, the bacteria will grow. This growth will continue until the production of by-product and diffusion of the nutrient lowers this ratio below a lower threshold, causing growth to stop. Growth will not resume until the diffusion of by-product raises the relative concentrations above an upper threshold that is distinct from the lower. Numerics in [6] reproduced the formation of distinctive concentric rings observed in experiments, however the question of the existence and uniqueness of solutions, as well as a thorough explanation of the mechanism of pattern formation, remained open.

Another application in developmental biology can be found, e.g., in [8], and an analysis of the corresponding stationary solutions in [9].

11.1.2 Setting of the Problem

In this chapter we will treat the following prototype problem:

$$u_t = \Delta u + f(u, v), \quad v = \mathcal{H}(\xi_0, u), \quad (x, t) \in Q_T, \tag{11.1}$$

$$u|_{t=0} = \varphi, \quad x \in Q, \tag{11.2}$$

$$\left. \frac{\partial u}{\partial v} \right|_{\partial' Q_T} = 0. \tag{11.3}$$

Here $Q \subset \mathbb{R}^n$ is a domain with smooth boundary, $Q_T := Q \times (0, T)$, where $T > 0$, $\partial' Q_T := \partial Q \times (0, T)$, u is a real-valued function on Q_T, and $\mathcal{H}(\xi_0, u)$ is a hysteresis operator defined as follows (see Fig. 11.1a). Fix two real numbers $\alpha < \beta$, an integer $\xi_0 \in \{-1, 1\}$, and two continuous functions $H_1 : (-\infty, \beta] \to \mathbb{R}$ and $H_{-1} : [\alpha, \infty) \to \mathbb{R}$ such that $H_1(u) \neq H_{-1}(u)$ for $u \in [\alpha, \beta]$. Define the sets

$$\Sigma_1 := \{(u, v) \in \mathbb{R}^2 \mid u \in (-\infty, \beta), v = H_1(u)\},$$

$$\Sigma_{-1} := \{(u, v) \in \mathbb{R}^2 \mid u \in (\alpha, \infty), v = H_{-1}(u)\}.$$

Definition 11.1.1 Let $u, v : [0, T] \to \mathbb{R}$, where u is a continuous function. We say that $v = \mathcal{H}(\xi_0, u)$ if the following hold:

1. $(u(t), v(t)) \in \Sigma_1 \cup \Sigma_{-1}$ for every $t \in [0, T]$.
2. If $u(0) \in (\alpha, \beta)$, then $v(0) = H_{\xi_0}(u(0))$.
3. If $u(t_0) \in (\alpha, \beta)$, then $v(t)$ is continuous in a neighorhood of t_0.

The operator $\mathcal{H}(\xi_0, u)$ is called the *non-ideal relay* and item 3 means that the non-ideal relay jumps up (or down) when $u = \alpha$ (or $u = \beta$). This definition is equivalent to the definitions of non-ideal relay found in [1, 10, 11]. If $\mathcal{H}(\xi_0, u)(t) = H_j(u(t))$, then we call $\xi(t) := j$ the configuration of $\mathcal{H}$ at the moment t, and we call ξ_0 the initial configuration. Now let $u : Q_T \to \mathbb{R}$ be a function of (x, t) and $\xi_0 : Q \to \{-1, 1\}$ a function of x, then $\mathcal{H}(\xi_0, u)(x, t)$ is defined in the same way by treating x as a parameter, i.e., there is a non-ideal relay at every $x \in Q$ with input $u(x, t)$, configuration $\xi(x, t)$, and initial configuration $\xi_0(x)$.

11.1.3 Set-Valued Hysteresis

First results on the well-posedness of (11.1)–(11.3) were obtained in [12, 13] for set-valued hysteresis, and their model problems are worth explaining in more detail. In both papers, the uniqueness of solutions as well as their continuous dependence on initial data remained open.

First we discuss the work of Visintin [13], which treats (11.1)–(11.3) for arbitrary $n \geq 1$ with $\mathcal{H}(\xi_0, u)$ replaced by a set-valued operator called a *completed relay* (see Fig. 11.1b). We still use the thresholds $\alpha < \beta$, and will consider constant hysteresis branches $H_1(u) \equiv 1$, and $H_{-1}(u) \equiv -1$. We also define the set $\Sigma_0 := \{(u, v) \in \mathbb{R}^2 \mid u \in [\alpha, \beta], v \in (-1, 1)\}$.

Definition 11.1.2 Let $u, v : [0, T] \to \mathbb{R}$, where u is a continuous function, and let $\xi_0 \in [-1, 1]$. We say $v \in \mathcal{H}_{\mathrm{Vis}}(\xi_0, u)$ if the following hold:

1. $(u(t), v(t)) \in \overline{\Sigma_1} \cup \overline{\Sigma_{-1}} \cup \Sigma_0$ for every $t \in [0, T]$.
2. If $u(0) \in (\alpha, \beta)$, then $v(0) = \xi_0$; if $u(0) = \alpha$ (or β), then $v(0) \in [\xi_0, 1]$ (or $v(0) \in [-1, \xi_0]$).

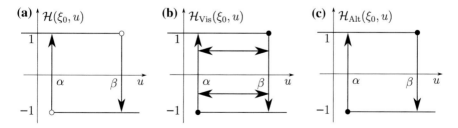

Fig. 11.1 The hysteresis operator with $H_1(u) \equiv 1$ and $H_{-1}(u) \equiv -1$

3. If $u(t_0) \in (\alpha, \beta)$, then $v(t)$ is constant in a neighborhood of t_0.
4. If $u(t_0) = \alpha$ (or β), then $v(t)$ is non-decreasing (or non-increasing) in a neighborhood of t_0.

By treating x as a parameter, $\mathcal{H}_{\text{Vis}}(\xi_0, u)$ is defined for $u : Q_T \to \mathbb{R}$ as we have done previously for $\mathcal{H}(\xi_0, u)$. Visintin [13] proved the existence of u and v such that the equation

$$u_t = \Delta u + v, \quad v \in \mathcal{H}_{\text{Vis}}(\xi_0, u),$$

with $n \geq 1$, Dirichlet boundary conditions, and initial data φ is satisfied in a weak sense in Q_T. Visintin [13] and more recently Aiki and Kopfova [14] proved the existence of solutions to modified versions of [6, 7], where the hysteretic discontinuity was a completed relay responding to a scalar input. A non-ideal relay with vector input, as in [6, 7], behaves almost identically to a non-ideal relay with scalar input, but for clarity of exposition we only consider scalar inputs in this chapter.

Let us now turn to the model hysteresis operator $\mathcal{H}_{\text{Alt}}(\xi_0, u)$ proposed by Alt in [12] (see Fig. 11.1c). We still consider $H_1(u) \equiv 1$ and $H_{-1}(u) \equiv -1$, and introduce the set

$$\tilde{\Sigma}_0 := \{(u, v) \in \mathbb{R}^2 \mid u = \alpha, v \in [-1, 1)\} \cup \{(u, v) \in \mathbb{R}^2 \mid u = \beta, v \in (-1, 1]\}.$$

Definition 11.1.3 Let $u, v : [0, T] \to \mathbb{R}$, where u is a continuous function, and let $\xi_0 \in \{-1, 1\}$. We say that $v \in \mathcal{H}_{\text{Alt}}(\xi_0, u)$ if the following hold:

1. $(u(t), v(t)) \in \Sigma_1 \cup \Sigma_{-1} \cup \tilde{\Sigma}_0$ for every $t \in [0, T]$.
2. If $u(0) \in [\alpha, \beta]$, then $v(0) = \xi_0$.
3. If $u(t_0) \in (\alpha, \beta)$, then $v(t)$ is constant in a neighborhood of t_0.
4. If $u(t_0) = \alpha$ (or β), then $v(t)$ is non-decreasing (or non-increasing) in a neighborhood of t_0.

One can define $\mathcal{H}_{\text{Alt}}(\xi_0, u)$ for $u : Q_T \to \mathbb{R}$ by treating x as a parameter as we did when defining $\mathcal{H}(\xi_0, u)$ and $\mathcal{H}_{\text{Vis}}(\xi_0, u)$.

To highlight the main difference between the completed relay $\mathcal{H}_{\text{Vis}}(\xi_0, u)$ and Alt's relay $\mathcal{H}_{\text{Alt}}(\xi_0, u)$, suppose that $\mathcal{H}_{\text{Vis}}(\xi_0, u)(t_0), \mathcal{H}_{\text{Alt}}(\xi_0, u)(t_0) \in (-1, 1)$ and $u(t_0) = \beta$ has a local maximum at time t_0. Then, as soon as u decreases, $\mathcal{H}_{\text{Alt}}$ jumps to -1, however $\mathcal{H}_{\text{Vis}}$ remains constant.

Let us introduce the notation $\{u = \alpha\} := \{(x, t) \in \overline{Q_T} \mid u(x, t) = \alpha\}$, with $\{u = \beta\}$ defined analogously. Alt's existence theorem can, omitting the technical assumptions, be stated in the following way. Let $n = 1$ and suppose $(\varphi, \xi_0) \in \overline{\Sigma_1} \cup \overline{\Sigma_{-1}}$. Then the following holds:

1. There exists u and v such that $v \in \mathcal{H}_{\text{Alt}}(\xi_0, u)$ a.e. in Q_T and

$$u_t = u_{xx} + v \quad \text{a.e. on } \{(x, t) \in Q_T \mid u(x, t) \notin \{\alpha, \beta\}\}.$$

2. We have

$$u_t = u_{xx} \quad \text{a.e. on } \{(x, t) \in Q_T \mid u(x, t) \in \{\alpha, \beta\}\},$$

$v \in [-1, 0]$ on $\{u = \beta\}$, and $v \in [0, 1]$ on $\{u = \alpha\}$.
3. Items 2–4 of Definition 11.1.3 hold in the following weak sense:
   For every $\psi \in C_0^\infty(Q \times [0, T))$ with $\psi \geq 0$ on $\{Q \times [0, T)\} \cap \{u = \alpha\}$ and $\psi \leq 0$ on $\{Q \times [0, T)\} \cap \{u = \beta\}$,

$$\int_{Q_T} (v - v_0)\psi_t \, dxdt \leq 0.$$

11.1.4   Slow-Fast Approximation

Equations of the type (11.1)–(11.3) are deeply connected with slow-fast systems where the variable v is replaced by a fast bistable ordinary differential equation with a small parameter $\delta > 0$

$$\delta v_t = g(u, v). \tag{11.4}$$

A typical example are the FitzHugh–Nagumo equations, where $g(u, v) = v - \frac{v^3}{3} - u$ and the hysteresis branches $H_1(u)$ and $H_{-1}(u)$ are the stable parts of the nullcline of g (see Fig. 11.2). The question of whether the hysteresis operator approximates the fast variable v as $\delta \to 0$ has been addressed for systems of ordinary differential equations (see, e.g., [15, 16] and further references in [17]), however the corresponding question for partial differential equations is still open.

11.1.5   Free Boundary Approach

Problem (11.1)–(11.3) with hysteresis has two distinct phases and a switching mechanism, hence it can be considered as a free boundary problem. First observe that the hysteresis $\mathcal{H}$ naturally segregates the domain into two subdomains depending on the value of $\xi(x, t)$. Denote

Fig. 11.2 a The nullcline of the S-shaped nonlinearity $g(u, v)$. **b** Hysteresis with nonconstant branches $H_1(u)$ and $H_{-1}(u)$

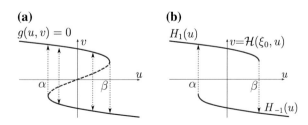

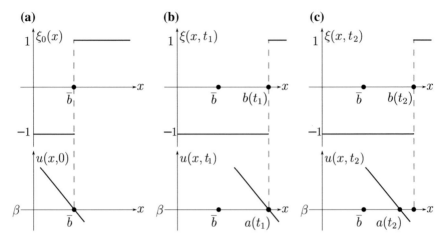

Fig. 11.3 An example of the hysteresis configuration ξ responding to an input u

$$Q_j := \{x \in Q \mid \xi_0(x) = j\}, \quad j = \pm 1. \tag{11.5}$$

Let us look at how the free boundary $\overline{Q_1} \cap \overline{Q_{-1}}$ can evolve for a simple example on the interval $Q = (0, 1)$. Consider a neighborhood U of $x \in Q$, and suppose at time $t = 0$, $Q_1 \cap U$ and $Q_{-1} \cap U$ are subintervals separated by a point $\overline{b} \in U$ (Fig. 11.3a). Let $u(x, t_0) > \beta$ for $x < \overline{b}$, $u(x, t_0) < \beta$ for $x > \overline{b}$, and let $x = a(t)$ be the unique solution of $u(x, t) = \beta$ in U. If at time $t_1 > 0$ the value of u at points $x > \overline{b}$ have already risen above β, then $\xi(x, t)$ has switched from 1 to -1. These are the points x such that $\overline{b} < x \leq a(t_1)$ (Fig. 11.3b). Now if at time $t_2 > t_1$ the value of u at the switched points has fallen below β again, $\xi(x, t)$ remains switched. These are the points x such that $a(t_2) < x < a(t_1)$ (Fig. 11.3c). More succinctly, $\xi(x, t) = -1$ if $x \leq b(t)$ and $\xi(x, t) = 1$ if $x > b(t)$, where $b(t) = \max_{0 \leq s \leq t} a(s)$.

The point of this example is to illustrate that the free boundary does not in general coincide with the points where u is equal to one of the threshold values. This is different from the two-phase parabolic obstacle problem (see, e.g., [18, 19]), which (11.1)–(11.3) reduces to if $\alpha = \beta$.

Assume the derivative $\varphi'(\overline{b})$ in the above example was non-vanishing on the boundary $\{\overline{b}\} = \overline{Q_1} \cap \overline{Q_{-1}}$. This is an example of *transverse* initial data, and whether the initial data is transverse or not will play an important role in the analysis of problem (11.1)–(11.3).

11.1.6   Overview

This chapter is organized in the following way.

In Sect. 11.2 we will investigate the well-posedness of (11.1)–(11.3) for *transverse* initial data. For $n = 1$ the existence of solutions and their continuous dependence on initial data was established in [11], uniqueness of the solution in [20] and the analogous results for systems of equations in [21]. Preliminary results for $n \geq 2$ were obtained in [22].

In Sect. 11.3 we consider the regularity of solutions u, in particular, whether the generalized derivatives $u_{x_i x_j}$ and u_t are uniformly bounded. We will summarize the results of [23], where the authors proved that these derivatives are locally bounded in a neighborhood of a point not on the free boundary. They also showed that this bound depends on the parabolic distance to the parts of the free boundary that do not contain the sets $\{u = \alpha\}$ or $\{u = \beta\}$.

In Sect. 11.4 we consider non-transverse data and the results of [24]. We will analyze a spatio-temporal pattern (called *rattling*) arising after spatial discretization of the reaction-diffusion equation and discuss its connection with the continuous model with hysteresis operators $\mathcal{H}$, $\mathcal{H}_{\mathrm{Vis}}$, and $\mathcal{H}_{\mathrm{Alt}}$.

11.2 Transverse Initial Data

11.2.1 Setting of a Model Problem

In this section we will discuss the well-posedness of problem (11.1)–(11.3) under the assumption that φ is transverse with respect to ξ_0, a notion which we will make precise shortly. In order to illustrate the main ideas, we will treat the following model problem in detail and then discuss generalizations at the end of this section (see Sect. 11.2.4). Let $h_{-1} \leq 0 \leq h_1$ be two constants, and let the hysteresis branches be given by $H_1(u) \equiv h_1$ and $H_{-1}(u) \equiv h_{-1}$. Consider the prototype problem

$$u_t = \Delta u + \mathcal{H}(\xi_0, u), \quad (x, t) \in Q_T, \tag{11.6}$$

$$u|_{t=0} = \varphi, \quad x \in Q, \tag{11.7}$$

$$\frac{\partial u}{\partial \nu}\bigg|_{\partial' Q_T} = 0. \tag{11.8}$$

We will treat $n = 1$ in Sect. 11.2.2 (see [11, 20]) and $n \geq 2$ (see [22]) in Sect. 11.2.3. Throughout this subsection we will always assume that φ and ξ_0 are *consistent* with each other, i.e., if $\varphi(x) < \alpha$ (or $\varphi(x) > \beta$), then $\xi_0(x) = 1$ (or $\xi_0(x) = -1$). In particular, this means that for every $x \in Q$, $\xi(x, t)$ is continuous from the right as a function of $t \in [0, T)$.

Since in general $\mathcal{H}(\xi_0, u) \in L_q(Q_T)$, we will look for solutions in the Sobolev space $W_q^{2,1}(Q_T)$ with $q > n + 2$. This is the space consisting of functions with two weak spatial derivatives and one weak time derivative from $L_q(Q_T)$ (see [25, Chap. 1]). If $u \in W_q^{2,1}(Q_T)$, then for every $t \in [0, T]$ the trace is well defined and

$u(\cdot, t) \in W_q^{2-2/q}(Q)$ (see, e.g., [25, p. 70]). To ensure that φ is regular enough to define the spatial transversality property, we henceforth fix a γ such that $0 < \gamma < 1 - (n+2)/q$. It follows that if $\varphi \in W_q^{2-2/q}(Q)$, then $\varphi \in C^\gamma(\overline{Q})$ and $\nabla \varphi \in (C^\gamma(\overline{Q}))^n$, where C^γ is the standard Hölder space (see [26, Sect. 4.6.1]).

The subspace $W_{q,N}^{2-2/q}(Q) \subset W_q^{2-2/q}(Q)$ of functions with homogeneous Neumann boundary conditions is a well-defined subspace, and in this section we always assume that $\varphi \in W_{q,N}^{2-2/q}(Q)$.

Definition 11.2.1 A solution to problem (11.6)–(11.8) on the time interval $[0, T)$ is a function $u \in W_q^{2,1}(Q_T)$ such that (11.6) is satisfied in $L_q(Q_T)$ and u satisfies (11.7) and (11.8) in terms of traces. A solution on $[0, \infty)$ is a function $u : Q \times [0, \infty) \to \mathbb{R}$ such that for any $T > 0$, $u|_{Q_T}$ is a solution in the sense just described.

We note that if $u \in W_q^{2,1}(Q_T)$, then $\mathcal{H}(\xi_0, u)$ is a measurable function on Q_T (see [1, Sect. 6.1]).

11.2.2 Case $n = 1$

Let $Q = (0, 1)$ and Q_j be given by (11.5).

Definition 11.2.2 Let $\varphi \in C^1(\overline{Q})$. We say φ is *transverse* with respect to ξ_0 if the following hold:

1. There is a $\overline{b} \in (0, 1)$ such that $Q_{-1} = \{x \mid 0 \le x \le \overline{b}\}$ and $Q_1 = \{x \mid \overline{b} < x \le 1\}$.
2. If $\varphi(\overline{b}) = \beta$, then $\varphi'(\overline{b}) < 0$.

An example of φ and ξ_0 satisfying Definition 11.2.2 is given in Fig. 11.3a.

Definition 11.2.3 A solution u is called *transverse* if for all $t \in [0, T]$, $u(\cdot, t)$ is transverse with respect to $\xi(\cdot, t)$.

Theorem 11.2.4 (See [11, Theorems 2.16 and 2.17]) *Suppose the initial data $\varphi \in W_{q,N}^{2-2/q}(Q)$ is transverse with respect to ξ_0. Then there is a $T > 0$ such that the following hold:*

1. *Any solution $u \in W_q^{2,1}(Q_T)$ of problem (11.6)–(11.8) is transverse.*
2. *There is at least one transverse solution $u \in W_q^{2,1}(Q_T)$ of problem (11.6)–(11.8).*
3. *If $u \in W_q^{2,1}(Q_T)$ is a transverse solution of problem (11.6)–(11.8), then it can be continued to a maximal interval of transverse existence $[0, T_{max})$, i.e., $u(x, T_{max})$ is not transverse or $T_{max} = \infty$.*

We will sketch the proof of Theorem 11.2.4, part 2, assuming that $\varphi(\overline{b}) = \beta$ and $\varphi'(\overline{b}) < 0$.

Let us define the closed, convex, bounded subset of $C[0, T]$

$$B := \{b \in C[0, T] \mid b(t) \in [0, 1], b(0) = \overline{b}\}.$$

For any $b_0 \in B$, define the function

$$F(x,t) := \begin{cases} h_{-1} & \text{if } 0 \le x \le b_0(t), \\ h_1 & \text{if } b_0(t) < x \le 1. \end{cases} \tag{11.9}$$

Let $u \in W_q^{2,1}(Q_T)$ be the solution to problem (11.6)–(11.8) with nonlinearity F in place of $\mathcal{H}(\xi_0, u)$. We claim that T can be chosen small enough such that the configuration $\xi(x,t)$ of $\mathcal{H}(\xi_0, u)$ is defined by a unique discontinuity point $b(t)$. Note that we do not yet claim that $F = \mathcal{H}(\xi_0, u)$.

To prove the claim, first fix $T_0 > 0$. It is a result of classical parabolic theory [25, Chap. 4] that for all $T \in [0, T_0]$

$$\|u\|_{C^\gamma(\overline{Q_T})} + \|u_x\|_{C^\gamma(\overline{Q_T})} \le C_1 \left(\|F\|_{L_q(Q_T)} + \|\varphi\|_{W_{q,N}^{2-2/q}(Q)} \right) \le C_2, \tag{11.10}$$

where $C_1, C_2, \dots > 0$ depend only on T_0 and q. The claim now follows from (11.10) with the help of the implicit function theorem.

Observe that u is a solution of problem (11.6)–(11.8) if $\mathcal{H}(\xi_0, u) = F$, i.e., $b_0 = b$. We therefore look for a fixed point of the map $\mathcal{R} : B \to B$, $\mathcal{R}(b_0) := b$.

Consider $b_{01}, b_{02} \in B$ and define F_1, F_2 via b_{01}, b_{02} similarly to (11.9), and let u_1, u_2 be the corresponding solutions. Observe that $F_1 \ne F_2$ only if

$$\min(b_{01}(t), b_{02}(t)) < x < \max(b_{01}(t), b_{02}(t)),$$

in particular,

$$\begin{aligned} \|u_1 - u_2\|_{C^\gamma(\overline{Q_T})} + \|u_{1x} - u_{2x}\|_{C^\gamma(\overline{Q_T})} &\le C_1 \|F_1 - F_2\|_{L_q(Q_T)}, \\ &\le C_3 \|b_{01} - b_{02}\|_{C[0,T]}^{1/q}. \end{aligned} \tag{11.11}$$

Applying (11.10) again, and using $\varphi'(\overline{b}) > 0$ and the implicit function theorem, we see that the left hand side of (11.11) bounds $\|a_1 - a_2\|_{C[0,T]}$. One can additionally show that $\|a_1 - a_2\|_{C[0,T]}$ bounds $\|b_1 - b_2\|_{C[0,T]}$, hence

$$\|b_1 - b_2\|_{C[0,T]} \le \|a_1 - a_2\|_{C[0,T]} \le C_4 \|b_{01} - b_{02}\|_{C[0,T]}^{1/q}. \tag{11.12}$$

In particular (11.12) shows that $\mathcal{R}$ is a continuous map on B. Moreover, one can use (11.10) to show that $\mathcal{R}(B)$ is bounded in $C^\gamma[0, T]$, and since $C^\gamma[0, T]$ is compactly embedded into $C[0, T]$, the Schauder fixed point theorem implies that $\mathcal{R}$ has a fixed point.

Theorem 11.2.5 (see [20, Theorem 2.2]) *If u_1 and u_2 are transverse solutions of problem (11.6)–(11.8) with the same φ, then $u_1 \equiv u_2$.*

We prove the theorem by expressing solutions as a convolution with the Green function $G(x, y, t, s)$ for the heat equation with Neumann boundary conditions.

Let us use this function to estimate the solution $w = u_1 - u_2$ of the heat equation with zero initial data, Neumann boundary conditions, and the right hand side $h = \mathcal{H}(\xi_0, u_1) - \mathcal{H}(\xi_0, u_2)$:

$$|w(x, t)| \leq \int_0^t \int_Q |G(x, y, t, s)| |h(y, s)| \, dy ds. \tag{11.13}$$

Also note that G satisfies the inequality (see, e.g., [27])

$$|G(x, y, t, s)| \leq \frac{C_1}{(t - s)^{1/2}}, \quad x, y \in Q, \, 0 \leq s < t, \tag{11.14}$$

where $C_1 > 0$ does not depend on x, y, t or s.

Similarly to the proof of Theorem 11.2.4, for every $s \leq t$ the integral of $|h(y, s)|$ over Q is bounded by $\|b_1 - b_2\|_{C[0,t]}$ and hence by $\|a_1 - a_2\|_{C[0,t]}$ and hence by $\|u_1 - u_2\|_{C(\overline{Q_t})}$. Combining this with (11.13) and (11.14), and taking the supremum over $(x, t) \in Q_T$ we get

$$\|w\|_{C(\overline{Q_T})} \leq C_2 \sqrt{T} \|w\|_{C(\overline{Q_T})},$$

where $C_2 > 0$ does not depend on T. Thus $w = 0$ for T small enough. A passage to arbitrary T is standard.

Theorem 11.2.6 (See [11, Theorem 2.9]) *Let $u \in W_q^{2,1}(Q_T)$ be a transverse solution of problem (11.6)–(11.8). If $\|\varphi - \varphi_n\|_{W_{q,N}^{2-2/q}(Q)} \to 0$ and $|\bar{b}_n - \bar{b}| \to 0$ as $n \to \infty$, then for sufficiently large n, problem (11.6)–(11.8) has a solution $u_n \in W_q^{2,1}(Q_T)$ with initial data φ_n and initial configuration ξ_{0n} defined via $\bar{b}_n$. Furthermore, $\|u_n - u\|_{W_q^{2,1}(Q_T)} \to 0$ as $n \to \infty$.*

The crux of the proof is showing that for sufficiently large n, all the solutions exist on the same time interval $[0, T]$. To this end we note that we have in fact given an explicit construction of T, and that this T depends on $\bar{b}$, $\|\varphi\|_{W_q^{2-2/q}(Q)}$, and if $\varphi(\bar{b}) = \beta$, also on $\varphi'(\bar{b})$. Hence for φ_n and $\bar{b}_n$ close enough to φ and $\bar{b}$ in their respective norms, the same T can be used.

11.2.3 Case $n \geq 2$

For the case $n \geq 2$ a notion of transversality has been studied in a model problem. For clarity we will define transversality for the case where the threshold β is adjoined to the free boundary between Q_1 and Q_{-1}, and α is not. In what follows, let int(A) denote the topological interior of a subset $A \subset Q$, and let $\{\varphi = \alpha\}$ be defined similarly to $\{u = \alpha\}$ but taking $x \in \overline{Q}$ instead of $(x, t) \in \overline{Q_T}$. In [22] the existence and

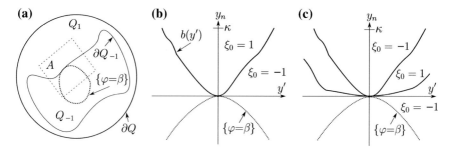

Fig. 11.4 An example of **a** the sets $Q_{\pm 1}$, **b** transverse data, and **c** non-transverse data

uniqueness of solutions were studied for initial data transverse in the following sense (see Fig. 11.4a, b, and recall that Q_j is given by (11.5)).

Definition 11.2.7 We say the function φ is transverse with respect to ξ_0 if the following hold:

1. Q_1 and Q_{-1} are measurable, $\partial Q_{-1} \subset Q, \partial Q_1 = \partial Q_{-1} \cup \partial Q$, and ∂Q_{-1} has zero Lebesgue measure.
2. $\varphi(x) < \beta$ for $x \in \text{int}(Q_1) \cup \partial Q$.
3. $\varphi(x) > \alpha$ for $x \in \overline{Q_{-1}}$.
4. If $x \in \{u = \beta\} \cap \partial Q_{-1}$, then there is a neighbourhood A of x, a set $A' \subset \mathbb{R}^{n-1}$, a $\kappa > 0$, and a map ψ such that

   (a) ψ is a composition of a translation and a rotation. and

$$\psi(A) = A' \times [-\kappa, \kappa], \quad \psi(x) = (0, 0).$$

   (b) There is a continuous function $\overline{b} : A' \to [-\kappa, \kappa]$ such that the configuration function $\xi_0 \circ \psi^{-1}$ in $\psi(A)$ (which we denote by $\xi_0(y', y_n), y' \in A')$ is given by

$$\xi_0(y', y_n) = \begin{cases} -1 & \text{if } -\kappa \leq y_n \leq \overline{b}(y'), \\ 1 & \text{if } \overline{b}(y') < y_n \leq \kappa. \end{cases}$$

   (c) $\varphi \circ \psi^{-1}$, which we write as $\varphi(y', y_n)$, satisfies $\varphi_{y_n}(0, 0) < 0$.

We observe that in Sect. 11.2.2, the boundary between Q_1 and Q_{-1} was a single point $\overline{b}$. But when $n \geq 2$, this boundary is assumed to have the structure of a continuous codimension 1 submanifold in a neighborhood of a point on the free boundary where φ takes a threshold value. Also note that for $n \geq 2$ non-transversality can be caused by the geometry of ∂Q_{-1} in addition to the possible degeneracy of $\nabla \varphi$ (see Fig. 11.4c and Sect. 11.2.4 for further discussion).

Theorem 11.2.8 (see [22, Theorems 3.18 and 3.19]) *Assume that* $n \geq 2$ *and* $\varphi \in W^{2-2/q}_{q,N}(Q)$ *is transverse with respect to* ξ_0. *Then there is a* $T > 0$ *such that any* solution $u \in W_q^{2,1}(Q_T)$ to problem (11.6)–(11.8) is transverse and there is at least one such solution. Moreover, if for some $T' > 0$, u_1 and u_2 are two transverse solutions to problem (11.6)–(11.8) on $Q_{T'}$, then $u_1 \equiv u_2$.

The main ideas of the proof are similar to those for the case $n = 1$. Since $(\varphi(y', \cdot), \xi_0(y', \cdot))$ is transverse in the 1d sense for every $y' \in A'$, one can prove continuity of a map $\mathcal{R}$ that now maps functions $u_0 \in C^\lambda(\overline{Q_T})$ ($\lambda < \gamma$) to solutions $\mathcal{R}(u_0) := u$ of problem (11.6)–(11.8) with the right hand side $\mathcal{H}(\xi_0, u_0)$. Estimate (11.10) implies that $u \in C^\gamma(\overline{Q_T})$, and the compactness of the embedding $C^\gamma(\overline{Q_T}) \subset C^\lambda(\overline{Q_T})$ and the Schauder fixed point theorem together imply that $\mathcal{R}$ has a fixed point in $C^\gamma(\overline{Q_T})$.

11.2.4 Generalizations and Open Problems

Let us list some generalizations for the case $n = 1$.

Change of topology. Suppose $u(x, t)$ becomes non-transverse at some time T in the sense of Definition 11.2.2. Then one of two possibilities arise. Either $u(x, T)$ has touched a threshold with zero spatial derivative at some point in $(0, 1)$, or this is not the case but $\lim_{t \to T} b(t) = 1$. In the latter case, one can continue the solution, and it remains unique, by redefining the problem effectively without hysteresis [11, Theorem 2.18]. We say that the topology of the hysteresis has changed at time T, in the sense that ξ transitions from piecewise constant to uniformly constant.

Continuous dependence on initial data. If u is a solution such that the topology has changed for some $t_1 < T$, then u need not continuously depend on the initial data since a sequence of approximating solutions u_n may become non-transverse at moments τ_n with $\tau_n < t_1$ and $\lim_{n \to \infty} \tau_n = t_1$ (the dashed line in Fig. 11.5). But

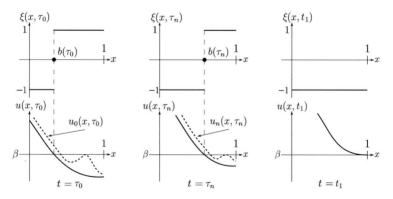

Fig. 11.5 A solution u (drawn as *solid lines* in the *lower picture*) and its configuration ξ (the *upper picture*) that remain transverse as a discontinuity of ξ disappears at time t_1. The *dashed line* in the *lower picture* is a series of non-transverse approximations u_n that become non-transverse at moments τ_n with $\tau_n < t_1$ and $\lim_{n \to \infty} \tau_n = t_1$

if we also assume that each u_n is a transverse solution, then solutions do depend continuously on their initial data.

Finite number of discontinuities. The results in Sect. 11.2.2 remain valid if the hysteresis topology is defined by finitely many discontinuity points. The hysteresis changing topology in the sense we described for one point of discontinuity corresponds to these points merging together in the general case (see Fig. 11.6).

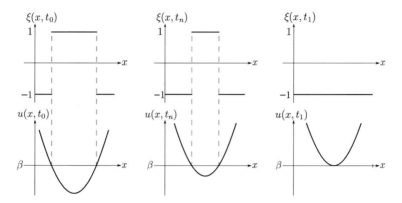

Fig. 11.6 Discontinuities merging as $t \to t_1$

General nonlinearity. The results in this section also hold for the more general problem (11.1)–(11.3). First one must assume that f is locally Lipschitz and dissipative (see [11, Condition 2.11]). With such an f, and if H_1 and H_{-1} are locally Hölder continuous, then transverse solutions exist and can be continued up to a maximal interval of transverse existence. If one additionally assumes that transverse solutions are unique, they can be shown to continuously depend on their initial data. To prove the uniqueness of solutions the authors of [20, 22] make the stronger assumption on H_1 and H_{-1}, namely that

$$|H_1(u_1) - H_1(u_2)| \leq \frac{M}{(\beta - u_1)^\sigma + (\beta - u_2)^\sigma}|u_1 - u_2|,$$

for u_1, u_2 in a left neighborhood of β, with $M > 0$ and $\sigma \in (0, 1)$, plus an analogous inequality for H_{-1} and a right neighborhood of α. This condition covers the case where H_1 and H_{-1} are the stable branches in the slow-fast approximation as in Fig. 11.2 (see the appendix of [20] for further discussion).

Systems of equations. In [21, Theorem 2.1], the results of Sect. 11.2.2 were generalized to systems of equations of the type in problem (11.1)–(11.3). It was also shown therein that problem (11.1)–(11.3) can be coupled to ordinary differential equations to cover the Hoppensteadt–Jäger model from [6, 7].

Let us conclude this subsection by discussing an open problem.

Open problem. In Fig. 11.4c, one can see that for every $y' \neq 0$, $(\varphi(y', \cdot), \xi_0(y', \cdot))$ is transverse in the 1d sense (with two discontinuties), but since the free boundary cannot be represented as a graph with codomain y_n at the point $y' = 0$, this initial data is not transverse. Whether Definition 11.2.7 can be generalized to include such cases is the subject of future work, and at this stage the authors strongly suspect that item 4 of Definition 11.2.7 can be replaced by the following statement: *if $x \in \{u = \beta\} \cap \partial Q_{-1}$, then $\nabla \varphi(x) \neq 0$.* In other words, the assumption that the free boundary is a graph is not necessary, and hence Fig. 11.4c would also be transverse. This question is intimately linked to the topology of the free boundary. Whether solutions can be continued to a maximal interval of existence and how to pose continuous dependence of initial data is unclear for the quite general conditions on Q_{-1} and Q_1 in Definition 11.2.7. These questions also apply to the case where $n = 1$ and ξ_0 has infinitely many discontinuities.

11.3 Regularity of Strong Solutions

To begin with let us discuss what we mean by regularity of solutions in this context. First observe that we cannot expect a classical solution since $\mathcal{H}$ has a jump discontinuity. Therefore the "optimal" regularity we expect is $W_\infty^{2,1}$. In this section we obtain $W_\infty^{2,1}$ "locally", for points $(x, t) \in Q_T$ outside of the static part of the free boundary. We will also assume the following condition:

Condition 11.3.1 $H_1(u) \equiv 1$ *and* $H_{-1}(u) \equiv -1$.

Let us introduce the notation $Q_T^{\pm 1} := \{(x, t) \mid \xi(x, t) = \pm 1\}$ and observe that u is smooth on the interior of $Q_T^{\pm 1}$.

The free boundary is defined as the set $\Gamma := \partial Q_T^1 \cap \partial Q_T^{-1}$. Moreover, we define $\Gamma_\alpha := \{u = \alpha\} \cap \Gamma$ and $\Gamma_\beta := \{u = \beta\} \cap \Gamma$. Note that both Γ_α and Γ_β have zero Lebesgue measure whenever u is a solution of problem (11.6)–(11.8). This follows from the fact that $u_t - \Delta u = 0$ a.e. on $\Gamma_\alpha \cup \Gamma_\beta$ and Condition 11.3.1 (see Alt's argument in the introduction and [12]).

The estimates we obtain will depend critically on the *static* part of the free boundary $\Gamma_v := \Gamma \backslash (\Gamma_\alpha \cup \Gamma_\beta)$. If $(x, t) \in \Gamma_v$, then $u(x, t) \neq \alpha, \beta$ and by continuity of u, $u(x, t \pm \tau) \neq \alpha, \beta$ for τ sufficiently small. This means $\xi(x, t \pm \tau) = \xi(x, t)$ and so if we draw the t-axis vertically as in Fig. 11.7, Γ_v looks like a vertical strip.

Next we recall the definition of a parabolic cylinder

$$P_r(x^0, t^0) := \{x \in \mathbb{R}^n \mid \|x^0 - x\|_{\mathbb{R}^n} < r\} \times (t^0 - r^2, t^0 + r^2), \quad r > 0.$$

We define the parabolic distance between (x^0, t^0) and a set $A \subset Q_T$ as

$$\text{dist}_p((x^0, t^0), A) := \sup\{r > 0 \mid P_r(x^0, t^0) \cap \{t \leq t^0\} \cap A = \emptyset\}.$$

This is all the notation we need to state the main result of [23].

Fig. 11.7 A possible scenario where $\Gamma \neq \Gamma_\alpha \cup \Gamma_\beta$ and Γ_ν appears. *White* and *grey* indicate the regions Q_T^1 and Q_T^{-1} respectively

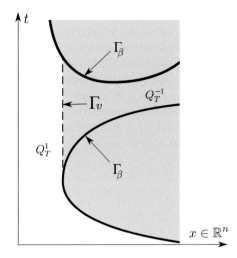

Theorem 11.3.2 (see [23, Theorem 2.3]) *We assume that* $n \geq 1$ *and* u *is a solution of problem* (11.6)–(11.8). *Then*

$$|u_t(x,t)| + \sum_{i,j=1}^{n} |u_{x_i x_j}(x,t)| \leq C(\rho_\nu, \rho_b, M), \quad a.e. \ (x,t) \in Q_T \backslash \overline{\Gamma_\nu},$$

where C *depends on* $\rho_\nu := dist_p((x,t), \Gamma_\nu)$, $\rho_b := dist_p((x,t), \partial' Q_T \cup (Q \times \{0\}))$, *and* $M := \sup_{(x,t) \in Q_T} |u(x,t)|$.

To explain the main ideas in the proof we define some further notation. Let $\Gamma_\alpha^0 = \Gamma_\alpha \cap \{\nabla u = 0\}$ and $\Gamma_\alpha^* = \Gamma_\alpha \backslash \Gamma_\alpha^0$, with Γ_β^0 and Γ_β^* defined similarly. Furthermore, define $\Gamma^0 = \Gamma_\alpha^0 \cup \Gamma_\beta^0$ and $\Gamma^* = \Gamma_\alpha^* \cup \Gamma_\beta^*$.

The crucial point in the proof is the quadratic growth estimate

$$\sup_{P_r(x,t)} |u - \beta| \leq C_1(\rho_\nu, \rho_b, M)r^2 \quad \text{for} \quad r \leq \min\{\rho_\nu, \rho_b\}, \tag{11.15}$$

and $(x,t) \in \Gamma_\beta^0$ (the estimate on Γ_α^0 is similar). The main tool for showing the quadratic bound (11.15) is the local rescaled version of the Caffarelli monotonicity formula, see [23, 28, 29].

Furthermore, the quadratic growth estimate (11.15) implies the corresponding linear bound for $|\nabla u|$

$$\sup_{P_r(x,t)} |\nabla u| \leq C_2(\rho_\nu, \rho_b, M)r \quad \text{for all} \quad r \leq \min\{\rho_\nu, \rho_b\}, \tag{11.16}$$

with $(x,t) \in \Gamma^0$. The dependence of C_1 and C_2 on the distance ρ_ν in (11.15) and (11.16) arises due to the monotonicity formula. Near Γ_ν neither the local rescaled version of Caffarelli's monotonicity formula nor its generalizations (such as the almost monotonicity formula) are applicable to the positive and negative parts of the spatial directional derivatives $D_e u$, with $e \in \mathbb{R}^n$.

Besides estimates (11.15) and (11.16), one also needs information about the behaviour of u_t near Γ^*. Although u_t may have jumps across the free boundary, one can show that u_t is a continuous function in a neighborhood of $(x, t) \in \Gamma^* \backslash \Gamma_v$. In addition, the monotonicity of the jumps of $\mathcal{H}(\xi_0, u)$ in the t-direction provides one-sided estimates of u_t near Γ_α and Γ_β. Combining these results with the observation that $u_t \leq 0$ on $\Gamma_\alpha^* \backslash \Gamma_v$, and $u_t \geq 0$ on $\Gamma_\beta^* \backslash \Gamma_v$ gives

$$\sup_{\Gamma^* \backslash \Gamma_v} |u_t| \leq C_3(\rho_b, M). \tag{11.17}$$

Inequalities (11.15)–(11.17) allow one to apply methods from the theory of free boundary problems (see, e.g., [18, 19]) and estimate $|u_t(x, t)|$ and $|u_{x_i x_j}(x, t)|$ for a.e. $(x, t) \in Q_T \backslash \overline{\Gamma_v}$.

11.4 Non-transverse Initial Data

11.4.1 Setting of a Problem

In this section we summarize the recent work [24], where the nontransverse case is analyzed for $x \in \mathbb{R}$, and indicate directions for further research. We will be interested in the behavior of solutions near one of the thresholds, say β. Therefore, we set $\alpha = -\infty$ and $\beta = 0$ (see Fig. 11.8) and assume that the initial data satisfy $\varphi(x) = -cx^2 + o(x^2)$ in a small neighborhood of the origin, $\varphi(x) < 0$ everywhere outside of the origin, $\xi_0(x) = -1$ for $x = 0$, and $\xi_0(x) = 1$ for $x \neq 0$. In particular, we assume $c > 0$. In this situation, the theorems in Sect. 11.2.2 are not applicable. Hence, to understand the dynamics of the solution near the origin, we approximate the continuous equation (11.6) by its spatial discretization and the initial data by the discrete quadratic function. Namely, we choose a grid step $\varepsilon > 0$, set $u_n^\varepsilon(t) := u(\varepsilon n, t)$, $n \in \mathbb{Z}$, and consider the system of infinitely many ordinary differential equations with hysteresis

Fig. 11.8 Hysteresis with thresholds $\alpha = -\infty$ and $\beta = 0$

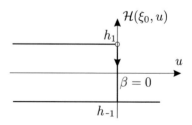

$$\frac{du_n^\varepsilon}{dt} = \frac{u_{n+1}^\varepsilon - 2u_n^\varepsilon + u_{n-1}^\varepsilon}{\varepsilon^2} + \mathcal{H}(u_n^\varepsilon), \quad t > 0, \ n \in \mathbb{Z}, \tag{11.18}$$

supplemented by the nontransverse (quadratic) initial data

$$u_n^\varepsilon(0) = -c(\varepsilon n)^2, \quad n \in \mathbb{Z}. \tag{11.19}$$

Here we do not explicitly indicate the dependence of $\mathcal{H}$ on ξ_0, assuming that $\mathcal{H}(u_n^\varepsilon)(t) = h_1$ if $u_n^\varepsilon(s) < 0$ for all $s \in [0, t]$ and $\mathcal{H}(u_n^\varepsilon)(t) = h_{-1}$ otherwise. As before, we assume that $h_{-1} \leq 0 < h_1$.

Due to [24, Theorem 2.5], problem (11.18), (11.19) admits a unique solution in the class of functions satisfying

$$\sup_{s\in[0,t]} |u_n^\varepsilon(s)| \leq A e^{B|n|}, \quad n \in \mathbb{Z}, \ t \geq 0,$$

with some $A = A(t, \varepsilon) \geq 0$ and $B = B(t, \varepsilon) \in \mathbb{R}$. Thus, we are now in a position to discuss the dynamics of solutions for each fixed grid step ε and analyze the limit $\varepsilon \to 0$.

First, we observe that ε in (11.18), (11.19) can be scaled out. Indeed, setting

$$u_n(t) := \varepsilon^{-2} u_n^\varepsilon(\varepsilon^2 t) \tag{11.20}$$

reduces problem (11.18), (11.19) to the equivalent one

$$\begin{cases} \dfrac{du_n}{dt} = u_{n+1} - 2u_n + u_{n-1} + \mathcal{H}(u_n), \quad t > 0, \ n \in \mathbb{Z}, \\ u_n(0) = -cn^2, \quad n \in \mathbb{Z}. \end{cases} \tag{11.21}$$

Using the comparison principle, it is easy to see that if $h_1 \leq 2c$, then $u_n(t) < 0$ for all $n \in \mathbb{Z}$ and $t > 0$ and, therefore, no switchings happen for $t > 0$. Let us assume that

$$h_{-1} \leq 0 < 2c < h_1. \tag{11.22}$$

It is easy to show that $u_n(t) \leq 0$ for all $n \in \mathbb{Z}$ and $t > 0$. However, some nodes can now reach the threshold $\beta = 0$ and switch the hysteresis. The main question is which nodes do this and according to which law.

11.4.2 Numerical Observations

The following pattern formation behavior is indicated by numerics (see Fig. 11.9). As time goes on, the spatial profile of $u_n(t)$ forms two symmetric hills propagating away from the origin. At the same time, the whole spatial profile oscillates up and

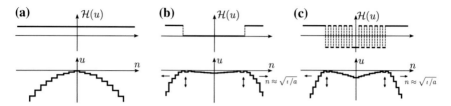

Fig. 11.9 Upper graphs represent spatial profiles of the hysteresis $\mathcal{H}(u_n)$ and lower graphs the spatial profiles of the solution u_n. **a** Nontransverse initial data. **b** Spatial profiles at a moment $t > 0$ for $h_{-1} = 0$. **c** Spatial profiles at a moment $t > 0$ for $h_{-1} = -h_1 < 0$

down (never exceeding the threshold $\beta = 0$) and touches the threshold $\beta = 0$ in such a way that

$$\lim_{j \to \infty} \frac{N_{\text{ns}}(j)}{N_{\text{s}}(j)} = \frac{|h_{-1}|}{h_1}, \tag{11.23}$$

where $N_{\text{s}}(j)$ and $N_{\text{ns}}(j)$ are integers denoting the number of nodes in the set $\{u_0, u_{\pm 1}, \ldots, u_{\pm j}\}$ that switch and do not switch, respectively, on the time interval $[0, \infty)$. In [24], such a spatio-temporal pattern was called *rattling*.

A more specific pattern occurs if $|h_{-1}|/h_1 = p_{\text{ns}}/p_{\text{s}}$, where p_{s} and p_{ns} are co-prime integers. In this case, for any j large enough, the set $\{u_{j+1}, \ldots, u_{j+p_{\text{s}}+p_{\text{ns}}}\}$ contains exactly p_{s} nodes that switch and p_{ns} nodes that do not switch on the time interval $[0, \infty)$.

If a node u_n switches on the time interval $[0, \infty)$, then we denote its switching moment by t_n; otherwise, set $t_n := \infty$. In particular, finite values of t_n characterize the propagation velocity of the two hills mentioned above. Numerics indicates that, for the nodes where t_n is finite, we have

$$t_n = an^2 + \begin{cases} O(\sqrt{n}) & \text{if } h_{-1} = 0, \\ O(n) & \text{if } h_{-1} < 0, \end{cases} \quad \text{as } n \to \infty, \tag{11.24}$$

and

$$|u_{k+1}(t) - u_k(t)| \le b, \quad |k| \le n, \ t \ge t_n, \ n = 0, 1, 2, \ldots, \tag{11.25}$$

where $a, b > 0$ do not depend on k and n. In particular, (11.24) and (11.25) mean that the hills propagate with velocity of order $t^{-1/2}$, while the cavity between the hills has a bounded steepness, which distinguishes the observed phenomenon from the "classical" traveling wave situation.

11.4.3 Rigorous Result

The recent work [24] provides a rigorous analysis of the rattling in the case $h_{-1} = 0$, where, according to (11.24), all the nodes are supposed to switch at time moments satisfying

$$t_n = an^2 + q_n, \quad |q_n| \le E\sqrt{n}, \tag{11.26}$$

where $E > 0$ does not depend on $n \in \mathbb{Z}$. In [24], the authors found the coefficient a and proved that if finitely many nodes $u_n(t)$, $n = 0, \pm 1, \cdots \pm n_0$, switch at time moments t_n satisfying (11.26), then all the nodes $u_n(t)$, $n \in \mathbb{Z}$, switch at time moments t_n satisfying (11.26) (see the rigorous statement below). One of the main tools in the analysis is the so-called discrete Green function $y_n(t)$ that is a solution of the problem

$$\begin{cases} \dot{y}_0 = \Delta y_0 + 1, & t > 0, \\ \dot{y}_n = \Delta y_n, & t > 0, \ n \ne 0, \\ y_n(0) = 0, & n \in \mathbb{Z}. \end{cases} \tag{11.27}$$

The important property of the discrete Green function is the following asymptotics proved in [30]:

$$y_n(t) = \sqrt{t} f\left(\frac{|n|}{\sqrt{t}}\right) + O\left(\frac{1}{\sqrt{t}}\right) \quad \text{as } t \to \infty, \tag{11.28}$$

where

$$f(x) := 2x \int_x^\infty \mathbb{Z}^{-2} h(\mathbb{Z}) \, d\mathbb{Z}, \quad h(x) := \frac{1}{2\sqrt{\pi}} e^{-\frac{x^2}{4}}, \tag{11.29}$$

and $O(\cdot)$ does not depend on $n \in \mathbb{Z}$.

Now if we (inductively) assume that the nodes $u_0, u_{\pm 1}, \dots u_{\pm(n-1)}$ switched at the moments satisfying (11.26), while no other nodes switched on the time interval $[0, t_{n-1}]$, then the dynamics of the node

$u_n(t)$ for $t \ge t_{n-1}$ (and until the next switching in the system occurs) is given by

$$u_n(t) = -cn^2 + (h_1 - 2c)t - h_1 \sum_{k=-(n-1)}^{n-1} y_{n-k}(t - t_k). \tag{11.30}$$

At the (potential) switching moment $t_n = an^2 + q_n$, the relations $t_k = ak^2 + q_k$ ($|k| \le n - 1$), equality (11.30), the Taylor formula, and asymptotics (11.28) yield

$$0 = -cn^2 + (h_1 - 2c)an^2 - h_1 \sum_{k=-(n-1)}^{n-1} y_{n-k}\left(a(n^2 - k^2)\right) + \text{l.o.t.}$$

$$= -cn^2 + (h_1 - 2c)an^2 - h_1 \sum_{k=-(n-1)}^{n-1} \sqrt{a(n^2 - k^2)} f\left(\frac{n-k}{\sqrt{a(n^2 - k^2)}}\right) \quad (11.31)$$

$$+ \text{l.o.t.}$$

$$= (-c + (h_1 - 2c)a - h_1 R_n(a)) n^2 + \text{l.o.t.},$$

where

$$R_n(a) := \sum_{k=-(n-1)}^{n-1} \frac{1}{n}\sqrt{a(1 - (k/n)^2)} f\left(\frac{1 - k/n}{\sqrt{a(1 - (k/n)^2)}}\right)$$

and "l.o.t." stands for lower order terms that we do not explicitly specify here. Note that $R_n(a)$ is the Riemann sum for the integral

$$I_f(a) := \int_{-1}^{1} \sqrt{a(1 - x^2)} f\left(\frac{1 - x}{\sqrt{a(1 - x^2)}}\right) dx. \quad (11.32)$$

Therefore, equality (11.31) can be rewritten as

$$0 = \left(-c + (h_1 - 2c)a - h_1 I_f(a)\right) n^2 + \text{l.o.t.} \quad (11.33)$$

It is proved in [24] that there exists a unique $a > 0$ for which the coefficient at n^2 in (11.33) vanishes. The most difficult part is to analyze the lower order terms in (11.33) that involve:

1. the remainders $q_0, q_{\pm 1}, \ldots, q_n$ from (11.26) arising from (11.30) via the application of the Taylor formula,
2. the remainder in the asymptotic (11.28) for the discrete Green function $y_n(t)$,
3. the remainders arising from approximating the integral $I_f(a)$ by the Riemann sum $R_n(a)$.

In particular, one has to prove that if $|q_j| \leq E\sqrt{|j|}$ for $j = 0, \pm 1, \ldots, \pm(n - 1)$, then the lower order terms vanish for a specified above and $|q_n| \leq E\sqrt{|n|}$. This allows one to continue the inductive scheme and (after an appropriate analysis of the nodes $u_{\pm(n+1)}(t), u_{\pm(n+2)}(t), \ldots$ for $t \in [t_{n-1}, t_n]$) complete the proof.

The rigourous formulation of the main result in [24] is as follows.

Theorem 11.4.1 (see [24, Theorem 3.2]) *Assume that* (11.22) *holds and that* $h_{-1} = 0$. *Let* $a = a(h_1/c) > 0$ *be a (unique) root of the equation*

$$-c + (h_1 - 2c)a - h_1 I_f(a) = 0 \quad (11.34)$$

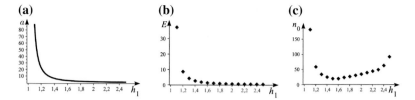

Fig. 11.10 Dependence on h_1 of the values of a, E, and $n_0(E)$ that fulfill assumptions (11.35) for $c = 1/2$. **a** The values of a are found as roots of (11.34). **b, c** The values of E and $n_0(E)$ are calculated for discrete values $h_1 = 1.1, 1.2, 1.3, \ldots, 2.5$

with $I_f(a)$ given by (11.32). *Then there exists a constant $E_0 = E_0(h_1, c, a) > 0$ and a function $n_0 = n_0(E) = n_0(E, h_1, c, a)$ (both explicitly constructed) with the following property. If*

> *finitely many nodes $u_0(t), u_1(t), \ldots, u_{n_0}(t)$ switch at moments t_n*
> *satisfying (11.26) with the above a and some $E \geq E_0$,*   (11.35)

then each node $u_n(t)$, $n \in \mathbb{Z}$, switches; moreover, the switching occurs at a time moment t_n satisfying (11.26) with a and E as in (11.35).

We note that the explicit formula (11.30) for the solution $u_n(t)$ allows one to verify the fulfillment of finitely many assumptions (11.35) numerically with an arbitrary accuracy for any given values of h_1 and c. The graphs in Fig. 11.10 taken from [24] represent the values of a, E, and $n_0(E)$ that fulfill assumption (11.35) for $c = 1/2$ and $h_1 = 1.1, 1.2, 1.3, \ldots, 2.5$.

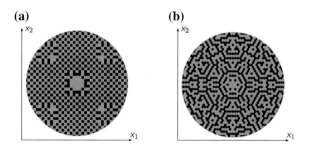

Fig. 11.11 A snapshot for a time moment $t > 0$ of a two-dimensional spatial profile of hysteresis taking values $h_1 > 2c > 0$ and $h_{-1} = -h_1 < 0$. The nontrasverse initial data is given by $\varphi(x) = -c(x_1^2 + x_2^2)$. *Grey* (*black*) squares or hexagons correspond to the nodes that have (not) switched on the time interval $[0, t]$. **a** Discretization on the square lattice. **b** Discretization on the triangular lattice

11.4.4 Open Problems

To conclude this section, we indicate several directions of further research in the nontransverse case.

Case $h_{-1} < 0$. In this case, one has to additionally prove a specific switching pattern (11.23). We expect that the tools developed in [24] will work for rational h_1/h_{-1}. The irrational case appears to be a much more difficult problem.

Multi-dimensional case. Numerics indicates that the behavior analogous to (11.23) occurs in higher spatial dimensions for different kinds of approximating grids. Figure 11.11 illustrates the switching pattern for a two-dimensional analog of problem (11.21), where the Laplacian is discretized on the square and triangular lattices, respectively.

Limit $\varepsilon \to 0$. We introduce the function

$$u^\varepsilon(x,t) := u_n^\varepsilon(t), \quad x \in [\varepsilon n - \varepsilon/2, \varepsilon n + \varepsilon/2), \ n \in \mathbb{Z},$$

(which is piecewise constant in x for every fixed t). Making the transformation inverse to (11.20) and assuming (11.23) and (11.24), we can deduce that, as $\varepsilon \to 0$, the function $u^\varepsilon(x,t)$ approximates a smooth function $u(x,t)$, which satisfies $u(x,t) = 0$ for $x \in (-\sqrt{t/a}, \sqrt{t/a})$. In other words, $u(x,t)$ sticks to the threshold line $\beta = 0$ on the expanding interval $x \in (-\sqrt{t/a}, \sqrt{t/a})$.

Similarly to $u^\varepsilon(x,t)$, we consider the function

$$H^\varepsilon(x,t) := \mathcal{H}(u_n^\varepsilon)(t), \quad x \in [\varepsilon n - \varepsilon/2, \varepsilon n + \varepsilon/2), \ n \in \mathbb{Z},$$

which is supposed to approximate the hysteresis $\mathcal{H}(u)(x,t)$ in (11.6). We see that the spatial profile of $H^\varepsilon(x,t)$ for $x \in (-\sqrt{t/a}, \sqrt{t/a})$ is a step-like function taking values h_1 and h_{-1} on alternating intervals of length of order ε. Hence, it has no pointwise limit as $\varepsilon \to 0$, but converges in a weak sense to the function $H(x,t)$ given by $H(x,t) = 0$ for $x \in (-\sqrt{t/a}, \sqrt{t/a})$ and $H(x,t) = h_1$ for $x \notin (-\sqrt{t/a}, \sqrt{t/a})$. We emphasize that $H(x,t)$ does not depend on h_{-1} (because a does not). On the other hand, if $h_{-1} < 0$, the hysteresis operator $\mathcal{H}(u)(x,t)$ in (11.6) cannot take value 0 by definition, which clarifies the essential difficulty with the well-posedness of the original problem (11.6) in the nontransverse case. To overcome the non-wellposedness, one need to allow the intermediate value 0 for the hysteresis operator, cf. the discussion of modified hysteresis operators due to Visintin and Alt in the introduction. A rigorous analysis of the limit $\varepsilon \to 0$ is an open problem, which may lead to a unique "physical" choice of an appropriate element in the multi-valued Visintin's hysteresis $\mathcal{H}_{\mathrm{Vis}}(\xi_0, u)$ in Definition 11.1.2.

Rattling in slow-fast systems. One may think that the rattling occurs exclusively due to the discontinuous nature or hysteresis. This is not quite the case. Consider an equation of type (11.6) with the hysteresis $\mathcal{H}(\xi_0, u)$ replaced by the solution v of a bistable ordinary differential equation of type (11.4), e.g.,

$$u_t = u_{xx} + v, \quad \delta v_t = g(u, v). \tag{11.36}$$

Numerical solution of system (11.36) with a nontransverse initial data $u(x, 0) = -cx^2 + o(x^2)$ and $v(x, 0) = H_1(\beta)$ near the origin reveals a behavior analogous to that for a spatially discrete system (see Fig. 11.12). As the spatial profile of $u(x, t)$ touches the threshold β at some point x_0, the spatial profile of $v(x, t)$ forms a peak-like transition layer around x_0 that rapidly converges to a plateau. Thus, as time goes on, the spatial profile of $v(x, t)$ converges to a step-like function taking values $H_1(\beta)$ and $H_{-1}(\beta)$ on alternating intervals, whose length tends to zero as $\delta \to 0$. A rigorous analysis of the limit $\delta \to 0$ is an open problem.

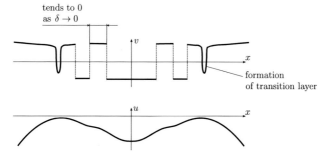

Fig. 11.12 Lower and upper graphs are spatial profiles of the solution $u(x, t)$ and $v(x, t)$, respectively, for problem (11.36) with initial data $u|_{t=0} = -cx^2 + o(x^2)$, $v|_{t=0} = H_1(\beta)$

Acknowledgments The authors are grateful for the support of the DFG project SFB 910 and the DAAD project G-RISC. The work of the first author was partially supported by the Berlin Mathematical School. The work of the second author was partially supported by the DFG Heisenberg programme. The work of the third author was partially supported by Chebyshev Laboratory (Department of Mathematics and Mechanics, St. Petersburg State University) under RF Government grant 11.G34.31.0026, JSC "Gazprom neft", by the Saint-Petersburg State University research grant 6.38.223.2014 and RFBR 15-01-03797a.

References

1. A. Visintin, *Differential Models of Hysteresis*. Applied Mathematical Sciences (Springer-Verglag, Berlin, 1994)
2. P. Krejčí, *Hysteresis Convexity and Dissipation in Hyperbolic Equations*. GAKUTO International series (Gattötoscho, 1996)
3. M. Brokate, J. Sprekels, *Hysteresis and Phase Transitions*. Applied Mathematical Sciences (Springer-Verlag, New York, 1996)
4. A. Visintin, Acta Applicandae Mathematicae **132**(1), 635 (2014)
5. A. Visintin, Discrete Contin. Dyn. Syst., Ser. S 8(4), 793 (2015)
6. F. Hoppensteadt, W. Jäger, in *Biological Growth and Spread. Lecture Notes in Biomathematics*, vol. 38, ed. by W. Jäger, H. Rost, P. Tautu (Springer, Berlin Heidelberg, 1980), pp. 68–81

7. F. Hoppensteadt, W. Jäger, C. Pöppe, in *Modelling of Patterns, in Space and Time. Lecture Notes in Biomathematics*, ed. by W. Jäger, J.D. Murray (Springer, Berlin Heidelberg, 1984), vol. 55, pp. 123–134
8. A. Marciniak-Czochra, Math. Biosci. **199**(1), 97 (2006)
9. A. Köthe, Hysteresis-driven pattern formation in reaction-diffusion-ode models. Ph.D. thesis, University of Heidelberg (2013)
10. M. Krasnosel'skii, M. Niezgodka, A. Pokrovskii, *Systems with Hysteresis* (Springer, Berlin, 2012)
11. P. Gurevich, S. Tikhomirov, R. Shamin, SIAM J. Math. Anal. **45**(3), 1328 (2013)
12. H.W. Alt, Control Cybern. **14**(1–3), 171 (1985)
13. A. Visintin, SIAM J. Math. Anal. **17**(5) (1986)
14. T. Aiki, J. Kopfová, in *Recent Advances in Nonlinear Analysis* (2008), pp. 1–10
15. P. Krejčí, J. Physics.: Conf. Ser. (22), 103 (2005)
16. E. Mischenko, N. Rozov, *Differential Equations with Small Parameters and Relaxation Oscillations* (Plenum, New York, 1980)
17. C. Kuehn, *Multiple Time Scale Dynamics, Applied Mathematical Sciences*, vol. 191 (Springer International Publishing, 2015)
18. D. Apushkinskaya, N. Uraltseva, St. Petersbg. Math. J. **25**(2), 195 (2014)
19. H. Shahgholian, N. Uraltseva, G.S. Weiss, Adv. Math. **221**(3), 861 (2009)
20. P. Gurevich, S. Tikhomirov, Nonlinear Anal. **75**(18), 6610 (2012)
21. P. Gurevich, S. Tikhomirov, Mathematica Bohemica (Proc. Equadiff 2013) **139**(2), 239 (2014)
22. M. Curran, Local well-poseness of a reaction-diffusion equation with hysteresis. Master's thesis, Fachbereich Mathematik und Informatik, Freie Universität Berlin (2014)
23. D. Apushkinskaya, N. Uraltseva, Interfaces and Free Boundaries **17**(1), 93 (2015)
24. P. Gurevich, S. Tikhomirov, arXiv:1504.02385 [math.AP] (2015)
25. O. Ladyzhenskaya, V. Solonnikov, N. Uraltseva, *Linear and Quasilinear Equations of Parabolic Type* (American Mathematical Society, Providence, Rohde Island, 1968)
26. H. Triebel, *Interpolation Theory, Function Spaces, Differential Operators*. Carnegie-Rochester Conference Series on Public Policy (North-Holland Publishing Company, 1978)
27. S. Ivasishen, Math. USSR-Sb (4), 461 (1981)
28. L. Caffarelli, S. Salsa, *A Geometric Approach to Free Boundary Problems*. Graduate Studies in Mathematics (American Mathematical Soc., 2005)
29. D. Apushkinskaya, H. Shahgholian, N. Uraltseva, J. Math. Sci. **115**(6), 2720 (2003)
30. P. Gurevich, arXiv:1504.02673 [math.AP] (2015)

Chapter 12
Deriving Effective Models for Multiscale Systems via Evolutionary Γ-Convergence

Alexander Mielke

Abstract We discuss possible extensions of the recently established theory of evolutionary Γ-convergence for gradient systems to nonlinear dynamical systems obtained by perturbation of a gradient systems. Thus, it is possible to derive effective equations for pattern forming systems with multiple scales. Our applications include homogenization of reaction-diffusion systems, the justification of amplitude equations for Turing instabilities, and the limit from pure diffusion to reaction-diffusion. This is achieved by generalizing the Γ-limit approaches based on the energy-dissipation principle or the evolutionary variational estimate.

12.1 Introduction

The theory of evolutionary Γ-convergence was developed for families of gradient systems $(X, \mathcal{E}_\varepsilon, \mathcal{R}_\varepsilon)_{\varepsilon \in [0,1]}$, which define the family of gradient flows

$$\mathrm{D}_{\dot{u}}\mathcal{R}_\varepsilon(u^\varepsilon, \dot{u}^\varepsilon) = -\mathrm{D}\mathcal{E}_\varepsilon(u^\varepsilon), \qquad u^\varepsilon(0) = u_\varepsilon^0.$$

The aim of the theory is to provide as general conditions as possible for the convergence of the energy functionals $\mathcal{E}_\varepsilon \rightsquigarrow \mathcal{E}_0$ and of the dissipation potentials $\mathcal{R}_\varepsilon \rightsquigarrow \mathcal{R}_0$ for $\varepsilon \to 0$, that still guarantee that the solutions $u^\varepsilon : [0, T] \to X$ converge to a solution $u^0 : [0, T] \to X$ of the limiting gradient flow as $\varepsilon \to 0$. We refer to the surveys [6, 23, 34–36]. We emphasize here that there are numerous much older works relating to the case that X is a Hilbert space H and $\mathcal{R}_\varepsilon(u, \dot{u}) = \frac{1}{2}\|\dot{u}\|_H^2$ is independent of ε such that only equation has the form $\dot{u} = -\mathrm{D}\mathcal{E}_\varepsilon(u)$ where A_ε is a maximal monotone operator, see [3, 7].

A. Mielke
Weierstraß-Institut für Angewandte Analysis und Stochastik,
Mohrenstraße 39, 10117 Berlin, Germany
e-mail: mielke@wias-berlin.de

© Springer International Publishing Switzerland 2016   235
E. Schöll et al. (eds.), *Control of Self-Organizing Nonlinear Systems*,
Understanding Complex Systems, DOI 10.1007/978-3-319-28028-8_12

Here we are interested in perturbed gradient systems, where we allow the energy functional $\mathcal{E}_\varepsilon$ to depend on the time $t \in [0, T]$ and the equation to contain a non-gradient term h_ε. We use the quadruple $(X, \mathcal{E}_\varepsilon, \mathcal{R}_\varepsilon, h_\varepsilon)$ to denote the perturbed gradient system, which then defines an evolutionary equation

$$D_{\dot{u}}\mathcal{R}_\varepsilon(u^\varepsilon, \dot{u}^\varepsilon) = -D\mathcal{E}_\varepsilon(u^\varepsilon) + h_\varepsilon(t, u^\varepsilon), \qquad u^\varepsilon(0) = u^0_\varepsilon. \qquad (12.1)$$

Here we understand that h_ε is a lower order perturbation of the gradient system obtained for $h_\varepsilon \equiv 0$. Thus, the hope is that it is possible to generalize the strong results on evolutionary convergence of gradient systems (see [23, 35]) to the perturbed case without adding too much technicalities.

Hence, there are two major motivations for considering perturbed gradient systems. On the one hand, there may be cases where a given system has a particular gradient structure $(\widehat{X}, \widehat{\mathcal{E}}_\varepsilon, \widehat{\mathcal{R}}_\varepsilon)$, but it may be easier to treat it as a perturbed gradient system $(X, \mathcal{E}_\varepsilon, \mathcal{R}_\varepsilon)$. We highlight this by looking at the reaction-diffusion system

$$\dot{u} = \mathrm{div}\left(a_\varepsilon(x)\nabla u\right) + \frac{c_\varepsilon(x)(1-uv)}{d_\varepsilon(x)+u+v}, \quad \dot{v} = \mathrm{div}\left(b_\varepsilon(x)\nabla v\right) + \frac{c_\varepsilon(x)(1-uv)}{d_\varepsilon(x)+u+v},$$

where $u, v > 0$ are densities and $a_\varepsilon, b_\varepsilon, c_\varepsilon$ and d_ε are positive ε-periodic coefficients. It was shown in [21] that this system is has a gradient system with

$$\widehat{\mathcal{E}}(u, v) = \int_\Omega \lambda_B(u) + \lambda_B(v)\,\mathrm{d}x \quad \text{with } \lambda_B(u) := u\log u - u + 1,$$

$$\widehat{\mathcal{R}}^*_\varepsilon(u, v, \mu, \nu) = \int_\Omega \left(\frac{a_\varepsilon}{2}|\nabla\xi|^2 + \frac{b_\varepsilon}{2}|\nabla v|^2 + C_\varepsilon(x, u, v)(\xi + v)^2\right)\mathrm{d}x,$$

where $c_\varepsilon(x, u, v) = \frac{c_\varepsilon(x)uv}{(d_\varepsilon(x)+u+v)(\log(uv)-1)} > 0$, and $\mathcal{R}^*_\varepsilon$ is the Legendre dual potential of $\mathcal{R}_\varepsilon$, see (12.15). However, doing a multiscale analysis for the limit $\varepsilon \to 0$ is very difficult because of the dependence of $\widehat{\mathcal{R}}^*_\varepsilon$ on u and v.

For a perturbed gradient structure we may choose the classical L^2 gradient structure for the leading terms and treat the reactions as perturbations, i.e.

$$\mathcal{E}_\varepsilon(u, v) = \int_\Omega \left(\frac{a_\varepsilon}{2}|\nabla u|^2 + \frac{b_\varepsilon}{2}|\nabla v|^2\right)\mathrm{d}x, \quad \mathcal{R}_\varepsilon(\dot{u}, \dot{v}) = \frac{1}{2}\|\dot{u}\|^2_{L^2} + \frac{1}{2}\|\dot{v}\|^2_{L^2},$$

and the perturbation $h_\varepsilon(x, u, v) = \frac{c_\varepsilon(x)(1-uv)}{d_\varepsilon(x)+u+v}(1, 1)^\top$. For such a system the limit $\varepsilon \to 0$ can be taken much more easily, see Sect. 12.5.1 and [27, 31].

On the other hand, the treatment of perturbed gradient systems is important, since the dynamics of pure gradient systems is completely different from perturbed ones. In gradient systems, typical solutions converge to local minimizer of the energy for $t \to \infty$. In a perturbed s gradient system, much more complicated dynamics can happen, like Hopf bifurcations or chaos, see e.g. [16].

Section 12.2 provides a priori estimates for the perturbed gradient system $(X, \mathcal{E}_\varepsilon,$ $\mathcal{R}_\varepsilon, h_\varepsilon)$. In additions to the standard conditions on gradient systems the new assumption is an estimate of the form $\mathcal{R}_\varepsilon^*(u, \frac{1}{c} h_\varepsilon(t, u)) \leq C\mathcal{E}(t, u)$ (cf. (12.3)). Based only on these simply estimates we provide two abstract results on evolutionary Γ-convergence.

The first result on evolutionary Γ-convergence for perturbed gradient systems is given in Theorem 4 and relies on the rather strong assumption of λ-convexity. For this we assume that X is a Hilbert space H, that the dissipation potentials have the quadratic form $\mathcal{R}_\varepsilon(u, \dot{u}) = \frac{1}{2} \langle \mathbb{G}_\varepsilon \dot{u}, \dot{u} \rangle$, and that there exists a $\lambda \in \mathbb{R}$ such that $u \mapsto \mathcal{E}_\varepsilon(u) - \lambda \mathcal{R}_\varepsilon(u)$ is convex for all $\varepsilon \in [0, 1]$. Otherwise the assumptions are rather weak, since the simple Γ-convergence of $\mathcal{E}_\varepsilon(t, \cdot) \xrightarrow{\Gamma} \mathcal{E}_0(t, \cdot)$ and pointwise convergence of $\mathcal{R}_\varepsilon$ are essentially sufficient.

The second result on evolutionary Γ-convergence for perturbed gradient systems relies on De Giorgi's energy-dissipation principle. It is much more flexible, since no λ-convexity is needed and $\mathcal{R}_\varepsilon$ can be much more general. The major new quantity in this approach is the dissipation functional

$$\mathfrak{D}_\varepsilon(u(\cdot)) := \int_0^T \left(\mathcal{R}_\varepsilon(u(t), \dot{u}(t)) + \mathcal{R}_\varepsilon^*\left(u(t), h_\varepsilon(t, u(t)) - D_u \mathcal{E}_\varepsilon(t, u(t))\right) \right) dt.$$

As a major assumption of the abstract result in Theorem 7 one needs the liminf estimate $\liminf_{\varepsilon \to 0} \mathfrak{D}_\varepsilon(u^\varepsilon(\cdot)) \geq \mathfrak{D}(u(\cdot))$ if $u^\varepsilon \rightharpoonup u$ in $W^{1,p}([0, T]; X)$.

In Sect. 12.5 we discuss a few possible applications of the general results. We first consider the classical question of homogenization of reaction-diffusion systems as a didactical example. There we treat the diffusion part as a gradient part associated with the convex and quadratic Dirichlet energy. Because of the semilinear structure, all the nonlinear reaction terms can be treated as non-gradient perturbations. We are able to apply the λ-convex theory and refer to [19] for a comparison of the strengths and weaknesses of the two different approaches discussed on the basis of the homogenization of a Cahn–Hilliard-type problem.

In Sect. 12.5.2 we reconsider the theory developed in [22] for pure gradient systems. There it was shown that the Ginzburg–Landau equations can be understood as the evolutionary Γ-limit of the suitable scaled Swift-Hohenberg equation. We discuss the usage of perturbed gradient systems to analyze a coupled system of Swift-Hohenberg equations introduced in [33].

Finally we speculate concerning the usage of evolutionary Γ-convergence to derive a nonlinear reaction-diffusion system from a single Fokker–Planck-type master equation of diffusion in physical space as well as along a chemical reaction path. This follows the spirit of [2, 20, 29, 30], where chemical reaction is understood as a limit of diffusion.

12.2   Energy Control and a Priori Estimates

As was announced earlier, we will consider the non-gradient term h_ε as a lower-order perturbation of the gradient system. Before specifying this, we fix the major properties of the energy functionals $\mathcal{E}_\varepsilon$. We assume that the reflexive and separable Banach space $\mathbf{Z}$ is compactly embedded into $\mathbf{X}$ and

$$\operatorname{dom}\mathcal{E}_\varepsilon := \{ (t, u) \mid \mathcal{E}_\varepsilon(t, u) < \infty \} = [0, T] \times \mathbf{D}_\varepsilon, \quad \mathbf{D}_\varepsilon := \operatorname{dom}\mathcal{E}_\varepsilon(0, \cdot), \quad (12.2a)$$

$$\exists\, c_0, \alpha > 0 \ \forall\, (\varepsilon, t, u) \in [0, 1] \times [0, T] \times \mathbf{X} : \ \mathcal{E}_\varepsilon(t, u) \geq c_0 \|u\|_{\mathbf{Z}}^\alpha, \quad (12.2b)$$

$$\exists\, \Lambda_{\mathrm{na}} \geq 0 \ \forall\, (\varepsilon, t, u) \in [0, 1] \times [0, T] \times \mathbf{D}_\varepsilon : \ |\partial_t \mathcal{E}_\varepsilon(t, u)| \leq \Lambda_{\mathrm{na}} \mathcal{E}_\varepsilon(t, u),$$
$$(12.2c)$$

where $\|u\|_{\mathbf{Z}} = \infty$ for $u \in \mathbf{X} \setminus \mathbf{Z}$. Note that the energies are only defined up to a constant, so we can choose $C = 0$ in the usual condition of coercivity $\mathcal{E}_\varepsilon(t, u) \geq c_0 \|u\|_{\mathbf{Z}}^\alpha - C$.

In this section we consider general dissipation potentials $\mathcal{R}_\varepsilon : \mathbf{X} \times \mathbf{X} \to [0, \infty]$, which means that $\mathcal{R}_\varepsilon(u, \cdot) : \mathbf{X} \to [0, \infty]$ is a lower semicontinuous and convex functional satisfying additionally $\mathcal{R}_\varepsilon(u, 0) = 0$. The first condition on the perturbation $h_\varepsilon : [0, T] \times \mathbf{X} \to \mathbf{X}^*$ is the following bound:

$$\exists\, \Lambda_{\mathrm{ng}} \geq 0, \ c \in\,]0, 1[\, \forall\, (\varepsilon, t, u) \in [0, 1] \times [0, T] \times \mathbf{X} : \qquad (12.3)$$

$$\mathcal{R}_\varepsilon^*\left(u, \frac{1}{c} h_\varepsilon(t, u)\right) \leq \frac{\Lambda_{\mathrm{ng}}}{c} \mathcal{E}(t, u).$$

Based on these assumptions we first derive a control of the energy $\mathcal{E}_\varepsilon$ for fixed w and along solutions $u : [0, T] \to \mathbf{X}$ of the perturbed gradient flow

$$\mathrm{D}_{\dot{u}} \mathcal{R}_\varepsilon(u, \dot{u}(t)) = -\mathrm{D}_u \mathcal{E}_\varepsilon(t, u(t)) + h_\varepsilon(t, u(t)) \text{ for a.a. } t \in [0, T]. \qquad (12.4)$$

Note that all the estimates are uniform in $\varepsilon \in [0, 1]$.

Proposition 1 *If* (12.2c) *holds, then for all* $(\varepsilon, s, t, w) \in [0, 1] \times [0, T]^2 \times \mathbf{D}_\varepsilon$:

$$\mathrm{e}^{-\Lambda_{\mathrm{na}}|t-s|} \mathcal{E}_\varepsilon(s, w) \leq \mathcal{E}_\varepsilon(t, w) \leq \mathrm{e}^{\Lambda_{\mathrm{na}}|t-s|} \mathcal{E}_\varepsilon(s, w). \qquad (12.5)$$

Assuming additionally (12.3) *and setting* $\Lambda := \Lambda_{\mathrm{na}} + \Lambda_{\mathrm{ng}}$, *every solution* $u : [0, T] \to \mathbf{X}$ *of* (12.4) *satisfies, for* $0 \leq s < t \leq T$, *the estimate*

$$\mathcal{E}_\varepsilon(t, u(t)) + \int_s^t (1 - c)\mathcal{R}_\varepsilon(u(r), \dot{u}(r))\, \mathrm{d}r \ \leq \mathrm{e}^{\Lambda(t-s)} \mathcal{E}_\varepsilon(s, u(s)). \qquad (12.6)$$

Proof Equation (12.5) follows by a simple Gronwall estimate based on (12.2c).

For the second result we apply $\langle \cdot, \dot{u} \rangle$ to (12.4) and use $\langle \mathrm{D}_{\dot{u}} \mathcal{R}_\varepsilon(u, \dot{u}), \dot{u} \rangle \geq \mathcal{R}_\varepsilon(u, \dot{u})$ and the chain rule for $\mathcal{E}_\varepsilon$ to obtain the energy estimate

$$\mathcal{E}_\varepsilon(t, u(t)) + \int_s^t \mathcal{R}_\varepsilon(u, \dot{u})\,\mathrm{d}r \le \mathcal{E}_\varepsilon(s, u(s)) + \int_s^t \partial_r \mathcal{E}_\varepsilon(r, u(r)) + \langle h_\varepsilon(r, u(r)), \dot{u}(r)\rangle\,\mathrm{d}r.$$

Estimating $\langle h_\varepsilon, \dot{u}(r)\rangle \le c\mathcal{R}_\varepsilon^*(u, \frac{1}{c}h_\varepsilon) + c\mathcal{R}_\varepsilon(u, \dot{u}) \le \Lambda_{\mathrm{ng}}\mathcal{E}_\varepsilon(t, u) + c\mathcal{R}_\varepsilon(u, \dot{u})$ we find the purely energetic a priori estimate

$$\mathcal{E}_\varepsilon(t, u(t)) + \int_s^t (1-c)\mathcal{R}_\varepsilon(u, \dot{u})\,\mathrm{d}r \le \mathcal{E}_\varepsilon(s, u(s)) + \int_s^t \Lambda\mathcal{E}_\varepsilon(r, u(r))\,\mathrm{d}r. \quad (12.7)$$

Neglecting $\mathcal{R}_\varepsilon$, Gronwall's estimate gives $\mathcal{E}_\varepsilon(t, u(t)) \le \mathrm{e}^{\Lambda(t-s)}\mathcal{E}_\varepsilon(s, u(s))$ for all $t \in [s, T]$. Next we replace t by r in the latter relation and insert it into the right-hand side of (12.7), which provides the assertion (12.6). □

The main point of this proposition is that we are able to derive uniform a priori estimates as follows:

Corollary 2 (Uniform a priori estimates) *Assume that the dissipation potentials are equicoercive:*

$$\exists\, c_R > 0, \; p > 1 \; \forall\, (\varepsilon, u, v) \in [0, 1] \times X^2 : \quad \mathcal{R}_\varepsilon(u, v) \ge c_R\|v\|_X^p, \quad (12.8)$$

and that the initial energies satisfy $\mathcal{E}_\varepsilon(0, u_\varepsilon^0) \le C_E < \infty$. Then, there exists $C_* < \infty$ such that the solutions $u_\varepsilon : [0, T] \to X$ of (12.4) satisfy

$$\|u_\varepsilon(\cdot)\|_{\mathrm{L}^\infty([0,T];Z)} + \|u_\varepsilon(\cdot)\|_{\mathrm{W}^{1,p}([0,T];X)} \le C. \quad (12.9)$$

Proof We use (12.6) for $s = 0$ and $t \le T$, where the right-hand side is estimated by $\mathrm{e}^{\Lambda T}C_E < \infty$. Now, the coercivity (12.2b) of $\mathcal{E}_\varepsilon$ gives the bound in $\mathrm{L}^\infty([0, T]; Z)$. Then, the coercivity (12.8) of $\mathcal{R}_\varepsilon$ gives the bound in $\mathrm{W}^{1,p}([0, T]; X)$. □

12.3   Perturbed Evolutionary Variational Estimate

In this section we consider a simple Hilbert-space setting, i.e. the dynamic space X is a Hilbert space H with norm $\|\cdot\|$, and the dissipation potentials $\mathcal{R}_\varepsilon$ are one-half of the square of Hilbert-space norms. Nevertheless, we do not work with one Hilbert space but with a family of norms:

$$\exists\, C > 0 \; \forall\, \varepsilon \in [0, 1] \; \exists\, \mathbb{G}_\varepsilon = \mathbb{G}_\varepsilon^* \in \mathrm{Lin}(H, H) : \quad (12.10)$$

$$\mathcal{R}_\varepsilon(v) = \frac{1}{2}\langle \mathbb{G}_\varepsilon v, v\rangle \text{ and } \frac{1}{2C}\|v\|^2 \le \mathcal{R}_\varepsilon(v) \le \frac{C}{2}\|v\|^2.$$

For the energies $\mathcal{E}_\varepsilon : [0, T] \times H \to \mathbb{R}_\infty$ we assume that they are uniformly λ-convex:

$$\exists \lambda_* \in \mathbb{R} \; \forall \, (\varepsilon, t) \in [0, 1] \times [0, T]: \quad \mathcal{E}_\varepsilon(t, \cdot) + \lambda_* \mathcal{R}_\varepsilon(\cdot) \text{ is convex on } \boldsymbol{H}; \qquad (12.11)$$

$$\mathcal{E}_\varepsilon(t, u_\theta) \leq (1 - \theta)\mathcal{E}_\varepsilon(t, u_0) + \theta \mathcal{E}_\varepsilon(t, u_1) + \lambda_* \theta(1 - \theta)\mathcal{R}_\varepsilon(u_1 - u_0),$$

where $u_\theta := (1 - \theta)u_0 + \theta u_1$. For sufficiently smooth $\mathcal{E}_\varepsilon$ condition (12.11) simply means

$$\mathcal{E}_\varepsilon(t, w) \geq \mathcal{E}_\varepsilon(t, u) + \langle \mathrm{D}\mathcal{E}_\varepsilon(t, u), w - u \rangle + \lambda_* \mathcal{R}_\varepsilon(w - u). \qquad (12.12)$$

For the non-gradient term $h_\varepsilon : [0, T] \times \boldsymbol{H} \to \boldsymbol{H}^*$ we assume that it is controlled by the gradient parts as in (12.3).

Here we do not address the question of existence and uniqueness of solutions, which we assume to hold. (For this one may additionally impose a global Lipschitz continuity of h_ε.) Our concern is the convergence of the solutions $u_\varepsilon : [0, T] \to \boldsymbol{H}$ for the perturbed gradient system $(\boldsymbol{H}, \mathcal{E}_\varepsilon, \mathcal{R}_\varepsilon, h_\varepsilon)$, i.e. u_ε satisfies (12.4).

Our next result provides a reformulation of this equation in terms of a *perturbed evolutionary variational estimate* (PEVE), which is a direct generalization of the metric theory in [1, 10], where $\Lambda_{na} = \Lambda_{ng} = 0$. Since it is a statement for fixed $\varepsilon \in [0, 1]$, we can drop the index ε here.

Proposition 3 *Assume that the assumptions (12.10), (12.2), (12.11), and (12.3) hold and set $\Lambda := \Lambda_{na} + \Lambda_{ng}$. Then, a function $u \in \mathrm{H}^1([0, T]; \boldsymbol{H}) \cap \mathrm{L}^\infty(0, T; \boldsymbol{Z})$ solves (12.4) if and only if (PEVE) holds:*

$$\forall \, 0 \leq s < t \; \forall \, w \in \boldsymbol{H}:$$

$$\mathrm{e}^{\lambda_*(t-s)}\mathcal{R}(u(t) - w) - \mathcal{R}(u(s) - w) + A_*^+(t - s)\mathcal{E}(t, u(t)) \qquad \text{(PEVE)}$$

$$\leq A_*^-(t - s)\mathcal{E}(t, w) - \int_s^t \mathrm{e}^{\lambda_*(r-s)} \langle h(r, u(r)), w - u(r) \rangle \, \mathrm{d}r,$$

where $A_*^\pm(r) = \left(\mathrm{e}^{\lambda_* r} - \mathrm{e}^{\mp \Lambda r}\right)/(\lambda_* \pm \Lambda)$ (giving $A_*^\pm(0) = 0$ and $(A_*^\pm)'(0) = 1$).

Proof We first show that (12.4) implies (PEVE). For this, we choose arbitrary w and apply $\langle \cdot, u(t) - w \rangle$ to (12.4) to obtain

$$\frac{\mathrm{d}}{\mathrm{d}t}\mathcal{R}(u(t) - w) \overset{(12.10)}{=} \langle \mathrm{D}\mathcal{R}(\dot{u}), u - w \rangle \overset{(12.4)}{=} \langle \mathrm{D}\mathcal{E}(t, u) - h(t, u), w - u \rangle$$

$$\overset{(12.12)}{\leq} \mathcal{E}(t, w) - \mathcal{E}(t, u) - \lambda_* \mathcal{R}(w - u) - \langle h(t, u), w - u \rangle.$$

Moving $-\lambda_* \mathcal{R}(w - u)$ to the left-hand side and multiplying by $\mathrm{e}^{\lambda_*(t-s)}$ we can integrate over $t \in [s, t_1]$. Renaming t and t_1 into r and t, respectively, we find

$$\mathrm{e}^{\lambda_*(t-s)}\mathcal{R}(u(t) - w) - \mathcal{R}(u(s) - w)$$

$$\leq \int_s^t \mathrm{e}^{\lambda_*(r-s)}\left(\mathcal{E}(r, w) - \mathcal{E}(r, u(r)) - \langle h(r, u(r)), w - u(r) \rangle\right) \mathrm{d}r.$$

From (12.5) we obtain $\mathcal{E}(r, w) \leq e^{A(t-r)}\mathcal{E}(t, w)$, and (12.6) implies $\mathcal{E}(r, u(r)) \geq e^{-A(t-r)}\mathcal{E}(t, u(t))$. Inserting this into the last estimate and doing the integration in $r \in [s, t]$ explicitly for the first two terms leads to the desired result (PEVE).

We now show that PEVE implies (12.4). For this we divide both sides by $t - s > 0$ and then take the limit $s \nearrow t$. Using $A_*^{\pm}(r)/r \to 1$ for $r \searrow 0$ we obtain the differential form again, namely

$$\frac{d}{dt}\mathcal{R}(u - w) = \langle \mathbb{G}\dot{u}, u - w \rangle = \langle D_u \mathcal{E}(t, u), w - u \rangle$$
$$\leq \mathcal{E}(t, w) - \mathcal{E}(t, u) - \lambda_* \mathcal{R}(u - w) + \langle h(t, u), u - w \rangle.$$

Keeping t fixed and inserting the test function $w = u(t) - \delta v$ with $\delta > 0$, we divide by δ first and then pass to the limit to obtain $\langle \dot{u}, v \rangle \leq \langle -D\mathcal{E}(t, u) + h(t, u), v \rangle$. Since v is arbitrary, we also have the opposite sign (replace v by $-v$), and (12.4) is established. $\qquad\square$

The above characterization of solutions of the perturbed gradient system $(\boldsymbol{H}, \mathcal{E}_\varepsilon, \mathcal{R}_\varepsilon, h_\varepsilon)$, which give rise to the evolution equation (12.4), allows us to formulate a result concerning evolutionary Γ-convergence. For this we use the notion of (strong) Γ-convergence of the energies, continuous convergence of the dissipation potentials, and strong convergence of the perturbations:

$$\mathcal{E}_\varepsilon \xrightarrow{\Gamma} \mathcal{E}_0, \quad \text{i.e.} \begin{cases} w_\varepsilon \to w \text{ in } \boldsymbol{H} \implies \liminf_{\varepsilon \to 0} \mathcal{E}_\varepsilon(t, w_\varepsilon) \geq \mathcal{E}_0(t, w_0), \\ \forall \widehat{w}_0 \; \exists \widehat{w}_\varepsilon \to \widehat{w}_0 \text{ in } \boldsymbol{H} : \; \mathcal{E}_\varepsilon(t, w_\varepsilon) \to \mathcal{E}_0(t, \widehat{w}_0); \end{cases} \tag{12.13a}$$

$$\mathcal{R}_\varepsilon \xrightarrow{C} \mathcal{R}_0, \quad \text{i.e. } w_\varepsilon \rightharpoonup w_0 \text{ in } \boldsymbol{Z} \implies \mathcal{R}_\varepsilon(w_\varepsilon) \to \mathcal{R}_0(w_0); \tag{12.13b}$$

$$w_\varepsilon \rightharpoonup w_0 \text{ in } \boldsymbol{Z} \implies h_\varepsilon(t, w_\varepsilon) \rightharpoonup h_0(t, w_0) \text{ in } \boldsymbol{H}^*. \tag{12.13c}$$

Concerning the static Γ-convergence in (12.13a) we refer to the standard textbooks [4, 6, 9]. In these statements the weak convergence in $\boldsymbol{Z}$ can be replaced by the more general and maybe more flexible statement of convergence within sublevels of $\mathcal{E}_\varepsilon$, namely $w_\varepsilon \to w_0$ in $\boldsymbol{H}$ and $\mathcal{E}_\varepsilon(t, w_\varepsilon) \leq C$. Clearly, the equicoercivity (12.2b) implies weak convergence in $\boldsymbol{Z}$.

The following result relies on PEVE and the a priori estimate provided in Corollary 2. The latter shows that the desired accumulating points exist, since the unit ball in $W^{1,p}([0, T]; \boldsymbol{Z})$ is weakly compact, i.e. converging subsequences as assumed in the following result always exist.

Theorem 4 (Evolutionary Γ-convergence via PEVE)   *Let the assumptions of Proposition 3 and (12.13) hold. If for a family of solutions $u^\varepsilon : [0, T] \to \boldsymbol{H}$ of (12.4) a subsequence $(u^{\varepsilon_k})_{k \in \mathbb{N}}$ satisfies*

$$\varepsilon_k \to 0 \quad \text{and} \quad u^{\varepsilon_k} \rightharpoonup u \text{ in } H^1([0, T]; \boldsymbol{H}),$$

then u is a solution of the limiting perturbed gradient system $(\boldsymbol{H}, \mathcal{E}_0, \mathcal{R}_0, h_0)$, i.e. u solves (12.4) for $\varepsilon = 0$.

Proof By the a priori estimate in Corollary 2 we can assume $u^{\varepsilon k} \rightharpoonup u$ in $\mathrm{H}^1([0, T]; \boldsymbol{H})$ and

$$\forall\, t \in [0, T]: \quad u^{\varepsilon k}(t) \rightharpoonup u(t) \text{ in } \boldsymbol{Z} \text{ and } u^{\varepsilon k}(t) \to u(t) \text{ in } \boldsymbol{H}.$$

We now exploit that the perturbed evolutionary variational estimate (PEVE) holds with λ_* and Λ independently of ε. For $0 \le s < t \le T$ and $w \in \boldsymbol{H}$ we have

$$
\begin{aligned}
\mathrm{e}^{\lambda_*(t-s)} \mathcal{R}_\varepsilon(u^{\varepsilon k}(t) - w) &- \mathcal{R}_\varepsilon(u^{\varepsilon k}(s) - w) + A_*^+(t - s)\mathcal{E}_\varepsilon(t, u^{\varepsilon k}(t)) \\
&\le A_*^-(t - s)\mathcal{E}_\varepsilon(t, w) - \int_s^t \mathrm{e}^{\lambda_*(r-s)} \langle h_\varepsilon(r, u^{\varepsilon k}(t)), w - u^{\varepsilon k}(r) \rangle \, \mathrm{d}r.
\end{aligned}
\tag{12.14}
$$

Fixing s and t we may now choose a suitable test function $w = w^{\varepsilon k}$, namely such that $w^{\varepsilon k} \to w^0$ and $\mathcal{E}(t, w^{\varepsilon k}) \to \mathcal{E}(t, w^0)$ (cf. (12.13a)). Note that the equicoercivity implies $w^{\varepsilon k} \rightharpoonup w^0$ in $\boldsymbol{Z}$.

Hence, we can pass to the limit inferior for $\varepsilon_k \to 0$ in (12.14). Indeed, on the left-hand side the first two terms converge to $\mathrm{e}^{\lambda_*(t-s)} \mathcal{R}_0(u(t) - w^0) - \mathcal{R}_0(u(s) - w^0)$ because of (12.13b), whereas the third term has a liminf bounded from below by $A_*^+(t - s)\mathcal{E}_0(t, u(t))$, where we use $A_*^+(t - s) > 0$. On the right-hand side the first term converges to $A_*^-(t - s)\mathcal{E}_0(t, w^0)$ by the choice of $w^{\varepsilon k}$, whereas the second term converges to $\int_s^t \mathrm{e}^{\lambda_*(r-s)} \langle h_0(r, u(r)), w^0 - u(r) \rangle \, \mathrm{d}r$ by strong convergence of $w^{\varepsilon k} - u^{\varepsilon k}(r)$ and weak convergence of $h_{\varepsilon_k}(r, u^{\varepsilon k}(r))$. Thus, since w^0 is arbitrary, (PEVE) is established for u, and by Proposition 3 we know that u is a solution of (12.4) for $\varepsilon = 0$. $\qquad\square$

12.4 De Giorgi's Energy-Dissipation Principle

To prepare for De Giorgi's reformulation of gradient flows in terms, we recall the following fact from convex analysis. For a convex function $\Psi : X \to \mathbb{R}_\infty := \mathbb{R} \cup \{\infty\}$ the Legendre–Fenchel dual $\Psi^* : X^* \to \mathbb{R}_\infty$ is defined via

$$\Psi^*(\xi) := \sup\{ \langle \xi, v \rangle - \Psi(v) \mid v \in X \}
\tag{12.15}$$

and the convex subdifferential via

$$\partial\Psi(v) = \{ \xi \in X^* \mid \Psi(w) \ge \Psi(v) + \langle \xi, w - v \rangle \text{ for all } w \in X \}.
\tag{12.16}$$

Proposition 5 *Let X be a reflexive Banach space and $\Psi : X \to \mathbb{R}_\infty$ be proper, convex, and lower semi-continuous. Then, the following holds:*
(A) Young-Fenchel estimate: $\forall\, (v, \xi) \in X \times X^* : \quad \Psi(v) + \Psi^*(\xi) \ge \langle \xi, v \rangle.$
(B) Fenchel equivalence ([13, 15]): for all $(v, \xi) \in X \times X^*$ we have

$$(i)\ \xi \in \partial\Psi(v) \iff (ii)\ v \in \partial\Psi^*(\xi) \iff (iii)\ \Psi(v) + \Psi^*(\xi) = \langle \xi, v \rangle.$$

We emphasize that the relation (i) is a relation in dual space X^*, (ii) is a relation in X, and (iii) is a relation in $\mathbb{R}$. Using (A), it is immediate that (iii) can be replaced by the estimate (iii)$'$ $\Psi(v) + \Psi^*(\xi) \leq \langle \xi, v \rangle$.

We can apply these equivalences with $\Psi(\cdot) = \mathcal{R}_\varepsilon(u, \cdot)$ to the formulation of the gradient flow associated with our perturbed gradient system $(X, \mathcal{E}_\varepsilon, \mathcal{R}_\varepsilon, h_\varepsilon)$ and obtain three equivalent formulations:

force balance $\mathrm{D}_{\dot{u}}\mathcal{R}_\varepsilon(u, \dot{u}) = -\mathrm{D}_u\mathcal{E}_\varepsilon(t, u) + h_\varepsilon(t, u)$;

rate equation $\dot{u} = \mathrm{D}_\xi \mathcal{R}_\varepsilon^*\big(u, -\mathrm{D}_u\mathcal{E}_\varepsilon(t, u) + h_\varepsilon(t, u)\big)$;

power balance $\mathcal{R}_\varepsilon(u, \dot{u}) + \mathcal{R}_\varepsilon^*(u, h_\varepsilon(t, u) - \mathrm{D}_u\mathcal{E}_\varepsilon(t, u)) = \langle h_\varepsilon(t, u) - \mathrm{D}_u\mathcal{E}_\varepsilon(t, u), \dot{u} \rangle$.

The main point is that a time-integrated version of the third formulation can be used to characterize solutions of perturbed gradient systems. For this we need an *abstract chain rule* for $\mathcal{E}_\varepsilon$. We say that $(X, \mathcal{E})$ satisfies the chain rule if for all $p \geq 1$ the following holds. If $u \in \mathrm{W}^{1,p}([0, T]; X)$, $\mathcal{E}(\cdot, u(\cdot)) \in \mathrm{L}^1([0, T])$, and $\mathrm{D}_u\mathcal{E}(\cdot, u(\cdot)) \in \mathrm{L}^{p*}([0, T]; X^*)$, then $t \mapsto \mathcal{E}(t, u(t))$ is absolutely continuous and

$$\frac{\mathrm{d}}{\mathrm{d}t}\mathcal{E}(t, u(t)) = \langle \xi(t), \dot{u}(t) \rangle + \partial_t\mathcal{E}(t, u(t)) \text{ a.e. in } [0, T]. \quad (12.17)$$

We refer to [26, 32] for general treatments and derivations of such abstract chain rules. Using this chain rule, we can integrate the power balance in time and replace $\langle \mathrm{D}_u\mathcal{E}_\varepsilon(t, u), \dot{u} \rangle$ by the difference of the initial and final energies plus an integral over $\partial_t\mathcal{E}_\varepsilon$. De Giorgi's energy-dissipation principle (EDP) states that this integrated version of the power estimate (iii)$'$ is equivalent to the force balance (12.4) for a.a. $t \in [0, T]$. Again we can drop the parameter $\varepsilon > 0$.

Theorem 6 (De Giorgi's EDP) *Assume that $(X, \mathcal{E})$ satisfies the chain rule* (12.17) *and that there exists $C, p > 1$ such that $(1 + \|\dot{u}\|^p)/C \leq \mathcal{R}(u, \dot{u}) \leq C(1 + \|\dot{u}\|^p)$. Then a function $u \in \mathrm{W}^{1,p}([0, T]; X)$ is a solution of the perturbed gradient system $(X, \mathcal{E}, \mathcal{R}, h)$ if and only if it satisfies the Upper Energy-Dissipation Estimate*

$$\mathcal{E}(T, u(T)) + \mathfrak{D}(u(\cdot)) \leq \mathcal{E}(0, u(0)) + \int_0^T \partial_t\mathcal{E}(t, u(t)) + \langle h(t, u(t)), \dot{u}(t) \rangle \, \mathrm{d}t,$$
$$\text{(UEDE)}$$

where De Giorgi's dissipation functional $\mathfrak{D}$ is given by

$$\mathfrak{D}(u(\cdot)) := \int_0^T \mathcal{R}(u(t), \dot{u}(t)) + \mathcal{R}^*\big(u(t), h(t, u(t)) - \mathrm{D}_u\mathcal{E}(t, u(t))\big) \, \mathrm{d}t. \quad (12.18)$$

This result is a simple generalization of [23, Theorem 3.3], where the proof for the case $h \equiv 0$ is given. We remark that the EDP relates the final energy $\mathcal{E}(T, u(T))$ plus the dissipated energy $\int_0^T \mathcal{R} + \mathcal{R}^* \, \mathrm{d}t$ to the initial energy $\mathcal{E}(0, u(0))$ plus the external work $\int_0^T \partial_t\mathcal{E}(t, u(t)) \, \mathrm{d}t$ and the work due to the non-gradient terms

$\int_0^T \langle h(t, u(t)), \dot{u}(t) \rangle \, dt$. It is sufficient to establish the UEDE, then by the chain rule one obtains an equality in UEDE giving the power balance.

The EDP is ideal for proving evolutionary Γ-convergence. In fact, it is the basis of the famous Sandier–Serfaty approach, see [34, 35]. For this we look at the ε-dependent UEDE:

$$\mathcal{E}_\varepsilon(T, u^\varepsilon(T)) + \mathfrak{D}_\varepsilon(u^\varepsilon(\cdot)) \leq \mathcal{E}_\varepsilon(0, u_\varepsilon^0) + \int_0^T \partial_t \mathcal{E}_\varepsilon(t, u^\varepsilon(t)) + \langle h_\varepsilon(t, u^\varepsilon(t)), \dot{u}^\varepsilon(t) \rangle \, dt.$$
$$(12.19)$$

The main importance of the EDP is that it involves the UEDE, which states that the final and the dissipated energies only need to have a good upper bound. Hence, in passing to the Γ-limit it will be sufficient to have good liminf estimates for these terms, while the right-hand side can be controlled by the well-preparedness of the initial conditions and proper assumptions on the power of the external forces $\partial_t \mathcal{E}_\varepsilon(t, u)$ and the power $\langle h_\varepsilon(t, u), \dot{u} \rangle$. The following result gives sufficient conditions for evolutionary Γ-convergence, in fact for "pE-convergence" in the sense of [23].

Theorem 7 (Evolutionary Γ-convergence via EDP) *Assume that the perturbed gradient systems $(X, \mathcal{E}_\varepsilon, \mathcal{R}_\varepsilon, h_\varepsilon)$ satisfy (12.2), (12.3), (12.8) and that*

$$\mathcal{E}_\varepsilon(t, \cdot) \xrightarrow{\Gamma} \mathcal{E}_0(t, \cdot) \quad \text{and} \quad \mathcal{E}_\varepsilon(0, u_\varepsilon^0) \to \mathcal{E}_0(0, u_0^0); \qquad (12.20a)$$

$$(X, \mathcal{E}_0) \text{ satisfies the chain rule}; \qquad (12.20b)$$

$$w_\varepsilon \xrightarrow{Z} w_0 \implies \left(\partial_t \mathcal{E}_\varepsilon(t, w_\varepsilon) \to \partial_t \mathcal{E}_0(t, w_0) \,\&\, h_\varepsilon(t, w_\varepsilon) \xrightarrow{Z^*} h_0(t, w_0)\right); \quad (12.20c)$$

$$\widehat{w}_\varepsilon(\cdot) \rightharpoonup \widehat{w}_0(\cdot) \text{ in } W^{1,p}([0, T]; X) \implies \mathfrak{D}_0(\widehat{w}_0) \leq \liminf_{\varepsilon \to 0} \mathfrak{D}_\varepsilon(\widehat{w}^\varepsilon). \qquad (12.20d)$$

If $u^\varepsilon : [0, T] \to X$ is a family of solutions for (12.4) with $u^\varepsilon(0) = u_\varepsilon^0$ and

$$\varepsilon_k \to 0 \quad \text{and} \quad u^{\varepsilon_k} \rightharpoonup u \text{ in } W^{1,p}([0, T]; X) \text{ as } k \to \infty,$$

then u is a solution for the perturbed system $(X, \mathcal{E}_0, \mathcal{R}_0, h_0)$ with $u(0) = u_0^0$.

The crucial and most difficult condition here is the liminf estimate for De Giorgi's dissipation potential, where $\mathcal{D}_0$ again must have the form (12.18). The liminf estimate is then sufficient, since the duality of $\mathcal{R}_0$ and $\mathcal{R}_0^*$ and the chain rule (12.20b) imply equality again.

Proof Because of the assumptions we can use the a priori estimates of Corollary 2 and may assume the additional convergences

$$\forall t \in [0, T]: \quad u^{\varepsilon_k}(t) \rightharpoonup u(t) \text{ in } Z \text{ and } u^{\varepsilon_k}(t) \to u(t) \text{ in } X.$$

Using the EDP in Theorem 6 we know that u^ε satisfies the UEDE (12.19). Using the assumptions (12.20a) and (12.20b) and the a priori estimates, we easily see that the right-hand side in (12.19) converges to $\mathcal{E}_0(0, u_0^0) + \int_0^T \partial_t \mathcal{E}_0(t, u(t)) + \langle h_0(t, u(t)), \dot{u}(t) \rangle \, dt$.

On the left we have $\mathcal{E}_0(T, u(T)) \leq \liminf_{\varepsilon_k \to 0} \mathcal{E}_{\varepsilon_k}(T, u^{\varepsilon_k}(T))$ and $\mathfrak{D}_0(u(\cdot)) \leq \liminf_{\varepsilon_k \to 0} \mathfrak{D}_{\varepsilon_k}(u^{\varepsilon_k}(\cdot))$. Thus, the UEDE for u with $\varepsilon = 0$ is established, and the EDP in Theorem 6 implies that u solves (12.4) for $\varepsilon = 0$. □

Based on the philosophy of this result, the notion of "EDP-convergence" was introduced in [20] by asking $\mathfrak{D}_\varepsilon \overset{\Gamma}{\to} \mathfrak{D}_0$ in $W^{1,p}([0, T]; X)$. This convergence is in fact much more than what is needed for evolutionary Γ-convergence. In principle, in (12.20d) it is sufficient to obtain the desired liminf estimate only along solutions. In contrast, EDP-convergence asks for a Γ-convergence along arbitrary functions. This is physically justified by fluctuation theory, which gives the proper justification of gradient structures, see e.g. [28].

Remark 8 A similar theory may be derived for perturbed gradient systems in the form

$$\dot{u} = D_\xi \mathcal{R}_\varepsilon^*\big(u, -D_u \mathcal{E}_\varepsilon(t, u)\big) + g_\varepsilon(t, u).$$

The corresponding energy-dissipation principle takes the form

$$\mathcal{E}_\varepsilon(T, u(T)) + \widehat{\mathfrak{D}}_\varepsilon(u) \leq \mathcal{E}_\varepsilon(0, u(0)) + \int_0^T \big(\partial_t \mathcal{E}_\varepsilon(t, u) + \langle D_u \mathcal{E}_\varepsilon(t, u), g_\varepsilon(t, u) \rangle\big) \, dt,$$

where $\widehat{\mathfrak{D}}_\varepsilon(u) = \int_0^T \big(\mathcal{R}_\varepsilon(u, \dot{u} - g_\varepsilon(t, u)) + \mathcal{R}_\varepsilon^*(u, -D_u \mathcal{E}_\varepsilon(u))\big) \, dt.$

We refer to [8, 11, 12] for the usage of this variational principle, where the term $\langle D_u \mathcal{E}_\varepsilon(t, u), g_\varepsilon(t, u) \rangle$ even disappears because of a Hamiltonian structure of g_ε.

12.5 Applications of Evolutionary Γ-Convergence

We provide a few possible applications of the two theories developed above.

12.5.1 Homogenization of Reaction-Diffusion System

We only discuss a few simple results, where we emphasize that scalar reaction-diffusion equations can easily be treated as unperturbed gradient systems. However, for general systems no gradient structure exists. We consider a vector $u = (u_1, \ldots, u_I) \in \mathbb{R}^I$ of concentrations depending on $(t, x) \in [0, T] \times \Omega$, where Ω is a bounded smooth domain in $\mathbb{R}^d$, which we may consider as a periodically structured solid, surface or interface. The reaction-diffusion system reads

$$M_\varepsilon(x)\dot{u} = \mathrm{div}\left(A_\varepsilon(x)\nabla u\right) - F_\varepsilon(x,u) \text{ in } \Omega, \quad A_\varepsilon(x)\nabla u \cdot \nu = 0 \text{ on } \partial\Omega. \quad (12.21)$$

Here M_ε, A_ε, and F_ε depend periodically on x in the form

$$M_\varepsilon(x) = \mathbb{M}\left(\frac{1}{\varepsilon}x\right), \quad A_\varepsilon(x) = \mathbb{A}\left(\frac{1}{\varepsilon}x\right), \quad F_\varepsilon(x,u) = \mathbb{F}\left(\frac{1}{\varepsilon}x,u\right),$$

where the functions $\mathbb{M}$, $\mathbb{A}$, and $\mathbb{F}$ are 1-periodic in the variable $y = \frac{1}{\varepsilon}x \in \mathbb{R}^d$, viz. $\mathbb{M}(y+k) = \mathbb{M}(y)$ for all $y \in \mathbb{R}^d$ and all $k \in \mathbb{Z}^d$.

We can apply the theory of perturbed gradient systems by using the spaces $X = L^2(\Omega; \mathbb{R}^I)$ and $Z = H^1(\Omega; \mathbb{R}^I)$ and the functionals

$$\mathcal{E}_\varepsilon(u) = \int_\Omega \frac{1}{2}\nabla u \cdot A_\varepsilon(x)\nabla u + \frac{1}{2}|u|^2 \, dx \text{ and } \mathcal{R}_\varepsilon(\dot{u}) = \int_\Omega \frac{1}{2}\dot{u} \cdot M_\varepsilon(x)\dot{u}\,dx.$$

For the perturbation h_ε we choose $h_\varepsilon(t,x,u) = u - F_\varepsilon(x,u)$.

In addition to the 1-periodicity, the main assumptions on the functions $\mathbb{M}$, $\mathbb{A}$, and $\mathbb{F}$ are the following. There exists $C, c_0 > 0$ such that

$$\mathbb{M} = \mathbb{M}^\top \in L^\infty(\mathbb{R}^d; \mathbb{R}^{I \times I}), \quad \xi \cdot \mathbb{M}(y)\xi \geq c_0|\xi|^2,$$
$$\mathbb{A} = \mathbb{A}^\top \in L^\infty(\mathbb{R}^d; \mathbb{R}^{(I \times d) \times (I \times d)}), \quad \varXi : \mathbb{A}(y)\varXi \geq c_0|\varXi|^2,$$
$$\mathbb{F}(\cdot,u) \in L^\infty(\mathbb{R}^d; \mathbb{R}^I), \quad |\mathbb{F}(y,u) - \mathbb{F}(y,\tilde{u})| \leq C|u - \tilde{u}|,$$

for all $u, \tilde{u} \in \mathbb{R}^I$, $y, \xi \in \mathbb{R}^d$, and $\varXi \in \mathbb{R}^{I \times d}$.

First we observe that the general assumptions (12.2) hold with $D_\varepsilon = Z = H^1(\Omega; \mathbb{R}^I)$, $\alpha = 2$, and $\Lambda_{na} = 0$. Moreover, (12.3) holds since

$$\mathcal{R}_\varepsilon^*(u, h_\varepsilon) \leq C_1\|h_\varepsilon\|_{L^2}^2 \leq C(1 + \|u\|_{L^2}^2) \leq \Lambda_{ng}\mathcal{E}_\varepsilon(u).$$

We now show that the theory developed in Sect. 12.3 for the perturbed evolutionary variational estimate holds. By the definition of $\mathcal{R}_\varepsilon$ it is quadratic on the Hilbert space $H = L^2(\Omega; \mathbb{R}^I)$, i.e. (12.10) holds. Moreover, $\mathcal{E}_\varepsilon$ is convex, so (12.11) holds with $\lambda_* = 0$.

To apply Theorem 4 we need to establish convergence for $\mathcal{E}_\varepsilon$, $\mathcal{R}_\varepsilon$, and h_ε. Strong Γ-convergence of $\mathcal{E}_\varepsilon$ in H (or similarly weak Γ-convergence in Z) holds with

$$\mathcal{E}_0(u) = \int_\Omega \left(\frac{1}{2}\nabla u : A_{\mathrm{eff}}\nabla u + \frac{1}{2}|u|^2\right)dx,$$

where the effective tensor follows from linear homogenization, see e.g. [5, 9]. Since weak convergence in $Z = H^1(\Omega; \mathbb{R}^I)$ implies strong convergence in $H = L^2(\Omega; \mathbb{R}^I)$, it is easy to show that $w_\varepsilon \rightharpoonup w_0$ in Z implies

$$\mathcal{R}_\varepsilon(w_\varepsilon) \to \mathcal{R}_0(w_0) = \int_\Omega \frac{1}{2} w_0 \cdot M_{\text{eff}} w_0 \, dx \quad \text{with } M_{\text{eff}} = \int_{[0,1]^d} \mathbb{M}(y) \, dy,$$

$$h_\varepsilon(w_\varepsilon) \to h_0(w_0) = w_0 - F_{\text{eff}}(w_0) \text{ in } \boldsymbol{H}, \quad \text{where } F_{\text{eff}}(w) = \int_{[0,1]^d} \mathbb{F}(y, w) \, dy.$$

We refer to [27] for the last convergence. Thus, assumption (12.13) is established, Theorem 4 is applicable, and the limiting perturbed gradient system $(\boldsymbol{H}, \mathcal{E}_0, \mathcal{R}_0, h_0)$ is identified by using A_{eff}, M_{eff}, and F_{eff} in its definition. In particular, the limiting perturbed gradient flow is given by the effective reaction-diffusion system

$$M_{\text{eff}} \dot{u} = \text{div}\left(A_{\text{eff}} \nabla u\right) - F_{\text{eff}}(u).$$

Of course, the above homogenization problem only serves as a didactical example, since the result is well known. However, the theory allows for significant generalizations. We first mention the homogenization of the Cahn-Hilliard equation in [19], where also a comparison between the two abstract approaches (PEVE versus EDP) is done. In [27, 31] the case of ε-dependent diffusion constants is two-scale convergence and proving strong convergence via a suitable Gronwall estimates.

12.5.2 Justification of Amplitude Equations

An application of the theory developed in Sect. 12.3 to the justification of amplitude equations is given in [22] for the case of pure gradient systems. The suitably rescaled fourth-order Swift-Hohenberg equation with periodic boundary condition on the circle $\mathbb{S}$ reads

$$\dot{w} = -\frac{1}{\varepsilon^2}\left(1 + \varepsilon^2 \partial_x^2\right)^2 w + \mu w + \beta \varepsilon w_x - w^3 \quad \text{on } \mathbb{S} := \mathbb{R}/_{2\pi\mathbb{Z}} \tag{12.22}$$

and is a gradient system for $\beta = 0$ on the Hilbert space $L^2(\mathbb{S})$ for the energy functional $\mathcal{F}_\varepsilon^{\text{SH}}(w) = \int_\mathbb{S} \frac{1}{2\varepsilon^2}(w + \varepsilon^2 w_{xx})^2 - \frac{\mu}{2} w^2 + \frac{1}{4} w^4 \, dx$ and the dissipation potential $\mathcal{R}^{\text{SH}}(\dot{w}) = \frac{1}{2}\|\dot{w}\|_{L^2}^2$. Here we show that the case $\beta \neq 0$ can be treated as a perturbed gradient system.

Because of the special form of the linear operator all typical solutions of (12.22) will spatially oscillate on the scale ε and are approximately of the form $w(t, x) \approx \text{Re}(A(t, x)e^{ix/\varepsilon})$. Using a suitable bijection $\mathbb{M}_\varepsilon$ between $L^2(\mathbb{S})$ and a proper subspace of $\boldsymbol{H} := L^2(\mathbb{S};)$, which satisfies $w = \text{Re}((\mathbb{M}_\varepsilon w)e^{ix/\varepsilon})$, one can define the amplitudes $A^\varepsilon = \mathbb{M}_\varepsilon w^\varepsilon \in \boldsymbol{H}$ and finds perturbed gradient systems $(\boldsymbol{H}, \mathcal{E}_\varepsilon, \mathcal{R}_\varepsilon, h_\varepsilon)$ with $\mathcal{E}_\varepsilon(\mathbb{M}_\varepsilon w) = \mathcal{F}_\varepsilon^{\text{SH}}(w)$, $\mathcal{R}_\varepsilon(\mathbb{M}_\varepsilon \dot{w}) = \mathcal{R}^{\text{SH}}(\dot{w})$, and the non-gradient part $h_\varepsilon(A) = \beta(iA + \varepsilon \partial_x A)/2$.

Using the theory developed in [22] (cf. Theorem 2.3 there with $\gamma = 0$) one can show that Theorem 4 applies with $\boldsymbol{Z} = H^1(\mathbb{S};)$, and we find evolutionary Γ-convergence to the perturbed gradient system $(\boldsymbol{H}, \mathcal{E}^{\text{GL}}, \mathcal{R}^{\text{GL}}, h_0)$ with

$$\mathcal{E}^{\mathrm{GL}}(A) = \int_{\mathbb{S}} \left(|A'|^2 - \frac{\mu}{4}|A|^2 + \frac{3}{32}|A|^4 \right) dx \text{ and } \mathcal{R}^{\mathrm{GL}}(\dot{A}) = \frac{1}{4}\|\dot{A}\|_{L^2}^2$$

and $h_0(A) = i\beta A/2$, which leads to the limiting perturbed gradient flow given by the Ginzburg-Landau equation

$$\dot{A} = 4A_{xx} + (\mu + i\beta)A - \tfrac{3}{4}|A|^2 A.$$

This result is not too surprising, since the perturbation introduced by $\beta \neq 0$ can be compensated by a rotation of the form $w(t, x) = \widetilde{w}(t, x - \varepsilon\beta t)$, which then transforms into a phase shift $A(t, x) = \widetilde{A}(t, x)e^{i\beta t}$ via $\mathbb{M}_\varepsilon$.

The theory for perturbed gradient systems can be used in much more general situations. We may consider a system of two Swift-Hohenberg equations with different critical wave lengths that are coupled in a non-gradient manner:

$$\dot{u} = -\frac{1}{\varepsilon^2}\left(1 + \varepsilon^2\partial_x^2\right)^2 u + \mu_1 u + (\eta + \beta)w - u^3,$$

$$\dot{w} = -\frac{1}{\varepsilon^2}\left(1 + \mu^2\varepsilon^2\partial_x^2\right)^2 w + \mu_2 w + (\eta - \beta)u - w^3.$$

We refer to [33] for this model in the case $\mu_1 = \mu_2$ and $\eta = 0$. Here u has the critical wave length $2\pi\varepsilon$ while that of w is $2\pi\mu\varepsilon$. The coupling between the two system occurs through a gradient term η or a non-gradient term β.

Thus, we can define the associated perturbed gradient system via

$$\boldsymbol{H} = L^2(\mathbb{S})^2, \quad \mathcal{R}^{\mathrm{cSH}}(\dot{u}, \dot{w}) = \frac{1}{2}\|\dot{u}\|_{L^2}^2 + \frac{1}{2}\|\dot{w}\|_{L^2}^2, \quad h(u, w) = \beta\begin{pmatrix} w \\ -u \end{pmatrix},$$

$$\mathcal{E}_\varepsilon^{\mathrm{cSH}}(u, w) = \int_{\mathbb{S}} \left(\frac{(u + \varepsilon^2 u_{xx})^2 + (w + \mu^2\varepsilon^2 w_{xx})^2}{2\varepsilon^2} + E(u, w) \right) dx$$

with $E(u, w) = -(\mu_1 u^2 + \mu_2 w^2)/2 - \eta uw + (u^4 + w^4)/4$. It is clear that the theory developed in Sect. 12.3 is principally applicable and that the induced limiting system for $\varepsilon \to 0$ will again be a perturbed gradient system given in terms of two possibly coupled Ginzburg–Landau equations. However, the critical bifurcations do no longer occur at $\mu_j = 0$. So, one needs to do a careful linear bifurcation analysis first. This and the justification of the arising amplitude equations will be the content of subsequent work.

12.5.3 From Diffusion to Reaction

In a series of papers it was shown that simple reactions can be understood as evolutionary Γ-limits of diffusion systems, if the occurrence of a reaction is measured moving along a reaction path. In particular, for an interchange reaction $A \rightleftharpoons B$ one should consider A and B are minima, which are separated by a saddle point. We refer to [2, 29, 30] for a series of papers along this spirit.

In [20] a systematic approach based on the energy-dissipation principle was developed allowing for a simultaneous treatment of diffusion in a physical domain $\Omega \in \mathbb{R}^d$ with points $x \in \Omega$ and the diffusion along the chemical reaction variable $y \in [0, 1] =: \Upsilon$. Denoting by $u(t, x, y)$ the concentration of particles one can write the master equation based on a gradient system, where the energy functional is the relative entropy with respect to the equilibrium state w_ε, namely

$$\mathcal{E}_\varepsilon(u) = \int_{\Omega \times \Upsilon} \lambda_B\big(u(x, y)/w_\varepsilon(y)\big)w_\varepsilon(y)\,\mathrm{d}y\,\mathrm{d}x \quad \text{with } \lambda_B(z) := z \log z - z + 1,$$

where the equilibrium state is $w_\varepsilon(y) = e^{-V(y)/\varepsilon}/Z_\varepsilon$ with $Z_\varepsilon = \int_\Upsilon e^{-V(y)/\varepsilon}\,\mathrm{d}y$. Here $y = 0$ corresponds to the pure state A, while $y = 1$ corresponds to the pure state B. We assume $V(0) = V(1) = 0$ and $0 < V(y) < 1 = V(1/2)$ for $y \in \Upsilon \setminus \{0, 1/2, 1\}$. The full state space X is the set $M(\Omega \times \Upsilon)$ of all non-negative Radon measures on $\Omega \times \Upsilon$.

Since in general the mass per particle can change during reactions we define a function $m : \Upsilon \to \mathbb{R}_{>0}$ such that the total mass $\int_{\Omega \times \Upsilon} m(y)u(t, x, y)\,\mathrm{d}y\,\mathrm{d}x$ is conserved. E.g. for the reaction $3\,O_2 \rightleftharpoons 2\,O_3$ one may set $m(0) = 2$, $m(1/2) = 1$, and $m(1) = 3$, where we assume that $y = 1/2$ corresponds to O_1. Using the function m we can define a dissipation potential $\mathcal{R}_\varepsilon$ via its Legendre dual

$$\mathcal{R}_\varepsilon^*(u, \xi) = \int_{\Omega \times \Upsilon} \frac{1}{2}\Big(\mu(y)|\nabla_x \xi|^2 + \tau_\varepsilon\Big[\frac{\partial_y \xi}{m(y)}\Big]^2\Big)u\,\mathrm{d}y\,\mathrm{d}x,$$

where μ is a possibly y-dependent spatial mobility and $\tau_\varepsilon \gg 1$ is the chemical mobility. The latter has to be scaled in a suitable manner to allow the particles to overcome the potential barrier of size $1/\varepsilon$ at $y = 1/2$.

Using that $D\mathcal{E}_\varepsilon(u) = \log(u/w_\varepsilon)$, the master equation (Kolmogorov's forward equation) for u is given via $\dot{u} = D_\xi \mathcal{R}_\varepsilon^*(u, -D\mathcal{E}_\varepsilon(u))$ and takes the explicit form

$$\dot{u} = \mu(y)\Delta_x u + \frac{\tau_\varepsilon}{m(y)}\partial_y\Big(u\,\partial_y\Big[\frac{\log u + V(y)/\varepsilon}{m(y)}\Big]\Big).$$

Generalizing the results in [20], where only the case $m \equiv 1$ was treated, it should be possible to show that the gradient systems $(M(\Omega \times \Upsilon), \mathcal{E}_\varepsilon, \mathcal{R}_\varepsilon)$ have the evolutionary Γ-limit $(M(\Omega \times \Upsilon), \mathcal{E}_0, \mathcal{R}_0)$, where the limit energy $\mathcal{E}_0$ is only finite if all the particles are in pure states $y = 0$ or $y = 1$, i.e.

$$\mathcal{E}_0(u) = \int_\Omega \Big(\lambda_B(c_0/c_0^*)c_0^* + \lambda_B(c_1/c_1^*)c_1^*\Big)\,\mathrm{d}x \quad \text{if } u = c_0\delta_{y=0} + c_1\delta_{y=1}$$

and $+\infty$ else. This means that we now have two concentrations c_0 and c_1 depending only on time t and the physical position $x \in \Omega$.

Fixing $m(1/2) = 1$ the limiting dissipation potential $\mathcal{R}_0^*$ takes the form

$$
\mathcal{R}_0^*(c_0, c_1; \eta_0, \eta_1) = \int_\Omega \left(\sum_{j=0}^{1} \frac{\mu_j c_j}{2} |\nabla_x \eta_j|^2 + k\big(c_0^{m_1} c_1^{m_0}\big)^{1/2} \mathfrak{S}^*(m_1 \eta_0 - m_0 \eta_1) \right) dx
$$

where $\mathfrak{S}^*(\eta) = 4(\cosh(\eta/2) - 1)$, $\mu_j = \mu(j)$, and $m_j = m(j)$. Thus, we expect evolutionary convergence to the nonlinear reaction-diffusion system

$$
\dot{c}_0 = \mu_0 \Delta_x c_0 + m_1 k\big((c_0/c_0^*)^{m_1} - (c_1/c_1^*)^{m_0}\big),
$$
$$
\dot{c}_1 = \mu_1 \Delta_x c_1 - m_0 k\big((c_0/c_0^*)^{m_1} - (c_1/c_1^*)^{m_0}\big).
$$

It is interesting to note that $\mathcal{R}_0^*$ is no longer quadratic in the chemical potentials η_j, but contains exponential terms through $\mathfrak{S}^*$. This seems to correspond nicely to the de Donder–Marcelin kinetics as described in [14, Definition 3.3], [17, Equation (11)], or [18, Equation (69)], and generalizes the usual quadratic fluctuation theory, cf. [28]. The importance of the function $\mathfrak{S}^*$ for fluctuations in reactions and jump processes was first highlighted in [25] based on large-deviation principles. Further discussions are found in [20, 24].

Acknowledgments The research was partially supported by the DFG within the SFB 910 (subproject A5) and by the ERC under AdG 267802 AnaMultiScale. The author is grateful to Matthias Liero, Sina Reichelt and Mark Peletier for stimulating discussions.

References

1. L. Ambrosio, N. Gigli, G. Savaré, *Gradient Flows in Metric Spaces and in the Space of Probability Measures*, Lectures in Mathematics ETH Zürich (Birkhäuser Verlag, Basel, 2005)
2. S. Arnrich, A. Mielke, M.A. Peletier, G. Savaré, M. Veneroni, Passing to the limit in a Wasserstein gradient flow: from diffusion to reaction. Calc. Var. Part. Diff. Eqns. **44**, 419–454 (2012)
3. H. Attouch, *Variational Convergence of Functions and Operators* (Pitman Advanced Publishing Program, Pitman, 1984)
4. A. Braides, *Γ-Convergence for Beginners* (Oxford University Press, 2002)
5. A. Braides, A handbook of Γ-convergence, in *Handbook of Differential Equations. Stationary Partial Differential Equations*, vol. 3, ed. by M. Chipot, P. Quittner (Elsevier, 2006)
6. A. Braides, *Local Minimization, Variational Evolution and Γ-Convergence*. Lecture Notes in Mathematics, vol. 2094 (Springer, 2013)
7. H. Brézis, *Opérateurs maximaux monotones et semi-groupes de contractions dans les espaces de Hilbert* (North-Holland Publishing Co., Amsterdam, 1973)
8. M. Buliga, G. de Saxcé, A symplectic Brezis–Ekeland–Nayroles principle. arXiv:1408.3102 (2014)
9. G. Dal Maso, *An Introduction to Γ-Convergence* (Birkhäuser Boston Inc., Boston, 1993)
10. S. Daneri, G. Savaré, Lecture notes on gradient flows and optimal transport, in *Optimal Transportation. Theory and Applications*, ed. by Y. Ollivier, H. Pajot, C. Villani (Cambridge University Press, 2014), pp. 100–144
11. M.H. Duong, A. Lamacz, M.A. Peletier, U. Sharma, Variational approach to coarse-graining of generalized gradient flows. arXiv:1507.03207v1 (2015)

12. M.H. Duong, M.A. Peletier, J. Zimmer, GENERIC formalism of a Vlasov-Fokker-Planck equation and connection to large-deviation principles. Nonlinearity **26**, 2951–2971 (2013)
13. I. Ekeland, R. Temam, *Convex Analysis and Variational Problems* (North Holland, 1976)
14. M. Feinberg, On chemical kinetics of a certain class. Arch. Rational Mech. Anal. **46**, 1–41 (1972)
15. W. Fenchel, On conjugate convex functions. Can. J. Math. **1**, 73–77 (1949)
16. B. Fiedler, P. Poláčik, Complicated dynamics of scalar reaction diffusion equations with a nonlocal term. Proc. Roy. Soc. Edinb. Sect. A **115**, 167–192 (1990)
17. A.N. Gorban, I.V. Karlin, V.B. Zmievskii, S.V. Dymova, Reduced description in the reaction kinetics. Phys. A **275**, 361–379 (2000)
18. M. Grmela, Multiscale equilibrium and nonequilibrium thermodynamics in chemical engineering. Adv. Chem. Eng. **39**, 75–128 (2010)
19. M. Liero, S. Reichelt, Homogenization of Cahn-Hilliard-type equations via evolutionary Γ-convergence. WIAS Preprint **2114**, (2015)
20. M. Liero, A. Mielke, M.A. Peletier, D.R.M. Renger, On microscopic origins of generalized gradient structures. Discr. Cont. Dynam. Systems Ser. S, (2015). Submitted (WIAS Preprint 2148)
21. A. Mielke, A gradient structure for reaction-diffusion systems and for energy-drift-diffusion systems. Nonlinearity **24**, 1329–1346 (2011)
22. A. Mielke, Deriving amplitude equations via evolutionary Γ-convergence. Discr. Cont. Dynam. Syst. Ser. A, **35**(6), (2015)
23. A. Mielke, On evolutionary Γ-convergence for gradient systems, in *Macroscopic and Large Scale Phenomena*, Lecture Notes in Applied Math. Mechanics, vol. 3, ed. by A. Muntean, J. Rademacher, A. Zagaris (Springer, 2015. Proceedings of Summer School in Twente University, June 2012. To appear (WIAS Preprint 1915), 55 p
24. A. Mielke, R.I.A. Patterson, M.A. Peletier, D.R.M. Renger, Non-equilibrium thermodynamical principles for nonlinear chemical reactions and systems with coagulation and fragmentation. WIAS Preprint **2168**, (2015)
25. A. Mielke, M.A. Peletier, D.R.M. Renger, On the relation between gradient flows and the large-deviation principle, with applications to Markov chains and diffusion. Potential Anal. **41**(4), 1293–1327 (2014)
26. A. Mielke, R. Rossi, G. Savaré, Nonsmooth analysis of doubly nonlinear evolution equations. Calc. Var. Part. Diff. Eqns. **46**(1–2), 253–310 (2013)
27. A. Mielke, S. Reichelt, M. Thomas, Two-scale homogenization of nonlinear reaction-diffusion systems with slow diffusion. Netw. Heterg. Mater. **9**(2), 353–382 (2014)
28. L. Onsager, S. Machlup, Fluctuations and irreversible processes. Phys. Rev. **91**(6), 1505–1512 (1953)
29. M.A. Peletier, G. Savarè, M. Veneroni, From diffusion to reaction via Γ-convergence. SIAM J. Math. Anal. **42**(4), 1805–1825 (2010)
30. M.A. Peletier, G. Savaré, M. Veneroni, Chemical reactions as Γ-limit of diffusion [revised reprint of [29]. SIAM Rev. **54**(2), 327–352 (2012)
31. S. Reichelt, Error estimates for nonlinear reaction-diffusion systems involving different diffusion length scales. J. Phys.: Conf. Ser. (2015). To appear (WIAS Preprint 2008)
32. R. Rossi, G. Savaré, Gradient flows of non convex functionals in Hilbert spaces and applications. ESAIM Control Optim. Calc. Var. **12**, 564–614 (2006)
33. D. Schüler, S. Alonso, A. Torcini, M. Bär, Spatio-temporal dynamics induced by competing instabilities in two asymmetrically coupled nonlinear evolution equations. Chaos **24**(4), 043142 (2014)
34. E. Sandier, S. Serfaty, Γ-convergence of gradient flows with applications to Ginzburg-Landau. Comm. Pure Appl. Math. **LVII**, 1627–1672 (2004)
35. S. Serfaty, Gamma-convergence of gradient flows on Hilbert spaces and metric spaces and applications. Discr. Cont. Dynam. Syst. Ser. A **31**(4), 1427–1451 (2011)
36. U. Stefanelli, The Brezis-Ekeland principle for doubly nonlinear equations. SIAM J. Control Optim. **47**(3), 1615–1642 (2008)

Chapter 13
Moment Closure—A Brief Review

Christian Kuehn

Abstract Moment closure methods appear in myriad scientific disciplines in the modelling of complex systems. The goal is to achieve a closed form of a large, usually even infinite, set of coupled differential (or difference) equations. Each equation describes the evolution of one "moment", a suitable coarse-grained quantity computable from the full state space. If the system is too large for analytical and/or numerical methods, then one aims to reduce it by finding a moment closure relation expressing "higher-order moments" in terms of "lower-order moments". In this brief review, we focus on highlighting how moment closure methods occur in different contexts. We also conjecture via a geometric explanation why it has been difficult to rigorously justify many moment closure approximations although they work very well in practice.

13.1 Introduction

The idea of moment-based methods is most easily explained in the context of stochastic dynamical systems. Abstractly, such a system generates a time-indexed sequence of random variables $x = x(t) \in \mathcal{X}$, say for $t \in [0, +\infty)$ on a given state space $\mathcal{X}$. Let us assume that the random variable x has a well-defined probability density function (PDF) $p = p(x, t)$. Instead of trying to study the full PDF, it is a natural step to just focus on certain moments $m_j = m_j(t)$ such as the mean, the variance, and so on, where $j \in \mathcal{J}$ and $\mathcal{J}$ is an index set and $\mathbb{M} = \{m_j : j \in \mathcal{J}\}$ is a fixed finite-dimensional space of moments. In principle, we may consider any moment space $\mathbb{M}$ consisting of a choice of coarse-grained variables approximating the full system, not just statistical moments. A typical moment-closure based study consists of four main steps:

C. Kuehn (✉)
Institute for Analysis and Scientific Computing, Vienna University
of Technology, 1040 Vienna, Austria
e-mail: ckuehn@asc.tuwien.ac.at

© Springer International Publishing Switzerland 2016   253
E. Schöll et al. (eds.), *Control of Self-Organizing Nonlinear Systems*,
Understanding Complex Systems, DOI 10.1007/978-3-319-28028-8_13

(S0) **Moment Space** *Select* the space $\mathbb{M}$ containing a hierarchy of moments m_j.

(S1) **Moment Equations** The next step is to derive evolution equations for the moments m_j. In the general case, such a system will be *high-dimensional* and fully *coupled*.

(S2) **Moment Closure** The large, often even infinite-dimensional, system of moment equations has to be *closed* to make it tractable for analytical and numerical techniques. In the general case, the closed system will be *nonlinear* and it will only *approximate* the full system of all moments.

(S3) **Justification and Verification** One has to justify, why the expansion made in step (S1) and the approximation made in step (S2) are useful in the context of the problem considered. In particular, the *choice* of the m_j and the *approximation properties* of the closure have to be answered.

Each of the steps (S0)–(S3) has its own difficulties. We shall not focus on (S0) as selecting what good 'moments' or 'coarse-grained' variables are creates its own set of problems. Instead, we consider some classical choices. (S1) is frequently a lengthy computation. Deriving relatively small moment systems tends to be a manageable task. For larger systems, computer algebra packages may help to carry out some of the calculations. Finding a good closure in (S2) is very difficult. Different approaches have shown to be successful. The ideas frequently include heuristics, empirical/numerical observations, physical first-principle considerations or a-priori assumptions. This partially explains, why mathematically rigorous justifications in (S3) are relatively rare and usually work for specific systems only. However, comparisons with numerical simulations of particle/agent-based models and comparisons with explicit special solutions have consistently shown that moment closure methods are an efficient tool. Here we shall also not consider (S3) in detail and refer the reader to suitable case studies in the literature.

Although moment closure ideas appear virtually across all quantitative scientific disciplines, a unifying theory has not emerged yet. In this review, several lines of research will be highlighted. Frequently the focus of moment closure research is to optimize closure methods with one particular application in mind. It is the hope that highlighting common principles will eventually lead to a better global understanding of the area.

In Sect. 13.2 we introduce moment equations more formally. We show how to derive moment equations via three fundamental approaches. In Sect. 13.3 the basic ideas for moment closure methods are outlined. The differences and similarities between different closure ideas are discussed. In Sect. 13.4 a survey of different applications is given. As already emphasized in the title of this review, we do not aim to be exhaustive here but rather try to indicate the common ideas across the enormous breadth of the area.

13.2 Moment Equations

The derivation of moment equations will be explained in the context of three classical examples. Although the examples look quite different at first sight, we shall indicate how the procedures are related.

13.2.1 Stochastic Differential Equations

Consider a probability space $(\Omega, \mathcal{F}, \mathbb{P})$ and let $W = W(t) \in \mathbb{R}^L$ be a vector of independent Brownian motions for $t \in \mathbb{R}$. A system of stochastic differential equations (SDEs) driven by $W(t)$ for unknowns $x = x(t) \in \mathbb{R}^N = \mathcal{X}$ is given by

$$\mathrm{d}x = f(x)\,\mathrm{d}t + F(x)\,\mathrm{d}W \tag{13.1}$$

where $f : \mathbb{R}^N \to \mathbb{R}^N$, $F : \mathbb{R}^N \to \mathbb{R}^{N \times L}$ are assumed to be sufficiently smooth maps, and we interpret the SDEs in the Itô sense [1, 2]. Alternatively, one may write (13.1) using white noise, i.e., via the generalized derivative of Brownian motion, $\xi := W'$ [1] as

$$x' = f(x) + F(x)\xi, \qquad ' = \frac{\mathrm{d}}{\mathrm{d}t}. \tag{13.2}$$

For the equivalent Stratonovich formulation see [3]. Instead of studying (13.1)–(13.2) directly, one frequently focuses on certain moments of the distribution. For example, one may make the *choice* to consider

$$m_j(t) := \langle x(t)^j \rangle = \langle x_1(t)^{j_1} \cdots x_N(t)^{j_N} \rangle, \tag{13.3}$$

where $\langle \cdot \rangle$ denotes the expected (or mean) value and $j \in \mathcal{J}$, $j = (j_1, \ldots, j_N)$, $j_n \in \mathbb{N}_0$, where $\mathcal{J}$ is a certain set of multi-indices so that $\mathbb{M} = \{m_j : j \in \mathcal{J}\}$. Of course, it should be noted that $\mathcal{J}$ can be potentially a very large set, e.g., for the cardinality of all multi-indices up to order J we have

$$\left| \left\{ j \in \mathbb{N}_0^N : |j| = \sum_n j_n \leq J \right\} \right| = \binom{J+N}{J} = \frac{(J+N)!}{J!N!}.$$

However, the main steps to derive evolution equations for m_j are similar for every fixed choice of J, N. After defining $m_j = m_j(t)$ (or any other "coarse-grained" variables), we may just differentiate m_j. Consider as an example the case $N = 1 = L$, and $\mathcal{J} = \{1, 2, \ldots, J\}$, where we write the multi-index simply as $j = j \in \mathbb{N}_0$. Then averaging (13.2) yields

$$m_1' = \langle x' \rangle = \langle f(x) \rangle + \langle F(x)\xi \rangle,$$

which illustrates the problem that we may never hope to express the moment equations explicitly for any nonlinear SDE if f and/or F are not expressible as convergent power series, i.e., if they are not analytic. The term $\langle F(x)\xi \rangle$ is not necessarily equal to zero for general nonlinearities F as $\int_0^t F(x(s)) \, dW(s)$ is only a *local* martingale under relatively mild assumptions [2]. Suppose we simplify the situation drastically by assuming a quadratic polynomial f and constant additive noise

$$f(x) = a_2 x^2 + a_1 x + a_0, \qquad F(x) \equiv \sigma \in \mathbb{R}. \tag{13.4}$$

Then we can actually use that $\langle \xi \rangle = 0$ and get

$$m_1' = \langle x' \rangle = a_2 \langle x^2 \rangle + a_1 \langle x \rangle + a_0 = a_2 m_2 + a_1 m_1 + a_0.$$

Hence, we also need an equation for the moment m_2. Using Itô's formula one finds the differential

$$d(x^2) = [2xf(x) + \sigma^2] \, dt + 2x\sigma \, dW$$

and taking the expectation it follows that

$$\begin{aligned} m_2' &= 2\langle a_2 x^3 + a_1 x^2 + a_0 x \rangle + \sigma^2 + \sigma \langle 2x\xi \rangle \\ &= 2(a_2 m_3 + a_1 m_2 + a_0 m_1) + \sigma^2, \end{aligned} \tag{13.5}$$

where $\langle 2x\xi \rangle = 0$ due to the martingale property of $\int_0^t 2x(s) \, dW_s$. The key point is that the ODE for m_2 depends upon m_3. The same problem repeats for higher moments and we get an infinite system of ODEs, even for the simplified case considered here. For a generic nonlinear SDE, the moment system is a fully-coupled infinite-dimensional system of ODEs. Equations at a given order $|\boldsymbol{j}| = J$ depend upon higher-order moments $|\boldsymbol{j}| > J$, where $|\boldsymbol{j}| := \sum_n j_n$.

Another option to derive moment equations is to consider the Fokker-Plank (or forward Kolmogorov) equation associated to (13.1)–(13.2); see [3]. It describes the probability density $p = p(x, t|x_0, t_0)$ of x at time t starting at $x_0 = x(t_0)$ and is given by

$$\frac{\partial p}{\partial t} = -\sum_{k=1}^{N} \frac{\partial}{\partial x_k}[pf] + \frac{1}{2} \sum_{i,k=1}^{N} \frac{\partial^2}{\partial x_i \partial x_k}[(FF^T)_{ik} p]. \tag{13.6}$$

Consider the case of additive noise $F(x) \equiv \sigma$, quadratic polynomial nonlinearity $f(x)$ and $N = 1 = L$ as in (13.4), then we have

$$\frac{\partial p}{\partial t} = -\frac{\partial}{\partial x}[(a_2 x^2 + a_1 x + a_0)p] + \frac{\sigma^2}{2} \frac{\partial^2 p}{\partial x^2}. \tag{13.7}$$

The idea to derive equations for m_j is to multiply (13.7) by x^j, integrate by parts and use some a-priori known properties or assumptions about p. For example, we have

$$m'_1 = \langle x' \rangle = \int_{\mathbb{R}} x \frac{\partial p}{\partial t} \, dx$$

$$= \int_{\mathbb{R}} -x \frac{\partial}{\partial x} [(a_2 x^2 + a_1 x + a_0) p] \, dx + \int_{\mathbb{R}} x \frac{\sigma^2}{2} \frac{\partial^2 p}{\partial x^2} \, dx.$$

If p and its derivative vanish at infinity, which is quite reasonable for many densities, then integration by parts gives

$$m'_1 = \int_{\mathbb{R}} [(a_2 x^2 + a_1 x + a_0) p] \, dx = a_2 m_2 + a_1 m_1 + a_0$$

as expected. A similar calculation yields the equations for other moments. Using the forward Kolmogorov equation generalizes in a relatively straightforward way to other Markov process, e.g., to discrete-time and/or discrete-space stochastic processes; in fact, many discrete stochastic processes have natural ODE limits [4–7]. In the context of Markov processes, yet another approach is to utilize the moment generating function or Laplace transform $s \mapsto \langle \exp[isx] \rangle$ (where $i := \sqrt{-1}$) to determine equations for the moments.

13.2.2 Kinetic Equations

A different context where moment methods are used frequently is kinetic theory [8–10]. Let $x \in \Omega \subset \mathbb{R}^N$ and consider the description of a gas via a single-particle density $\varrho = \varrho(x, t, v)$, which is nonnegative and can be interpreted as a probability density if it is normalized; in fact, the notational similarity between p from Sect. 13.2.1 and the one-particle density ϱ is deliberate. The pair $(x, v) \in \Omega \times \mathbb{R}^N$ is interpreted as position and velocity. A kinetic equation is given by

$$\frac{\partial \varrho}{\partial t} + v \cdot \nabla_x \varrho = Q(\varrho), \tag{13.8}$$

where $\nabla_x = \left(\frac{\partial}{\partial x_1}, \ldots, \frac{\partial}{\partial x_N} \right)^{\mathsf{T}}$, suitable boundary conditions are assumed, and $\varrho \mapsto Q(\varrho)$ is the collision operator acting only on the v-variable at each $(x, t) \in \mathbb{R}^N \times [0, +\infty)$ with domain $\mathcal{D}(Q)$. For example, for short-range interaction and hard-sphere collisions [11] one would take for a function $v \mapsto G(v)$ the operator

$$Q(G)(v) = \int_{\mathbb{S}^{N-1}} \int_{\mathbb{R}^N} \|v - w\| [G(w^*)G(v^*) - G(v)G(w)] \, dw \, d\psi$$

where $v^* = \frac{1}{2}(v + w + \|v - w\|\psi)$, $w^* = \frac{1}{2}(v + w + \|v - w\|\psi)$ for $\psi \in \mathbb{S}^{N-1}$ and $\mathbb{S}^{N-1}$ denotes the unit sphere in $\mathbb{R}^N$. We denote velocity averaging by

$$\langle G \rangle = \int_{\mathbb{R}^N} G(v) \, dv,$$

where the overloaded notation $\langle \cdot \rangle$ is again deliberately chosen to highlight the similarities with Sect. 13.2.1. It is standard to make several assumptions about the collision operator such as the conservation of mass, momentum, energy as well as local entropy dissipation

$$\langle Q(G) \rangle = 0, \quad \langle v Q(G) \rangle = 0, \quad \langle \|v\|^2 Q(G) \rangle = 0, \quad \langle \ln(G) Q(G) \rangle \le 0. \quad (13.9)$$

Moreover, one usually assumes that the steady states of (13.8) are Maxwellian (Gaussian-like) densities of the form

$$\rho_*(v) = \frac{q}{(2\pi\theta)^{N/2}} \exp\left(-\frac{\|v - v_*\|^2}{2\theta}\right), \quad (q, \theta, v_*) \in \mathbb{R}^+ \times \mathbb{R}^+ \times \mathbb{R}^N \quad (13.10)$$

and that Q commutes with certain group actions [8] implying symmetries. Note that the physical constraints (13.9) have important consequences, e.g., entropy dissipation implies the local dissipation law

$$\frac{\partial}{\partial t} \langle \varrho \ln \varrho - \varrho \rangle + \nabla_x \cdot \langle v(\varrho \ln \varrho - \varrho) \rangle = \langle \ln \varrho \, Q(\varrho) \rangle \le 0. \quad (13.11)$$

while mass conservation implies the local conservation law

$$\frac{\partial}{\partial t} \langle \varrho \rangle + \nabla_x \cdot \langle v \varrho \rangle = 0 \quad (13.12)$$

with similar local conservation laws for momentum and energy. The local conservation law indicates that it could be natural, similar to the SDE case above, to multiply the kinetic equation (13.8) by polynomials and then average. Let $\{m_j = m_j(v)\}_{j=1}^J$ be a basis for a J-dimensional space of polynomials $\mathbb{M}$. Consider a column vector $M = M(v) \in \mathbb{R}^J$ containing all the basis elements so that every element $m \in \mathbb{M}$ can be written as $m = \alpha^\top M$ for some vector $\alpha \in \mathbb{R}^J$. Then it follows

$$\frac{\partial}{\partial t} \langle \varrho M \rangle + \nabla_x \cdot \langle v \varrho M \rangle = \langle Q(\varrho) M \rangle \quad (13.13)$$

by multiplying and averaging. This is exactly the same procedure as for the forward Kolmogorov equation for the SDE case above. Observe that (13.13) is a J-dimensional set of moment equations when viewed component-wise. This set is usually not closed. We already see by looking at the case $M \equiv v$ that the second term in (13.13) will usually generate higher-order moments.

13.2.3 Networks

Another common situation where moment equations appear are network dynamical systems. Typical examples occur in epidemiology, chemical reaction networks and socio-economic models. Here we illustrate the moment equations [12–15] for the classical susceptible-infected-susceptible (SIS) model [16] on a fixed network; for remarks on adaptive networks see Sect. 13.4. Given a graph of K nodes, each node can be in two states, infected I or susceptible S. Along an SI-link infections occur at rate τ and recovery of infected nodes occurs at rate γ. The entire (microscopic) description of the system is then given by all potential configurations $x \in \mathbb{R}^N = \mathcal{X}$ of non-isomorphic graph configurations of S and I nodes. Even for small graphs, N can be extremely large since already just all possible node configurations without considering the topology of the graph are 2^K. Therefore, it is natural to consider a coarse-grained description. Let $m_I = \langle I \rangle = \langle I \rangle(t)$ and $m_S = \langle S \rangle = \langle S \rangle(t)$ denote the average number of infected and susceptibles at time t. From the assumptions about infection and recovery rates we formally derive

$$\frac{dm_S}{dt} = \gamma m_I - \tau \langle SI \rangle, \tag{13.14}$$

$$\frac{dm_I}{dt} = \tau \langle SI \rangle - \gamma m_I, \tag{13.15}$$

where $\langle SI \rangle =: m_{SI}$ denotes the average number of SI-links. In (13.14) the first term describes that susceptibles are gained proportional to the number of infected times the recovery rate γ. The second term describes that infections are expected to occur proportional to the number of SI-links at the infection rate τ. Equation (13.15) can be motivated similarly. However, the system is not closed and we need an equation for $\langle SI \rangle$. In addition to (13.14)–(13.15), the result [14, Theorem 1] states that the remaining second-order motif equations are given by

$$\frac{dm_{SI}}{dt} = \gamma(m_{II} - m_{SI}) + \tau(m_{SSI} - m_{ISI} - m_{SI}), \tag{13.16}$$

$$\frac{dm_{II}}{dt} = -2\gamma m_{II} + 2\tau(m_{ISI} + m_{SI}), \tag{13.17}$$

$$\frac{dm_{SS}}{dt} = 2\gamma m_{SI} - 2\tau m_{SSI}, \tag{13.18}$$

where we refer also to [12, 13]; it should be noted that (13.16)–(13.18) does not seem to coincide with a direct derivation by counting links [17, (9.2)–(9.3)]. In any case, it is clear that third-order motifs must appear, e.g., if we just look at the motif ISI then an infection event generates two new II-links so the higher-order topological motif structure does have an influence on lower-order densities. If we pick the second-order space of moments

$$\mathbb{M} = \{m_I, m_S, m_{SI}, m_{SS}, m_{II}\} \tag{13.19}$$

the Eqs. (13.14)–(13.15) and (13.16)–(13.18) are not closed. We have the same
problems as for the SDE and kinetic cases discussed previously. The derivation of the
SIS moment equations can be based upon formal microscopic balance considerations.
Another option is write the discrete finite-size SIS-model as a Markov chain with
Kolmogorov equation

$$\frac{dx}{dt} = Px, \tag{13.20}$$

which can be viewed as an ODE of 2^K equations given by a matrix P. One defines
the moments as averages, e.g., taking

$$\langle I \rangle(t) := \sum_{k=0}^{K} kx^{(k)}(t), \qquad \langle S \rangle(t) := \sum_{k=0}^{K} (K-k)x^{(k)}(t),$$

where $x^{(k)}(t)$ are all states with k infected nodes at time t. Similarly one can define
higher moments, multiply the Kolmogorov equation by suitable terms, sum the equation as an analogy to the integration presented in Sect. 13.2.2, and derive the moment
equations [14]. For any general network dynamical systems, moment equations can
usually be derived. However, the choice which moment (or coarse-grained) variables
to consider is far from trivial as discussed in Sect. 13.4.

13.3 Moment Closure

We have seen that moment equations, albeit being very intuitive, do suffer from the
drawback that the number of moment equations tends to grow rapidly and the exact
moment system tends to form an infinite-dimensional system given by

$$\begin{aligned}
\frac{dm_1}{dt} &= h_1(m_1, m_2, \ldots), \\
\frac{dm_2}{dt} &= h_2(m_2, m_3, \ldots), \\
\frac{dm_3}{dt} &= \cdots,
\end{aligned} \tag{13.21}$$

where we are going to assume from now on the even more general case $h_j = h_j(m_1, m_2, m_3, \ldots)$ for all j. In some cases, working with an infinite-dimensional
system of moments may already be preferable to the original problem. We do not
discuss this direction further and instead try to close (13.21) to obtain a finite-
dimensional system. The idea is to find a mapping H, usually expressing the higher-
order moments in terms of certain lower-order moments of the form

$$H(m_1, \ldots, m_\kappa) = (m_{\kappa+1}, m_{\kappa+2}, \ldots) \tag{13.22}$$

for some $\kappa \in \mathcal{J}$, such that (13.21) yields a closed system

$$
\begin{aligned}
\frac{dm_1}{dt} &= h_1(m_1, m_2, \ldots, m_\kappa, H(m_1, \ldots, m_\kappa)), \\
\frac{dm_2}{dt} &= h_2(m_1, m_2, \ldots, m_\kappa, H(m_1, \ldots, m_\kappa)), \\
\vdots\;\; &= \qquad\qquad\qquad \vdots \\
\frac{dm_\kappa}{dt} &= h_\kappa(m_1, m_2, \ldots, m_\kappa, H(m_1, \ldots, m_\kappa)).
\end{aligned}
\tag{13.23}
$$

The two main questions are

(Q1) How to find/select the mapping H?
(Q2) How well does (13.23) approximate solutions of (13.21) and/or of the original dynamical system from which the moment equations (13.21) have been derived?

Here we shall focus on describing the several answers proposed to (Q1). For a general nonlinear system, (Q2) is extremely difficult and Sect. 13.3.4 provides a geometric conjecture why this could be the case.

13.3.1 Stochastic Closures

In this section we focus on the SDE (13.1) from Sect. 13.2.1. However, similar principles apply to all incarnations of the moment equations we have discussed. One possibility is to *truncate* [18] the system and neglect all moments higher than a certain order, which means taking

$$
H(m_1, \ldots, m_\kappa) = (0, 0, \ldots). \tag{13.24}
$$

Albeit being rather simple, the advantage of (13.24) is that it is trivial to implement and does not work as badly as one may think at first sight for many examples. A variation of the theme is to use the method of steady-state of moments by setting

$$
\begin{aligned}
0 &= h_{\kappa+1}(m_1, m_2, \ldots, m_\kappa, m_{\kappa+1}, \ldots), \\
0 &= h_{\kappa+2}(m_1, m_2, \ldots, m_\kappa, m_{\kappa+1}, \ldots), \\
\vdots\;\; &= \qquad\qquad \vdots
\end{aligned}
\tag{13.25}
$$

and try to solve for all higher-order moments in terms of $(m_1, m_2, \ldots, m_\kappa)$ in the algebraic equations (13.25). As we shall point out in Sect. 13.3.4, this is nothing but the quasi-steady-state assumption in disguise. Similar ideas as for zero and steady-sate moments can also be implemented using central moments and cumulants [18].

Another common idea for moment closure principles is to make an a priori assumption about the distribution of the solution. Consider the one-dimensional SDE example ($N = 1 = L$) and suppose $x = x(t)$ is normally distributed. For a normal distribution with mean zero and variance v^2, we know the moments

$$\langle x^j \rangle = v^j (j-1)!!, \quad \text{if } j \text{ is even}, \qquad \langle x^j \rangle = 0, \quad \text{if } j \text{ is odd}, \tag{13.26}$$

so one closure method, the so-called *Gaussian (or normal) closure*, is to set

$$m_j = 0 \quad \text{if } j \geq 3 \text{ and } j \text{ is odd},$$
$$m_j = (m_2)^{j/2} (j-1)!! \quad \text{if } j \geq 4 \text{ and } j \text{ is even}.$$

A similar approach can be implemented using central moments. If x turns out to deviate substantially from a Gaussian distribution, then one has to question whether a Gaussian closure is really a good choice. The Gaussian closure principle is one choice of a wide variety of distributional closures. For example, one could assume the moments of a *lognormal distribution* [19] instead

$$x \sim \exp[\tilde{\mu} + \tilde{v}\tilde{x}], \ \tilde{x} \sim \mathcal{N}(0, 1), \ \Rightarrow \langle x^j \rangle = m_j = \exp\left[j\tilde{\mu} + \frac{1}{2} j^2 \tilde{v}^2 \right] \tag{13.27}$$

where '$\sim$' means 'distributed according to' a given distribution and $\mathcal{N}(0, 1)$ indicates the standard normal distribution. Solving for $(\tilde{\mu}, \tilde{v})$ in (13.27) in terms of (m_1, m_2) yields a moment closure $(m_3, m_4, \ldots) = H(m_1, m_2)$. The same principle also works for discrete state space stochastic process, using a-prior distribution assumption. A typical example is the *binomial closure* [20] and mixtures of different distributional closure have also been considered [21, 22].

13.3.2 Physical Principle Closures

In the context of moment equations of the form (13.13) derived from kinetic equations, a typical moment closure technique is to consider a constrained closure based upon a postulated physical principle. The constraints are usually derived from the original kinetic equation (13.8), e.g., if it satisfies certain symmetries, entropy dissipation and local conservation laws, then the closure for the moment equations should aim to capture these properties somehow. For example, the assumption

$$\text{span}\{1, v_1, \ldots, v_N, \|v\|^2\} \subset \mathbb{M}$$

turns out to be necessary to recover conservation laws [8], while assuming that the space $\mathbb{M}$ is invariant under suitable transformations is going to preserve symmetries. However, even by restricting the space of moments to preserve certain physical assumptions, this usually does not constraint the moments enough to get a closure. Following [8] suppose that the single-particle density is given by

$$\varrho = \mathfrak{M}(\alpha) = \exp[\alpha^\top M(v)], \qquad m = m(v) \in \mathbb{M} \text{ s.t. } m(v) = \alpha^\top M(v) \tag{13.28}$$

for some *moment densities* $\alpha = \alpha(x, t) \in \mathbb{R}^J$. Using (13.28) in (13.13) leads to

$$\frac{\partial}{\partial t}\langle \mathfrak{M}(\alpha)M \rangle + \nabla_x \cdot \langle v\mathfrak{M}(\alpha)M \rangle = \langle Q(\mathfrak{M}(\alpha))M \rangle. \qquad (13.29)$$

Observe that we may view (13.29) as a system of J equations for the J unknowns α. Hence, one has formally achieved closure. The question is what really motivates the exponential ansatz (13.28). Introduce new variables $\eta = \langle \mathfrak{M}(\alpha)M \rangle$ and define a function

$$H(\eta) = -\langle \mathfrak{M}(\alpha) \rangle + \alpha^\top \eta$$

and one may show that $\alpha = [D_\eta H](\eta)$. It turns out [8] that $H(\eta)$ can be computed by solving the entropy minimization problem

$$\min_{\varrho}\{\langle \varrho \ln \varrho - \varrho \rangle : \langle M\varrho \rangle = \eta\} = H(\eta), \qquad (13.30)$$

where the constraint $\langle M\varrho \rangle = \eta$ prescribes certain moments; we recall that $M = M(v)$ is the fixed vector containing the moment space basis elements and the relation $\alpha = [D_\eta H](\eta)$ holds. From a statistical physics perspective, it may be more natural to view (13.30) as an entropy maximization problem [23] by introducing another minus sign. Therefore, the choice of the exponential function in the ansatz (13.28) does not only guarantee non-negativity but it was developed as it is the Legendre transform of the so-called entropy density $\varrho \mapsto \varrho \ln \varrho - \varrho$ so it naturally relates to a physical optimization problem [8].

To motivate further why using a closure motivated by entropy corresponds to certain physical principles, let us consider the 'minimal' moment space

$$\mathbb{M} = \mathrm{span}\{1, v_1, \ldots, v_N, \|v\|^2\}$$

The closure ansatz (13.28) can be facilitated using the vector $M(v) = (1, v_1, \ldots, v_N, \|v\|^2)$ but then [24] the ansatz is related to the Maxwellian density (13.10) since

$$\rho_*(v) = \exp[\alpha^\top M(v)], \quad \alpha = \left(\ln\left(\frac{q}{(2\pi\theta)^{3/2}}\right) - \frac{\|v_*\|}{2\theta}, \frac{v_*}{\theta}, -\frac{1}{2\theta}\right)^\top$$

but Maxwellian densities are essentially Gaussian-like densities and we again have a *Gaussian closure*. Using a Gaussian closure implies that the moment equations become the Euler equations of gas dynamics, which can be viewed as a mean-field model near equilibrium for the mesoscopic single-particle kinetic equation (13.8), which is itself a limit of microscopic equations for each particle [25, 26].

Taking a larger moment space $\mathbb{M}$ one may also get the Navier-Stokes equation as a limit [8], and this hydrodynamic limit can even be justified rigorously under certain assumptions [27]. This clearly shows that moment closure methods can link physical theories at different scales.

13.3.3 *Microscopic Closures*

Since there are limit connections between the microscopic level and macroscopic moment equations, it seems plausible that starting from an individual-based network model, one may motivate moment closure techniques. Here we shall illustrate this approach for the SIS-model from Sect. 13.2.3. Suppose we start at the level of first-order moments and let $\mathbb{M} = \{m_I, m_S\}$. To close (13.14)–(13.15) we want a map

$$m_{SI} = H(m_I, m_S). \tag{13.31}$$

If we view the density of the I nodes and S nodes as very *weakly correlated* random variables then a first guess is to use the approximation

$$m_{SI} = \langle SI \rangle \approx \langle S \rangle \langle I \rangle = m_S m_I. \tag{13.32}$$

Plugging (13.32) into (13.14)–(13.15) yields the *mean-field* SIS model

$$\begin{aligned} m'_S &= \gamma m_I - \tau m_S m_I, \\ m'_I &= \tau m_S m_I - m_I. \end{aligned} \tag{13.33}$$

The mean-field SIS model is one of the simplest examples where one clearly sees that although the moment equations are *linear* ODEs, the moment-closure ODEs are frequently *nonlinear*. It is important to note that (13.32) is not expected to be valid for all possible networks as it ignores the graph structure. A natural alternative is to consider

$$m_{SI} = \langle SI \rangle \approx m_d \langle S \rangle \langle I \rangle = m_d m_S m_I, \tag{13.34}$$

where m_d is the mean degree of the given graph/network. Hence it is intuitive that (13.32) is valid for a complete graph in the limit $K \to \infty$ [15].

If we want to find a closure similar to the approximation (13.32) for second-order moments with $\mathcal{M}$ as in (13.19), then the classical choice is the *pair-approximation* [28–30]

$$m_{abc} \approx \frac{m_{ab} m_{bc}}{m_b}, \qquad a, b, c \in \{S, I\} \tag{13.35}$$

which just means that the density of triplet motifs is given approximately by counting certain link densities that form the triplet. In (13.35) we have again ignored pre-factors from the graph structure such as the mean excess degree [12, 17]. As before, the assumption (13.35) is neglecting certain correlations and provides a mapping

$$(m_{SSI}, m_{ISI}) = H(m_{II}, m_{SS}, m_{SI}) = \left(\frac{m_{SS} m_{SI}}{m_S}, \frac{m_{SI} m_{SI}}{m_S} \right) \tag{13.36}$$

and substituting (13.36) into (13.16)–(13.18) yields a system of five closed nonlinear ODEs. Many other paradigms for similar closures exist. The idea is to use the interpretation of the moments and approximate certain higher-order moments based upon certain assumptions for each moment/motif. In the cases discussed here, this means neglecting certain *correlation terms* from random variables. At least on a formal level, this is approach is related to the other closures we have discussed. For example, forcing maximum entropy means minimizing correlations in the system while assuming a certain distribution for the moments just means assuming a particular correlation structure of mixed moments.

13.3.4 Geometric Closure

All the moment closure methods described so far, have been extensively tested in many practical examples and frequently lead to very good results; see Sect. 13.4. However, regarding the question (Q2) on approximation accuracy of moment closure, no completely general results are available. To make progress in this direction I conjecture that a high-potential direction is to consider moment closures in the context of geometric invariant manifold theory. There is very little mathematically rigorous work in this direction [31] although the relevance [32, 33] is almost obvious.

Consider the abstract moment equations (13.21). Let us assume for illustration purposes that we know that (13.21) can be written as a system

$$
\begin{aligned}
\frac{dm_1}{dt} &= h_1(m_1, m_2, \ldots, m_\kappa, m_{\kappa+1}, m_{\kappa+2}, \ldots), \\
\frac{dm_2}{dt} &= h_2(m_1, m_2, \ldots, m_\kappa, m_{\kappa+1}, m_{\kappa+2}, \ldots), \\
\vdots \;\; &= \qquad \vdots \\
\frac{dm_\kappa}{dt} &= h_\kappa(m_1, m_2, \ldots, m_\kappa, m_{\kappa+1}, m_{\kappa+2}, \ldots). \\
\frac{dm_{\kappa+1}}{dt} &= \frac{1}{\varepsilon} h_{\kappa+1}(m_1, m_2, \ldots, m_\kappa, m_{\kappa+1}, m_{\kappa+2}, \ldots). \\
\frac{dm_{\kappa+2}}{dt} &= \frac{1}{\varepsilon} h_{\kappa+2}(m_1, m_2, \ldots, m_\kappa, m_{\kappa+1}, m_{\kappa+2}, \ldots). \\
\vdots \;\; &= \qquad \vdots
\end{aligned}
\tag{13.37}
$$

where $0 < \varepsilon \ll 1$ is a small parameter and each of the component functions of the vector field h is of order $\mathcal{O}(1)$ as $\varepsilon \to 0$. Then (13.37) is a fast-slow system [34, 35] with fast variables $(m_{\kappa+1}, m_{\kappa+2}, \ldots)$ and slow variables $(m_1, \ldots, m_\kappa)$. The classical *quasi-steady-state assumption* [36] to reduce (13.37) to a lower-dimensional system is to take

$$
0 = \frac{dm_{\kappa+1}}{dt}, \qquad 0 = \frac{dm_{\kappa+2}}{dt}, \qquad \cdots .
$$

This generates a system of differential-algebraic equations and if we can solve the algebraic equations

$$
0 = h_{\kappa+1}(m_1, m_2, \ldots), \qquad 0 = h_{\kappa+2}(m_1, m_2, \ldots), \qquad \cdots
\tag{13.38}
$$

via a mapping H as in (13.22) we end up with a closed system of the form (13.23).

The quasi-steady-state approach hides several difficulties that are best understood geometrically from the theory of normally hyperbolic invariant manifolds, which is well exemplified by the case of fast-slow systems. For fast-slow systems, the algebraic equations (13.38) provide a representation of the *critical manifold*

$$C_0 = \{(m_1, m_2, \ldots) : h_j = 0 \text{ for } j > \kappa, j \in \mathbb{N}\}.$$

However, it is crucial to note that, despite its name, C_0 is not necessarily a manifold but in general just an algebraic variety. Even if we assume that C_0 is a manifold and we would be able to find a mapping H of the form (13.22), this mapping is generically only possible *locally* [34, 37]. Even if we assume in addition that the mapping is possible globally, then the dynamics on C_0 given by (13.22) does not necessarily approximate the dynamics of the full moment system for $\varepsilon > 0$. The relevant property to have a dynamical approximation is *normal hyperbolicity*, i.e., the 'matrix'

$$\left(\frac{\partial h_j}{\partial m_l} \right) \Bigg|_{C_0}, \qquad j, l \in \{\kappa + 1, \kappa + 2, \ldots\}$$

has no eigenvalues with zero real parts; in fact, this matrix is just the total derivative of the fast variables restricted to points on C_0 but for moment equations it is usually infinite-dimensional. Even if we assume in addition that C_0 is normally hyperbolic, which is a very strong and *non-generic* assumption for a fast-slow system [34, 35], then the dynamics given via the map H is only the *lowest-order approximation*. The correct full dynamics is given on a *slow manifold*

$$C_\varepsilon = \{(m_{\kappa+1}, m_{\kappa+2}, \ldots) = H(m_1, m_2, \ldots, m_\kappa) + \mathcal{O}(\varepsilon)\} \qquad (13.39)$$

so H is only correct up to order $\mathcal{O}(\varepsilon)$. This novel viewpoint on moment closure shows why it is probably quite difficult [38] to answer the approximation question (Q2) since for a general nonlinear system, the moment equations will only admit a closure via an explicit formula *locally* in the phase space of moments. One has to be very lucky, and probably make very effective use of special structures [39, 40] in the dynamical system, to obtain any *global* closure. Local closures are also an interesting direction to pursue [41].

13.4   Applications and Further References

Historically, applications of moment closure can at least be traced back to the classical Kirkwood closure [42] as well as statistical physics applications, e.g., in the Ising model [43]. The Gaussian (or normal) closure has a long history as well [44]. In mechanical applications and related nonlinear vibrations questions, stochastic mechanics models have been among the first where moment closure techniques for stochastic processes have become standard tools [45, 46] including the idea to just discard higher-order moments [47]. By now, moment closure methods have permeated practically all natural sciences as evidenced by the classical books [48, 49]. For SDEs, moment closure methods have not been used as intensively as one may guess but see [50].

For kinetic theory, closure methods also have a long history, particularly starting from the famous Grad 13-moment closure [51, 52], and moment methods have become fundamental tools in gas dynamics [53]. One particularly important application for kinetic-theory moment methods is the modelling of plasmas [54, 55]. In general, it is quite difficult to study the resulting kinetic moment equations analytically [56, 57] but many numerical approaches exist [58–61]. Of course, the maximum entropy closure we have discussed is not restricted to kinetic theory [62] and maximum entropy principles appear in many contexts [63–67].

One area where moment closure methods are employed a lot recently is mathematical biology. For example, the pair approximation [12] and its variants [68] are frequently used in various models including lattice models [69–74], homogeneous networks [75, 76] and many other network models [77–80]. Several closures have also included higher-order moments [81, 82] and truncation ideas are still used [83–85]. Applications to various different setups for epidemic spreading are myriad [85, 86]. A typical benchmark problem for moment methods in biology is the stochastic logistic equation [87–93]. Furthermore, spatial models in epidemiology and ecology have been a focus [94–97]. There are several survey and comparison papers with a focus on epidemics application and closure-methods available [13, 98–100]. There is also a link from mathematical biology and moment closure to transport and kinetic equations [101, 102], e.g., in applications of cell motion [103]. Also physical constraints, as we have discussed for abstract kinetic equations, play a key role in biology, e.g., trying to guarantee non-negativity [86].

Another direction is network dynamics [104], where moment closure methods have been used very effectively are adaptive, or co-evolutionary, networks with dynamics of and on the network [30, 105]. Moment equations are one reason why one may hope to describe self-organization of adaptive networks [106] by low-dimensional dynamical systems models [107]. Applications include opinion formation [108, 109] with a focus on the classical voter model [110–112]; see [113] for a review of closure methods applied to the voter model. Other applications are found again in epidemiology [114–120] and in game theory [121–123]. The maximum entropy-closure we introduced for kinetic equations has also been applied in the context of complex networks [124] and spatial network models in biology [125]. An overview of the use of the pair approximation, several models, and the relation to master equations can be found in [126]. It has also been shown that in many cases low-order or mean-field closures can still be quite effective [127].

On the level of moment equations in network science, one has to distinguish between purely moment or motif-based choices of the space $\mathbb{M}$ and the recent proposal to use heterogeneous degree-based moments. For example, instead of just tracking the moment of a node density, one also characterizes the degree distribution [128] of the node via new moment variables [129]. Various applications of heterogeneous moment equations have been investigated [130, 131].

Another important applications are stochastic reaction networks [132–134], where the mean-field reaction-rate equations are not accurate enough [135]. A detailed computation of moment equations from the master equation of reaction-rate models is given in [136]. In a related area, turbulent combustion models are investigated using moment closure [137–141]. For turbulent combustion, one frequently considers so-called conditional moment closures where one either conditions upon the flow being turbulent or restricts moments to certain parts of phase space; see [142] for a very detailed review.

Further applications we have not focused on here can be found in genetics [143], client-server models in computer science [144, 145], mathematical finance [146], systems biology [147], estimating transport coefficients [148], neutron transport [149], and radiative transport problems [150, 151]. We have also not focused on certain methods to derive moment equations including moment-generating functions [152–154], Lie-algebraic methods [155], and factorial moment expansions [156].

In summary, it is clear that many different areas are actively using moment closure methods and that a cross-disciplinary approach could yield new insights on the validity regimes of various methods. Furthermore, it is important to emphasize again that only a relatively small snapshot of the current literature has been given in this review and a detailed account of all applications of moment closure methods would probably fill many books.

Acknowledgments I would like to thank the Austrian Academy of Science (ÖAW) for support via an APART Fellowship and the EU/REA for support via a Marie-Curie Integration Re-Integration Grant. Support by the Collaborative Research Center 910 of the German Science Foundation (DFG) to attend the "International Conference on Control of Self-Organizing Nonlinear Systems" in 2014 is also gratefully acknowledged. Furthermore, I would like to thank Thomas Christen, Thilo Gross, Thomas House and an anonymous referee for very helpful feedback on various preprint versions of this chapter.

References

1. L. Arnold, *Stochastic Differential Equations: Theory and Applications* (Wiley, 1974)
2. O. Kallenberg, *Foundations of Modern Probability*, 2nd edn. (Springer, New York, NY, 2002)
3. C. Gardiner, *Stochastic Methods*, 4th edn. (Springer, Berlin Heidelberg, Germany, 2009)
4. T. Kurtz, J. Appl. Prob. **7**(1), 49 (1970)
5. T. Kurtz, Stoch. Proc. Appl. **6**(3), 223 (1978)
6. R. Darling, J. Norris, Prob. Surv. **5**, 37 (2008)
7. A. Bátkai, I. Kiss, E. Sikolya, P. Simon, Netw. Heterog. Media **7**, 43 (2012)
8. C. Levermore, J. Stat. Phys. **83**(5), 1021 (1996)
9. C. Cercignani, *Mathematical Methods in Kinetic Theory* (Springer, 1969)
10. P. Krapivsky, S. Redner, E. Ben-Naim, *A Kinetic View of Statistical Physics* (CUP, 2010)
11. S. Mischler, C. Mouhot, Invent. Math. **193**(1), 1 (2013)
12. M. Keeling, Proc. R. Soc. London B **266**(1421), 859 (1999)
13. D. Rand, CWI Quarterly **12**(3), 329 (1999)
14. M. Taylor, P. Simon, D. Green, T. House, I. Kiss, J. Math. Biol. **64**, 1021 (2012)
15. P. Simon, M. Taylor, I. Kiss, J. Math. Biol. **62**(4), 479 (2011)

16. O. Diekmann, J. Heesterbeek, *Mathematical Epidemiology of Infectious Diseases: Model Building, Analysis and Interpretation* (Wiley, 2000)
17. A.L. Do, T. Gross, in *Adaptive Networks: Theory, Models, and Applications*, ed. by T. Gross, H. Sayama (Springer, 2009), pp. 191–208
18. L. Socha, *Linearization Methods for Stochastic Dynamic Systems* (Springer, 2008)
19. A. Ekanayake, L. Allen, Stoch. Anal. Appl. **28**, 907 (2010)
20. I. Kiss, P. Simon, Bull. Math. Biol. **74**(7), 1501 (2012)
21. I. Krishnarajah, A. Cook, G. Marion, G. Gibson, Bull. Math. Biol. **67**(4), 855 (2005)
22. I. Krishnarajah, A. Cook, G. Marion, G. Gibson, Math. Biosci. **208**(2), 621 (2007)
23. E. Jaynes, Proc. IEEE **70**(9), 939 (1982)
24. C. Levermore, W. Morokoff, SIAM J. Appl. Math. **59**(1), 72 (1998)
25. C. Cercignani, *The Boltzmann Equation and Its Applications* (Springer, 1988)
26. H. Spohn, Rev. Mod. Phys. **52**(3), 569 (1980)
27. F. Golse, L. Saint-Raymond, Invent. Math. **155**(1), 81 (2004)
28. M. Keeling, D. Rand, A. Morris, Proc. R. Soc. B **264**(1385), 1149 (1997)
29. M. Keeling, K. Eames, J.R. Soc, Interface **2**(4), 295 (2005)
30. T. Gross, C.D. D'Lima, B. Blasius, Phys. Rev. Lett. **96**, (208701) (2006)
31. J. Stark, P. Iannelli, S. Baigent, Nonl. Anal. **47**, 753 (2001)
32. U. Dieckmann, R. Law, in *The Geometry of Ecological Interactions: Simplifying Spatial Complexity*, ed. by U. Dieckmann, R. Law, J. Metz (CUP, 2000), pp. 412–455
33. J. Pacheco, A. Traulsen, M. Nowak, Phys. Rev. Lett. **97**, (258103) (2006)
34. C. Kuehn, *Multiple Time Scale Dynamics* (Springer, 2015). 814 pp
35. C. Jones, in *Dynamical Systems (Montecatini Terme, 1994), Lect. Notes Math.*, vol. 1609 (Springer, 1995), pp. 44–118
36. L. Segel, M. Slemrod, SIAM Rev. **31**(3), 446 (1989)
37. N. Fenichel, J. Differ. Equ. **31**, 53 (1979)
38. A. Hasofer, M. Grigoriu, J. Appl. Mech. **62**(2), 527 (1995)
39. D. Pelinovsky, V. Zharnitsky, arXiv:1505.03354 (2015), pp. 1–45
40. C.D. Genio, T. House, Phys. Rev. E **88**(4), 040801 (2013)
41. G. Böhme, T. Gross, Phys. Rev. E **83**, (035101) (2011)
42. J. Kirkwood, J. Chem. Phys. **3**(5), 300 (1935)
43. R. Kikuchi, Phys. Rev. **81**(6), 988 (1951)
44. P. Whittle, J.R. Stat, Soc. B **19**(2), 268 (1957)
45. V. Bolotin, *Random Vibrations of Elastic Systems* (Springer, 1984)
46. R. Ibrahim, *Parametric Random Vibration* (Dover, 2008)
47. J. Richardson, in *Stochastic Processes in Mathematical Physics and Engineering*, ed. by R. Bellman (AMS, 1964), pp. 290–302
48. N. van Kampen, *Stochastic Processes in Physics and Chemistry* (North-Holland, 2007)
49. G. Adomian, *Stochastic Systems* (Academic Press, 1983)
50. R. Bobryk, J. Math. Anal. Appl. **329**(1), 703 (2007)
51. H. Grad, Comm. Pure Appl. Math. **2**(4), 331 (1949)
52. H. Struchtrup, M. Torrilhon, Phys. Fluids **15**(9), 2668 (2003)
53. H. Struchtrup, *Macroscopic Transport Equations for Rarefied Gas Flows* (Springer, 2005)
54. R. Robson, R. White, Z. Petrović, Rev. Mod. Phys. **77**(4), 1303 (2005)
55. G. Hammett, F. Perkins, Phys. Rev. Lett. **64**(25), 3019 (1990)
56. L. Desvillettes, Arch. Rat. Mech. Anal. **123**(4), 387 (1993)
57. T. Elmroth, Arch. Rat. Mech. Anal. **82**(1), 1 (1983)
58. C. Groth, J. McDonald, Cont. Mech. Thermodyn. **21**(6), 467 (2009)
59. C. Levermore, W. Morokoff, B. Nadiga, Phys. Fluids **10**(12), 3214 (1998)
60. J. McDonald, C. Groth, Cont. Mech. Thermodyn. **25**(5), 573 (2013)
61. M. Torrilhon, H. Struchtrup, J. Fluid Mech. **513**, 171 (2004)
62. A. Singer, J. Chem. Phys. **121**(8), 3657 (2004)
63. R. Abramov, J. Comput. Phys. **226**(1), 621 (2007)
64. A. Rangan, D. Cai, Phys. Rev. Lett. **96**(17), (178101) (2006)

65. J. Cernohorsky, S. Bludman, Astrophys. J. **433**, 250 (1994)
66. I. Csiszar, Ann. Stat. **19**(4), 2032 (1991)
67. L. Borland, F. Pennini, A. Plastino, A. Plastino, Eur. J. Phys. B **12**(2), 285 (1999)
68. C. Bauch, Math. Biosci. **198**, 217 (2005)
69. K. Sato, H. Matsuda, A. Sasaki, J. Math. Biol. **32**(3), 215 (1994)
70. J. Filipe, G. Gibson, Phil. Trans. R. Soc. London B **353**(1378), 2153 (1998)
71. J. Filipe, G. Gibson, Bull. Math. Biol. **63**(4), 603 (2001)
72. S. Ellner, J. Theor. Biol. **210**(4), 435 (2001)
73. M. Nakamaru, H. Matsuda, Y. Iwasa, J. Theor. Biol. **184**(1), 65 (1997)
74. H. Matsuda, N. Ogita, A. Sasaki, K. Sato, Prog. Theor. Phys. **88**(6), 1035 (1992)
75. T. Petermann, P.D.L. Rios, J. Theor. Biol. **229**(1), 1 (2004)
76. G. Rozhnova, A. Nunes, Phys. Rev. E **79**(4), (041922) (2009)
77. S. Bansal, B. Grenfell, L. Meyers, J. R. Soc. Interface **4**, 879 (2007)
78. D. Rand, in *Advanced Ecological Theory*, ed. by J. McGlade (Wiley, 1994), pp. 100–142
79. S. Risau-Gusman, D. Zanette, J. Theor. Biol. **257**, 52 (2009)
80. E. Volz, L. Meyers, Proc. R. Soc. B **274**, 2925 (2007)
81. T. House, G. Davies, L. Danon, M. Keeling, Bull. Math. Biol. **71**(7), 1693 (2009)
82. M. Keeling, J. Theor. Biol. **205**(2), 269 (2000)
83. B. Bolker, S. Pacala, Theor. Popul. Biol. **52**(3), 179 (1997)
84. B. Bolker, S. Pacala, Am. Nat. **153**(6), 575 (1999)
85. K. Hausken, J. Moxnes, Math. Comput. Mod. Dyn. Syst. **16**(6), 555 (2010)
86. D. Hiebeler, Bull. Math. Biol. **68**, 1315 (2006)
87. M. Bartlett, J. Gower, P. Leslie, Biometrika **47**(1), 1 (1960)
88. J. Matis, T. Kiffe, Biometrics **52**, 980 (1996)
89. A. Singh, J. Hespanha, Bull. Math. Biol. **69**(6), 1909 (2007)
90. J. Matis, T. Kiffe, Theor. Popul. Biol. **56**(2), 139 (1999)
91. I. Nåsell, Theor. Popul. Biol. **63**(2), 159 (2003)
92. I. Nåsell, Theor. Popul. Biol. **64**(2), 233 (2003)
93. T. Newman, J. Ferdy, C. Quince, Theor. Popul. Biol. **65**(2), 115 (2004)
94. O. Ovaskainen, S. Cornell, Proc. Natl. Acad. Sci. USA **103**(34), 12781 (2006)
95. R. Law, U. Dieckmann, in *The Geometry of Ecological Interactions: Simplifying Spatial Complexity*, ed. by U. Dieckmann, R. Law, J. Metz (CUP, 2000), pp. 252–270
96. P.A. Noël, B. Davoudi, R. Brunham, L.D. amd B. Pourbohloul, Phys. Rev. E **79**, (026101) (2009)
97. M. Martcheva, H. Thieme, T. Dhirasakdanon, J. Math. Biol. **53**, 642 (2006)
98. B. Bolker, S. Pacala, S. Levin, in *The Geometry of Ecological Interactions: Simplifying Spatial Complexity*, ed. by U. Dieckmann, R. Law, J. Metz (CUP, 2000), pp. 388–411
99. J. Miller, I. Kiss, Math. Mod. Nat. Phenom. **9**, 4 (2014)
100. D. Murrell, U. Dieckmann, R. Law, J. Theor. Biol. **229**, 421 (2004)
101. T. Hillen, Discr. Cont. Dyn. Syst. B **4**, 961 (2004)
102. T. Hillen, Discr. Cont. Dyn. Syst. B **5**(2), 299 (2005)
103. T. Hillen, J. Math. Biol. **53**(4), 585 (2006)
104. M. Porter, J. Gleeson, *Frontiers in Applied Dynamical Systems: Reviews and Tutorials*, arXiv:1403.7663 (2014), pp. 1–32
105. T. Gross, B. Blasius, J. R. Soc. Interface **5**, 259 (2008)
106. S. Bornholdt, T. Rohlf, Phys. Rev. Lett. **84**(26), 6114 (2000)
107. C. Kuehn, Phys. Rev. E **85**(2), 026103 (2012)
108. C. Nardini, B. Kozma, A. Barrat, Phys. Rev. Lett. **100**, (158701) (2008)
109. D. Kimura, Y. Hayakawa, Phys. Rev. E **78**, (016103) (2008)
110. V. Sood, S. Redner, Phys. Rev. Lett. **94**, 178701 (2005)
111. E. Pugliese, C. Castellano, Europhys. Lett. **88**, (58004) (2009)
112. F. Vazquez, V. Eguíluz, New J. Phys. **10**, (063011) (2008)
113. G. Demirel, F. Vazquez, G. Böhme, T. Gross, Phys. D **267**, 68 (2014)
114. T. Gross, I. Kevrekidis, Europhys. Lett. **82**, (38004) (2008)

115. L. Shaw, I. Schwartz, Phys. Rev. E **77**, (066101) (2008)
116. L. Shaw, I. Schwartz, Phys. Rev. E **81**, 046120 (2010)
117. M. Taylor, T. Taylor, I. Kiss, arXiv:1110.4000v1 (2011)
118. V. Marceau, P.A. Noël, L. Hébert-Dufresne, A. Allard, L. Dubé, Phys. Rev. E **82**, (036116) (2010)
119. C. Kuehn, J. Nonlinear Sci. **23**(3), 457 (2013)
120. D. Zanette, S. Risau-Gusmán, J. Biol. Phys. **34**, 135 (2008)
121. G. Demirel, P. Prizak, P. Reddy, T. Gross, Eur. Phys. J. B **84**, 541 (2011)
122. F. Feng, T. Wu, L. Wang, Phys. Rev. E **79**, (036101) (2009)
123. C.D. Genio, T. Gross, New J. Phys. **13**, (103038) (2011)
124. T. Rogers, J. Stat. Mech. **2011**, P05007 (2011)
125. M. Raghib, N. Hill, U. Dieckmann, J. Math. Biol. **62**, 605 (2011)
126. J. Gleeson, Phys. Rev. X **3**(2), 021004 (2013)
127. J. Gleeson, S. Melnik, J. Ward, M. Porter, P. Mucha, Phys. Rev. E **85**(2), 026106 (2012)
128. J. Gleeson, Phys. Rev. Lett. **107**, (068701) (2011)
129. K. Eames, M. Keeling, Proc. Nat. Acad. Sci. USA **99**(20), 13330 (2002)
130. J. Lindquist, J. Ma, P.V. den Driessche, F. Willeboordse, J. Math. Biol. **62**(2), 143 (2011)
131. H. Silk, G. Demirel, M. Homer, T. Gross, New J. Phys. **16**(9), (093051) (2014)
132. B. Barzel, O. Biham, Phys. Rev. Lett. **106**, (150602) (2011)
133. B. Barzel, O. Biham, Phys. Rev. E **86**, (031126) (2012)
134. C.G. Uribe, G. Verghese, J. Chem. Phys. **126**(2), 024109 (2007)
135. C. Lee, K.H. Kim, P. Kim, J. Chem. Phys. **130**, (134107) (2009)
136. S. Engblom, Appl. Math. Comput. **180**(2), 498 (2006)
137. R. Bilger, Phys. Fluids A: Fluid Dyn. **5**(2), 436 (1993)
138. M. Roomina, R. Bilger, Combust. Flame **125**(3), 1176 (2001)
139. M. Mortensen, R. Bilger, Combust. Flame. **156**(1), 62 (2009)
140. A. Klimenko, Phys. Fluids **7**(2), 446 (1995)
141. S. Navarro-Martinez, A. Kronenbuerg, F.D. Mare, Flow, Turbul. Combust. **75**(1), 245 (2005)
142. A. Klimenko, R. Bilger, Prog. Energy Combust. Sci. **25**(6), 595 (1999)
143. E. Baake, T. Hustedt, Markov Proc. Relat. Fields **17**, 429 (2011)
144. M. Guenther, J. Bradley, in *Computer Performance Engineering* (Springer, 2011), pp. 87–101
145. M. Guenther, A. Stefanek, J. Bradley, in *Computer Performance Engineering* (Springer, 2013), pp. 32–47
146. H. Singer, Comput. Stat. **21**(3), 385 (2006)
147. C. Gillespie, I.E.T. Syst, Biol. **3**(1), 52 (2009)
148. T. Christen, F. Kassubek, J. Phys. D: Appl. Phys. **47**, 363001 (2014)
149. T. Brunner, J. Holloway, J. Quant. Spectr. Rad. Transfer **69**(5), 543 (2001)
150. M. Frank, B. Dubroca, A. Klar, J. Comput. Phys. **218**(1), 1 (2006)
151. H. Struchtrup, Ann. Phys. **257**(2), 111 (1997)
152. E. Volz, J. Math. Biol. **56**, 293 (2008)
153. T. House, M. Keeling, J. R. Soc. Interface **8**, 67 (2011)
154. J. Miller, J. Math. Biol. **62**(3), 349 (2011)
155. T. House, Bull. Math. Biol. **77**(4), 646 (2015)
156. B. Blaszczyszyn, Stoch. Proc. Appl. **56**(2), 321 (1995)

Part II
Concepts of Applications

Chapter 14
Feedback Control in Quantum Transport

Clive Emary

Abstract Quantum transport is the study of the motion of electrons through nano-scale structures small enough that quantum effects are important. In this contribution I review recent theoretical proposals to use the techniques of quantum feedback control to manipulate the properties of electron flows and states in quantum-transport devices. Quantum control strategies can be grouped into two broad classes: measurement-based control and coherent control, and both are covered here. I discuss how measurement-based techniques are capable of producing a range of effects, such as noise suppression, stabilisation of nonequillibrium quantum states and the realisation of a nano-electronic Maxwell's demon. I also describe recent results on coherent transport control and its relation to quantum networks.

14.1 Introduction

Feedback control of quantum-mechanical systems is a rapidly emerging topic [1, 2], developed most fully in the field of quantum optics [3]. Only recently have these ideas been extended to quantum transport, a field which looks to understand and control the motion of electrons through structures on the nano-scale [4]. The aim of this contribution is to review these recent developments.

Broadly speaking, quantum feedback strategies may usefully be classified into two types:

- *Measurement-based control*, where the quantum system is subject to measurements, the classical information from which forms the basis of the feedback loop;
- *Coherent control*, where the system, the controller and their interconnections are phase coherent such that the information flow in the feedback loop is of quantum information [5].

C. Emary (✉)
Department of Physics and Mathematics, University of Hull,
Hull HU6 7RX, UK

C. Emary
School of Mathematics and Statistics, Newcastle University,
Newcastle upon Tyne, Tyne and Wear NE1 7RU, UK

© Springer International Publishing Switzerland 2016   275
E. Schöll et al. (eds.), *Control of Self-Organizing Nonlinear Systems*,
Understanding Complex Systems, DOI 10.1007/978-3-319-28028-8_14

Mirroring the situation in optics, most of the work to-date on feedback in quantum transport has been within the measurement-based paradigm. In Sect. 14.2 here, we discuss a number of different measurement-based schemes and the physical results they can produce. Initial studies of coherent control in quantum transport have recently been performed and these are discussed in Sect. 14.3.

The scope of this review is limited to feedback schemes in which the target of the feedback is the electrons involved in the transport process. We shall not discuss situations where a transport device, such as a quantum point contact or single-electron transistor, acts as the readout stage in the feedback control of closed quantum system, e.g. a charge qubit. Such schemes have been extensively discussed elsewhere [6–9].

14.2 Measurement-Based Control

The basic idea behind measurement-based control in quantum transport is sketched in Fig. 14.1. Generically, the transport system we are looking to control is a small quantum system in the Coulomb blockade regime, weakly coupled to leads across which a potential difference is applied. In this regime, transport takes place via a series of discrete "jumps" in which electrons tunnel into or out of the system.

Our aim is to control some aspect of this process, be it the statistical properties of the current flow or the electronic states inside the device, through the establishment on a feedback loop based on the real-time detection of the electronic jumps using e.g. a quantum point contact (QPC) [10, 11]. The information gained from this electron

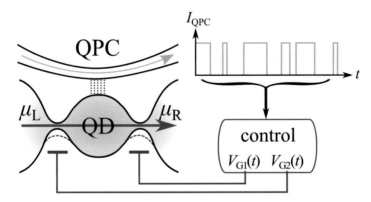

Fig. 14.1 Schematic of a measurement-based feedback control scheme applied to transport through a quantum dot (QD). The QD is connected to reservoirs (indicated by their chemical potentials μ_L and μ_R) and the *arrow* indicates current flow. The occupation status of the QD is detected with a quantum point contact (QPC), whose current gives rise to the time trace, *top right*. This information is then processed by control circuitry that modulates the gate potentials $V_{G1}(t)$ and $V_{G2}(t)$ in response, and in doing so alters the tunnel rates of electrons through the QD. In this way, a feedback loop is set up to control aspects of charge transfer through the dot

counting is processed and used to e.g. manipulate the gate voltages that serve to define the various properties of the transport system such as the tunnel coupling between dot and reservoirs. It should be noted that, although the system in question may be a quantum-mechanical one, the feedback loop here is an entirely classical affair.

14.2.1 Counting Statistics Formalism

The class of systems outlined above readily admits a description in terms of quantum master equations. The full-counting statistics (FCS) formalism [12], as applied to master equations [13], provides a convenient way to calculate transport properties, as well as motivate in a very physical way, various feedback schemes.

Let us consider a transport system described by a Markovian master equation,

$$\dot{\rho}(t) = \mathcal{W}\rho(t), \tag{14.1}$$

with $\rho(t)$ the reduced density matrix of the system at time t, and $\mathcal{W}$ the Liouvillian superoperator of the system [14]. In a weak-coupling approach, $\mathcal{W}$ describes a simple rate equation with transitions between system eigenstates. Alternatively, in the infinite bias limit [15], $\mathcal{W}$ defines a quantum master equation of Lindblad form with explicit unitary system dynamics and tunneling described in the local basis.

Irrespective of its precise form, the Liouvillian can be decomposed into terms that describe jump processes and those that do not. Let us focus in on a single, particular jump process and decompose the Liouvillian as

$$\mathcal{W} = \mathcal{W}_0 + \mathcal{J}, \tag{14.2}$$

where $\mathcal{J}$ is the superoperator describing the jump process in question, and where $\mathcal{W}_0$ describes the remaining evolution without jumps of this kind. Defining $\rho^{(n)}(t)$ as the density matrix of the system conditioned on n jumps of this type having occurred, the original master equation can be transformed into the number-resolved master equation [16]

$$\dot{\rho}^{(n)}(t) = \mathcal{W}_0\rho^{(n)}(t) + \mathcal{J}\rho^{(n-1)}(t). \tag{14.3}$$

Through definition of the Fourier transform $\rho(\chi; t) \equiv \sum_n e^{in\chi}\rho^{(n)}(t)$, we obtain the "counting-field-resolved" master equation

$$\dot{\rho}(\chi; t) = \mathcal{W}(\chi)\rho(\chi; t); \qquad \mathcal{W}(\chi) = \mathcal{W}_0 + e^{i\chi}\mathcal{J}. \tag{14.4}$$

This equation forms the basis of FCS calculations in master-equation approaches. Generalisation to counting more than one type of transition is straightforward.

14.2.2 Wiseman-Milburn Control and the Stabilisation of Non-equillibrium Pure States

Perhaps the simplest quantum-control scheme, well understood in quantum optics, is that due to Wiseman and Milburn [3, 17]. In essence, this scheme monitors the quantum jumps of the system and, directly after each, applies a fixed control operation to the system, assumed to act instantaneously. With such a control loop in place, counting and controlling a particular jump process, the dynamics of the system are still described by a master equation of the form Eq. (14.4), but with the original Liouvillian being replaced by its controlled counterpart

$$W(\chi) \rightarrow W_C(\chi) = W_0 + e^{i\chi} \mathcal{C}\mathcal{J}, \tag{14.5}$$

where $\mathcal{C}$ is the super-operator describing the control operation. This operation is typically a unitary operation acting on the system, but could also include non-unitary elements (with possible changes to the counting field structure, e.g. [18]).

Pöltl et al. [19] considered the application of Wiseman-Milburn control to transport models and demonstrated that it could be used to stabilise (in the sense to be defined below) a certain class of system state. They considered a generic infinite-bias two-lead transport model with internal coherences, restricted to the zero or one charge sectors. In this case the Liouvillian can be written as $W = W_0 + \mathcal{J}_L + \mathcal{J}_R$ with $\mathcal{J}_L$ describing electron tunneling in from the left and $\mathcal{J}_R$ describing tunneling out to the right. The "no-jump" part of the Liouvillian can then be written in terms of a non-Hermitian Hamiltonian:

$$W_0\rho = -i\left\{\widetilde{H}\rho - \rho\widetilde{H}^\dagger\right\}. \tag{14.6}$$

This Hamiltonian has eigenstates $\widetilde{H}|\psi_j\rangle = \varepsilon_j|\psi_j\rangle$ and $\langle\widetilde{\psi}_j|\widetilde{H} = \varepsilon_j\langle\widetilde{\psi}_j|$, which, in general, are non-adjoint. Pöltl et al. introduced a control operator $\mathcal{C}$ conditioned on the *incoming* jumps of the electrons and defined to rotate the post-jump state of the electron into one of the eigenstates $|\psi_j\rangle$. Since these states do not evolve under the action of W_0, an electron in state $|\psi_j\rangle$ will remain in it until it tunnels out. The dynamics of the system with control can therefore described by a simple two-level model with effective Liouvillian (in the basis of populations of empty and $|\psi_j\rangle$ states)

$$W_C^{(j)} = \begin{pmatrix} -\Gamma_L & \gamma_R^{(j)} \\ \Gamma_L & -\gamma_R^{(j)} \end{pmatrix}, \tag{14.7}$$

where Γ_L is the original rate of tunneling into the system and $\gamma_R^{(j)} = -2\,\mathrm{Im}(\varepsilon_j)$ is the new effective outgoing rate. In the limit of high in-tunneling rate, $\Gamma_L \rightarrow \infty$, the system spends the majority of the time in state $|\psi_j\rangle$ and the system is thus *stabilized* in this state. The state $|\psi_j\rangle$ is a *pure* state and thus very different from the stationary state of the system without control, which is typically mixed. Furthermore, due to the non-equillibrium character of the effective Hamiltonian, these states are also distinct from the eigenstates of the original system Hamiltonian.

Reference [19] applied these ideas to a non-equillbrium charge qubit consisting of a double quantum dot with coherent interdot tunneling. The stationary states of this model are mixed, to a greater or lesser degree, depending on the interdot tunnel coupling. By using the above feedback scheme with the appropriate choice of unitary feedback operator, it was shown to be possible to stabilise states over the complete surface of the Bloch sphere.

This model was also used to illustrate the effects of the control scheme on the current flowing through the device. When exact stabilisation takes place and the system is governed by Eq. (14.7), the FCS of the system naturally reduces to that of a two-level system. In the limit $\Gamma_L \to \infty$, these statistics become Poissonian, with all cumulants equal. This contrasts strongly with the FCS of the double quantum dot without control or with control parameters that do not lead to stabilisation. Thus, measurement of the output FCS can be used as part of a further (classical) feedback loop to isolate the stabilising control operation by minimising the distance between the system FCS distribution and that of a two-level system. The inverse problem of how to find formally the control operation that stablises a particular state was discussed for the general case in Ref. [9].

14.2.3   Current-Regulating Feedback

Historically, the first feedback control protocol to be proposed in quantum transport was that due to Brandes [20], who considered a feedback loop which served to modify the various elements of Eq. (14.3) such that they inherited a dependence on the number of jumps to have occurred:

$$\dot{\rho}^{(n)}(t) = \mathcal{W}_0^{(n)} \rho^{(n)}(t) + \mathcal{J}^{(n)} \rho^{(n-1)}(t). \tag{14.8}$$

In particular, Brandes considered that the new elements in Eq. (14.8) were the same as without feedback, but multiplied by analytic functions of the form

$$f[q_n(t)]; \quad q_n(t) \equiv I_0 t - n; \quad f[0] = 1. \tag{14.9}$$

Here, the quantity $q_n(t)$ describes the deviation between the actual number of charges to have flowed through the system, n, and a reference value, defined in terms of a reference current, I_0. In the linear feedback case, we have $f(x) = 1 + gx$ with g a small, dimensionless feedback parameter. Any rate multiplied by this function will increase if the actual charge transferred lags behind the reference, and decrease if it is in excess.

The results obtained with this current-regulating feedback are exemplified by the simple model of a unidirectional tunnel junction where $\mathcal{J}^{(n)} = -\mathcal{W}_0^{(n)} = \Gamma \{1 + g(I_0 t - n)\}$ is scalar. Without control, this is a Poisson process and all cumu-

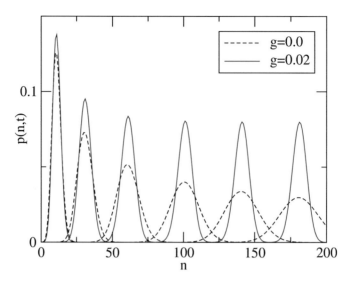

Fig. 14.2 The effect of current-governing feedback. Here is shown the distribution of the number of electrons n transferred through a tunnel junction at times $t = 30, 60, 100, 140, 180$ ($\Gamma = 1$). Shown are results for the case without ($g = 0$) and with linear ($g = 0.02$) feedback. Inclusion of the current-governing feedback leads to a freezing of the distribution. From Ref. [20]

lants of the transfered-charge distribution are equal: $C_k(t) = \Gamma t$. In particular, the width of the distribution grows linearly with time: $C_2(t) \equiv \langle n^2 \rangle(t) - \langle n \rangle^2(t) = \Gamma t$. With control in place, and I_0 set as the mean current without control, the first cumulant is unaltered. However, the second cumulant becomes

$$C_2(t) = \langle n^2 \rangle(t) - \langle n \rangle^2(t) = \frac{1}{2g} \left(1 - e^{-2g\Gamma t}\right). \tag{14.10}$$

Thus, the width of the FCS distribution no longer grows in time, but rather tends towards a fixed value. Figure 14.2 shows the FCS distribution with and without control and illustrates that feedback of this type leads to a freezing of the shape of the distribution. In the long-time limit, then, the relative fluctuation in the electron number, becomes vanishing small since $C_2(t)/C_1(t) \sim (2g\Gamma t)^{-1}$. This effect was also shown to hold for higher-dimensional models, in particular the single-electron transistor.

One potential application of this effect might be in the control of single-electron current sources [21], where reduction of the fluctuations in electron current is essential for the realisation of a useful quantum-mechanical definition of the ampere [22]. In this context, Fricke et al. have demonstrated the current locking of two electron pumps through a feedback mechanism based on the charging an mesoscopic island between them [23].

Whilst the number-dependence of Eq. (14.8) was originally considered to be the result of some external classical feedback circuit, subsequent work has shown that this kind of dependence can also arise from microscopic considerations [24]. Recently, Brandes has shown how the feedback coupling of Eq. (14.9) can arise autonomously in a series of interacting transport channels [25].

14.2.4 Piecewise-Constant Feedback and Maxwell's Demon

The idea of piecewise-constant feedback is best illustrated by direct consideration of the Maxwell's demon proposal of Ref. [18]. The set-up is exactly as in Fig. 14.1 with the quantum dot restricted to just two states ('empty' and 'full') and with the two reservoirs at finite bias and temperature.

The population of the QD is monitored in real time and, with piecewise-constant feedback, we apply one configuration of top-gate voltages when the QD is occupied, and a different set when it is empty. The electrons thus experience a different Liouvillian depending on whether the QD is empty or full. As shown in Refs. [18, 26], however, it is possible to describe the evolution of the system in terms of a single Liouvillian which, in the current case, written in the basis populations (empty, full), reads:

$$\mathcal{W}_{\text{fb}}^I(\chi_L, \chi_R) = \mathcal{W}_E(\chi_L, \chi_R) \begin{pmatrix} 1 & 0 \\ 0 & 0 \end{pmatrix} + \mathcal{W}_F(\chi_L, \chi_R) \begin{pmatrix} 0 & 0 \\ 0 & 1 \end{pmatrix}. \quad (14.11)$$

Here $\mathcal{W}_E$ is the Liouvillian when system in empty, and $\mathcal{W}_F(\chi_L, \chi_R)$ the Liouvillian when full. Both left (χ_L) and right (χ_R) counting fields are required.

In Ref. [18] it was assumed that only the dot-lead tunneling rates (and not, for example, the position of the energy level in the dot) are changed by the feedback loop. It can easily be seen how this arrangement might lead to a Maxwell's demon if we imagine that when the dot is empty, we completely close off tunneling to the right; and when the dot is full, we reverse the situation, and close tunneling to the left. In this situation, irrespective of bias or temperature, electrons will be preferentially transported from left to right through the QD. Even with less extreme modulation of the barriers, this scheme was shown to still drive a current against an applied bias, and thus extract work. In the classical limit, changing the barriers in the above fashion performs no work on the system, and thus, the current flow arises from the information gain of the feedback loop. This then is equivalent to Maxwell's demon. Two further feedback schemes, both based on Wiseman-Milburn style instantaneous control pulses, were also considered and shown to also give rise to the demon effect.

Subsequently, Esposito and Schaller [27] have formalised the notion of "Maxwell-demon feedbacks" and studied their thermodynamics. A physical, autonomous implementation of these ideas was discussed in Ref. [28], where the demon was realised by a second quantum dot connected to an independent electron reservoir.

We note also that a similar proposal involving a three-junction electron pump has also been proposed [29].

Tilloy et al. [30] have described a different method for generating a current through a transport device at zero bias that combines feedback control with the quantum Zeno effect [31]. They considered a double quantum dot in which the occupation of individual dots is continuously monitored. Conditioned on the outcome of this measurement, the measurement strength is changed: if the electron is found in the right-hand dot, the measurement strength is increased relative to the other possibilities. Then, due to the coherent coupling between the dots, the quantum Zeno effect means that the tunneling rate to the left dot will be suppressed whilst the rate for the electron to leave to the right lead remains untouched. The result is a flow of electrons through the double quantum dot to the right. Unlike in the single-dot Demon considered above, which may be perceived as a classical effect, the Zeno-based effect is irreducibly quantum-mechanical in nature.

14.2.5 Feedback Control with Delay

The preceding schemes have all assumed that the control operation is effected on the system immediately after the detection of a jump. In reality, however, there will always be a delay, of a time τ say, between detection and actuation. Wiseman considered the effects of delay on the class of feedback schemes outlined in Sect. 14.2.2 and gave modifications to Eq. (14.5) correct to first order in the delay time [17]. In Ref. [32], I showed that Wiseman's result actually holds for arbitrary delay time, providing one makes an additional "control-skipping" assumption, which means that if a jump occurs within the delay-time of an earlier jump, then the control operation for the first jump is discarded. With this assumption, it is straightforward to show that the delay-controlled system still obeys a master equation, but now a nonMarkovian one. The appropriate replacement for the kernel reads

$$\mathcal{W}(\chi) \to \mathcal{W}_{DC}(\chi, z) = \mathcal{W}_0 + \mathcal{D}(\chi, z)\mathcal{J}e^{i\chi}, \qquad (14.12)$$

with

$$\mathcal{D}(\chi, z) = 1 + [\mathcal{C} - 1]e^{(\mathcal{W}_0 - z)\tau}, \qquad (14.13)$$

the delayed control operation. In these expressions, z is the variable conjugate to time in the Laplace transform. In the time domain, we obtain the delayed nonMarkovian master equation

$$\dot{\rho}(t) = \mathcal{W}\rho(t) + (\mathcal{C} - 1)e^{\mathcal{W}_0\tau}\mathcal{J}\rho(t - \tau)\theta(t - \tau), \qquad (14.14)$$

in which the time evolution of the density matrix $\rho(t)$ depends not only on the state of the system at time t but also at previous time $t - \tau$.

Reference [32] considered the effect of delay on the state-stabilisation protocol of Ref. [19] and the Maxwell's demon of Ref. [18]. The influence of delay on the current governor was also discussed in [20]. In all these cases, the effects of delay are deleterious, but some effect of the control loop persists in the presence of delay. The influence of delay on the thermodynamics of Wiseman-Milburn feedback was studied in Ref. [33].

14.3   Coherent Control

Coherent feedback seeks to control a quantum system without the additional disturbance produced by the measurement step in measurement-based control. Various forms of coherent control have been discussed in the literature, e.g. Refs. [5, 34, 35]. However, the only type currently proposed for quantum transport [36, 37] is the *quantum feedback network* [35, 38–43], and this is the work we describe here.

14.3.1   Quantum Feedback Networks

In contrast to the measurement-based case, the quantum-feedback-network approach of Ref. [36] assumes that the system is strongly coupled to the leads, that the motion of the electrons through system and controller is phase coherent and that electron-electron interactions can be neglected. In this limit, transport can be described by Landauer-Büttiker theory [44].

Reference [36] considered that the system to be controlled was a four-terminal device (see Fig. 14.3), whose scattering matrix could be written in block form as

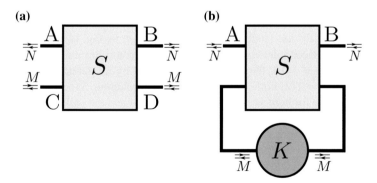

Fig. 14.3   a At the centre of the quantum feedback network discussed in Ref. [36] is a four-terminal device with scattering matrix S. Four leads are labeled A through D: A and B possess N bidirectional channels and leads C and D possess M. **b** The feedback loop is realised by connecting leads C and D together via a controller with scattering matrix, K. Figure taken from Ref. [36]

$$S = \begin{pmatrix} S_{\mathrm{I}} & S_{\mathrm{II}} \\ S_{\mathrm{III}} & S_{\mathrm{IV}} \end{pmatrix}. \tag{14.15}$$

Here, for example, block S_{I} describes all scattering processes between leads A and B, S_{II} describes all scattering processes starting in leads C and D and ending in leads A and B, and so on. All leads support states travelling in both directions. The feedback network is then formed by connecting leads C and D together through a controller with scattering matrix K. By considering all possible paths between leads A and B, the joint scattering matrix for the system-controller network is found to be:

$$S_{\mathrm{fb}} = S_{\mathrm{I}} + S_{\mathrm{II}} \frac{1}{1 - K\,S_{\mathrm{IV}}} K\,S_{\mathrm{III}}. \tag{14.16}$$

One key result stemming from this is that if the control scattering matrix, K, has the same dimension as the output matrix, S_{fb}, we can rearrange Eq. (14.16) to give

$$K = \frac{1}{S_{\mathrm{IV}} + S_{\mathrm{III}}\,(S_{\mathrm{fb}} - S_{\mathrm{I}})^{-1}\,S_{\mathrm{II}}}. \tag{14.17}$$

Thus, given an arbitrary original system matrix, S, we can obtain any desired output S_{fb} by choosing the control operator as in Eq. (14.17). This was dubbed "ideal control" in Ref. [36].

14.3.2 Conductance Optimisation

As an example of the use of the feedback network described by Eq. (14.16), Ref. [36] studied the optimisation of the conductance of chaotic quantum dots. The dots were modeled with $4N \times 4N$ scattering matrices taken from random matrix theory [45]. The number of active control channels in the controller was set as $1 \leq M \leq N$ and the elements of K chosen to maximise the conductance of the system-controller network. Results for the feedback network are shown in Fig. 14.4, and compared with those for a second network where system and controller are placed in series. Without control ($M = 0$), the conductance is given by the random-matrix-theory result $G/(NG_0) = 1/2$, with G_0 the conductance quantum. When $M = N$, ideal control is possible for both series and feedback setups and the ballistic conductance $G/(NG_0) = 1$ is obtained. For $0 < M < N$ there is a monotonic increase in the conductance for both series and feedback geometries. However, it is the feedback loop that offers the greater degree of conductance increase. The calculations of Ref. [36] also indicate that the feedback-loop geometry is also more robust under the influence of decoherence.

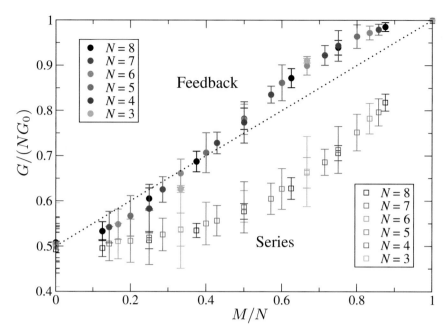

Fig. 14.4 Optimised conductance G of an ensemble of chaotic quantum dots under coherent control plotted as a function of the ratio of control to output dimension, M/N. The conductance is given in units of the ballistic conductance, NG_0, with $G_0 = 2e^2/h$. Results are shown for both feedback (*solid circles*) and series (*open squares*) configurations. Increasing the controller dimension increases the conductance gain towards the ideal-control limit of $G = NG_0$ at $M = N$. The feedback geometry outperforms its series counterpart for all $0 < M < N$. From Ref. [36]

14.4 Conclusion

In the majority of the proposals reviewed here, the target of the control has been the current flowing through the device. We have seen ways in which either the magnitude of the current, as in the Maxwell-demon and coherent-control proposals, or its statistical properties, as in the governor, can be modified by feedback. The exception to this was the proposal in Sect. 14.2.2, where the target of the control was the nonequillibrium states of the electrons inside the system itself. The manipulation of these states, however, also had a knock-on effect for current, which proved useful in diagnosing the effectiveness of the control procedure.

Of these proposals, the current-governor and Maxwell demon are the most hopeful candidates for experimental realisation in the foreseeable future. Indeed, a Maxwell's demon, similar in many respects to the one described here, has recently been realised in the single-electron box [46]. Such schemes are practicable because, although they take place in quantum-confined nanostructures, they do not rely on quantum coherence for their operation. The time-scales involved can therefore be relatively slow: in the FCS experiments of Ref. [10], for example, the QD was very weakly coupled to the reservoirs so that the typical time between tunnel events was of the order of a millisecond. Given this sort of timescale, it should be possible to build a control circuit fast enough to enact the required operations with a speed approximating the instantaneous ideal. By way of contrast, the pure-state stabilisation proposal of Ref. [19] can only function if the control loop operates on a timescale faster than the coherence time of the system being controlled. For charge coherences, this time is $\sim$1 ns [47], rending the construction of the feedback loop a considerable challenge. Progress could perhaps be made by controlling spin, rather than charge, degrees of freedom, since spin coherence times in QDs are far longer. We note that all such questions of external-circuits timescale are side-stepped by coherent-control protocols such as that described in Sect. 14.3. Here, however, the challenge is to go beyond abstract analysis and find appropriate physical systems to act as useful controllers.

References

1. J.E. Gough, Phil. Trans. R. Soc. A **370**, 5241 (2012)
2. J. Zhang, Y. Xi Liu, R.B. Wu, K. Jacobs, F. Nori, arXiv:1407.8536 (2014)
3. H.M. Wiseman, G.J. Milburn, Quantum Measurement and Control, 1st edn. (Cambridge University Press, 2009)
4. Y.V. Nazarov, Y.M. Blanter, Quantum Transport: Introduction to Nanoscience (Cambridge University Press, 2009)
5. S. Lloyd, Phys. Rev. A **62**, 022108 (2000)
6. R. Ruskov, A.N. Korotkov, Phys. Rev. B **66**, 041401 (2002)
7. A.N. Korotkov, Phys. Rev. B **71**, 201305 (2005)
8. G. Kießlich, G. Schaller, C. Emary, T. Brandes, Phys. Rev. Lett. **107**, 050501 (2011)
9. G. Kießlich, C. Emary, G. Schaller, T. Brandes, New J. Phys. **14**(12), 123036 (2012)
10. S. Gustavsson, R. Leturcq, B. Simovič, R. Schleser, T. Ihn, P. Studerus, K. Ensslin, D.C. Driscoll, A.C. Gossard, Phys. Rev. Lett. **96**, 076605 (2006)
11. T. Fujisawa, T. Hayashi, R. Tomita, Y. Hirayama, Science **312**, 1634 (2006)
12. L.S. Levitov, H. Lee, G.B. Lesovik, J. Math. Phys. **37**, 4845 (1996)
13. D.A. Bagrets, Y.V. Nazarov, Phys. Rev. B **67**, 085316 (2003)
14. T. Brandes, Phys. Rep. **408**, 315 (2005)
15. S.A. Gurvitz, Y.S. Prager, Phys. Rev. B **53**, 15932 (1996)
16. R.J. Cook, Phys. Rev. A **23**, 1243 (1981)
17. H.M. Wiseman, Phys. Rev. A **49**, 2133 (1994)
18. G. Schaller, C. Emary, G. Kiesslich, T. Brandes, Phys. Rev. B **84**, 085418 (2011)
19. C. Pöltl, C. Emary, T. Brandes, Phys. Rev. B **84**, 085302 (2011)
20. T. Brandes, Phys. Rev. Lett. **105**, 060602 (2010)
21. J.P. Pekola, O.P. Saira, V.F. Maisi, A. Kemppinen, M. Möttönen, Y.A. Pashkin, D.V. Averin, Rev. Mod. Phys. **85**, 1421 (2013)
22. S. Giblin, M. Kataoka, J. Fletcher, P. See, T. Janssen, J. Griffiths, G. Jones, I. Farrer, D. Ritchie, Nat. Commun. **3**, 930 (2012)
23. L. Fricke, F. Hohls, N. Ubbelohde, B. Kaestner, V. Kashcheyevs, C. Leicht, P. Mirovsky, K. Pierz, H.W. Schumacher, R.J. Haug, Phys. Rev. B **83**, 193306 (2011)
24. M. Schubotz, T. Brandes, Phys. Rev. B **84**, 075340 (2011)
25. T. Brandes, Phys. Rev. E **91**, 052149 (2015)
26. G. Schaller, Open Quantum Systems Far from Equilibrium (Springer International Publishing, 2014)
27. M. Esposito, G. Schaller, Europhys. Lett. **99**, 30003 (2012)

28. P. Strasberg, G. Schaller, T. Brandes, M. Esposito, Phys. Rev. Lett. **110**, 040601 (2013)
29. D.V. Averin, M. Möttönen, J.P. Pekola, Phys. Rev. B **84**, 245448 (2011)
30. A. Tilloy, M. Bauer, D. Bernard, EPL (Europhysics Letters) 107(2), 20010 (2014)
31. W.M. Itano, D.J. Heinzen, J.J. Bollinger, D.J. Wineland, Phys. Rev. A **41**, 2295 (1990)
32. C. Emary, Phil. Trans. R. Soc. A **371**, 20120468 (2013)
33. P. Strasberg, G. Schaller, T. Brandes, M. Esposito, Phys. Rev. E **88**, 062107 (2013)
34. H. Mabuchi, Phys. Rev. A **78**, 032323 (2008)
35. G. Zhang, M. James, IEEE Trans. Autom. Control **56**, 1535 (2011)
36. C. Emary, J. Gough, Phys. Rev. B **90**, 205436 (2014)
37. J. Gough, Phys. Rev. E **90**, 062109 (2014)
38. J.E. Gough, R. Gohm, M. Yanagisawa, Phys. Rev. A **78**, 062104 (2008)
39. J. Gough, M. James, Commun. Math. Phys. **287**, 1109 (2009)
40. J. Gough, M. James, IEEE Trans. Autom. Control **54**, 2530 (2009)
41. M. James, H. Nurdin, I. Petersen, IEEE Trans. Autom. Control **53**, 1787 (2008)
42. H.I. Nurdin, M.R. James, I.R. Petersen, Automatica **45**, 1837 (2009)
43. G. Zhang, M.R. James, Chin. Sci. Bull. **57**, 2200 (2012)
44. Y.M. Blanter, M. Büttiker, Phys. Rep. **336**, 1 (2000)
45. C.W.J. Beenakker, Rev. Mod. Phys. **69**, 731 (1997)
46. J.V. Koski, V.F. Maisi, J.P. Pekola, D.V. Averin, Proc. Natl. Acad. Sci. USA **111**, 13786 (2014)
47. T. Fujisawa, T. Hayashi, S. Sasaki, Rep. Prog. Phys. **69**, 759 (2006)

Chapter 15
Controlling the Stability of Steady States in Continuous Variable Quantum Systems

Philipp Strasberg, Gernot Schaller and Tobias Brandes

Abstract For the paradigmatic case of the damped quantum harmonic oscillator we present two measurement-based feedback schemes to control the stability of its fixed point. The first scheme feeds back a Pyragas-like time-delayed reference signal and the second uses a predetermined instead of time-delayed reference signal. We show that both schemes can reverse the effect of the damping by turning the stable fixed point into an unstable one. Finally, by taking the classical limit $\hbar \to 0$ we explicitly distinguish between inherent quantum effects and effects, which would be also present in a classical noisy feedback loop. In particular, we point out that the correct description of a classical particle conditioned on a noisy measurement record is given by a non-linear *stochastic* Fokker-Planck equation and *not* a Langevin equation, which has observable consequences on average as soon as feedback is considered.

15.1 Introduction

Continuous variable quantum systems are quantum systems whose algebra is described by two operators $\hat{x}$ and $\hat{p}$ (usually called position and momentum), which obey the commutation relation $[\hat{x}, \hat{p}] = i\hbar$. Such systems constitute an important class of quantum systems. They do not only describe the quantum mechanical analogue of the motion of classical heavy particles in an external potential, but they also arise, e.g., in the quantization of the electromagnetic field. Understanding them is important, e.g., in quantum optics [1], for purposes of quantum information processing [2, 3], or in the growing field of optomechanics [4]. Furthermore, due to the pioneering work of Wigner and Weyl, such systems have a well-defined classical limit and can be used to understand the transition from the quantum to the classical world [5].

P. Strasberg (✉) · G. Schaller · T. Brandes
Institut für Theoretische Physik, Technische Universität Berlin,
Hardenbergstr. 36,
10623 Berlin, Germany
e-mail: phist@physik.tu-berlin.de

© Springer International Publishing Switzerland 2016   289
E. Schöll et al. (eds.), *Control of Self-Organizing Nonlinear Systems*,
Understanding Complex Systems, DOI 10.1007/978-3-319-28028-8_15

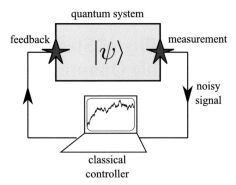

Fig. 15.1 Generic sketch for a closed-loop feedback scheme, in which we wish to control the dynamics of a given quantum system. Note that the feedback loop itself is actuated by a classical controller, i.e., the information after the measurement is *classical* (it is a number and not an operator). Nevertheless, to obtain the correct dynamics of the quantum system one needs to pay additional attention to the measurement and feedback step

To each quantum system there is an operator associated, called the Hamiltonian $\hat{H}$, which describes its energy and determines the dynamics of the system if it is isolated. However, in reality each system is an *open* system, i.e., it interacts with a large environment (we call it the bath) or other degrees of freedom (e.g., external fields). Since the bath is so large that we cannot describe it in detail, it induces effects like damping, dissipation or friction, which will eventually bring the system to a steady state. Classically as well as quantum mechanically it is often important to be able to counteract such irreversible behaviour, for instance, by applying a suitably designed feedback loop.

In the quantum domain, however, feedback control faces additional challenges compared to the classical world [6], see also Fig. 15.1. Each closed loop control scheme starts by measuring a certain output of the system and tries to feed the so gained information back into the system by adjusting some system parameters to influence its dynamics. In the quantum world—due to the measurement postulate of quantum mechanics and the associated "collapse of the wavefunction"—the measurement itself significantly disturbs the system and thus, it already influences the dynamics of the system. If one does not take this fact correctly into account, one easily arrives at wrong conclusions. Nevertheless, beautiful experiments have shown that quantum feedback control is invaluable to protect quantum information and to stabilize non-classical states of light and matter in various settings, see e.g. Refs. [7–12] for a selection of pioneering work in this field.

In this contribution we will apply two measurement based control schemes to a simple quantum system, the damped harmonic oscillator (HO), by correctly taking into account measurement and feedback noise at the quantum level (Sects. 15.3 and 15.4). These schemes will reverse the effect of dissipation and—to the best of our knowledge—have not been considered in this form elsewhere. However, we will see that our treatment is conceptually very close to a classical noisy feedback loop.

With this contribution we thus also hope to provide a bridge between quantum and classical feedback control. For pedagogical reasons we will therefore first present the necessary technical ingredients (continuous quantum measurement theory, quantum feedback theory and the phase space formulation of quantum mechanics) in Sect. 15.2. Due to a limited amount of space we cannot derive them here, but we will try to make them as plausible as possible. Section 15.5 is then devoted to a thorough discussion of the classical limit of our results showing which effects are truly quantum and which can be also expected in a classical feedback loop. In the last section we will give an outlook about possible applications and extensions of our feedback loop.

15.2   Preliminary

15.2.1   The Damped Quantum Harmonic Oscillator

We will focus only on the damped HO in this paper, but we will discuss extensions and applications of our scheme to other systems in Sect. 15.6. Using a canonical transformation we can rescale position and momentum such that the Hamiltonian of the HO reads $\hat{H} = \omega(\hat{p}^2 + \hat{x}^2)/2$ with $[\hat{x}, \hat{p}] = i\hbar$. Introducing linear combinations of position and momentum, called the annihilation operator $\hat{a} \equiv (\hat{x} + i\hat{p})/\sqrt{2\hbar}$ and its hermitian conjugate, the creation operator $\hat{a}^\dagger$, we can express the Hamiltonian as $\hat{H} = \hbar\omega(\hat{a}^\dagger\hat{a} + 1/2)$. Note that we explicitly keep Planck's contant $\hbar$ to take the classical limit ($\hbar \to 0$) later on.

The state of the HO is described by a density matrix $\hat{\rho}$, which is a positive, hermitian operator with unit trace: $\mathrm{tr}\hat{\rho} = 1$. If the HO is coupled to a large bath of different oscillators at a temperature T, it is possible to derive a so-called master equation (ME) for the time evolution of the density matrix [1, 6, 13]:

$$\frac{\partial}{\partial t}\hat{\rho}(t) = -\frac{i}{\hbar}[\hat{H}, \hat{\rho}(t)] + \kappa(1 + n_B)\mathcal{D}[\hat{a}]\hat{\rho}(t) + \kappa n_B\mathcal{D}[\hat{a}^\dagger]\hat{\rho}(t). \qquad (15.1)$$

Here, we introduced the dissipator $\mathcal{D}$, which is defined for an arbitrary operator $\hat{o}$ by its action on the density matrix: $\mathcal{D}[\hat{o}]\hat{\rho} \equiv \hat{o}\hat{\rho}\hat{o}^\dagger - \{\hat{o}^\dagger\hat{o}, \hat{\rho}\}/2$ where $\{\hat{a}, \hat{b}\} \equiv \hat{a}\hat{b} + \hat{b}\hat{a}$ denotes the anti-commutator. Furthermore, $\kappa > 0$ is a rate of dissipation characterizing how strong the time evolution of the system is effected by the bath and n_B denotes the Bose-Einstein distribution, $n_B \equiv (e^{\beta\hbar\omega} - 1)^{-1}$, where $\beta \equiv 1/T$ is the inverse temperature (we set $k_B \equiv 1$). For later purposes we abbreviate the whole ME (15.1) by

$$\frac{\partial}{\partial t}\hat{\rho}(t) \equiv \mathcal{L}_0\hat{\rho}(t), \qquad (15.2)$$

where the "superoperator" $\mathcal{L}_0$ is often called the Liouvillian and the subscript 0 refers to the fact that this is the ME for the *free* time evolution of the HO *without* any measurement or feedback performed on it [6]. Furthermore, it will turn out to be convenient to introduce a superoperator notation for the commutator and anti-commutator:

$$\mathcal{C}[\hat{o}]\hat{\rho} \equiv [\hat{o}, \hat{\rho}], \quad \mathcal{A}[\hat{o}]\hat{\rho} \equiv \{\hat{o}, \hat{\rho}\}. \tag{15.3}$$

One easily verifies that the time evolution of the expectation values of position $\langle \hat{x} \rangle(t) \equiv \text{tr}\{\hat{x}\hat{\rho}(t)\}$ and momentum $\langle \hat{p} \rangle(t) \equiv \text{tr}\{\hat{p}\hat{\rho}(t)\}$ is

$$\frac{d}{dt}\langle x \rangle(t) = \omega\langle p \rangle(t) - \frac{\kappa}{2}\langle x \rangle(t), \tag{15.4}$$

$$\frac{d}{dt}\langle p \rangle(t) = -\omega\langle x \rangle(t) - \frac{\kappa}{2}\langle p \rangle(t),$$

as for the classical damped harmonic oscillator. More generally speaking, for an arbitrary dynamical system these equations describe the generic situation for a two-dimensional stable steady state $(x^*, p^*) = (0, 0)$ within the linear approximation around $(0, 0)$. For $\kappa < 0$ this would describe an unstable steady state, but physically we can only allow for positive κ. The positivity of κ is mathematically required by Lindblad's theorem [14, 15] to guarantee that Eq. (15.1) describes a valid time evolution of the density matrix.[1]

Finally, a reader unfamiliar with this subject might find it instructive to verify that the canonical equilibrium state

$$\hat{\rho}_{\text{eq}} \sim e^{-\beta\hat{H}} \sim e^{-\beta\hbar\omega\hat{a}^\dagger\hat{a}} \tag{15.5}$$

is a steady state of the total ME (15.1) as it is expected from arguments of equilibrium statistical mechanics.

15.2.2 Continuous Quantum Measurements

In introductory courses on quantum mechanics (QM) one only learns about projective measurements, which yield the maximum information but are also maximally invasive in the sense that they project the total state $\hat{\rho}$ onto a single eigenstate. QM, however, also allows for much more general measurement procedures [6]. For our purposes, so-called continuous quantum measurements are most suited. They arise by considering very weak (i.e., less invasive) measurements, which are repeatedly performed on the system. In the limit where the time between two measurements goes to zero and the measurement becomes infinitely weak, we end up with a contin-

---

[1]The situation of an unstable fixed point would be modeled by exchanging the operators $\hat{a}$ and $\hat{a}^\dagger$ in the dissipators. This would correspond to a negative κ in the equation for the mean position and momentum. The feedback schemes presented here also work in that case.

uous quantum measurement scheme. For a quick introduction see Ref. [16]. Using their notation, one needs to replace $k \mapsto \gamma/(4\hbar)$ to obtain our results.

For our purposes we want to continuously measure (or "monitor") the position of the HO. Details of how to model the system-detector interaction can be found elsewhere [17–21]. Here, we restrict ourselves to showing the results and we try to make them plausible afterwards. If we neglect any contribution from $\mathcal{L}_0$ (Eq. (15.2)) for the moment, the time evolution of the density matrix due to the measurement of $\hat{x}$ is [6, 16–21]

$$\frac{\partial}{\partial t}\hat{\rho}(t) = -\frac{\gamma}{4\hbar}\mathcal{C}^2[\hat{x}]\hat{\rho}(t) \equiv \mathcal{L}_{\text{meas}}\hat{\rho}(t), \tag{15.6}$$

i.e., it involves a double commutator with the position operator $\hat{x}$ as defined in Eq. (15.3). Here, the new parameter γ has the physical dimension of a rate and quantifies the strength of the measurement. For $\gamma = 0$ we thus recover the case without any measurement. It is instructive to have a look how the matrix elements of $\hat{\rho}(t)$ evolve in the measurement basis $|x\rangle$ of the position operator $\hat{x} = \int dx\, x |x\rangle\langle x|$:

$$\frac{\partial}{\partial t}\langle x|\hat{\rho}(t)|x'\rangle = -\frac{\gamma}{4\hbar}(x - x')^2\langle x|\hat{\rho}(t)|x'\rangle. \tag{15.7}$$

We thus see that the off-diagonal elements (or "coherences") are exponentially damped whereas the diagonal elements (or "populations") remain unaffected. This is exactly what we would expect from a weak quantum measurement: the density matrix is perturbed only slightly but finally, in the long-time limit, it becomes diagonal in the measurement basis. Note that in case of a standard projective measurement scheme, the coherences would *instantaneously vanish*.

The ME (15.6) is, however, only half of the story because it tells us only about the average time evolution of the system, i.e., about the whole ensemble $\hat{\rho}$ averaged over all possible measurement records. The distinguishing feature of closed-loop control (as compared to open-loop control) is, however, that we want to influence the system based on a *single* (and not ensemble) measurement record. We denote the density matrix conditioned on a certain measurement record by $\hat{\rho}_c$ and call it the conditional density matrix. Its classical counterpart would be simply a conditional probability distribution.

In QM, even in absence of classical measurement errors, each single measurement record is necessarily noisy due to the inherent probabilistic interpretation of measurement outcomes in QM. The measurement signal $I(t)$ associated to the continuous position measurement scheme above can be shown to obey the stochastic process [6, 16]

$$dI(t) = \langle\hat{x}\rangle_c(t)dt + \sqrt{\frac{\hbar}{2\gamma\eta}}dW(t). \tag{15.8}$$

Here, by $\langle \hat{x} \rangle_c(t)$ we denoted the expectation value with respect to the conditional density matrix, i.e., $\langle \hat{x} \rangle_c(t) \equiv \text{tr}\{\hat{x}\hat{\rho}_c(t)\}$. Furthermore, $dW(t)$ is the Wiener increment. According to the standard rules of stochastic calculus, it obeys the relations [6, 16]

$$\mathbb{E}[dW(t)] = 0, \quad dW(t)^2 = dt \qquad (15.9)$$

where $\mathbb{E}[\dots]$ denotes a (classical) ensemble average over all noisy realizations. Furthermore, we have introduced a new parameter $\eta \in [0, 1]$, which is used to model the efficiency of the detector [6, 16, 20] with $\eta = 1$ corresponding to the case of a perfect detector.

Finally, we need to know how the state of the system evolves conditioned on a certain measurement record. This evolution is necessarily stochastic due to the stochastic measurement record. The so-called stochastic ME (SME) turns out to be given by [6, 16]

$$\hat{\rho}_c(t + dt) = \hat{\rho}_c(t) + \mathcal{L}_{\text{meas}}\hat{\rho}_c(t)dt + \sqrt{\frac{\gamma \eta}{2\hbar}}\mathcal{A}[\hat{x} - \langle \hat{x} \rangle_c(t)]\hat{\rho}_c(t)dW(t). \quad (15.10)$$

Because it will turn out to be useful, we have written the SME in an "incremental form" by explicitly using differentials as one would also do for numerical simulations. By definition we regard the quantity $[\hat{\rho}_c(t + dt) - \hat{\rho}_c(t)]/dt$ as being equivalent to $\partial_t \hat{\rho}_c(t)$. Using Eq. (15.8) we can express the SME alternatively as

$$\hat{\rho}_c(t + dt) = \hat{\rho}_c(t) + \mathcal{L}_{\text{meas}}\hat{\rho}_c(t)dt + \frac{\gamma \eta}{\hbar}\mathcal{A}[\hat{x} - \langle \hat{x} \rangle_c(t)]\hat{\rho}_c(t)[dI(t) - \langle \hat{x} \rangle_c(t)dt],$$
$$(15.11)$$

which explicitly demonstrates how our knowledge about the state of the system changes conditioned on a given measurement record $I(t)$.[2] We remark that the SME for $\hat{\rho}_c(t)$ is nonlinear in $\hat{\rho}_c(t)$, due to the fact that this in an equation of motion for a *conditional* density matrix.

To obtain the ME (15.6) for the average evolution, we only need to average the SME (15.10) over all possible measurement trajectories. In fact, it can be shown (see [6, 16]) that Eq. (15.10) has to be interpreted within the rules of Itô stochastic calculus (as well as all the following stochastic equations unless otherwise mentioned) such that

$$\mathbb{E}[\hat{\rho}_c(t)dW(t)] = \mathbb{E}[\hat{\rho}_c(t)]\mathbb{E}[dW(t)] = 0 \qquad (15.12)$$

holds. Defining $\hat{\rho}(t) \equiv \mathbb{E}[\hat{\rho}_c(t)]$, one can readily verify that the SME (15.10) yields on average Eq. (15.6).

---

[2]We explicitly adopt a Bayesian probability theory point of view in which probabilities (or more generally the density matrix $\hat{\rho}$) describe only (missing) human information. Especially, different observers (with possibly different access to measurement records) would associate *different* states $\hat{\rho}$ to the *same* system.

Taking the free evolution of the HO into account, Eq. (15.1), the total stochastic evolution of the system obeys

$$\hat{\rho}_c(t + dt) = \left\{ 1 + (\mathcal{L}_0 + \mathcal{L}_{\text{meas}})dt + \sqrt{\frac{\gamma\eta}{2\hbar}}\mathcal{A}[\hat{x} - \langle\hat{x}\rangle_c(t)]dW(t) \right\} \hat{\rho}_c(t). \quad (15.13)$$

Note that there are no "mixed terms" from the free evolution and the evolution due to the measurement to lowest order in dt. Furthermore, we remark that a solution of a SME is called a *quantum trajectory* in the literature [6, 13, 22].

15.2.3 Direct Quantum Feedback

In the following we will consider a form of quantum feedback control, which is sometimes called direct quantum feedback control because the measurement signal is directly fed back into the system (possibly with a delay) without any additional post-processing of the signal as, e.g., filtering or parameter estimation [21]. Direct quantum feedback based on a continuous measurement scheme was developed by Wiseman and Milburn [23–25]. Experimentally, the idea would be to continuously adjust a parameter of the Hamiltonian based on the measurement outcome (15.8) to control the dynamics of the system. Theoretically, we define the feedback control superoperator $\mathcal{F}$

$$[\dot{\rho}_c(t)]_{\text{fb}} = \mathcal{F}\hat{\rho}_c(t) \equiv -\frac{i}{\hbar}\frac{dI(t)}{dt}\mathcal{C}[\hat{z}]\hat{\rho}_c(t), \quad (15.14)$$

which describes a change of the free system Hamiltonian $\hat{H} = \omega(\hat{p}^2 + \hat{x}^2)/2$ to a new effective Hamiltonian $\omega(\hat{p}^2 + \hat{x}^2)/2 + \frac{dI(t)}{dt}\hat{z}$ containing a new term proportional to the measurement result and an arbitrary hermitian operator $\hat{z}$ (with units of $\dot{x}$). Here, we neglected any delay and assumed an instantaneous feedback of the measurement signal, but a delay can be easily incorporated, too, see Sect. 15.3.

Because the action of the feedback superoperator $\mathcal{F}$ on the system was merely postulated, we do not a priori know whether we have to interpret it according to the Itô or Stratonovich rules of stochastic calculus, but it turns out that only the latter interpretation gives senseful results [6, 23, 24]. Then, the effect of the feedback on the total time evolution of the system (including the measurement and free time evolution) can be found by exponentiating Eq. (15.14) [6, 23, 24]

$$\hat{\rho}_c(t + dt) = e^{\mathcal{F}dt}\left\{ 1 + \mathcal{L}_0dt + \mathcal{L}_{\text{meas}}dt + \sqrt{\frac{\gamma\eta}{2\hbar}}\mathcal{H}[\hat{x}]dW(t) \right\} \hat{\rho}_c(t) \quad (15.15)$$

and this equation is again of Itô type. Note that by construction this equation assures that the feedback step happens *after* the measurement as it must due to causality. Now, expanding $e^{\mathcal{F}dt}$ to first order in dt with $dI(t)$ from Eq. (15.8) (note that this requires to expand the exponential function up to *second* order due to the contribution from $dW(t)^2 = dt$) and using the rules of stochastic calculus gives the effective SME under feedback control:

$$\hat{\rho}_c(t + dt) = \hat{\rho}_c(t) + dt \left\{ \mathcal{L}_0 + \mathcal{L}_{\text{meas}} - \frac{i}{2\hbar} \mathcal{C}[\hat{z}] \mathcal{A}[\hat{x}] - \frac{1}{4\hbar\gamma\eta} \mathcal{C}^2[\hat{z}] \right\} \hat{\rho}_c(t)$$

$$(15.16)$$

$$+ dW(t) \left\{ \sqrt{\frac{\gamma\eta}{2\hbar}} \mathcal{A}[\hat{x} - \langle\hat{x}\rangle_c(t)] - \frac{i}{\hbar} \sqrt{\frac{\hbar}{2\gamma\eta}} \mathcal{C}[\hat{z}] \right\} \hat{\rho}_c(t).$$

If we take the ensemble average over the measurement records, we obtain the effective feedback ME

$$\frac{\partial}{\partial t} \hat{\rho}(t) = \left\{ \mathcal{L}_0 + \mathcal{L}_{\text{meas}} - \frac{i}{2\hbar} \mathcal{C}[\hat{z}] \mathcal{A}[\hat{x}] - \frac{1}{4\hbar\gamma\eta} \mathcal{C}^2[\hat{z}] \right\} \hat{\rho}(t) \qquad (15.17)$$

or more explicitly for our model

$$\frac{\partial}{\partial t} \hat{\rho}(t) = -\frac{i}{\hbar} \left\{ [\hat{H}, \hat{\rho}(t)] + \frac{1}{2} [\hat{z}, \hat{x}\hat{\rho}(t) + \hat{\rho}(t)\hat{x}] \right\}$$

$$+ \kappa(1 + n_B) \mathcal{D}[\hat{a}] \hat{\rho}(t) + \kappa n_B \mathcal{D}[\hat{a}^\dagger] \hat{\rho}(t) \qquad (15.18)$$

$$- \frac{\gamma}{4\hbar} [\hat{x}, [\hat{x}, \hat{\rho}(t)]] - \frac{1}{4\hbar\gamma\eta} [\hat{z}, [\hat{z}, \hat{\rho}(t)]].$$

Note that this equation is again linear in $\hat{\rho}(t)$ as it must be for a consistent statistical interpretation.

Before we give a short review about the last technically ingredient we need, which is rather unrelated to the previous content, we give a short summary. We have introduced the ME (15.1) for a HO of frequency ω, which is damped at a rate κ due to the interaction with a heat bath at inverse temperature β. We then started to continuously monitor the system at a rate γ with a detector of efficiency η. This procedure gave rise to a SME (15.13) conditioned on the measurement record (15.8). Finally, we applied feedback control by instantaneously changing the system Hamiltonian using the operator $\hat{z}$, which resulted in the effective ME (15.17).

15.2.4 Quantum Mechanics in Phase Space

The phase space formulation of QM is an equivalent formulation of QM, in which one tries to treat position and momentum on an equal footing (in contrast, in the Schrödinger formulation one has to work either in the position or ("exclusive or") momentum representation). By its design, phase space QM is very close to the classical phase space formulation of Hamiltonian mechanics and it is a versatile tool for a number of problems. For a more thorough introduction the reader is referred to Refs. [1, 5, 22, 26–28].

The central concept is to map the density matrix $\hat{\rho}$ to an object called the Wigner function:

$$W(x, p) \equiv \frac{1}{\hbar\pi} \int_{-\infty}^{\infty} dy \langle x - y|\hat{\rho}|x + y\rangle e^{2ipy/\hbar}. \tag{15.19}$$

The Wigner function is a quasi-probability distribution meaning that it is properly normalized, $\int dx dp W(x, p) = 1$, but can take on negative values. The expectation value of any function $F(x, p)$ in phase space can be computed via

$$\langle F(x, p) \rangle = \int_{\mathbb{R}^2} dx dp F(x, p) W(x, p) = \mathrm{tr}\{\hat{f}(\hat{x}, \hat{p})\hat{\rho}\}, \tag{15.20}$$

where the associated operator-valued observable $\hat{f}(\hat{x}, \hat{p})$ can be obtained from $F(x, p)$ via the Wigner-Weyl transform [5, 22, 28]. Roughly speaking this transformation symmetrizes all operator valued expressions. For instance, if $F(x, p) = xp$, then $\hat{f}(\hat{x}, \hat{p}) = (\hat{x}\hat{p} + \hat{p}\hat{x})/2$.

Each ME for a continuous variable quantum system can now be transformed to a corresponding equation of motion for the Wigner function. This is done by using certain correspondence rules between operator valued expressions and their phase space counterpart, e.g.,

$$\hat{x}\hat{\rho} \leftrightarrow \left(x + \frac{i\hbar}{2}\frac{\partial}{\partial p}\right) W(x, p), \tag{15.21}$$

which can be verified by applying Eq. (15.19) to $\hat{x}\hat{\rho}$ and some algebraic manipulations [22, 27].

The big advantage of the phase space formulation of QM is now that many MEs (namely those which can be called "linear") transform into an ordinary Fokker-Planck equation (FPE), for which many solution techniques are known [29]. We denote the general FPE for two variables (x, p) as

$$\frac{\partial}{\partial t} W(x, p, t) = \left\{-\nabla^T \cdot \mathbf{d} + \frac{1}{2}\nabla^T \cdot D \cdot \nabla\right\} W(x, p, t) \tag{15.22}$$

where $\nabla^T \equiv (\partial_x, \partial_p)$, the dot denotes a matrix product, $\mathbf{d}$ is the drift vector and D the diffusion matrix. It is then straightforward to confirm that the ME (15.1) corresponds to a FPE with

$$\mathbf{d}_x = \omega p - \frac{\kappa}{2}x, \quad \mathbf{d}_p = -\omega x - \frac{\kappa}{2}p, \tag{15.23}$$

$$D_{xx} = D_{pp} = \kappa\hbar\frac{1 + 2n_B}{2}, \quad D_{xp} = D_{px} = 0.$$

The SME (15.10) instead transforms to an equation for the conditional Wigner function $W_c(x, p, t)$ and reads

$$W_c(x, p, t + dt) = \left\{ 1 + dt \frac{\gamma\hbar}{4} \frac{\partial^2}{\partial p^2} + dW(t) \sqrt{\frac{2\gamma\eta}{\hbar}} [x - \langle x \rangle_c] \right\} W_c(x, p, t).$$

(15.24)

This does not have the standard form of a FPE. The additional term, however, does not cause any trouble in the interpretation of the Wigner function because we can still confirm that $\int dx dp\, W_c(x, p, t) = 1$.

Finally, we point out that the transition from quantum to classical physics is mathematically accomplished by the limit $\hbar \to 0$ [5]. Physically, of course, we do not have $\hbar = 0$ but the classical action of the particles motion becomes large compared to $\hbar$. We will discuss the classical limit of our equations in detail in Sect. 15.5.

15.3 Feedback Scheme I

The first feedback scheme we consider is the quantum analogue of the classical scheme considered in Ref. [30]. There the authors used a time delayed reference signal of the form $\langle \hat{x} \rangle(t) - \langle \hat{x} \rangle(t - \tau)$ to control the stability of the fixed point.[3] In our case we have to use the nosiy signal (15.8), i.e., we perform feedback based on

$$\delta I(t, \tau) \equiv [I(t) - I(t - \tau)]dt$$

(15.25)

$$= [\langle \hat{x} \rangle(t) - \langle \hat{x} \rangle_\tau(t)]dt + \sqrt{\frac{\hbar}{2\gamma\eta}} [dW(t) - dW_\tau(t)]$$

Here, a subscript τ indicates a shift of the time argument, i.e., $f_\tau(t) \equiv f(t - \tau)$. Due to this special form such feedback schemes are sometimes called Pyragas-like feedback schemes [33]. It should be noted however that we do not have a chaotic system here and we do not want to stabilize an unstable periodic orbit. In this respect, our feedback scheme is still an *invasive* feedback scheme, because the feedback-generated force does not vanish even if our goal to reverse the effect of the damping was achieved. We emphasize that such feedback schemes are widely used in classical control theory to influence the behaviour of, e.g., chaotic systems or complex networks [34, 35] and quite recently, there has been also a considerable interest to explore its quantum implications [36–43]. However, except of the feedback scheme in Ref. [41], the feedback schemes above were designed as *all-optical* or *coherent*

---

[3]In fact, in Ref. [30] they did not only feed back the results from a position measurement, but also from a momentum measurement. The simultaneous weak measurement of position and momentum can be also incorporated into our framework [17, 31, 32], but this would merely add additional terms without changing the overall message.

control schemes, in which the system is not subjected to an explicit measurement, but the environment is suitable engineered such that it acts back on the system in a very specific way. We will compare our scheme (which is based on explicit measurements) with these schemes towards the end of this section.

To see how our feedback scheme influences our system, we can still use Eq. (15.15) together with the measurement signal (15.25). Choosing $\hat{z} = k\hat{p}$ with $k \in \mathbb{R}$ and using that $dW(t)dW_\tau(t) = 0$ for $\tau \neq 0$ we obtain the SME

$$\hat{\rho}_c(t + dt) = \{1 + dt[\mathcal{L}_0 + \mathcal{L}_{\text{meas}}]\}\,\hat{\rho}_c(t)$$

$$- dt\left\{\frac{ik}{2\hbar}\mathcal{C}[\hat{p}]\mathcal{A}[\hat{x} - \langle x\rangle_c(t)] - \frac{ik}{\hbar}[\langle\hat{x}\rangle(t) - \langle\hat{x}\rangle_\tau(t)]\mathcal{C}[\hat{p}] - \frac{k^2}{2\hbar\gamma\eta}\mathcal{C}^2[\hat{p}]\right\}\hat{\rho}_c(t)$$

(15.26)

$$+ dW(t)\sqrt{\frac{\gamma\eta}{2\hbar}}\mathcal{A}[\hat{x} - \langle x\rangle_c(t)]\hat{\rho}_c(t) - \frac{ik}{\sqrt{2\hbar\gamma\eta}}[dW(t) - dW_\tau(t)]\mathcal{C}[\hat{p}]\hat{\rho}_c(t).$$

It is important to emphasize that also *time-delayed noise* enters the equation of motion for $\hat{\rho}_c(t)$. Because we do not know what $\mathbb{E}[\hat{\rho}_c(t)dW_\tau(t)]$ is in general, there is a priori no ME for the average time evolution of $\hat{\rho}(t)$. Approximating $\mathbb{E}[\hat{\rho}_c(t)dW_\tau(t)] \approx 0$ yields nonsense (the resulting ME would not even be linear in $\hat{\rho}$). This is, however, not a quantum feature and is equally true for classical feedback control based on a noisy, time-delayed measurement record (also see Sect. 15.5).

Due to the fact that there is no average ME, we are in principle doomed to simulated the SME (15.26) and average afterwards. However, as it turns out Eq. (15.26) can be transformed into a stochastic FPE whose solution is expected to be a Gaussian probability distribution. We will then see that the covariances indeed evolve *deterministicly*. Furthermore, it is possible to analytically deduce the equation of motion for the mean values on average. Within the Gaussian approximation we then have full knowledge about the evolution of the system.

Using the results from Sect. 15.2.4 we obtain

$$W_c(x, p, t + dt) = W_c(x, p, t) + dt\left(-\nabla^T \cdot \mathbf{d} + \frac{1}{2}\nabla^T \cdot D \cdot \nabla\right)W_c(x, p, t)$$

(15.27)

$$+ \left\{\sqrt{\frac{2\gamma\eta}{\hbar}}dW(x - \langle x\rangle_c) - k\sqrt{\frac{\hbar}{2\gamma\eta}}(dW - dW_\tau)\frac{\partial}{\partial x}\right\}W_c(x, p, t)$$

with the nonvanishing coefficients

$$\mathbf{d}_x = \omega p - \frac{\kappa}{2}x + k[x - \langle x\rangle_{c,\tau}(t)], \quad \mathbf{d}_p = -\omega x - \frac{\kappa}{2}p, \tag{15.28}$$

$$D_{xx} = \kappa\hbar\frac{1 + 2n_B}{2} + \frac{\hbar k^2}{\gamma\eta}, \quad D_{pp} = \kappa\hbar\frac{1 + 2n_B}{2} + \frac{\hbar\gamma}{2}. \tag{15.29}$$

We introduce the conditional covariances by

$$V_{x,c} \equiv \langle x^2 \rangle_c - \langle x \rangle_c^2, \quad V_{p,c} \equiv \langle p^2 \rangle_c - \langle p \rangle_c^2, \quad C_c \equiv \langle xp \rangle_c - \langle x \rangle_c \langle p \rangle_c \quad (15.30)$$

where we dropped already any time argument for notational convenience (we keep the subscript τ to denote the time-delay though). The time-evolution of the conditional means is then given by

$$d\langle x \rangle_c = \left\{ \omega \langle p \rangle_c - \frac{\kappa}{2} \langle x \rangle_c + k(\langle x \rangle_c - \langle x \rangle_{c,\tau}) \right\} dt \quad (15.31)$$

$$+ k \sqrt{\frac{\hbar}{2\gamma \eta}} (dW - dW_\tau) + \sqrt{\frac{2\gamma \eta}{\hbar}} V_{x,c} dW,$$

$$d\langle p \rangle_c = \left\{ -\omega \langle x \rangle_c - \frac{\kappa}{2} \langle p \rangle_c \right\} dt + \sqrt{\frac{2\gamma \eta}{\hbar}} C_c dW \quad (15.32)$$

Note that—for a stochastic simulation of these equations—we are required to simulate the equations for the covariances (Eqs. (15.35)–(15.37)), too. Interestingly, however, because the time-delayed noise enters only additively, we can also average these equations to obtain the unconditional evolution of the mean values directly:

$$\frac{d}{dt} \langle x \rangle = \omega \langle p \rangle - \frac{\kappa}{2} \langle x \rangle + k(\langle x \rangle - \langle x \rangle_\tau), \quad (15.33)$$

$$\frac{d}{dt} \langle p \rangle = -\omega \langle x \rangle - \frac{\kappa}{2} \langle p \rangle. \quad (15.34)$$

These equations are exactly the same as the classical equations in Ref. [30] if one considers only position measurements. Hence, we can successfully reproduce the classical feedback scheme *on average*. Unfortunately the treatment of delay differential equations is very complicated and our goal is not to study these equations in detail now. However, the reasoning why we can turn a stable fixed point into an unstable one goes like this: for $k = 0$ we clearly have a stable fixed point but for $k \gg \kappa$ we might neglect the term $-\frac{\kappa}{2} \langle x \rangle$ for a moment. If we choose $\tau = \pi/\omega$ (corresponding to half of a period of the undamped HO), we see that the "feedback force" $k(\langle x \rangle - \langle x \rangle_\tau)$ is always positive if $\langle x \rangle > 0$ and negative if $\langle x \rangle < 0$ (we assume $k > 0$). Hence, by looking at the differential equation it follows that the feedback term generates a drift "outwards", i.e., away from the fixed point $(0, 0)$, which at some point also cannot be compensated anymore by the friction of the momentum $-\frac{\kappa}{2} \langle p \rangle$. From the numerics, see Fig. 15.2, we infer that the critical feedback strength, which turns the stable fixed point into an unstable one is $k \geq \frac{\kappa}{2}$, also see Ref. [30] for a more detailed discussion of the domain of control.

Turning to the time evolution of the conditional covariances we obtain[4]

---

[4]Pay attention to the fact that we are using an Itô stochastic differential equation where the ordinary chain rule of differentiation does not apply. Instead, we have for instance for the stochastic change of the position variance $dV_{x,c} = d\langle x^2 \rangle_c - 2\langle x \rangle_c d\langle x \rangle_c - (d\langle x \rangle_c)^2$.

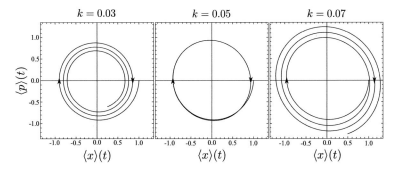

Fig. 15.2 Parametric plot of $(\langle x \rangle, \langle p \rangle)(t)$ as a function of time $t \in [0, 20]$ for different feedback strengths k based on Eqs. (15.33) and (15.34). The initial condition is $(\langle x \rangle, \langle p \rangle)(t) = (1, 0)$ for $t \leq 0$ and the other parameters are $\omega = 1, \kappa = 0.1$ and $\tau = \pi$. Note that the trajectory for $k = \kappa$ is not a perfect circle due to the asymmetric feedback, which is only applied to the x-coordinate and not to p

$$dV_{x,c} = \left\{ -\kappa V_{x,c} + 2\omega C_c - \frac{2\gamma \eta}{\hbar} V_{x,c}^2 + \kappa \hbar \frac{1 + 2n_B}{2} \right\} dt$$
$$+ \sqrt{\frac{2\gamma \eta}{\hbar}} \left\langle (x - \langle x \rangle_c)^3 \right\rangle_c dW, \tag{15.35}$$

$$dV_{p,c} = \left\{ -\kappa V_{p,c} - 2\omega C_c - \frac{2\gamma \eta}{\hbar} C_c^2 + \kappa \hbar \frac{1 + 2n_B}{2} + \frac{\hbar \gamma}{2} \right\} dt$$
$$+ \sqrt{\frac{2\gamma \eta}{\hbar}} \left\langle (x - \langle x \rangle_c)(p - \langle p \rangle_c)^2 \right\rangle_c dW, \tag{15.36}$$

$$dC_c = \left\{ \omega(V_{p,c} - V_{x,c}) - \kappa C_c - \frac{2\gamma \eta}{\hbar} V_{x,c} C_c \right\} dt$$
$$+ \sqrt{\frac{2\gamma \eta}{\hbar}} \left\langle (x - \langle x \rangle_c)^2 (p - \langle p \rangle_c) \right\rangle_c dW. \tag{15.37}$$

Unfortunately, we see that all the stochastic terms proportional to dW involve third order cumulants, which would in turn require to deduce equations for them as well. However, if we assume that the state of our system is Gaussian, these terms vanish due to the fact that third order cumulants of a Gaussian are zero. In fact, the assumption of a Gaussian state seems reasonable[5]: first of all, if the system is already Gaussian, it will also remain Gaussian for all times because then the Eqs. (15.31) and (15.32) as well as Eqs. (15.35)–(15.37) form a closed set. Second, even if we start with a non-Gaussian distribution, the state is expected to rapidly evolve to a Gaussian due to the continuous position measurement and the environmentally induced decoherence and

---

[5] We remark that a Gaussian state in QM, i.e., a system described by a Gaussian Wigner function, might still exhibit true quantum features like entanglement or squeezing [2, 3].

dissipation [44]. Then, the time evolution of the conditional covariances becomes indeed deterministic, i.e., the covariances (but not the means) behave identically in each single realization of the experiment:

$$\frac{d}{dt}V_{x,c} = -\kappa V_{x,c} + 2\omega C_c - \frac{2\gamma\eta}{\hbar}V_{x,c}^2 + \kappa\hbar\frac{1+2n_B}{2},\tag{15.38}$$

$$\frac{d}{dt}V_{p,c} = -\kappa V_{p,c} - 2\omega C_c - \frac{2\gamma\eta}{\hbar}C_c^2 + \kappa\hbar\frac{1+2n_B}{2} + \frac{\hbar\gamma}{2},\tag{15.39}$$

$$\frac{d}{dt}C_c = \omega(V_{p,c} - V_{x,c}) - \kappa C_c - \frac{2\gamma\eta}{\hbar}V_{x,c}C_c.\tag{15.40}$$

Thus, we can fully solve the conditional dynamics of the system by first solving the ordinary differential equations for the covariances and then, using this solution, we can integrate the stochastic equations (15.31) and (15.32) for the means.

Because the time evolution of the conditional covariances is the same for the second feedback scheme, we will discuss them in more detail in Sect. 15.4. Here, we just want to emphasize that we cannot simply average the conditional covariances to obtain the unconditional ones, i.e., $\mathbb{E}[V_{x,c}] \neq V_x \equiv \int dx dp x^2 W(x, p) - [\int dx dp x W(x, p)]^2$ in general. In fact, the conditional and unconditional covariances can behave very differently, see Sect. 15.4.

Finally, let us say a few words about our feedback scheme in comparison with the coherent control schemes in Refs. [36–40, 42, 43], which are designed for quantum optical systems and use an external mirror to induce an intrinsic time-delay in the system dynamics. Clearly, the advantage of the coherent control schemes is that they do not introduce additional noise because they avoid any explicit measurement. On the other hand, in our feedback loop we have the freedom to choose the feedback strength k at our will, which allows us to truely reverse the effect of dissipation. In fact, due to simple arguments of energy conservation, the coherent control schemes can only fully reverse the effect of dissipation if the external mirrors are *perfect*. Otherwise the overall system and controller is still loosing energy at a finite rate such that the system ends up in the same steady state as without feedback. Thus, as long as the coherent control loop does not have access to any external sources of energy, it is only able to counteract dissipation on a transient time-scale except one allows for perfect mirrors, which in turn would make it unnessary to introduce any feedback loop at all in our situation. It should be noted, however, that for transient time-scales coherent feedback might have strong advantages or it might be the case that one is not primarily interested in the prevention of dissipation (in fact, in Ref. [40] they use the control loop to *speed up* dissipation). The question whether one scheme is superior to the other is thus, in general, undecidable and needs a thorough case to case analysis.

15.4   Feedback Scheme II

We wish to present a second feedback scheme, in which we replace the time-delayed signal by a fixed reference signal such that no time-delayed noise enters the description and hence, we are not forced to work with a SME like Eq. (15.26). The measurement signal we wish to couple back is thus of the form

$$\delta I(t) = [I(t) - x^*(t)]dt = [\langle x \rangle(t) - x^*(t)]dt + \sqrt{\frac{\hbar}{2\gamma\eta}} dW(t) \qquad (15.41)$$

and our aim is to *synchronize* the motion of the HO with the external reference signal $x^*(t)$. Choosing $\hat{z} = k\hat{p}$ and using Eq. (15.15) yields the SME

$$\hat{\rho}_c(t + dt) = \hat{\rho}_c(t)$$

$$+ dt \left\{ \mathcal{L}_0 + \mathcal{L}_{\text{meas}} - \frac{ik}{2\hbar} \mathcal{C}[\hat{p}]\{\mathcal{A}[\hat{x}] - 2x^*(t)\} - \frac{k^2}{4\hbar\gamma\eta} \mathcal{C}^2[\hat{p}] \right\} \hat{\rho}_c(t)$$

$$\qquad (15.42)$$

$$+ dW(t) \left\{ \sqrt{\frac{\gamma\eta}{2\hbar}} \mathcal{H}[\hat{x}] - \frac{ik}{\sqrt{2\hbar\gamma\eta}} \mathcal{C}[\hat{p}] \right\} \hat{\rho}_c(t).$$

The associated FPE (15.22) for the conditional Wigner function is given by

$$W_c(x, p, t + dt) = W_c(x, p, t) + dt \left(-\nabla^T \cdot \mathbf{d} + \frac{1}{2} \nabla^T \cdot D \cdot \nabla \right) W_c(x, p, t)$$

$$\qquad (15.43)$$

$$+ dW(t) \left[\sqrt{\frac{2\gamma\eta}{\hbar}} (x - \langle x \rangle_c) - k \sqrt{\frac{\hbar}{2\gamma\eta}} \frac{\partial}{\partial x} \right] W_c(x, p, t)$$

with the nonvanishing coefficients

$$\mathbf{d}_x = \omega p - \frac{\kappa}{2} x + k[x - x^*(t)], \quad \mathbf{d}_p = -\omega x - \frac{\kappa}{2} p, \qquad (15.44)$$

$$D_{xx} = \kappa\hbar \frac{1 + 2n_B}{2} + \frac{\hbar k^2}{2\gamma\eta}, \quad D_{pp} = \kappa\hbar \frac{1 + 2n_B}{2} + \frac{\hbar\gamma}{2}.$$

Because we have no time-delayed noise here, the average, unconditional evolution of the system can be simply obtained by dropping all terms proportional to the noise $dW(t)$ due to Eq. (15.12). We thus have a fully Markovian feedback scheme here.

The equation of motion for the conditional means are

$$d\langle x\rangle_c = \left\{\omega\langle p\rangle - \frac{\kappa}{2}\langle x\rangle + k[\langle x\rangle - x^*(t)]\right\} dt \qquad (15.45)$$

$$+ \left(\sqrt{\frac{2\gamma\eta}{\hbar}}V_{x,c} + k\sqrt{\frac{\hbar}{2\gamma\eta}}\right) dW(t),$$

$$d\langle p\rangle_c = \left\{-\omega\langle x\rangle - \frac{\kappa}{2}\langle p\rangle\right\} dt + \sqrt{\frac{2\gamma\eta}{\hbar}}C_c dW(t) \qquad (15.46)$$

from which the average evolution directly follows:

$$\frac{d}{dt}\langle x\rangle = \omega\langle p\rangle - \frac{\kappa}{2}\langle x\rangle + k[\langle x\rangle - x^*(t)], \qquad (15.47)$$

$$\frac{d}{dt}\langle p\rangle = -\omega\langle x\rangle - \frac{\kappa}{2}\langle p\rangle. \qquad (15.48)$$

Again, it is not our purpose to investigate these equations in detail, but we will only focus on the special situation $k = \kappa/2$ and $x^*(t) = -y_0\cos(\omega t)$. Then,

$$\frac{d}{dt}\langle x\rangle = \omega\langle p\rangle + \frac{y_0\kappa}{2}\cos(\omega t), \qquad (15.49)$$

$$\frac{d}{dt}\langle p\rangle = -\omega\langle x\rangle - \frac{\kappa}{2}\langle p\rangle. \qquad (15.50)$$

These equations look very similar to the classical differential equation of an externally forced harmonic oscillator.[6] However, it is important to emphasize that we do not have an open-loop control scheme here although it looks like it at the average level of the means. The asymptotic solution of Eqs. (15.49) and (15.50) is given by

$$\lim_{t\to\infty}\langle x\rangle(t) = y_0\cos(\omega t) + \frac{\kappa y_0}{2\omega}\sin(\omega t), \qquad (15.51)$$

$$\lim_{t\to\infty}\langle p\rangle(t) = -y_0\sin(\omega t). \qquad (15.52)$$

Within the weak-coupling regime it is natural to assume that $\kappa/\omega \ll 1$ and we asymptotically obtain a circular motion $(\langle x\rangle, \langle p\rangle)(t) \approx y_0(\cos\omega t, -\sin\omega t)$. It is worth to stress that we always reach the asymptotic solution independent of the chosen initial condition, also see Fig. 15.3. As a consequence, the limit cycle given by Eqs. (15.51) and (15.52) is stable. In contrast, the equations of motion for the first scheme are completely scale-invariant, i.e., an arbitrary scaling of the form $(\langle\tilde{x}\rangle, \langle\tilde{p}\rangle) \equiv \alpha(\langle x\rangle, \langle p\rangle)$, $\alpha \in \mathbb{R}$, leaves the Eqs. (15.33) and (15.34) unchanged and the effect of the feedback depends on the initial condition.

---

[6]Indeed, if we would choose the feedback operator $\hat{z} = k\hat{x}$, the resulting differential equations for $\langle x\rangle$ and $\langle p\rangle$ would exactly resemble the differential equation of a classical harmonic oscillator with sinusoidal driving force.

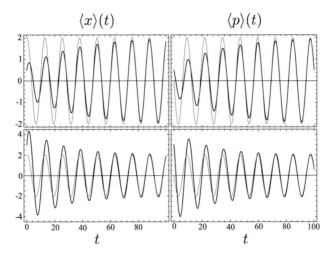

Fig. 15.3 Plot of the average mean values $\langle x \rangle(t)$ and $\langle p \rangle(t)$ as a function of time t (*black, thick lines*) compared to the asymptotic solution given by Eqs. (15.51) and (15.52) (*red, thin lines*). The *upper panel* corresponds to the initial condition $(\langle x \rangle, \langle p \rangle)(0) = (1/2, 1/2)$ and the lower one to $(\langle x \rangle, \langle p \rangle)(0) = (3, 3)$. Other parameters are $\omega = 1$, $\kappa = 1/4$ and $y_0 = 2$

Within the Gaussian assumption, the conditional covariances (i.e., the covariances an observer *with* access to the measurement result would associate to the state of the system) evolve as for the first scheme according to

$$\frac{d}{dt}V_{x,c} = -\kappa V_{x,c} + 2\omega C_c - \frac{2\gamma\eta}{\hbar}V_{x,c}^2 + \kappa\hbar\frac{1+2n_B}{2}, \qquad (15.53)$$

$$\frac{d}{dt}V_{p,c} = -\kappa V_{p,c} - 2\omega C_c - \frac{2\gamma\eta}{\hbar}C_c^2 + \kappa\hbar\frac{1+2n_B}{2} + \frac{\hbar\gamma}{2}, \qquad (15.54)$$

$$\frac{d}{dt}C_c = \omega(V_{p,c} - V_{x,c}) - \kappa C_c - \frac{2\gamma\eta}{\hbar}V_{x,c}C_c. \qquad (15.55)$$

In contrast, the unconditional covariances (which an observer *without* access to the measurement results would associate to the state of the system) obey

$$\frac{d}{dt}V_x = (2k - \kappa)V_x + 2\omega C + \kappa\hbar\frac{1+2n_B}{2} + \frac{\hbar k^2}{2\gamma\eta}, \qquad (15.56)$$

$$\frac{d}{dt}V_p = -\kappa V_p - 2\omega C + \kappa\hbar\frac{1+2n_B}{2} + \frac{\hbar\gamma}{2}, \qquad (15.57)$$

$$\frac{d}{dt}C = \omega(V_p - V_x) + (k - \kappa)C. \qquad (15.58)$$

Comparing both sets of equations, the most striking difference is that Eqs. (15.53)–(15.55) are nonlinear differential equations whereas Eqs. (15.56)–(15.58) are linear. Especially note the term proportional to $-\frac{2\gamma\eta}{\hbar}V_{x,c}^2$ in Eq. (15.53), which tends to

squeeze the wavepacket in the x-direction. This is the effect of the continuous measurement performed on the system, which tends to localize the state. However, if we
average over (or, equivalently, ignore) the measurement results, this effect is missing.
Furthermore, note that Eqs. (15.53)–(15.55) do not contain the parameter k, which
quantifies how strongly we feed back the signal.

Solving Eqs. (15.53)–(15.55) for its steady state is possible, but the exact expressions are extremely lengthy. However, to see how the continuous measurement influences the conditional covariances we will have a look at the special case of no damping ($\kappa = 0$). To appreciate this case we remind us that the steady state covariances of a
damped HO for no measurement and no feedback are given by $V_x = V_p = \hbar(n_B + \frac{1}{2})$
and $C = 0$.[7] Especially, at zero temperature ($n_B = 0$), we have a minimum uncertainty wave packet satisfying the lower bound of the Heisenberg uncertainty relation,
$V_x V_p = \hbar^2/4$. Now, for $\kappa = 0$, we can expand the conditional covariances in powers
of γ:

$$\lim_{t \to \infty} V_{x,c}(t) = \frac{\hbar}{2\sqrt{\eta}} - \frac{\sqrt{\eta}\hbar\gamma^2}{16\omega^2} + \mathcal{O}(\gamma^3), \qquad (15.59)$$

$$\lim_{t \to \infty} V_{p,c}(t) = \frac{\hbar}{2\sqrt{\eta}} + \frac{3\sqrt{\eta}\hbar\gamma^2}{16\omega^2} + \mathcal{O}(\gamma^3), \qquad (15.60)$$

$$\lim_{t \to \infty} C_c(t) = \frac{\hbar\gamma}{4\omega} + \mathcal{O}(\gamma^3). \qquad (15.61)$$

We see that the uncertainty in position is reduced at the expense of an increased
uncertainty in momentum. This is exactly what we have to expect from a position
measurement due to Heisenberg's uncertainty principle. Note that this effect is weak-
ened for larger frequencies ω of the oscillator because the continuous measurement
has problems to "follow" the state appropriately. Furthermore, an imperfect detector ($\eta < 1$) will always increase the variances. Finally, we remark that these results
become meaningless in the strict limit $\gamma \to 0$ because in that case there is simply *no*
conditional dynamics. This becomes also clear by looking at Eq. (15.56), in which
the last term diverges in this limit because we would feed back an infinitely noisy
signal.

Thus, an observer with access to the measurement record would associate very
different covariances to the system in comparison to an observer without that knowledge. However, a detailed discussion of the time evolution of the covariances is
beyond the scope of the present paper. Instead, we find it more interesting to discuss the relationship between the present quantum feedback scheme and its classical
counterpart, to which we turn now.

---

[7]We can obtain this result by computing the steady state of Eqs. (15.56)–(15.58) where we first
send $k \to 0$ and then $\gamma \to 0$.

15.5 Classical Limit

To take the classical limit in our case we have to use a little trick because simply taking $\hbar \to 0$ does not yield the correct result. In fact, for $\hbar \to 0$ we have from Eq. (15.8) that $I(t) = \langle x \rangle_c(t)$, which only makes sense if we can observe the particle with infinite accuracy, i.e., its conditional probability distribution is a delta function with respect to the position x. We explicitly wish, however, to model a *noisy* classical measurement. We thus additionally demand that $\eta \to 0$. More specifically, we set $\eta \equiv \hbar/\sigma$ with σ finite such that Eq. (15.8) becomes

$$dI(t) = \langle x \rangle_c(t)dt + \sqrt{\frac{\sigma}{2\gamma}}dW(t) \qquad (15.62)$$

and we remark that the case $\sigma \to 0$ corresponds to an error-free measurement.

The FPE (15.22) for the free evolution of the HO with drift vector and diffusion matrix from Eq. (15.23) becomes for $\hbar \to 0$ (note that the Bose-Einstein distribution n_B contains $\hbar$ as well and needs to be expanded)

$$\frac{\partial}{\partial t}P(x, p, t) \equiv \mathcal{L}_0^{\text{cl}} P(x, p, t) \qquad (15.63)$$

$$= \left\{ -\nabla^T \cdot \begin{pmatrix} \omega p - \frac{\kappa}{2}x \\ -\omega x - \frac{\kappa}{2}x \end{pmatrix} + \frac{\kappa}{2\beta\omega}\nabla^T \cdot \nabla \right\} P(x, p, t).$$

To emphasize the fact that the Wigner function $W(x, p)$ becomes an ordinary probability distribution in the classical limit, we denoted it by $P(x, p)$. As expected, we see that Eq. (15.63) corresponds to a FPE for a Brownian particle in a harmonic potential where position and momentum are both damped (usually one considers only the momentum to be damped [29]). This peculiarity is a consequence of an approximation made in deriving the ME (15.1), which is known as the secular or rotating-wave approximation. Nevertheless, one easily confirms that the canonical equilibrium state $P_{\text{eq}} \sim \exp[-\beta\omega(p^2 + x^2)/2]$ is a steady state of this FPE as it must be.

Next, it turns out to be interesting to discuss the classical limit of the SME (15.13) describing the free evolution plus the influence of the continuous noisy measurement. Using Eq. (15.24) we obtain an equation for the conditional probability distribution $P_c(x, p)$

$$P_c(x, p, t + dt) = \left\{ 1 + \mathcal{L}_0^{\text{cl}}dt + \sqrt{\frac{2\gamma}{\sigma}}(x - \langle x \rangle_c)dW(t) \right\} P_c(x, p, t). \quad (15.64)$$

This is a *stochastic* FPE, which is nonlinear in P_c. It describes how our state of knowledge changes if we take into account the measurement record (15.62). However, averaging Eq. (15.64) over all measurement results yields Eq. (15.63), which reflects the fact that a classical measurement does not perturb the system.[8] This is in contrast to the quantum case where the average evolution is still influenced by $\mathcal{L}_{\text{meas}}$, see Eq. (15.6). Hence, the term $\mathcal{L}_{\text{meas}}$ in Eq. (15.6) is *purely* of quantum orgin and it describes the effect of decoherence on a quantum state under the influence of a measurement. This effect is absent in a classical world. Exactly the same equation and the same conclusions were already derived by Milburn following a different route [45].

The impact of these conclusions is, however, much more severe if one additionally considers feedback. As we will now show, applying feedback based on the use of the stochastic FPE (15.64), *does* indeed yield observable consequences even *on average*. Please note that trying to model the present situation by a classical Langevin equation is nonsensical. If we would use a Langevin equation to describe our state of knowlegde about the system, we would implicitly ascribe an objective reality to the fact that there *is* a definite position x_0 and momentum p_0 of the particle corresponding to a probability distribution $\delta(x - x_0)\delta(p - p_0)$. This is, however, *not true* from the point of view of the observer who has to apply feedback control based on incomplete information (i.e., the noisy measurement record). Results we would obtain from a Langevin equation treatment can only be recovered in the limit of an error-free measurement, i.e., for $\sigma \to 0$, as we will demonstrate in Appendix 15.6.

For simplicity we will only have a look at the second feedback scheme from Sect. 15.4 because we can directly obtain the classical limit for the average evolution from Eq. (15.43). The same situation is, however, also encountered by considering the first scheme. Taking $\hbar \to 0$ in Eq. (15.43) together with the coefficients (15.44) then yields the FPE

$$\frac{\partial}{\partial t} P(x, p, t) = \left\{ \mathcal{L}_0^{\text{cl}} - \partial_x k[x - x^*(t)] + \frac{k^2 \sigma}{2\gamma} \partial_x^2 \right\} P(x, p, t). \qquad (15.65)$$

The first term to the correction of $\mathcal{L}_0^{\text{cl}}$ is the term one would expect for a noiseless feedback loop, too. The second, however, only arises due to the noisy measurement (we see that it vanishes for $\sigma \to 0$) and causes an additional diffusion in the x direction simply due to the fact that the observer applies a slightly wrong feedback control compared to the "perfect" situation without measurement errors.

We thus conclude this section by noting that the treatment of continuous noisy classical measurements faces similar challenges as in the quantum setting. On average the measurement itself does not influence the classical dynamics, but we see that we obtain new terms even on average if we use this measurement to perform feedback control. Most importantly, because a feedback loop has to be implemented by the observer who has access to the measurement record, it is in general not possible to model this situation with a Langevin equation. Furthermore, we remark that the situation is expected to be even more complicated for time-delayed feedback, where no average description is a priori possible.

---

[8]This is true at least in our context. In principle, it is of course possible to construct classical measurements, which perturb the system, too [6].

15.6 Summary and Outlook

Because we discussed the meaning of our results already during the main text in detail, we will only give a short summary together with a discussion on possible extensions and applications.

We have used two simple feedback schemes, which are known to change the stability of a steady state obtained from linearizing a dynamical system around that fixed point in the classical case. For the simple situation of a damped quantum HO we have seen that *on average* we obtain the same dynamics for the mean values as expected from a classical treatment and thus, classical control strategies might turn out to be very useful in the quantum realm, too.

However, the fact that a classical control scheme works so well in the quantum regime depends on two crucial assumptions. First of all, we have used a linear system (the HO). Having a non-linear system Hamiltonian (e.g., a Hamiltonian with a quartic potential $\sim \hat{x}^4$) would complicate the treatment because already the equations for the mean values would contain higher order moments, as e.g. $\langle x^3 \rangle$ in case of the quartic oscillator. Simply factorizing them as $\langle x^3 \rangle \approx \langle x \rangle^3$ would imply that we are already using a classical approximation. However, as in the classical treatment, where the equations of motion are obtained from linearizing a (potentially non-linear) dynamical system around the fixed point, it might also be possible in the quantum regime to neglect non-linear terms in the vicinity of the fixed point. Whether or not this is possible crucially depends on the localization of the state in phase space, i.e., on its covariances. Here, continuous quantum measurements can actually turn out to be helpful because they tend to localize the wavefunction and counteract a possible spreading of the state.

The second important assumption we used was that we restricted ourselves to continuous variable quantum systems. The reason why we obtained simple equations of motion is related to the commutation relation $[\hat{x}, \hat{p}] = i\hbar$, which we implicitly used to obtain the evolution equation for the Wigner function. Formally, phase space methods are also possible for other quantum systems, but the maps are much more complicated [27]. For such systems the methods presented here might be useful under certain special assumptions, but in general one should expect them to fail.

Acknowledgments PS wishes to thank Philipp Hövel, Lina Jaurigue and Wassilij Kopylov for helpful discussions about time-delayed feedback control. Financial support by the DFG (SCHA 1646/3-1, SFB 910, and GRK 1558) is gratefully acknowledged.

Appendix

We want to show that the stochastic FPE, which in general describes the *incomplete* state of knowledge of an observer, reduces to a Langevin equation in the error-free limit, i.e., in the limit in which we have indeed *complete* knowledge about the state of the system. Because we are only interested in a proof of principle here, we will consider the simplified situation of an overdamped particle.[9] The Langevin equation of an overdamped Brownian particle in an external potential $U(x)$ is usually given as (see, e.g., [29])

$$\dot{x}(t) = -\frac{2}{\kappa}U'(x) + \sqrt{\frac{T}{\kappa}}\xi(t) \qquad (15.66)$$

with $U'(x) \equiv \frac{\partial U(x)}{\partial x}$ and the Gaussian white noise $\xi(t) \equiv \frac{dW(t)}{dt}$. Furthermore, note that our friction constant is $\frac{\kappa}{2}$ and not—as it is often denoted—γ because we use γ already for the measurement rate. Now, it is important to remark that at this point Eq. (15.66) simply describes a convenient *numerical tool to simulate* a stochastic process. By the mathematical rules of stochastic calculus it is guaranteed that the Langevin equation gives the same averages as the corresponding FPE, i.e., the ensemble average $\mathbb{E}[f(x)]$ over all noisy trajectories for some function $f(x)$ is equal to the expectation value $\langle f(x)\rangle$ taken with respect to the solution of the FPE.

In Sect. 15.5 we have suggested that the correct state of the system based on a noisy position measurement is given by the stochastic FPE (15.64), which for an overdamped particle becomes (see also Ref. [45])

$$P_c(x, t+dt) = \left\{ 1 + \mathcal{L}_0^{cl}dt + \sqrt{\frac{2\gamma}{\sigma}}(x - \langle x\rangle_c)dW(t) \right\} P_c(x, t) \qquad (15.67)$$

with [29]

$$\mathcal{L}_0^{cl} = \frac{\partial}{\partial x}\left(\frac{2U'(x)}{\kappa} + \frac{\partial}{\partial x}\frac{2T}{\kappa} \right). \qquad (15.68)$$

Furthermore, we have also claimed that the parameter σ in (15.62) quantifies the error of the measurement. This suggests that we should be able to recover the Langevin Eq. (15.66) in the limit $\sigma \to 0$ in which we can observe the particle with infinite precision.

To show this we compute the expectation value of the position according to Eq. (15.67):

$$d\langle x\rangle_c(t) = -\frac{2}{\kappa}\langle U'(x)\rangle_c dt + \sqrt{\frac{2\gamma}{\sigma}}V_c(t)dW(t) \qquad (15.69)$$

where $V_c = \langle x^2\rangle_c - \langle x\rangle_c^2$ denotes the variance of the particles position. Because the conditional variance enters this equation, we compute its time evolution, too:

---

[9]The complete description of an underdamped particle (i.e., a particle descibed by its position x and momentum p), which is based on a continuous measurement of its position x alone, Eq. (15.62), faces the additional challenge that we have to first estimate the momentum p based on the noisy measurement results.

$$dV_c = -\frac{4}{\kappa}\left[\langle xU'(x)\rangle_c(t) - \langle x\rangle_c(t)\langle U'(x)\rangle_c(t)\right]dt + \frac{4T}{\kappa}dt - \frac{2\gamma}{\sigma}V_c^2 dt \quad (15.70)$$

$$+ \sqrt{\frac{2\gamma}{\sigma}}\left\langle[x - \langle x\rangle_c(t)]^3\right\rangle dW(t).$$

To make analytical progress we now need two assumptions. First, we will assume that $P_c(x, t)$ is a Gaussian probability distribution. In fact, because the measurement tends to localize the probability distribution and it is itself modeled as a Gaussian process, this assumption seems reasonable. In addition, we expect $P_c(x, t)$ to become a delta-distribution in the limit $\sigma \to 0$, which is a Gaussian distribution, too. This assumption allows us to drop the stochastic term in Eq. (15.70). Second, within the variance of $P_c(x, t)$ we assume that we can expand $U(x)$ in a Taylor series and approximate it by a quadratic function $ax^2 + bx + c$. This implies $\langle xU'(x)\rangle_c(t) \approx a\langle x^2\rangle_c(t) + b\langle x\rangle_c(t)$. In fact, this assumption seems also reasonable because we expect the measurement to be precise enough such that we can locally resolve the evolution of the particle sufficiently well (especially for small σ); or to put it differently: a measurement only makes sense if the conditional variance V_c of $P_c(x, t)$ is small enough. Using this approximation, too, we can then write Eq. (15.70) as

$$\frac{d}{dt}V_c(t) = \frac{4\kappa}{2} - \frac{4a}{\kappa}V_c - \frac{2\gamma}{\sigma}V_c^2. \quad (15.71)$$

The only physical steady state solution of this equation is

$$\lim_{t\to\infty} V_c(t) = \frac{a\sigma}{\gamma\kappa}\left(\sqrt{1 + \frac{2T\gamma\kappa}{a^2\sigma}} - 1\right). \quad (15.72)$$

Inserting this into Eq. (15.69) yields

$$d\langle x\rangle_c(t) = -\frac{2}{\kappa}\langle U'(x)\rangle_c dt + \frac{a}{\kappa}\sqrt{\frac{2\sigma}{\gamma}}\left(\sqrt{1 + \frac{2T\gamma\kappa}{a^2\sigma}} - 1\right)dW(t). \quad (15.73)$$

In this equation we can take the limit $\sigma \to 0$ such that

$$d\langle x\rangle_c^0(t) = -\frac{2}{\kappa}\langle U'(x)\rangle_c^0 dt + \sqrt{\frac{4T}{\kappa}}dW(t), \quad (15.74)$$

where we introduced a superscript 0 on all expectation values to denote the error-free limit. This equation looks already very similar to the LE (15.66). In fact, mathematically *this is the LE* since from Eq. (15.62) we can see that the measurement result becomes for $\sigma \to 0$ $dI(t) = \langle x\rangle_c^0(t)dt$. This implies that the measurement result is deterministic and not stochastic anymore, which is only compatible if the associated probability distribution $P_c(x, t)$ is a delta distribution $\delta(x - x^*)$ where x^* describes the *true instantaneous position* of the particle without any uncertainty.

Then, Eq. (15.69) becomes

$$dx^*(t) = -\frac{2}{\kappa}U'(x^*)dt + \sqrt{\frac{4T}{\kappa}}dW(t). \tag{15.75}$$

Now, this equation is *not just a numerical tool, but describes real physical objectivity* because x^* coincides with the observed position in the lab. This distinction might seem very nitpicking, but it is of crucial importance if we want to perform feedback based on incomplete information.

References

1. M.O. Scully, M.S. Zubairy, *Quantum Optics* (Cambridge University Press, Cambridge, 1997)
2. S.L. Braunstein, P. van Loock, Rev. Mod. Phys. **77**, 513 (2005)
3. C. Weedbrook, S. Pirandola, R. García-Patrón, N.J. Cerf, T.C. Ralph, J.H. Shapiro, S. Lloyd, Rev. Mod. Phys. **84**, 621 (2012)
4. M. Aspelmeyer, T.J. Kippenberg, F. Marquardt, Rev. Mod. Phys. **86**, 1391 (2014)
5. L.D. Faddeev, O.A. Yakubovskii, *Lectures on Quantum Mechanics for Mathematics Students*, Student Mathematical Library, vol. 47. (American Mathematical Society, 2009)
6. H.M. Wiseman, G.J. Milburn, *Quantum Measurement and Control* (Cambridge University Press, Cambridge, 2010)
7. W.P. Smith, J.E. Reiner, L.A. Orozco, S. Kuhr, H.M. Wiseman, Phys. Rev. Lett. **89**, 133601 (2002)
8. P. Bushev, D. Rotter, A. Wilson, F. Dubin, C. Becher, J. Eschner, R. Blatt, V. Steixner, P. Rabl, P. Zoller, Phys. Rev. Lett. **96**, 043003 (2006)
9. G.G. Gillett, R.B. Dalton, B.P. Lanyon, M.P. Almeida, M. Barbieri, G.J. Pryde, J.L. OBrien, K.J. Resch, S.D. Bartlett, A.G. White. Phys. Rev. Lett. **104**, 080503 (2010)
10. C. Sayrin, I. Dotsenko, X. Zhou, B. Peaudecerf, T. Rybarczyk, S. Gleyzes, P. Rouchon, M. Mirrahimi, H. Amini, M. Brune, J.M. Raimond, S. Haroche, Nature (London) **477**, 73 (2011)
11. R. Vijay, C. Macklin, D.H. Slichter, S.J. Weber, K.W. Murch, R. Naik, A.N. Korotkov, I. Siddiqi, Nature (London) **490**, 77 (2012)
12. D. Riste, C.C. Bultink, K.W. Lehnert, L. DiCarlo, Phys. Rev. Lett. **109**, 240502 (2012)
13. H.J. Carmichael, *An Open Systems Approach to Quantum Optics* (Lecture Notes, Springer, Berlin, 1993)
14. G. Lindblad, Commun. Math. Phys. **48**, 119 (1976)
15. V. Gorini, A. Kossakowski, E.C.G. Sudarshan, J. Math. Phys. **17**, 821 (1976)
16. K. Jacobs, D.A. Steck, Contemp. Phys. **47**, 279 (2006)
17. A. Barchielli, L. Lanz, G.M. Prosperi, Nuovo Cimento B **72**, 79 (1982)
18. A. Barchielli, L. Lanz, G.M. Prosperi, Found. Phys. **13**, 779 (1983)
19. C.M. Caves, G.J. Milburn, Phys. Rev. A **36**, 5543 (1987)
20. H.M. Wiseman, G.J. Milburn, Phys. Rev. A **47**, 642 (1993)
21. A.C. Doherty, K. Jacobs, Phys. Rev. A **60**, 2700 (1999)
22. C. Gardiner, P. Zoller, *Quantum Noise* (Springer-Verlag, Berlin Heidelberg, 2004)
23. H.M. Wiseman, G.J. Milburn, Phys. Rev. Lett. **70**, 548 (1993)
24. H.M. Wiseman, G.J. Milburn, Phys. Rev. A **49**, 1350 (1994)
25. H.M. Wiseman, Phys. Rev. A **49**, 2133 (1994)
26. M. Hillery, R.F. O'Connell, M.O. Scully, E. Wigner, Phys. Rep. **106**, 121 (1984)
27. H.J. Carmichael, *Statistical Methods in Quantum Optics 1: Master Equations and Fokker-Planck Equations* (Springer, New York, 1999)

28. C. Zachos, D. Fairlie, T. Curtright (eds.). *Quantum mechanics in phase space: an overview with selected papers* (World Scientific, vol. **34**, 2005)
29. H. Risken, *The Fokker-Planck Equation: Methods of Solution and Applications* (Springer, Berlin Heidelberg, 1984)
30. P. Hövel, E. Schöll, Phys. Rev. E **72**, 046203 (2005)
31. E. Arthurs, J.L. Kelly, Bell Syst. Tech. J. **44**, 725 (1965)
32. A.J. Scott, G.J. Milburn, Phys. Rev. A **63**, 042101 (2001)
33. K. Pyragas, Phys. Lett. A **170**, 421 (1992)
34. E. Schöll, H.G. Schuster (eds.), *Handbook of Chaos Control* (Wiley-VCH, Weinheim, 2007)
35. V. Flunkert, I. Fischer, E. Schöll (eds.). Dynamics, control and information in delay-coupled systems: an overview. Phil. Trans. R. Soc. A, **371**, 1999 (2013)
36. S.J. Whalen, M.J. Collett, A.S. Parkins, H.J. Carmichael, Quantum Electronics Conference & Lasers and Electro-Optics (CLEO/IQEC/PACIFIC RIM), IEEE pp. 1756–1757 (2011)
37. A. Carmele, J. Kabuss, F. Schulze, S. Reitzenstein, A. Knorr, Phys. Rev. Lett. **110**, 013601 (2013)
38. F. Schulze, B. Lingnau, S.M. Hein, A. Carmele, E. Schöll, K. Lüdge, A. Knorr, Phys. Rev. A **89**, 041801(R) (2014)
39. S.M. Hein, F. Schulze, A. Carmele, A. Knorr, Phys. Rev. Lett. **113**, 027401 (2014)
40. A.L. Grimsmo, A.S. Parkins, B.S. Skagerstam, New J. Phys. **16**, 065004 (2014)
41. W. Kopylov, C. Emary, E. Schöll, T. Brandes, New J. Phys. **17**, 013040 (2015)
42. A.L. Grimsmo, Phys. Rev. Lett. **115**, 060402 (2015)
43. J. Kabuss, D.O. Krimer, S. Rotter, K. Stannigel, A. Knorr, A. Carmele, arXiv:1503.05722
44. W.H. Zurek, S. Habib, J.P. Paz, Phys. Rev. Lett. **70**, 1187 (1993)
45. G.J. Milburn, Quantum Semiclass. Opt. **8**, 269 (1996)

Chapter 16
Chimera States in Quantum Mechanics

Victor Manuel Bastidas, Iryna Omelchenko, Anna Zakharova,
Eckehard Schöll and Tobias Brandes

Abstract Classical chimera states are paradigmatic examples of *partial synchronization patterns* emerging in nonlinear dynamics. These states are characterized by the spatial coexistence of two dramatically different dynamical behaviors, i.e., synchronized and desynchronized dynamics. Our aim in this contribution is to discuss signatures of chimera states in quantum mechanics. We study a network with a ring topology consisting of N coupled quantum Van der Pol oscillators. We describe the emergence of chimera-like quantum correlations in the covariance matrix. Further, we establish the connection of chimera states to quantum information theory by describing the quantum mutual information for a bipartite state of the network.

16.1 Introduction

Self-organization is one of the most intriguing phenomena in nature. Currently, there is a plethora of studies concerning pattern formation and the emergence of spiral waves, Turing structures, synchronization patterns, etc. In classical systems of coupled nonlinear oscillators, the phenomenon of chimera states, which describes the spontaneous emergence of coexisting synchronized and desynchronized dynamics in networks of identical elements, has recently aroused much interest [1]. These intriguing spatio-temporal patterns were originally discovered in models of coupled phase oscillators [2, 3]. The last decade has seen an increasing interest in chimera states in dynamical networks [4–12]. It was shown that they are not limited to phase oscillators, but can be found in a large variety of different systems including time-discrete maps [13], time-continuous chaotic models [14], neural systems [15–17], Van der Pol oscillators [18], and Boolean networks [19].

Chimera states were found also in systems with higher spatial dimensions [7, 20–23]. New types of these peculiar states having multiple incoherent regions

V.M. Bastidas · I. Omelchenko (✉) · A. Zakharova · E. Schöll · T. Brandes
Institut für Theoretische Physik, Technische Universität Berlin,
Hardenbergstr. 36, 10623 Berlin, Germany
e-mail: omelchenko@itp.tu-berlin.de

E. Schöll et al. (eds.), *Control of Self-Organizing Nonlinear Systems*,
Understanding Complex Systems, DOI 10.1007/978-3-319-28028-8_16

[15, 17, 24–26], as well as amplitude-mediated [27, 28], and pure amplitude chimera states [29] were discovered.

The nonlocal coupling has been usually considered as a necessary condition for chimera states to evolve in systems of coupled oscillators. Recent studies have shown that even global all-to-all coupling [28, 30–32], as well as more complex coupling topologies allow for the existence of chimera states [12, 33–37]. Furthermore, time-varying network structures can give rise to alternating chimera states [38].

Possible applications of chimera states in natural and technological systems include the phenomenon of unihemispheric sleep [39], bump states in neural systems [40, 41], epileptic seizure [42], power grids [43], or social systems [44]. Many works considering chimera states have been based on numerical results. A deeper bifurcation analysis [45] and even a possibility to control chimera states [46, 47] were obtained only recently.

The experimental verification of chimera states was first demonstrated in optical [48] and chemical [49, 50] systems. Further experiments involved mechanical [51], electronic [52, 53] and electrochemical [54, 55] oscillator systems, Boolean networks [19], the optical comb generated by a passively mode-locked quantum dot laser [56], and superconducting quantum interference devices [57].

While synchronization of classical oscillators has been well studied since the early observations of Huygens in the seventeenth century [58], synchronization in quantum mechanics has only very recently become a focus of interest. For example, quantum signatures of synchronization in a network of globally coupled Van der Pol oscillators have been investigated [59, 60]. Related works focus on the dynamical phase transitions of a network of nanomechanical oscillators with arbitrary topologies characterized by a coordination number [61], and the semiclassical quantization of the Kuramoto model by using path integral methods [62].

Contrary to classical mechanics, in quantum mechanics the notion of phase-space trajectory is not well defined. As a consequence, one has to define new measures of synchronization for continuous variable systems like optomechanical arrays [61]. These measures are based on quadratures of the coupled systems and allow one to extend the notion of phase synchronization to the quantum regime [63]. Additional measures of synchronization open intriguing connections to concepts of quantum information theory [64, 65], such as decoherence-free subspaces [66], quantum discord [67], entanglement [68, 69], and mutual information [70]. Despite the intensive theoretical investigation of quantum signatures of synchronized states, up to now the quantum manifestations of chimera states are still unresolved.

Recently, we have studied the emergence of chimera states in a network of coupled quantum Van der Pol oscillators [71], and here we review this work. Unlike in previous studies [72], we address the fundamental issue of the dynamical properties of chimera states in a continuous variable system. Considering the chaotic nature of chimera states [9], we study the short-time evolution of the quantum fluctuations at the Gaussian level. This approach allows us to use powerful tools of quantum information theory to describe the correlations in a nonequilibrium state of the system. We show that quantum manifestations of the chimera state appear in the covariance matrix and are related to bosonic squeezing, thus bringing these signatures into the realm of observability in trapped ions [59], optomechanical arrays [61], and driven-dissipative Bose-Einstein condensates [73, 74]. We find that the chimera states can be characterized in terms of Rényi quantum mutual information [75]. Our results reveal that the mutual information for a chimera state lies between the values for synchronized and desynchronized states, which extends in a natural way the definition of chimera states to quantum mechanics.

16.2 Nonlinear Quantum Oscillators

In this section we describe a recent theoretical proposal to realize a quantum analogue of the Van der Pol oscillator [59, 60].

16.2.1 The Classical Van der Pol Oscillator

The classical Van der Pol oscillator is given by the equation of motion [76]

$$\ddot{Q} + \omega_0^2 Q - \epsilon(1 - 2Q^2)\dot{Q} = 0, \tag{16.1}$$

where $Q \in \mathbb{R}$ is the dynamical variable, ω_0 is the linear frequency, and $\epsilon > 0$ is the nonlinearity parameter. One important aspect of this equation is that the interplay between negative damping proportional to $-\dot{Q}$ and nonlinear damping $Q^2\dot{Q}$ leads to the existence of self-sustained limit cycle oscillations. Similarly to the method discussed in Ref. [59], we consider a transformation into a rotating frame $Q(t) = 2^{-1/2}(\alpha(t)e^{i\omega_0 t} + \alpha^*(t)e^{-i\omega_0 t})$ with a slowly varying complex amplitude $\alpha = 2^{-1/2}(Q + iP)$, and Q and $P = \dot{Q}$ denote position and conjugate momentum, respectively. In the rotating frame, one can neglect fast oscillating terms in Eq. (16.1) as long as the condition $\epsilon \ll 1$ holds. This enables us to obtain an effective amplitude equation which has the form of a Stuart-Landau equation

$$\dot{\alpha}(t) = \frac{\epsilon}{2}(1 - |\alpha(t)|^2)\alpha(t) \tag{16.2}$$

describing the dynamics of the oscillator. In the stationary state, i.e., when $\dot{\alpha}(t) = 0$, it admits the existence of a limit cycle defined by $|\alpha(t)|^2 = 1$.

16.2.2 The Quantum Van der Pol Oscillator

To obtain a quantum analogue of the Van der Pol oscillator, we require a mechanism to inject energy into the system in a linear way (negative damping) and to induce nonlinear losses. Such features can be accomplished by using trapped ions setups [77], as proposed in Ref. [59]. In the case of trapped ions, the dynamic degrees of freedom are described by means of bosonic creation and annihilation operators $a^\dagger$ and a, respectively. The dissipative dynamics is governed by the Lindblad master equation

$$\dot{\rho}(t) = 2\kappa_1 \left(a^\dagger \rho a - \frac{1}{2}\{\rho, aa^\dagger\} \right) + 2\kappa_2 \left(a^2 \rho(a^\dagger)^2 - \frac{1}{2}\{\rho, (a^\dagger)^2 a^2\} \right), \quad (16.3)$$

in a rotating frame with frequency ω_0, where ρ is the density matrix. The rates κ_1 and κ_2 describe one-photon and two-photon dissipative processes, respectively [59]. These dissipative processes are the quantum analogue of negative damping and nonlinear losses.

In the high-photon density limit $\langle a^\dagger a \rangle = |\alpha|^2 \gg 1$, one can describe bosonic quantum fluctuations $\tilde{a}$, $\tilde{a}^\dagger$ about the mean field α. This approach enables us to study the time evolution of the quantum fluctuations, which is influenced by the mean field solution. Correspondingly, at mean-field level, the system resembles the classical behavior of the Van der Pol oscillator in the $\epsilon \ll 1$ limit. Unfortunately, to obtain analytical results we must confine ourselves to the study of Gaussian quantum fluctuations. This implies some limitations in the description of the long-time dynamics of the fluctuations. For example, even if one prepares the system in a coherent state at $t = 0$: $\rho(0) = |\alpha(0)\rangle\langle\alpha(0)|$, i.e., a bosonic Gaussian state, there are quantum signatures of the classical limit cycle leading to non-Gaussian effects. Therefore, within the framework of a Gaussian description, one is able to describe only short-time dynamics, where the non-Gaussian effects are negligible.

16.2.3 Gaussian Quantum Fluctuations and Semiclassical Trajectories

Let us begin by considering the decomposition $a(t) = \tilde{a} + \alpha(t)$ of the bosonic operator a in terms of the quantum fluctuations $\tilde{a}$ and the mean field α. For completeness, in the appendix we calculate explicitly the Gaussian quantum fluctuations for the quartic oscillator. To formalize this procedure from the perspective of the master equation [78, 79], we define the density matrix in the co-moving frame $\rho_\alpha(t) = \hat{D}^\dagger[\alpha(t)]\rho(t)\hat{D}[\alpha(t)]$, where $\hat{D}[\alpha(t)] = \exp\left[\alpha(t)\tilde{a}^\dagger - \alpha^*(t)\tilde{a}\right]$ is a displacement operator. In the co-moving frame, we obtain a master equation with Liouville operators $\hat{\mathcal{L}}_1$ and $\hat{\mathcal{L}}_2$

$$\dot{\rho}_\alpha(t) = -\frac{i}{\hbar}[\hat{H}^{(\alpha)}(t), \rho_\alpha(t)] + 2\kappa_1 \hat{D}^\dagger[\alpha(t)]\left(\tilde{a}^\dagger \rho \tilde{a} - \frac{1}{2}\{\rho, \tilde{a}\tilde{a}^\dagger\} \right) \hat{D}[\alpha(t)]$$

$$+ 2\kappa_2 \hat{D}^\dagger[\alpha(t)]\left(\tilde{a}^2 \rho(\tilde{a}^\dagger)^2 - \frac{1}{2}\{\rho, (\tilde{a}^\dagger)^2 \tilde{a}^2\} \right) \hat{D}[\alpha(t)]$$

$$\equiv -\frac{i}{\hbar}[\hat{H}^{(\alpha)}(t), \rho_\alpha(t)] + \hat{\mathcal{L}}_1 \rho_\alpha + \hat{\mathcal{L}}_2 \rho_\alpha, \quad (16.4)$$

where we have defined $\hat{H}^{(\alpha)}(t) = -i\hbar \hat{D}^\dagger [\alpha(t)] \partial_t \hat{D} [\alpha(t)]$ and the anticommutator $\{\hat{A}, \hat{B}\} = \hat{A}\hat{B} + \hat{B}\hat{A}$. In the co-moving frame, the coherent dynamics is generated by the Hamiltonian

$$\hat{H}^{(\alpha)}(t) = -\frac{i\hbar}{2}[\dot{\alpha}(t)\alpha^*(t) - \alpha(t)\dot{\alpha}^*(t)] - i\hbar[\dot{\alpha}(t)\tilde{a}^\dagger - \dot{\alpha}^*(t)\tilde{a}]. \tag{16.5}$$

A next step in the calculation of the Gaussian quantum fluctuations is to expand the Liouville operators $\hat{\mathcal{L}}_1$ and $\hat{\mathcal{L}}_2$ in terms of the quantum fluctuations. We begin by considering the Liouville operator $\hat{\mathcal{L}}_1$, which preserves the Gaussian character of the initial state $\rho(0) = |\alpha(0)\rangle\langle\alpha(0)|$. By using elementary properties of the displacement operator we can decompose the dissipative term into coherent and incoherent parts:

$$\hat{\mathcal{L}}_1 \rho_\alpha = 2\kappa_1 \hat{D}^\dagger [\alpha(t)] \left(\tilde{a}^\dagger \rho \tilde{a} - \frac{1}{2}\{\rho, \tilde{a}\tilde{a}^\dagger\} \right) \hat{D} [\alpha(t)]$$

$$= 2\kappa_1 \left(\tilde{a}^\dagger \rho_\alpha \tilde{a} - \frac{1}{2}\{\rho_\alpha, \tilde{a}\tilde{a}^\dagger\} \right) - \frac{i}{\hbar}\left[i\hbar\kappa_1 \alpha \tilde{a}^\dagger, \rho_\alpha\right] - \frac{i}{\hbar}\left[-i\hbar\kappa_1 \alpha^* \tilde{a}, \rho_\alpha\right]. \tag{16.6}$$

In the calculation of the dissipative term proportional to κ_2, one needs to be particularly careful, because it causes non-Gaussian effects due to two-photon processes. Interestingly, one can decompose the Liouville operator $\hat{\mathcal{L}}_2$ into coherent terms given by commutators of an effective Hamiltonian with the density operator and terms preserving the Gaussian character of the initial state. In addition, we also obtain explicitly the non-Gaussian contributions:

$$\hat{\mathcal{L}}_2 \rho_\alpha = 2\kappa_2 \hat{D}^\dagger [\alpha(t)] \left(\tilde{a}^2 \rho (\tilde{a}^\dagger)^2 - \frac{1}{2}\{\rho, (\tilde{a}^\dagger)^2 \tilde{a}^2\} \right) \hat{D} [\alpha(t)]$$

$$= 2\kappa_2 \left(\tilde{a}^2 \rho_\alpha (\tilde{a}^\dagger)^2 - \frac{1}{2}\{\rho_\alpha, (\tilde{a}^\dagger)^2 \tilde{a}^2\} \right) + 4\kappa_2 \alpha^* \left(\tilde{a}^2 \rho_\alpha \tilde{a}^\dagger - \frac{1}{2}\{\rho_\alpha, \tilde{a}^\dagger \tilde{a}^2\} \right)$$

$$+ 4\kappa_2 \alpha \left(\tilde{a} \rho_\alpha (\tilde{a}^\dagger)^2 - \frac{1}{2}\{\rho_\alpha, (\tilde{a}^\dagger)^2 \tilde{a}\} \right) + 8\kappa_2 |\alpha|^2 \left(\tilde{a} \rho_\alpha \tilde{a}^\dagger - \frac{1}{2}\{\rho_\alpha, \tilde{a}^\dagger \tilde{a}\} \right)$$

$$- \frac{i}{\hbar}\left[i\hbar\kappa_2 (\alpha^*)^2 \tilde{a}^2, \rho_\alpha\right] - \frac{i}{\hbar}\left[-i\hbar\kappa_2 \alpha^2 (\tilde{a}^\dagger)^2, \rho_\alpha\right]$$

$$- \frac{i}{\hbar}\left[2i\hbar\kappa_2 \alpha (\alpha^*)^2 \tilde{a}, \rho_\alpha\right] - \frac{i}{\hbar}\left[-2i\hbar\kappa_2 \alpha^* \alpha^2 \tilde{a}^\dagger, \rho_\alpha\right]. \tag{16.7}$$

In the semiclassical high-density limit $|\alpha|^2 \gg 1$, one can safely neglect the effect of the non-Gaussian terms in Eqs. (16.6) and (16.7). Similarly to Refs. [78, 79], to define the position $\alpha(t)$ of the co-moving frame, we require vanishing linear terms in the coherent part of the master equation (16.4). This is achieved as long as the condition

$$\dot{\alpha}(t) = \kappa_1 \alpha(t) - 2\kappa_2 \alpha(t)|\alpha(t)|^2 \tag{16.8}$$

is satisfied. After neglecting such terms, we obtain the master equation

$$\dot{\rho}_\alpha(t) = -i\kappa_2 \left[i(\alpha^*)^2 \tilde{a}^2 - i\alpha^2 (\tilde{a}^\dagger)^2, \rho_\alpha \right] + 2\kappa_1 \left(\tilde{a}^\dagger \rho_\alpha \tilde{a} - \frac{1}{2} \{ \rho_\alpha, \tilde{a} \tilde{a}^\dagger \} \right)$$

$$+ 8\kappa_2 |\alpha(t)|^2 \left(\tilde{a} \rho_\alpha \tilde{a}^\dagger - \frac{1}{2} \{ \rho_\alpha, \tilde{a}^\dagger \tilde{a} \} \right). \tag{16.9}$$

One can interpret this procedure from a geometrical point of view. An initial coherent state $\rho(0) = |\alpha(0)\rangle \langle \alpha(0)|$ corresponds to the vacuum $\rho_\alpha(0) = |0\rangle \langle 0|$ in the co-moving frame centered at $\alpha(0)$. On the other hand, $\alpha(0)$ plays the role of the initial condition for the classical equation of motion Eq. (16.8). The solution $\alpha(t)$ of Eq. (16.8) not only gives us the position of the co-moving frame, but it is also responsible for the emergence of time dependent rates in the master equation and time dependent squeezing. In the stationary limit, however, the classical equations of motion exhibit self-sustained oscillations with amplitude $|\alpha(t)|^2 = \kappa_1/2\kappa_2$. In this limit, the master equation ceases to have time-dependent coefficients.

16.3   Quantum Description of a Network of Coupled Van der Pol Oscillators

We consider a quantum network consisting of a ring of N coupled Van der Pol oscillators. Such a network can be described by the master equation for the density matrix $\rho(t)$

$$\dot{\rho} = -\frac{i}{\hbar}[\hat{H}, \rho] + 2 \sum_{l=1}^{N} \left[\kappa_1 \mathcal{D}(a_l^\dagger) + \kappa_2 \mathcal{D}(a_l^2) \right], \tag{16.10}$$

where $a_l^\dagger$, a_l are creation and annihilation operators of bosonic particles and $\mathcal{D}(\hat{O}) = \hat{O}\rho\hat{O}^\dagger - \frac{1}{2}(\hat{O}^\dagger\hat{O}\rho + \rho\hat{O}^\dagger\hat{O})$ describes dissipative processes with rates $\kappa_1, \kappa_2 > 0$, which we have discussed in Sect. 16.2.2. In addition, we have imposed periodic boundary conditions $a_l = a_{l+N}$ for the bosonic operators. In contrast to Ref. [59], we consider a nonlocal coupling between the oscillators. Therefore, the Hamiltonian in the interaction picture reads $\hat{H} = \hbar \sum_{l\neq m=1}^{N} K_{l,m}(a_l^\dagger a_m + a_l a_m^\dagger)$, where $K_{l,m} = \frac{V}{2d}\Theta(d - |l - m|)$ is the coupling matrix of the network and $\Theta(x)$ is the Heaviside step function. This kind of coupling implies that Eq. (16.10) has a rotational S^1 symmetry as discussed in Ref. [29]. One can include counter-rotating terms such as $a_l^\dagger a_m^\dagger$ in the coupling, but this would lead to symmetry breaking.

This definition implies that the coupling is zero if the distance $|l - m|$ between the lth and mth the nodes is bigger than the coupling range d. On the other hand, if $|l - m| < d$, then $K_{l,m} = \frac{V}{2d}$. In the particular case $d = N/2$ one has all-to-all coupling and recovers the results of Ref. [59].

Now we compare Eq. (16.10) with the general form of the Lindblad master equation [80]

$$\dot{\rho}(t) = -\frac{i}{\hbar}[\hat{H}, \rho] + \sum_{\mu} \gamma_{\mu} \left(\hat{L}_{\mu} \rho \hat{L}_{\mu}^{\dagger} - \frac{1}{2}\{\rho, \hat{L}_{\mu}^{\dagger} \hat{L}_{\mu}\} \right) \tag{16.11}$$

with Lindblad operators $\hat{L}_{\mu}$. This enables us to introduce an effective Hamiltonian which describes the dynamics between quantum jumps

$$\hat{H}_{\text{eff}} = \hat{H} - \frac{i\hbar}{2} \sum_{\mu} \gamma_{\mu} \hat{L}_{\mu}^{\dagger} \hat{L}_{\mu}. \tag{16.12}$$

In the case of the master equation Eq. (16.10), the effective Hamiltonian reads

$$\hat{H}_{\text{eff}} = -i\hbar\kappa_1 \sum_{l=1}^{N} (a_l^{\dagger} a_l + 1) - i\hbar\kappa_2 \sum_{l=1}^{N} \hat{n}_l(\hat{n}_l - 1)$$

$$+ \hbar \sum_{l \neq m = 1}^{N} K_{l,m}(a_l^{\dagger} a_m + a_l a_m^{\dagger}), \tag{16.13}$$

where $\hat{n}_l = a_l^{\dagger} a_l$. The Hamiltonian Eq. (16.13) describes a Bose-Hubbard model with long range interactions, where on-site energies and chemical potential are complex. This kind of model arises naturally in the context of driven-dissipative Bose-Einstein condensation [73, 74].

16.3.1 Gaussian Quantum Fluctuations and Master Equation

We define the expansion $b_l(t) = \hat{D}^{\dagger}[\boldsymbol{\alpha}(t)] a_l \hat{D}[\boldsymbol{\alpha}(t)] = \tilde{a}_l + \alpha_l(t)$, where $\hat{D}[\boldsymbol{\alpha}(t)] = \exp\left[\boldsymbol{\alpha}(t) \cdot \hat{\tilde{\boldsymbol{a}}}^{\dagger} - \boldsymbol{\alpha}^*(t) \cdot \hat{\tilde{\boldsymbol{a}}}\right]$, $\boldsymbol{\alpha}(t) = [\alpha_1(t), \ldots, \alpha_N(t)]$ and $\hat{\tilde{\boldsymbol{a}}} = (\tilde{a}_1, \ldots, \tilde{a}_N)$ as in Ref. [81]. In this work we consider the semiclassical regime, where the magnitude of the mean field $\alpha_l(t)$ is larger than the quantum fluctuations $\tilde{a}_l$ as in Refs. [78, 79]. By using the expansion of the master equation about the mean-field $\boldsymbol{\alpha}(t)$ described in the previous section, we obtain a master equation for the density operator in a co-moving frame $\rho_{\alpha}(t) = \hat{D}^{\dagger}[\boldsymbol{\alpha}(t)] \rho(t) \hat{D}[\boldsymbol{\alpha}(t)]$

$$\dot{\rho}_{\alpha} \approx -\frac{i}{\hbar}[\hat{H}_{Q}^{(\alpha)}, \rho_{\alpha}] + 2 \sum_{l=1}^{N} \left[\kappa_1 \mathcal{D}(\tilde{a}_l^{\dagger}) + 4\kappa_2 |\alpha_l|^2 \mathcal{D}(\tilde{a}_l)\right]. \tag{16.14}$$

In addition, the coherent dynamics of the fluctuations are governed by the Hamiltonian

$$\hat{H}_Q^{(\alpha)}(t) = \hbar \sum_{l=1}^{N} \left(i\kappa_2 (\alpha_l^*)^2 \tilde{a}_l^2 - i\kappa_2 \alpha_l^2 (\tilde{a}_l^\dagger)^2 + \sum_{r=1}^{N} K_{l,l+r} (\tilde{a}_l^\dagger \tilde{a}_{l+r} + \tilde{a}_l \tilde{a}_{l+r}^\dagger) \right)$$

$$+ \hbar \sum_{l=1}^{N} \left(-i[\dot{\alpha}_l(t)\tilde{a}_l^\dagger - \dot{\alpha}_l^*(t)\tilde{a}_l] + i\kappa_1 \alpha_l \tilde{a}_l^\dagger - i\kappa_1 \alpha_l^* \tilde{a} \right)$$

$$+ \hbar \sum_{l=1}^{N} \left(2i\kappa_2 \alpha_l (\alpha_l^*)^2 \tilde{a}_l - 2i\kappa_2 \alpha_l^* \alpha_l^2 \tilde{a}_l^\dagger \right)$$

$$+ \hbar \sum_{r=1}^{N} K_{l,l+r} (\alpha_{l+r} \tilde{a}_l^\dagger + \alpha_{l+r}^* \tilde{a}_l + \alpha_l^* \tilde{a}_{l+r} + \alpha_l \tilde{a}_{l+r}^\dagger). \tag{16.15}$$

To obtain the equations for the mean field, the linear terms in the expansion Eq. (16.15) must vanish, which leads to the equation

$$\dot{\alpha}_l(t) = \alpha_l(t)(\kappa_1 - 2\kappa_2 |\alpha_l(t)|^2) - i \sum_{s \neq l}^{N} K_{l,s} \alpha_s(t) \tag{16.16}$$

with a similar equation for $\dot{\alpha}_l^*(t)$. Finally, by using the master equation (16.14) we can calculate the equations of motion as follows

$$\frac{d\langle \tilde{a}_i \rangle}{dt} = \text{tr}[\tilde{a}_i \dot{\rho}_\alpha(t)] = \kappa_1 \langle \tilde{a}_i \rangle - 4\kappa_2 |\alpha_i|^2 \langle \tilde{a}_i \rangle - 2\kappa_2 \alpha_i^2 \langle \tilde{a}_i^\dagger \rangle - i \sum_{s \neq i}^{N} K_{i,s} \langle \tilde{a}_s \rangle. \tag{16.17}$$

16.3.2 Relation to the Continuum Limit and Linearization

To understand the meaning of the Gaussian quantum fluctuations we discuss the continuum limit of the classical equations of motion Eq. (16.16). In the continuum limit, $N \to \infty$, the complex variable $\alpha_l(t) = r_l(t)e^{i\phi_l(t)}$ can be described by means of a complex field $\alpha(x, t) = |\alpha(x, t)|e^{i\phi(x,t)}$, where x is the continuous version of the index l. Correspondingly, $|\alpha(x, t)|$ and $\phi(x, t)$ represent the amplitude and phase fields, respectively [2].

In the continuum limit, a ring of N coupled nodes can be described by means of a classical field $\alpha(x, t)$ defined on a circle of length L, where x is the position coordinate. In addition, if one introduces the continuum version $K(x - y)$ of the coupling matrix $K_{l,m}$, the dynamics of such a field can be described by the equation of motion

$$\frac{\partial \alpha(x,t)}{\partial t} = \alpha(x,t)(\kappa_1 - 2\kappa_2|\alpha(x,t)|^2) - i\int_0^L dy\, K(x-y)\alpha(y,t), \quad (16.18)$$

which is the continuum limit of Eq. (16.16) and resembles the field equation discussed in Ref. [2]. In particular, if we assume the amplitude $|\alpha(x,t)| = r_0$ to be constant after the system is trapped into the limit cycle, we obtain a differential equation for the phases

$$i\frac{\partial \phi(x,t)}{\partial t} = (\kappa_1 - 2\kappa_2 r_0^2) - i\int_0^L dy\, K(x-y)e^{-i[\phi(x,t)-\phi(y,t)]}. \quad (16.19)$$

Following the method described in Ref. [2], one can introduce a mean field

$$r(x,t)e^{i\Theta(x,t)} = \int_0^L dy\, K(x-y)e^{i\phi(y,t)}. \quad (16.20)$$

This method works well in the case of phase chimeras. However, in the case of amplitude-mediated chimeras [27, 28], one requires to study both phase $\phi(x,t)$ and amplitude $|\alpha(x,t)|$ fields.

In order to have a better understanding of the quantum fluctuations, let us linearize the equation of motion for the field Eq. (16.18) about a solution $\alpha_0(x,t)$. For this purpose, let us consider the decomposition of the field $\alpha(x,t) = \alpha_0(x,t) + \tilde{a}(x,t)$, where $\tilde{a}(x,t)$ is a small perturbation such that $|\alpha_0(x,t)| \ll |\tilde{a}(x,t)|$. Now let us assume that we expand Eq. (16.18) up to first order in the perturbation. After some algebraic manipulations we obtain

$$\frac{\partial \tilde{a}(x,t)}{\partial t} = \kappa_1 \tilde{a}(x,t) - 4\kappa_2|\alpha_0(x,t)|^2\tilde{a}(x,t)$$
$$- 2\kappa_2[\alpha_0(x,t)]^2\tilde{a}^*(x,t) - i\int_0^L dy\, K(x-y)\tilde{a}(y,t). \quad (16.21)$$

One can observe that this equation is precisely the continuum limit of the equations of motions Eq. (16.17) for the expectation values of the quantum fluctuations. In particular, $\alpha_0(x,t)$ plays the role of the mean field $\alpha_l(t)$ and $\tilde{a}(x,t)$ is the continuum limit of the expectation value $\langle \tilde{a}_l(t) \rangle$.

Now the role of the Gaussian quantum fluctuations is clear: By neglecting non-Gaussian contributions in the master equation [78, 79], one constructs the master equation (16.14) governing the evolution of the quantum fluctuations. Due to the chaotic nature of the classical chimera states [9], one expects giant quantum fluctuations about the semiclassical trajectories [82]. As a consequence, the Gaussian approximation, i.e., the master equation Eq. (16.14) fails to describe the long-time dynamics.

16.3.3   The Gutzwiller Ansatz and the Master Equation

In this section we describe the different methods to tackle the emergence of chimera states in the quantum regime. Due to the nature of the nodes of the network, one has to truncate the Hilbert space up to a certain occupation number n_t of the oscillator. This implies that if one has a network with N nodes, one has to solve a system of $(n_t + 1)^{2N}$ coupled differential equations for the elements $\rho_{n,m}$ of the density matrix as follows from Eq. (16.10). Chimera states usually emerge in networks consisting at least of $N = 40$ nodes. This means that if one truncates the Hilbert space of the oscillator up to $n_t = 2$ one has to solve a system of 3^{80} coupled differential equations. In the incoherent regime of the network one expects vanishing coherences. As a consequence of this one has to solve only the evolution of the populations, which involves the solution of 3^{40} ordinary differential equations. From the previous analysis we conclude that the complete solution of the master equation (16.10) is not possible in order to find the quantum signatures of synchronization and even to describe the incoherent regime. Therefore one has to invoke alternative methods of solution as we describe below.

We start by considering the Gutzwiller ansatz $\rho(t) = \bigotimes_{l=1}^{N} \rho_l(t)$ discussed in Refs. [59, 61]. By inserting this in the master equation Eq. (16.10), we obtain a self-consistent system of equations for the density matrix of the lth site

$$\dot{\rho}_l(t) = -\frac{i}{\hbar}[\hat{H}_l, \rho_l] + 2\kappa_1 D(a_l^\dagger) + \kappa_2 D(a_l^2), \qquad (16.22)$$

where we define the self-consistent local Hamiltonian

$$\hat{H}_l = \Gamma_l a_l^\dagger + \Gamma_l^* a_l. \qquad (16.23)$$

Motivated by the original approach of Kuramoto [2] and a recent work [68], we have defined the mean field $\Gamma_l = \hbar \sum_{r=1}^{N} K_{l,l+r} \langle a_{l+r} \rangle$, which resembles the order parameter Eq. (16.20). This order parameter takes into account the contributions of the quantum coherences. In contrast to the complete solution of Eq. (16.10), the Gutzwiller ansatz allows one to obtain the solution of the problem with polynomial resources. More specifically, instead of solving 3^{2N} equations, one has to solve $3^2 N$ equations if one truncates the bosonic Hilbert space at $n_t = 2$. The minimal size of a chain that supports chimeras is of the order of $N = 40$, therefore one has to solve only 360 equations of motion.

For completeness, we discuss briefly the equations of motion derived from Eq. (16.10)

$$\frac{d\langle a_l \rangle}{dt} = \text{tr}[a_l \dot{\rho}(t)] = \kappa_1 \langle a_l \rangle - 2\kappa_2 \langle a_l^\dagger a_l^2 \rangle - i \sum_{r=1}^{N} K_{l,l+r} \langle a_{l+r} \rangle. \qquad (16.24)$$

Interestingly, the equations of motion Eq. (16.17) constitute a particular case of
Eq. (16.24), because if one only considers the contributions of the Gaussian fluctuations, one obtains a natural way to factorize expectation values [59, 61].

16.4 Classical Chimera States and Phase-Space Methods

The equations of motion Eq. (16.16) resemble a system of coupled Stuart-Landau
oscillators [29]. The solution $\boldsymbol{\alpha}(t)$ of the equations of motion Eq. (16.16), provides
us information about the dynamics of the mean field. Such a mean field plays a
fundamental role in the study of the master equation Eq. (16.14). Within the Gaussian
approximation, the mean field is responsible for coherent effects such as squeezing
in Eq. (16.15). In addition, the amplitude $|\alpha_l(t)|$ determines the dissipation rates
which appear in Eq. (16.14). Therefore to investigate the evolution of the density
matrix, we require the time evolution $\boldsymbol{\alpha}(t)$. In this section we show that the mean
field exhibits chimera-like dynamics. By using phase-space methods, we investigate
quantum signatures of these states.

16.4.1 Emergence of Classical Chimera States

In the case of the uncoupled system $V = 0$, i.e., $K_{l,m} = 0$, a single Van der Pol oscillator [59, 60] exhibits a limit cycle with radius $r_0 = \sqrt{\frac{\kappa_1}{2\kappa_2}} = 1.58$ for the parameters
$\kappa_2 = 0.2\kappa_1$. For convenience, in the coupled system we consider initial conditions
at $t = 0$ in such a way that each oscillator has the same amplitude $|\alpha_l(0)| \approx 1.58$.
In addition, we consider phases drawn randomly from a Gaussian distribution in
space $\phi_l(0) = \frac{\theta}{\sqrt{2\pi}\sigma} \exp[-\frac{(l-\mu)^2}{2\sigma^2}]$, where $-24\pi < \theta < 24\pi$ is a random number,
$\mu = N/2$ and $\sigma = 9$. Figure 16.1a shows the initial conditions. In terms of the coordi-
nates $\alpha_l(t) = \frac{Q_l(t)+iP_l(t)}{\sqrt{2\hbar}}$, the initial conditions must satisfy $\sqrt{Q_l^2(0) + P_l^2(0)} \approx 2.24$,
which defines the green circle in Fig. 16.1b.

Besides the description of chimera states, we also discuss completely synchronized and completely desynchronized solutions. To obtain such solutions, we consider different coupling strengths V. However, for every case, we restrict ourselves
to the same initial conditions as in Fig. 16.1.

From our previous discussion in Sect. 16.3.1, the classical equations of motion
Eq. (16.16) must be satisfied in order to investigate the evolution of the master
equation Eq. (16.14) in the co-moving frame. In the polar representation $\alpha_l(t) =
r_l(t)e^{i\phi_l(t)}$, the equations of motion couple amplitude $r_l(t)$ and phase $\phi_l(t)$ of the indi-
vidual oscillators. We numerically solve Eq. (16.16) for a network of $N = 50$ coupled
oscillators with coupling range $d = 10$. We consider initial conditions $|\alpha_l(t_0)| \approx r_0$,
where $r_0 = 1.58$, and phases drawn randomly from a Gaussian distribution in space.

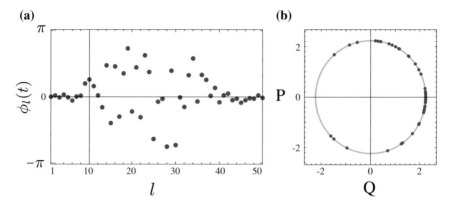

Fig. 16.1 Initial conditions used in the simulations. **a** Initial distribution of the phases $\phi_l(0)$ drawn randomly from a Gaussian distribution in space. **b** Phase-space representation of the initial conditions for the oscillators. The *green circle* represents the limit cycle with radius $|\alpha_l(0)| \approx 1.58$, where $\alpha_l(t) = \frac{Q_l(t) + iP_l(t)}{\sqrt{2\hbar}}$. Parameters: $\hbar = 1$, $d = 10$, $\kappa_2 = 0.2\kappa_1$, and $N = 50$

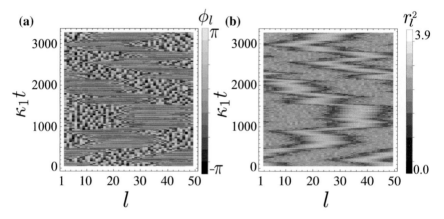

Fig. 16.2 Space-time representation of the classical chimera state. We consider a representation $\alpha_l(t) = r_l(t)e^{i\phi_l(t)}$ of the individual oscillators in terms of amplitude $r_l(t)$ and phase $\phi_l(t)$. **a** Depicts the time evolution of the phase chimera, where we represent the phases $\{\phi_l(t)\}$. Similarly, **b** depicts the amplitudes $\{r_l^2(t)\}$. *Parameters*: $d = 10$, $\kappa_2 = 0.2\kappa_1$, $V = 1.2\kappa_1$, and $N = 50$ [71]

Figure 16.2 depicts the time evolution of a classical chimera state. In Fig. 16.2a we show the space-time representation of the phases $\phi_l(t)$ of the individual oscillators. One can observe that for a fixed time, there is a domain of synchronized oscillators that coexists with a domain of desynchronized motion, which is a typical feature of chimera states. Besides the phase, also the amplitude exhibits chimera dynamics as we show in Fig. 16.2b. One can observe that the width of the synchronized region changes with time. Similarly, the center of mass of the synchronized region moves randomly along the ring. Chimera states with these features have been reported in the literature and are referred to as *breathing* and *drifting* chimeras [9].

16.4.2 Solution of the Fokker-Planck Equation

As discussed in the previous section, the classical equations of motion Eq. (16.16) exhibit a chimera state. By using the knowledge we have about the mean field $\boldsymbol{\alpha}(t)$ in the semiclassical limit $|\alpha_l| \gg 1$, we can study the time evolution of the quantum fluctuations $\tilde{a}(t)$ in the co-moving frame by solving the master equation Eq. (16.14). With this aim, we consider the pure coherent state as an initial density matrix $\rho(t_0) = \bigotimes_{l=1}^{N} |\alpha_l(t_0)\rangle\langle\alpha_l(t_0)|$, where $|\alpha_l(t_0)| \approx 1.58$ and we choose the phases as in the left panel of Fig. 16.3. This initial condition corresponds to a fixed time $t_0 = 3000/\kappa_1$ in Fig. 16.2. In the co-moving frame, such initial condition reads $\rho_\alpha(t_0) = \bigotimes_{l=1}^{N} |0_l\rangle\langle 0_l|$. For convenience, let us consider a representation of the bosonic operators $a_l = (\hat{q}_l + i\hat{p}_l)/\sqrt{2\hbar}$ and $\tilde{a}_l = (\hat{\tilde{q}}_l + i\hat{\tilde{p}}_l)/\sqrt{2\hbar}$ in terms of position and momentum operators $\hat{q}_l$ and $\hat{p}_l$, respectively. We also introduce the complex variables $z_l = (q_l + ip_l)/\sqrt{2\hbar}$, $\tilde{z}_l = (\tilde{q}_l + i\tilde{p}_l)/\sqrt{2\hbar}$, which allow us to define the coordinates $z^T = (z_1, \ldots, z_N)$ in the laboratory frame and $\tilde{z}^T = (\tilde{z}_1, \ldots, \tilde{z}_N)$ in the co-moving frame. These coordinates are related via $z = \boldsymbol{\alpha}(t) + \tilde{z}$. The variables q_l, $\tilde{q}_l$ and p_l, $\tilde{p}_l$ denote position and conjugate momentum, respectively.

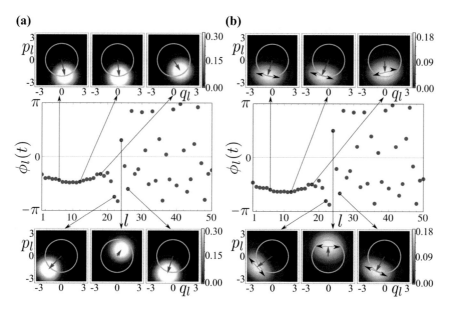

Fig. 16.3 Quantum signatures of the classical chimera state. **a** Snapshot of the phase chimera depicted in Fig. 16.2 at $\kappa_1 t_0 = 3000$. We consider an initial density matrix $\rho(t_0)$ which is a tensor product of coherent states centered around the positions of the individual oscillators as depicted in the insets (Husimi function). **b** After a short-time interval $\kappa_1 \Delta t = 0.5$, quantum correlations appear in the form of squeezing (*black double arrows* in the insets). Parameters: $d = 10$, $\kappa_2 = 0.2\kappa_1$, $V = 1.2\kappa_1$, and $N = 50$ [71]

Now let us define the Wigner representation of the density operator $\rho_\alpha(t)$ [80]:

$$W_\alpha(\tilde{z}) = \int \frac{d^{2N}\boldsymbol{\lambda}}{\pi^{2N}} e^{-\boldsymbol{\lambda}\cdot\tilde{z}^* + \boldsymbol{\lambda}^*\cdot\tilde{z}} \operatorname{tr}\left[\rho_\alpha(t)e^{-\boldsymbol{\lambda}\cdot\hat{\bar{a}}^\dagger + \boldsymbol{\lambda}^*\cdot\hat{\bar{a}}}\right], \tag{16.25}$$

where $\boldsymbol{\lambda} = (\lambda_1,\ldots,\lambda_N)$ denote the integration variables. The Husimi function $Q(z) = \frac{1}{\pi}\langle z|\rho(t)|z\rangle$ is intimately related to the Wigner function via the transformation [80]

$$Q_\alpha(\tilde{z}) = \frac{2}{\pi}\int W_\alpha(\tilde{x})e^{-2|\tilde{z}-\tilde{x}|^2}d^{2N}\tilde{x}. \tag{16.26}$$

The Husimi function can be obtained numerically after solving the master equation by using the Gutzwiller ansatz [59, 61] as we discussed in Sect. 16.3.3. The insets in the right panel of Fig. 16.3 depict the Husimi functions of the individual nodes after a short evolution time $\Delta t = 0.5/\kappa_1$. One can observe that even if one prepares the system in a separable state, quantum fluctuations arise in the form of bosonic squeezing of the oscillators [80]. In the insets of Fig. 16.3, the arrows indicate the direction perpendicular to the squeezing direction for the individual oscillators. For oscillators within the synchronized region, the squeezing occurs almost in the same direction. In contrast, the direction of squeezing is random for oscillators in the desynchronized region, which reflects the nature of the chimera state.

By using standard techniques of quantum optics [80], the master equation Eq. (16.14) can be represented as a Fokker-Planck equation for the Wigner function Eq. (16.25), which depends on the mean field solution of Eq. (16.16) and contains information of the chimera state

$$\frac{\partial W_\alpha}{\partial t} = \sum_{l=1}^{N}\left[2\kappa_2(\alpha_l^*)^2\partial_{\tilde{z}_l^*}\tilde{z}_l + (4\kappa_2|\alpha_l|^2 - \kappa_1)\partial_{\tilde{z}_l}\tilde{z}_l + \left(2\kappa_2|\alpha_l|^2 + \frac{\kappa_1}{2}\right)\partial^2_{\tilde{z}_l,\tilde{z}_l^*}\right]W_\alpha$$
$$- i\frac{V}{2d}\sum_{l=1}^{N}\sum_{\substack{m=l-d\\m\neq l}}^{l+d}(\partial_{\tilde{z}_m}\tilde{z}_l^* - \partial_{\tilde{z}_l}\tilde{z}_l)W_\alpha + \text{H.c.} \tag{16.27}$$

For convenience, we consider the Wigner representation $W_\alpha(\tilde{\boldsymbol{R}},t)$ of the density operator $\rho_\alpha(t)$ in terms of the new variables $\tilde{\boldsymbol{R}}^T = (\tilde{q}_1,\tilde{p}_1,\ldots,\tilde{q}_N,\tilde{p}_N)$. Correspondingly, the Fokker-Planck equation can be written also in terms of the quadratures $\tilde{q}_l$ and $\tilde{p}_l$

$$\frac{\partial W_\alpha}{\partial t} = -\sum_{i=1}^{2N}\mathscr{A}_{ij}(t)\partial_{\tilde{R}_i}(\tilde{R}_j W_\alpha) + \frac{1}{2}\sum_{i=1}^{2N}\mathscr{B}_{ij}(t)\partial^2_{\tilde{R}_i,\tilde{R}_j}W_\alpha. \tag{16.28}$$

Although this Fokker-Planck equation has time-dependent coefficients, one can derive an exact solution [80]

$$W_\alpha(\tilde{\boldsymbol{R}}, t) = \frac{\exp\left(-\frac{1}{2}\tilde{\boldsymbol{R}}^T \cdot \mathscr{C}^{-1}(t) \cdot \tilde{\boldsymbol{R}}\right)}{(2\pi)^N \sqrt{\det \mathscr{C}(t)}}, \tag{16.29}$$

where the covariance matrix $\mathscr{C}(t)$ is a solution of the differential equation $\dot{\mathscr{C}}(t) = \mathscr{A}(t)\mathscr{C}(t) + \mathscr{C}(t)\mathscr{A}^T(t) + \mathscr{B}(t)$. The matrix elements

$$\mathscr{C}_{ij} = \langle \frac{1}{2}\left(\hat{\tilde{R}}_i\hat{\tilde{R}}_j + \hat{\tilde{R}}_j\hat{\tilde{R}}_i\right)\rangle_\alpha - \langle\hat{\tilde{R}}_i\rangle_\alpha\langle\hat{\tilde{R}}_j\rangle_\alpha \tag{16.30}$$

of the covariance matrix contain information about the correlations between quantum fluctuations $\hat{\tilde{R}}_{2l-1} = \hat{\tilde{q}}_l$ and $\hat{\tilde{R}}_{2l} = \hat{\tilde{p}}_l$. The angular brackets $\langle \hat{O}\rangle_\alpha = \mathrm{tr}(\rho_\alpha\hat{O})$ denote the expectation value of an operator $\hat{O}$ calculated with the density matrix ρ_α. The solution $W_\alpha(\tilde{\boldsymbol{R}}, t)$ corresponds to a Gaussian distribution centered at the origin in the co-moving frame. In the laboratory frame, the Wigner function is centered at the classical trajectory $\boldsymbol{\alpha}(t)$. However, due to the chaotic nature of the classical chimera state [9], our exact solution is just valid for short-time evolution.

16.5   Chimera-Like Quantum Correlations in the Covariance Matrix

Now let us study the consequences of the exact solution for the short-time evolution of the Wigner function, where the time scale is given by the inverse of the linear dissipation rate κ_1. Once we obtain the solution of the equations of motion Eq. (16.16), we can find the corresponding covariance matrix $\mathscr{C}(t)$. As we have defined in the introduction, a chimera state is characterized by the coexistence in space of synchronized and desynchronized dynamics. Therefore, in order to understand the quantum manifestations of a chimera state, we need to study also quantum signatures of synchronized and desynchronized dynamics.

   Although we consider different coupling strengths V, we use the same initial conditions as in Fig. 16.1 to obtain the chimera, and completely synchronized and desynchronized states. In order to obtain the snapshot of the chimera state depicted in Fig. 16.4a, we let the system evolve up to a time $\kappa_1 t_0 = 3000.5$ for a coupling strength $V = 1.2$. Correspondingly, to obtain the snapshot of the synchronized solution shown in Fig. 16.4b, we let the system evolve a time $\kappa_1 t_{\mathrm{Syn}} = 25.5$ for $V = 1.6$. Finally, the snapshot of the desynchronized state in Fig. 16.4c is obtained after a time evolution $\kappa_1 t_{\mathrm{desyn}} = 8000.5$ for $V = 0.8$. The left column of Fig. 16.4 show snapshots of the phases for (a) chimera, (b) completely synchronized, and (c) completely desynchronized mean-field solutions of Eq. (16.16). The central column of Fig. 16.4 depicts the corresponding covariance matrices after a short evolution time $\Delta t = 0.5/\kappa_1$. For every plot, we have initialized the system at time t_i as a tensor product of coherent states $|\alpha_l(t_i)\rangle$ centered at the positions $\alpha_l(t_i)$ of the individual oscillators. As a consequence, the covariance matrix at the initial time is diago-

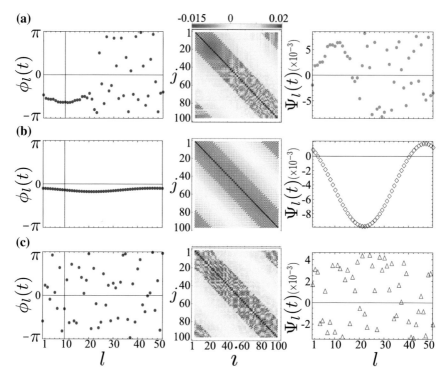

Fig. 16.4 Quantum fluctuations after a short-time evolution. Similarly to Fig. 16.3, we consider an initial density matrix $\rho(t_i)$ which is a tensor product of coherent states centered around the classical positions of the oscillators. Snapshots of the phase (*left column*) and covariance matrices (*central column*) after short-time evolution $\kappa_1 \Delta t = 0.5$ of the states: **a** chimera for $V = 1.2\kappa_1$, **b** synchronized state for $V = 1.6\kappa_1$, and **c** desynchronized state for $V = 0.8\kappa_1$. *Right column*: Weighted spatial average $\Psi_l(t)$ of the covariance matrix for the states shown in **a**, **b** and **c**, respectively. *Parameters*: $d = 10$, $\kappa_2 = 0.2\kappa_1$, and $N = 50$ [71]

nal $\mathscr{C}_{2l-1,2l-1}(t_i) = \langle \hat{\bar{q}}_l^2 \rangle_\alpha = \hbar/2$ and $\mathscr{C}_{2l,2l}(t_i) = \langle \hat{\bar{p}}_l^2 \rangle_\alpha = \hbar/2$, which reflects the Heisenberg uncertainty principle because $\langle \hat{\bar{q}}_l \rangle_\alpha = \langle \hat{\bar{p}}_l \rangle_\alpha = 0$.

After a short evolution time, quantum correlations are built up due to the coupling between the oscillators, and the covariance matrix exhibits a nontrivial structure which is influenced by the mean field solution. For example, the central panel of Fig. 16.4a shows a matrix plot of the covariance matrix corresponding to a chimera state obtained from the same initial condition as in Fig. 16.3. The covariance matrix acquires a block structure, where the upper 40×40 block (corresponding to nodes $l = 1, \ldots, 20$) shows a regular pattern matching the synchronized region of the chimera state. Similarly, the lower 60×60 block shows an irregular structure which corresponds to the desynchronized dynamics of the oscillators $l = 21, \ldots, 50$. In a similar fashion, Fig. 16.4b, c show the matrix $\mathscr{C}$ for completely synchronized and desynchronized states, respectively. In the case of a chimera state, this coincides with the results shown in Fig. 16.3, where the squeezing direction of the oscillators is related to the classical solution. In order to quantify these observations we define the weighted correlation as

$$\Psi_l(t) = \frac{V}{2d} \sum_{\substack{m=l-d \\ m \neq l}}^{l+d} \mathscr{C}_{2l,2m}(t). \tag{16.31}$$

This spatial average highlights the structure of the covariance matrix. The right column of Fig. 16.4 shows $\Psi_l(t)$ for (a) chimera, (b) synchronized, and (c) desynchronized states. The chimera state exhibits a regular and an irregular domain, exactly as the classical chimera does.

16.6 Quantum Mutual Information and Chimera States

Now let us consider a partition of the network into spatial domains of size L and $N - L$, which we call *Alice* (A) and *Bob* (B), respectively. This partition can be represented by considering a decomposition of the covariance matrix

$$\mathscr{C}(t) = \begin{pmatrix} \mathscr{C}_A(t) & \mathscr{C}_{AB}(t) \\ \mathscr{C}_{AB}^T(t) & \mathscr{C}_B(t) \end{pmatrix} \tag{16.32}$$

To study the interplay between synchronized and desynchronized dynamics, which is characteristic of a chimera state, we propose the use of an entropy measure [64, 65, 70]. Of particular interest is the Rényi entropy $S_\mu(\rho) = (1 - \mu)^{-1} \ln \mathrm{tr}(\rho^\mu)$, $\mu \in \mathbb{N}$, of the density matrix ρ, which is discussed in Ref. [75]. The Rényi entropy is of enormous interest in quantum information theory because it defines a family of additive entropies and enables us to quantify the amount of information of a quantum state ρ [75]. In particular, it converges to the von Neumann-Shannon entropy for $\mu \to 1$ and it is related to the purity of the quantum state for $\mu = 2$.

In terms of the Wigner representation of ρ_α, the Rényi entropy for $\mu = 2$ reads $S_2(\rho_\alpha) = -\ln\left[\int W_\alpha^2(\tilde{R}, t) d^{2N}\tilde{R}\right]$. Now let us consider the bipartite Gaussian state $\rho_{AB} = \rho_\alpha$ composed of *Alice* and *Bob* and define the tensor product $\rho_{Ref} = \rho_A \otimes \rho_B$ of the two marginals ρ_A and ρ_B.

To measure Gaussian Rényi-2 mutual information $\mathcal{I}_2(\rho_{A:B}) = S_2(\rho_A) + S_2(\rho_B) - S_2(\rho_{AB})$, we require the calculation of the relative sampling entropy between the total density matrix ρ_{AB} and the reference state ρ_{Ref} as shown in Ref. [75]. This leads to a formula $\mathcal{I}_2(\rho_{A:B}) = \frac{1}{2} \ln (\det \mathscr{C}_A \det \mathscr{C}_B / \det \mathscr{C})$ in terms of the covariance matrix Eq. (16.32). Figure 16.5a shows the variation of $\mathcal{I}_2(\rho_{A:B})$ as a function of the size L of the partition after an evolution time $\Delta t = 0.5/\kappa_1$. One can observe that for a

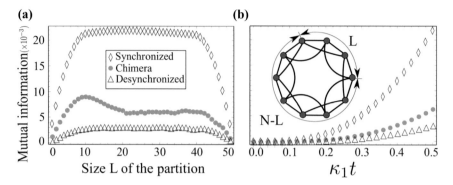

Fig. 16.5 Rényi quantum mutual information for the states shown in Fig. 16.4. The *green dots, blue diamonds*, and *purple triangles* represent the chimera, synchronized, and desynchronized states, respectively. **a** Gaussian Rényi-2 mutual information $\mathcal{I}_2(\rho_{A:B})$ as a function of the size L of *Alice* after an evolution time $\Delta t = 0.5/\kappa_1$. **b** The time evolution of the mutual information during the time interval Δt for a fixed size $L_c = 20$. *Inset*: Scheme of the nonlocally coupled network. *Parameters*: $d = 10$, $\kappa_2 = 0.2\kappa_1$, and $N = 50$ [71]

chimera state the mutual information is asymmetric as a function of L and there is a critical size $L_c = 20$, where a dramatic change of the correlations occurs.

Now let us consider the chimera state shown in Fig. 16.3, and let us consider a partition where the size of *Alice* is $L_c = 20$. Figure 16.5b shows the time evolution of mutual information for such a state. In addition, by using the same partition as for the chimera state, we calculate the mutual information for the synchronized and desynchronized states depicted in Fig. 16.4b, c, respectively. Our results reveal that the chimera state has a mutual information which lies between the values for synchronized and desynchronized states. This resembles the definition of a chimera state given at the beginning of the article.

16.7 Summary and Outlook

We have shown that quantum signatures of chimera states appear in the covariance matrix and in measures of mutual information. To quantify the structure of the covariance matrix, we have introduced a weighted spatial average of the quantum correlation, which reveals the nature of the classical trajectory, i.e., chimera, completely synchronized, or completely desynchronized state. The mutual information for a bipartite state $\mathcal{I}_2(\rho_{A:B})$ extends the definition of a chimera to the quantum regime and highlights the relation to quantum information theory. A possible experimental realization of our model could be carried out by means of trapped ions [77], as it was suggested in Ref. [59]. Other experimental possibilities include SQUID metamaterials [57] and Bose-Einstein condensation in the presence of dissipation and external driving [73, 74, 83]. In this context our approach is particularly interesting, because the continuum limit of the mean field Eq. (16.16) is a complex Ginzburg-Landau equation, which is nonlocal in space [2]. In this sense, our linearized master equation Eq. (16.14) enables us to study the Bogoliubov excitations above the mean field solution.

Acknowledgments V.M. Bastidas thanks L. M. Valencia and Y. Sato. The authors acknowledge inspiring discussions with J. Cerrillo, S. Restrepo, G. Schaller, and P. Strasberg. This work was supported by DFG in the framework of SFB 910.

Appendix

In this appendix we discuss the Gaussian quantum fluctuations of the quartic harmonic oscillator

$$H = \frac{p^2}{2m} + \frac{m\omega q^2}{2} + \frac{\lambda}{4}q^4 = \omega\left(a^\dagger a + \frac{1}{2}\right) + \lambda\left(\frac{1}{4m\omega}\right)^2 (a^\dagger + a)^4. \quad (16.33)$$

For simplicity, we consider units in such a way that $\hbar = 1$. As in the main text, we consider the decomposition $a = \tilde{a} + \alpha(t)$, where $\tilde{a}$ describes quantum fluctuations and $\alpha(t)$ is the mean field. We now consider the time-dependent displacement operator [81]

$$D[\alpha(t)] = \exp\left[\alpha(t)a^\dagger - \alpha^*(t)a\right] = \exp\left[\alpha(t)\tilde{a}^\dagger - \alpha^*(t)\tilde{a}\right]$$
$$= \exp\left[-\frac{|\alpha(t)|^2}{2}\right]\exp\left[\alpha(t)\tilde{a}^\dagger\right]\exp\left[-\alpha^*(t)\tilde{a}\right]. \quad (16.34)$$

Under this Gauge transformation, the Schrödinger equation $i\partial_t|\Psi(t)\rangle = H|\Psi(t)\rangle$ is transformed to $i\partial_t|\Psi_\alpha(t)\rangle = H^{(\alpha)}(t)|\Psi_\alpha(t)\rangle$, where $|\Psi_\alpha(t)\rangle = D^\dagger[\alpha(t)]|\Psi(t)\rangle$ and

$$\hat{H}^{(\alpha)}(t) = D^\dagger[\alpha(t)](H - i\partial_t)D[\alpha(t)]. \quad (16.35)$$

For later purposes, we need to use the identity

$$iD^\dagger[\alpha(t)]\partial_t D[\alpha(t)] = \frac{i}{2}[\dot{\alpha}(t)\alpha^*(t) - \alpha(t)\dot{\alpha}^*(t)] + i[\dot{\alpha}(t)\tilde{a}^\dagger - \dot{\alpha}^*(t)\tilde{a}]. \quad (16.36)$$

After some algebraic manipulations we can write

$$\hat{H}^{(\alpha)}(t) = \frac{\lambda}{16}\left(\frac{1}{m\omega}\right)^2 (\tilde{a}^\dagger + \tilde{a})^4 + \frac{1}{2}\left(\frac{1}{m\omega}\right)^2 (\tilde{a}^\dagger + \tilde{a})^3 \mathrm{Re}[\alpha(t)]$$

$$+ \omega\tilde{a}^\dagger\tilde{a} + \frac{3\lambda}{2}\left(\frac{1}{m\omega}\right)^2 (\tilde{a}^\dagger + \tilde{a})^2 (\mathrm{Re}[\alpha(t)])^2$$

$$- \mathrm{i}[\dot{\alpha}(t)\tilde{a}^\dagger - \dot{\alpha}^*(t)\tilde{a}] + \omega(\alpha^*\tilde{a} + \alpha\tilde{a}^\dagger) + 2\lambda\left(\frac{1}{m\omega}\right)^2 (\tilde{a}^\dagger + \tilde{a})(\mathrm{Re}[\alpha(t)])^3$$

$$- \frac{\mathrm{i}}{2}[\dot{\alpha}(t)\alpha^*(t) - \alpha(t)\dot{\alpha}^*(t)] + \omega|\alpha(t)|^2 + \lambda\left(\frac{1}{m\omega}\right)^2 (\mathrm{Re}[\alpha(t)])^4. \tag{16.37}$$

To study the quantum fluctuations about a semiclassical trajectory, we assume that $|\alpha(t)| \gg 1$. To obtain the quadratic fluctuations we must neglect the non-Gaussian terms in Eq. (16.37). In addition, the center of the co-moving frame $\alpha(t)$ must satisfy the condition

$$\dot{\alpha}(t) = -\mathrm{i}\left(\omega\alpha(t) + 2\lambda\left(\frac{1}{m\omega}\right)^2 (\mathrm{Re}[\alpha(t)])^3\right), \tag{16.38}$$

which corresponds to the classical equations of motion. This condition is satisfied if the linear terms in the quantum fluctuations $\tilde{a}$ of Eq. (16.37) vanish [78, 79]. We can go a step further and define the classical Hamiltonian function

$$H(\alpha, \alpha^*) = \omega|\alpha(t)|^2 + \lambda\left(\frac{1}{m\omega}\right)^2 (\mathrm{Re}[\alpha(t)])^4$$

$$= \omega\alpha^*(t)\alpha(t) + \lambda\left(\frac{1}{m\omega}\right)^2 \left[\frac{\alpha^*(t) + \alpha(t)}{2}\right]^4, \tag{16.39}$$

from which we obtain the equations of motion Eq. (16.38). In so doing we define the Poisson bracket $\{F(\alpha, \alpha^*), G(\alpha, \alpha^*)\} = -\mathrm{i}(\partial_\alpha F \partial_{\alpha^*} G - \partial_\alpha G \partial_{\alpha^*} F)$. By using this definition we obtain the equations of motion Eq. (16.38) as $\dot{\alpha}(t) = -\mathrm{i}\partial_{\alpha^*} H(\alpha, \alpha^*)$.

Now the role of each one of the terms in Eq. (16.37) is clear. In contrast, the quadratic terms give us the first quantum corrections about the semiclassical trajectory. To study these quantum fluctuations we need to study the quadratic Hamiltonian

$$\hat{H}_Q^{(\alpha)}(t) = \omega\tilde{a}^\dagger\tilde{a} + \frac{3\lambda}{2}\left(\frac{1}{m\omega}\right)^2 (\tilde{a}^\dagger + \tilde{a})^2 (\mathrm{Re}[\alpha(t)])^2 + L(\alpha, \alpha^*), \tag{16.40}$$

where $L(\alpha, \alpha^*) = -\frac{\mathrm{i}}{2}[\dot{\alpha}(t)\alpha^*(t) - \alpha(t)\dot{\alpha}^*(t)] + H(\alpha, \alpha^*)$ is the Lagrangian.

References

1. M.J. Panaggio, D.M. Abrams, Nonlinearity **28**, R67 (2015)
2. Y. Kuramoto, D. Battogtokh, Nonlin. Phen. Complex Syst. **5**, 380 (2002)
3. D.M. Abrams, S.H. Strogatz, Phys. Rev. Lett. **93**, 174102 (2004)
4. C.R. Laing, Phys. D **238**, 1569 (2009)
5. A.E. Motter, Nat. Phys. **6**, 164 (2010)
6. O.E. Omel'chenko, M. Wolfrum, Y.L. Maistrenko, Phys. Rev. E **81**, 065201 (2010)
7. O.E. Omel'chenko, M. Wolfrum, S. Yanchuk, Y.L. Maistrenko, O. Sudakov, Phys. Rev. E **85**, 036210 (2012)
8. E.A. Martens, Chaos **20**, 043122 (2010)
9. M. Wolfrum, O.E. Omel'chenko, S. Yanchuk, Y.L. Maistrenko, Chaos **21**, 013112 (2011)
10. T. Bountis, V. Kanas, J. Hizanidis, A. Bezerianos, Eur. Phys. J. Sp. Top. **223**, 721 (2014)
11. C.R. Laing, Phys. Rev. E **81**, 066221 (2010)
12. I. Omelchenko, A. Provata, J. Hizanidis, E. Schöll, P. Hövel, Phys. Rev. E **91**, 022917 (2015)
13. I. Omelchenko, Y.L. Maistrenko, P. Hövel, E. Schöll, Phys. Rev. Lett. **106**, 234102 (2011)
14. I. Omelchenko, B. Riemenschneider, P. Hövel, Y.L. Maistrenko, E. Schöll, Phys. Rev. E **85**, 026212 (2012)
15. I. Omelchenko, O.E. Omel'chenko, P. Hövel, E. Schöll, Phys. Rev. Lett. **110**, 224101 (2013)
16. J. Hizanidis, V. Kanas, A. Bezerianos, T. Bountis, Int. J. Bif. Chaos **24**, 1450030 (2014)
17. A. Vüllings, J. Hizanidis, I. Omelchenko, P. Hövel, New J. Phys. **16**, 123039 (2014)
18. I. Omelchenko, A. Zakharova, P. Hövel, J. Siebert, E. Schöll, Chaos **25**, 083104 (2015)
19. D.P. Rosin, D. Rontani, N.D. Haynes, E. Schöll, D.J. Gauthier, Phys. Rev. E **90**, 030902(R) (2014)
20. S.-I. Shima, Y. Kuramoto, Phys. Rev. E **69**, 036213 (2004)
21. E.A. Martens, C.R. Laing, S.H. Strogatz, Phys. Rev. Lett. **104**, 044101 (2010)
22. M.J. Panaggio, D.M. Abrams, Phys. Rev. Lett. **110**, 094102 (2013)
23. M.J. Panaggio, D.M. Abrams, Phys. Rev. E **91**, 022909 (2015)
24. G.C. Sethia, A. Sen, F.M. Atay, Phys. Rev. Lett. **100**, 144102 (2008)
25. Yu. Maistrenko, A. Vasylenko, O. Sudakov, R. Levchenko, V. Maistrenko, Int. J. Bif. Chaos **24**(8), 1440014 (2014)
26. J. Xie, E. Knobloch, H.-C. Kao, Phys. Rev. E **90**, 022919 (2014)
27. G.C. Sethia, A. Sen, G.L. Johnston, Phys. Rev. E **88**, 042917 (2013)
28. G.C. Sethia, A. Sen, Phys. Rev. Lett. **112**, 144101 (2014)
29. A. Zakharova, M. Kapeller, E. Schöll, Phys. Rev. Lett. **112**, 154101 (2014)
30. A. Yeldesbay, A. Pikovsky, M. Rosenblum, Phys. Rev. Lett. **112**, 144103 (2014)
31. L. Schmidt, K. Krischer, Phys. Rev. Lett. **114**, 034101 (2015)
32. F. Böhm, A. Zakharova, E. Schöll, K. Lüdge, Phys. Rev. E **91**, 040901(R) (2015)
33. T.-W. Ko, G.B. Ermentrout, Phys. Rev. E **78**, 016203 (2008)
34. M. Shanahan, Chaos **20**, 013108 (2010)
35. C.R. Laing, K. Rajendran, I.G. Kevrekidis, Chaos **22**, 013132 (2012)
36. N. Yao, Z.-G. Huang, Y.-C. Lai, Z. Zheng, Scientific Reports **3**, 3522 (2013)
37. Y. Zhu, Z. Zheng, J. Yang, Phys. Rev. E **89**, 022914 (2014)
38. A. Buscarino, M. Frasca, L.V. Gambuzza, P. Hövel, Phys. Rev. E **91**, 022817 (2015)
39. N.C. Rattenborg, C.J. Amlaner, S.L. Lima, Neurosci. Biobehav. Rev. **24**, 817 (2000)
40. C.R. Laing, C.C. Chow, Neural Comput. **13**, 1473 (2001)
41. H. Sakaguchi, Phys. Rev. E **73**, 031907 (2006)
42. A. Rothkegel, K. Lehnertz, New J. Phys. **16**, 055006 (2014)
43. A.E. Filatova, A.E. Hramov, A.A. Koronovskii, S. Boccaletti, Chaos **18**, 023133 (2008)
44. J.C. Gonzalez-Avella, M.G. Cosenza, M. San Miguel, Phys. A **399**, 24 (2014)
45. O.E. Omel'chenko, Nonlinearity **26**, 2469 (2013)
46. J. Sieber, O.E. Omel'chenko, M. Wolfrum, Phys. Rev. Lett. **112**, 054102 (2014)
47. C. Bick, E. Martens, New J. Phys. **17**, 033030 (2015)

48. A. Hagerstrom, T.E. Murphy, R. Roy, P. Hövel, I. Omelchenko, E. Schöll, Nat. Phys. **8**, 658 (2012)
49. M.R. Tinsley, S. Nkomo, K. Showalter, Nat. Phys. **8**, 662 (2012)
50. S. Nkomo, M.R. Tinsley, K. Showalter, Phys. Rev. Lett. **110**, 244102 (2013)
51. E.A. Martens, S. Thutupalli, A. Fourrière, O. Hallatschek, Proc. Nat. Acad. Sci. **110**, 10563 (2013)
52. L. Larger, B. Penkovsky, Y.L. Maistrenko, Phys. Rev. Lett. **111**, 054103 (2013)
53. L.V. Gambuzza, A. Buscarino, S. Chessari, L. Fortuna, R. Meucci, M. Frasca, Phys. Rev. E **90**, 032905 (2014)
54. L. Schmidt, K. Schönleber, K. Krischer, V. Garcia-Morales, Chaos **24**, 013102 (2014)
55. M. Wickramasinghe, I.Z. Kiss, PLoS ONE **8**, e80586 (2013)
56. E.A. Viktorov, T. Habruseva, S.P. Hegarty, G. Huyet, B. Kelleher, Phys. Rev. Lett. **112**, 224101 (2014)
57. N. Lazarides, G. Neofotistos, G.P. Tsironis, Phys. Rev. B **91**, 054303 (2015)
58. M. Kapitaniak, K. Czolczynski, P. Perlikowski, A. Stefanski, T. Kapitaniak, Phys. Rep. **517**, 1 (2012)
59. T.E. Lee, H.R. Sadeghpour, Phys. Rev. Lett. **111**, 234101 (2013)
60. S. Walter, A. Nunnenkamp, C. Bruder, Phys. Rev. Lett. **112**, 094102 (2014)
61. M. Ludwig, F. Marquardt, Phys. Rev. Lett. **111**, 073603 (2013)
62. I. Hermoso de Mendoza, L.A. Pachón, J. Gómez-Gardeñes, D. Zueco, Phys. Rev. E **90**, 052904 (2014)
63. A. Mari, A. Farace, N. Didier, V. Giovannetti, R. Fazio, Phys. Rev. Lett. **111**, 103605 (2013)
64. C. Weedbrook, S. Pirandola, R. García-Patrón, N.J. Cerf, T.C. Ralph, J.H. Shapiro, S. Lloyd, Rev. Mod. Phys. **84**, 621 (2012)
65. S.L. Braunstein, P. van Loock, Rev. Mod. Phys. **77**, 513 (2005)
66. G. Manzano, F. Galve, G. L. Giorgi, E. Hernández-García, R. Zambrini, Sci. Rep. **3** (2013)
67. G. Manzano, F. Galve, R. Zambrini, Phys. Rev. A **87**, 032114 (2013)
68. T.E. Lee, C.K. Chan, S. Wang, Phys. Rev. E **89**, 022913 (2014)
69. G.L. Giorgi, F. Plastina, G. Francica, R. Zambrini, Phys. Rev. A **88**, 042115 (2013)
70. V. Ameri, M. Eghbali-Arani, A. Mari, A. Farace, F. Kheirandish, V. Giovannetti, R. Fazio, Phys. Rev. A **91**, 012301 (2015)
71. V. Bastidas, I. Omelchenko, A. Zakharova, E. Schöll, T. Brandes, Phys. Rev. E **92**, 062924 (2015)
72. D. Viennot, L. Aubourg, Phys. Lett. A 380 (2016)
73. L.M. Sieberer, S.D. Huber, E. Altman, S. Diehl, Phys. Rev. Lett. **110**, 195301 (2013)
74. U.C. Täuber, S. Diehl, Phys. Rev. X **4**, 021010 (2014)
75. G. Adesso, D. Girolami, A. Serafini, Phys. Rev. Lett. **109**, 190502 (2012)
76. B. Van der Pol, Radio Rev. **1**, 701 (1920)
77. D. Leibfried, R. Blatt, C. Monroe, D. Wineland, Rev. Mod. Phys. **75**, 281 (2003)
78. D.A. Rodrigues, A.D. Armour, Phys. Rev. Lett. **104**, 053601 (2010)
79. N. Lörch, J. Qian, A. Clerk, F. Marquardt, K. Hammerer, Phys. Rev. X **4**, 011015 (2014)
80. H.J. Carmichael, *Statistical Methods in Quantum Optics 2: Non-Classical Fields* (Springer Science & Business Media, 2009)
81. K.E. Cahill, R.J. Glauber, Phys. Rev. **177**, 1857 (1969)
82. A. Altland, F. Haake, Phys. Rev. Lett. **108**, 073601 (2012)
83. J. Kasprzak, M. Richard, S. Kundermann, A. Baas, P. Jeambrun, J. Keeling, F. Marchetti, M. Szymańska, R. Andre, J. Staehli et al., Nature **443**(7110), 409 (2006)

Chapter 17
Multirhythmicity for a Time-Delayed FitzHugh-Nagumo System with Threshold Nonlinearity

Lionel Weicker, Lars Keuninckx, Gaetan Friart, Jan Danckaert and Thomas Erneux

Abstract A time-delayed FitzHugh-Nagumo (FHN) system exhibiting a threshold nonlinearity is studied both experimentally and theoretically. The basic steady state is stable but distinct stable oscillatory regimes may coexist for the same values of parameters (multirhythmicity). They are characterized by periods close to an integer fraction of the delay. From an asymptotic analysis of the FHN equations, we show that the mechanism leading to those oscillations corresponds to a limit-point of limit-cycles. In order to investigate their robustness with respect to noise, we study experimentally an electrical circuit that is modeled mathematically by the same delay differential equations. We obtain quantitative agreements between numerical and experimental bifurcation diagrams for the different coexisting time-periodic regimes.

17.1 Introduction

Excitable systems play important roles in biology and medicine. Phenomena such as the transmission of impulses between neurons, the cardiac arrhythmia, the aggregation of amoebas, the appearance of organized structures in the cortex of egg cells, all derive from the activity of excitable media [1–3]. The classical example of an excitable phenomenon is the firing of a nerve. According to the Hodgkin and Huxley (HH) equations [1, 4] a sub-threshold depolarization dies away monotonically, but a super-threshold depolarization initiates a spike potential. FitzHugh and

L. Weicker (✉) · G. Friart · T. Erneux
Optique Nonlinéaire Théorique, Université Libre de Bruxelles,
Campus Plaine, CP 231, 1050 Bruxelles, Belgium
e-mail: lionel.weicker@centralesupelec.fr

L. Weicker · L. Keuninckx · J. Danckaert
Applied Physics Research Group (APHY), Vrije Universiteit Brussel,
1050 Brussels, Belgium

L. Weicker
LMOPS, CentraleSupélec, Université Paris-Saclay, 57070 Metz, France

L. Weicker
LMOPS, CentraleSupélec, Université de Lorraine, 57070 Metz, France

© Springer International Publishing Switzerland 2016
E. Schöll et al. (eds.), *Control of Self-Organizing Nonlinear Systems*,
Understanding Complex Systems, DOI 10.1007/978-3-319-28028-8_17

Nagumo (FHN) [5, 6] later formulated a simplified version of the HH equations that describes the essential features of the nerve impulse in terms of two differential equations. The phase-plane analysis of the possible trajectories clarifies the conditions for excitability. A successful spike is generated only if a perturbation from the rest state surpasses a critical threshold.

The effects of time delays in neurosystems have recently attracted a lot of attention [7]. Delays are inherent in neuronal networks due to finite conduction velocities and synaptic transmission. Small neurons transmit over short distances <1 mm at velocities <2 m/s. Large neurons transmit over longer distances (cm to meters) at velocities of 10–100 m/s [8]. Specific synchronization or desynchronization patterns are essential for neural functioning and they have been investigated by formulating network models. Early studies considered coupled phase oscillator systems [9–12] that allowed analytical results. Biologically more realistic network models are now explored showing how time delays affect the structural heterogeneity of the network [13–16]. While most studies concentrated on populations of coupled limit-cycle oscillators, work has also been done on coupled excitable units. The case of two delayed coupled FHN systems has been examined in detail showing that stable periodic oscillations may coexist with a stable steady state [17–19]. The bifurcation phenomenon is fully induced by the delay τ and represents a new form of oscillatory synchronization exhibiting a period close to 2τ. Physically, the delayed coupling allows the sequential spiking of the two cells by controlling the timing of each pulse. In [20], we applied asymptotic techniques appropriate for slow-fast systems and constructed periodic solutions of a two delayed-coupled FHN system. We found that in addition to the 2τ-periodic solution, there exist periodic solutions of period $2\tau/n$ where $n = 1, 2, \ldots$ for the same values of the parameters. We experimentally investigated their robustness with respect to noise by designing an electronic circuit that simulates our two coupled FHN system. In this chapter, we consider only one FHN system subject to a delayed feedback and wonder if a stable periodic solution may still be an alternate to a stable steady state. This question was recently raised by Hövel (Sect. 6.3 in [21]). His work was motivated by earlier studies of Schöll and coworkers [22–24] who explored the effect of the delayed feedback on noise-induced oscillations. Here, we deliberately consider a delayed FHN problem where no Hopf bifurcation is possible even in the presence of a delayed feedback. We however anticipate that a periodic solution may appear through a limit-point of limit-cycles. Specifically, we consider the following FHN equations

$$\varepsilon x' = -x - y + H[x(t - \tau) - a], \qquad (17.1)$$
$$y' = x, \qquad (17.2)$$

where $H(x)$ is the Heaviside step function. $0 < a < 1/2$ is a threshold parameter for the onset of pulses. $\varepsilon \ll 1$ is a small parameter which implies that x is fast compared to y. $\tau = O(1)$ is the delay of the feedback. The presence of a threshold nonlinearity means that Eqs. (17.1) and (17.2) can effectively be treated as a piece-wise linear system. The study of piece-wise FHN systems has allowed for important advances in the understanding of excitable systems when diffusion is included [1, 25, 26] and/or delay [26, 27].

The multiplicity of periodic solutions of delay differential equations (DDE) is not a new phenomenon for oscillators subject to a delayed feedback. It has been shown for specific problems where the steady state is unstable that the period of the limit-cycle oscillations exhibits multiple hysteresis loops as the delay increases [28]. It is a generic phenomenon for a large class of DDEs [29]. Here we consider a slow-fast system with a stable steady state and with an arbitrary delay. We have found numerically that the coexistence of periodic solutions persist even for small delays $[\tau = O(\varepsilon)]$.

The steady state $(x, y) = (0, 0)$ is a stable focus whatever the value of τ. By contrast to the analysis in [20], we do not immediately take advantage of the small parameter ε but construct a periodic solution by combining two partial solutions valid for $x(t - \tau) < a$ and $x(t - \tau) > a$, respectively. We obtain transcendental equations for key properties of the solution that we then analyze in terms of parameter a. We show that the bifurcation mechanism for their emergence is a limit-point of limit-cycles.

The organization of the chapter is as follows. In Sect. 17.2.1, we investigate Eqs. (17.1) and (17.2) numerically and highlight the multirhythmicity phenomenon. More precisely, we observe stable periodic solutions characterized by different periods for the same values of the fixed parameters. To this end, we use different initial functions for the delayed variable x. In Sect. 17.2.2, we construct a time-periodic solution of Eqs. (17.1) and (17.2), and numerically determine bifurcation diagrams for all the periodic regimes found in Sect. 17.2.1. The bifurcation mechanism is then investigated analytically in Sect. 17.2.3 where we explore the limit $\varepsilon \to 0$. Section 17.3 is devoted to experiments on an electronic circuit described by Eqs. (17.1) and (17.2). We show that the periodic solutions predicted theoretically are robust to noise and we compare quantitatively experimental and numerical bifurcation diagrams. Finally, we summarize our main results in Sect. 17.4 and discuss how a specific regime has been selected in real optical devices.

17.2   Theory

17.2.1   Numerical Observations

Equations (17.1) and (17.2) admit only one steady state solution $(x, y) = (0, 0)$. This state is always stable but under specific initial conditions and parameter values, we observe different coexisting stable time-periodic solutions (multirhythmicity). See Fig. 17.1. The different regimes have been obtained using the following initial conditions

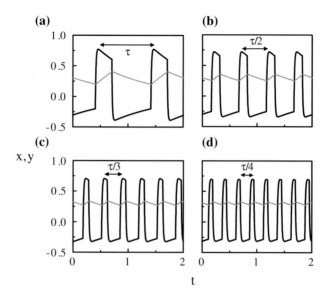

Fig. 17.1 Time traces for $x(t)$ (*black*) and $y(t)$ (*gray*) obtained numerically from Eqs. (17.1) and (17.2) with the initial conditions given by (17.3) and (17.4). The values of the fixed parameters are $\tau = 1$, $\varepsilon = 10^{-2}$, and $a = 0.2$. From (**a**) to (**d**), the figure shows different periodic solutions characterized by their period $T_n \simeq \tau/n$ $(n = 1, \ldots, 4)$

$$x(t) = \cos\left(\frac{2\pi nt}{\tau}\right) \quad (-\tau < t \leq 0), \tag{17.3}$$

$$y(0) = 0, \tag{17.4}$$

where n is an integer. Increasing n leads to periodic solutions with smaller periods and smaller orbits in the phase-plane. Figure 17.2 shows the τ-periodic limit-cycle in the phase-plane. It consists of two slowly varying parts following the slow manifold (broken lines)

$$y = -x + H(x - a) \tag{17.5}$$

connected by two fast transition layers at nearly constant values of y.

Each periodic solution is characterized by its period $T_n \simeq \tau/n$. Oscillations with smaller periods have also been found but are not shown for clarity. The extrema of x do not change significantly as n increases. The change is more dramatic if we examine the extrema of y and the bifurcation diagram is shown in Fig. 17.3a in terms of the extrema of y using a as the bifurcation parameter. The bifurcation diagram is obtained by a continuation method i.e., we integrate Eqs. (17.1) and (17.2) for a large interval of time changing a by steps and by using the previous solution as the new initial function.

We observe that the different solutions exist up to a critical value a_n^c. Beyond this value, the system either jumps to a periodic state exhibiting a larger period or to the

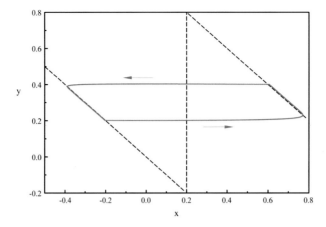

Fig. 17.2 Limit-cycle in the (x, y) phase-plane together with the slow manifold (17.5) (*broken lines*). Same values of the fixed parameters as in Fig. 17.1 and $n = 1$

stable steady state. The critical values a_n^c are listed in Table 17.1. We note that the a_n^c decreases as n increases which suggests that a large number of coexisting regimes are more likely to be found if a is close to 0. Table 17.1 also indicates the values obtained experimentally using an electronic circuit (see Sect. 17.3). The agreement is clearly quantitative. The bifurcation diagrams suggests that each branch of periodic solutions terminates at a limit point of limit-cycles. In order to verify this bifurcation mechanism, we construct an analytical solution in the next section.

17.2.2   Construction of the Periodic Solution

The numerical time integrations depend on the basins of attractions of each periodic solution as well as their linear stability properties. Transients can be very long near the limit points of periodic solutions which limit the accuracy of our numerical solutions. In this section, we propose an alternative method based on an analytical construction of each periodic solution.

Because Eqs. (17.1) and (17.2) are piecewise linear ordinary differential equations, we may construct a periodic solution by connecting two separate expressions valid for $x(t - \tau) < a$ and $x(t - \tau) > a$, respectively. We obtain strongly nonlinear transcendental equations for different unknown quantities. They will be solved numerically for a fixed value of ε and analytically in the limit ε small. Figure 17.4 shows the periodic solution of Fig. 17.1b for both $x(t)$ and $x(t - \tau)$. Time $t = 0$ is chosen as the time where $x(t - \tau) - a$ becomes positive causing a sudden increase of $x(t)$. Time $t = t_1$ is defined as the time where $x(t - \tau) - a$ becomes negative now causing a decrease of $x(t)$. Time $t = t_2$ is the total period where $x(t - \tau) - a$ is again positive. We assume that the period is given by

Fig. 17.3 Extrema of y as functions of parameter a for each T_n-periodic solution ($T_n \simeq \tau/n$) for $n = 1$ (*black*), $n = 2$ (*red*), $n = 3$ (*blue*), and $n = 4$ (*orange*). The critical points a_n^c mark the point where the T_n-periodic solution is no more observed. **a** Bifurcation diagrams of the stable periodic solutions of Eqs. (17.1) and (17.2). The fixed parameters are $\tau = 1$ and $\varepsilon = 10^{-2}$. **b** Bifurcation diagrams obtained experimentally using the electronic circuit described in Sect. 17.3. **c** Extrema of y obtained from the analytical construction of the periodic solutions (see Sect. 17.2.2) for the same values of the fixed parameters

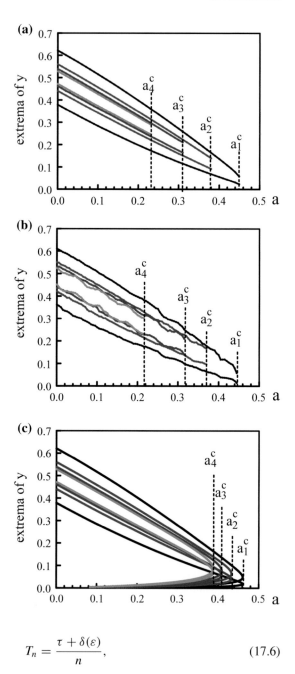

$$T_n = \frac{\tau + \delta(\varepsilon)}{n}, \tag{17.6}$$

where $\delta/n = O(\varepsilon)$ is defined as the small correction of τ/n. Consequently,

$$x(t - \tau) = x(t - nT_n + \delta) = x(t + \delta). \tag{17.7}$$

Table 17.1 a_n^c obtained numerically and experimentally for the T_n-periodic regimes

n	a_n^c numerical	a_n^c experimental
1	0.45	0.45
2	0.38	0.38
3	0.31	0.31
4	0.23	0.22

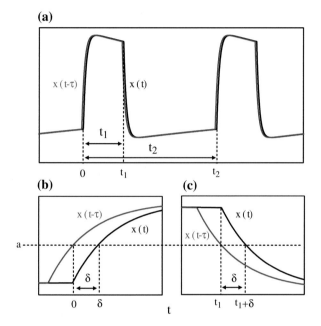

Fig. 17.4 Time series corresponding to Fig. 17.1b showing both $x(t)$ and $x(t-\tau)$. Times $t = 0, t_1$, and t_2 mark the points where $x(t-\tau) - a$ first becomes positive, then negative, and again positive. **b** and **c** are blow-ups of the fast transition layers at $t = 0$ and t_1, respectively

Equation (17.7) implies that $x(-\tau) = x(\delta)$ and $x(t_1 - \tau) = x(t_1 + \delta)$, as illustrated in Fig. 17.4b and c. In addition to the amplitude and waveform of the solution, we need to determine t_1 and δ.

17.2.2.1 $0 < t < t_1$

During the time interval $0 < t < t_1$, $x(t-\tau) > a$ and the Heaviside function is equal to 1 (see Fig. 17.4b). The equations for x and y are then given by

$$\varepsilon x' = -x - y + 1, \ y' = x. \tag{17.8}$$

They admit the solutions

$$x = Ae^{\lambda_+ t} + Be^{\lambda_- t}, \tag{17.9}$$

$$y = 1 - (1 + \varepsilon\lambda_+)Ae^{\lambda_+ t} - (1 + \varepsilon\lambda_-)Be^{\lambda_- t}, \tag{17.10}$$

where A and B are integration constants and

$$\lambda_\pm = \frac{-1 \pm \sqrt{1 - 4\varepsilon}}{2\varepsilon}. \tag{17.11}$$

17.2.2.2 $t_1 < t < t_2$

During the time interval $t_1 < t < t_2$, $x(t - \tau) < a$ and the Heaviside function is equal to 0 (see Fig. 17.4c). The equations for x and y now are

$$\varepsilon x' = -x - y, \ y' = x \tag{17.12}$$

and admit the solutions

$$x = Ce^{\lambda_+(t-t_1)} + De^{\lambda_-(t-t_1)}, \tag{17.13}$$

$$y = -(1 + \varepsilon\lambda_+)Ce^{\lambda_+(t-t_1)} - (1 + \varepsilon\lambda_-)De^{\lambda_-(t-t_1)}, \tag{17.14}$$

where C and D are two new constants of integration. Our problem depends on seven unknowns namely A, B, C, D, δ, t_1, and t_2. To obtain additional equations for these unknowns, we apply connection conditions. The details are relegated in the Appendix. We obtain a single equation relating t_1 and δ given by

$$0 = \frac{2 - e^{\lambda_+ t_{21}} - e^{\lambda_+ t_1}}{1 - e^{\lambda_+ t_2}} + \left(\frac{e^{\lambda_- t_{21}} - 2 + e^{\lambda_- t_1}}{1 - e^{\lambda_- t_2}} \right) e^{(\lambda_- - \lambda_+)\delta}. \tag{17.15}$$

If we fix t_1 and n, we may determine δ numerically from Eq. (17.15) using the dichotomy method. From (17.47), we then evaluate t_2. The coefficients A and B are obtained from (17.44) and (17.39), respectively. The value of a is computed by using (17.45). By taking the derivatives of (17.10) and (17.14), we determine the extrema of y for a single value of t_1 and n. If we apply the same procedure for different values of t_1 and n, we obtain bifurcation diagrams where a is the bifurcation parameter.

Figure 17.3c represents the extrema of y as function of the parameter a for different time-periodic regimes ($n = 1, 2, 3$ and 4). We clearly note that they emerge from limit-points of limit-cycles located at $a = a_n^c$. Comparing Fig. 17.3a and c, we note that the extrema obtained numerically by integrating Eqs. (17.1) and (17.2) start to slightly deviate from the extrema obtained by the analytical construction in the vicinity of the limit points. Moreover, there are significant differences between analytical and numerical estimates of the a_n^c. It suggests a possible change of stability of the branches of periodic solutions near the limit points although we didn't find any numerical evidence of a secondary bifurcation to quasi-periodic oscillations. It is most likely that the periodic solutions are weakly stable in the vicinity of the limit points.

17.2.3 The Limit $\varepsilon \to 0$

In order to further progress analytically, we investigate the asymptotic limit $\varepsilon \to 0$. For $\tau = O(1)$, $\varepsilon \to 0$ and $a \to a_n^c$, we note that $|\lambda_+| = O(1)$, $|\lambda_-| = O(\varepsilon^{-1})$, $t_1 = O(|\varepsilon \ln(\varepsilon)|) \ll 1$, and $t_2 = (\tau + \delta)/n = O(1)$. The two exponentials $\exp(\lambda_- t_{21})$ and $\exp(\lambda_+ t_{21})$ are then $O\left[\exp(-\varepsilon^{-1})\right]$ small quantities and Eq. (17.50) reduces to

$$1 - \lambda_+ t_1 + \frac{\lambda_+^2 t_1^2}{2} \frac{1 + e^{\lambda_+ \tau/n}}{1 - e^{\lambda_+ \tau/n}} + \left(-2 + e^{\lambda_- t_1}\right) e^{(\lambda_- - \lambda_+)\delta} = 0. \qquad (17.16)$$

From (17.16), we extract δ as

$$
\begin{aligned}
\delta &= \frac{1}{(\lambda_- - \lambda_+)} \ln\left(\frac{1 - \lambda_+ t_1}{2 - e^{\lambda_- t_1}}\right), \\
&\simeq \frac{\ln(2)}{(\lambda_+ - \lambda_-)}. \qquad (17.17)
\end{aligned}
$$

From (17.49) and using (17.16) and (17.17) in order to eliminate $\exp\left[(\lambda_- - \lambda_+)\delta\right]$ and $\exp(-\lambda_+ \delta)$, we obtain

$$a = \frac{1}{\varepsilon\left[\lambda_+(1 - \ln(2)) - \lambda_-\right]}\left[\frac{1}{2} + \lambda_+ t_1\left(\frac{1}{1 - e^{\lambda_+ \tau/n}} - \frac{1}{2}\right) - \frac{e^{\lambda_- t_1}}{4}\right]. \qquad (17.18)$$

We wish to find a limit point of limit-cycles. To this end, we analyze the condition $da/dt_1 = 0$. From Eq. (17.18), we find

$$t_{1c} = \frac{1}{\lambda_-} \ln\left[2\frac{\lambda_+}{\lambda_-}\left(\frac{1 + e^{\lambda_+ \tau/n}}{1 - e^{\lambda_+ \tau/n}}\right)\right]. \qquad (17.19)$$

Using $\lambda_+ \simeq -1$ and $\lambda_- \simeq -\varepsilon^{-1} + 1$, we simplify Eqs. (17.18) and (17.19), and obtain

$$a = \frac{1}{[1 - 2\varepsilon + \varepsilon \ln(2)]}\left[t_1\left(\frac{1}{2} - \frac{1}{1 - e^{-\tau/n}}\right) + \frac{1}{2} - \frac{e^{-\varepsilon^{-1} t_1}}{4}\right], \qquad (17.20)$$

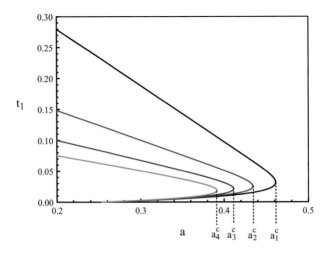

Fig. 17.5 The different *curves* represent the interval of time t_1 as functions of a for $n = 1$ (*black*), $n = 2$ (*red*), $n = 3$ (*blue*), and $n = 4$ (*orange*). They are obtained from Eq. (17.20) which is the leading approximation of the exact conditions for ε small. For each periodic solution, we note two branches that emerge from a limit point

and

$$t_{1c} = -\varepsilon \ln \left[2\varepsilon \left(\frac{1 + e^{-\tau/n}}{1 - e^{-\tau/n}} \right) \right]. \tag{17.21}$$

The critical values a_n^c are obtained by inserting (17.21) into (17.20). Figure 17.5 is obtained from Eq. (17.20) and represents the time t_1 as function of a for $n = 1, 2, 3$, and 4. As in Fig. 17.3c, we observe that oscillations of period τ/n beyond a_n^c are no more possible. Moreover, we note that the critical values obtained by evaluating numerically the analytical conditions valid for arbitrary ε or their approximations for ε small are very close. They are listed in the Table 17.2.

Table 17.2 a_n^c obtained analytically (exactly and in the limit $\varepsilon \to 0$) for the T_n-periodic regimes

n	a_n^c exact	a_n^c ($\varepsilon \to 0$)
1	0.462	0.461
2	0.434	0.433
3	0.411	0.409
4	0.391	0.388

17.3 Experiments

17.3.1 Circuit

In order to test the experimental accessibility of our theoretical results, we have build a nonlinear electronic circuit that simulate our FHN system (see Fig. 17.6).

Assuming the values of R_1, R_2 and R_3 are sufficiently close to each other, the evolution equations for V_x and V_y are given by

$$R_2 C_2 \frac{dV_x}{dt} = -V_x - V_y - V_z + 3V_{ref}, \tag{17.22}$$

$$\frac{dV_y}{dt} = \frac{V_x}{R_4 C_4} - \frac{V_{ref}}{R_4 C_4} + \frac{V_{ref}}{R_5 C_4} - \frac{V_y}{R_5 C_4}, \tag{17.23}$$

while

$$V_z = -V_{ref} H(V_x(T - T_D) - V_a) + V_{ref}.$$

The Heaviside function is accomplished by comparator U_{3a}. The circuit was built on a 'breadboard' and connected to a commercially available Digilent Nexys 2 Field Programmable Gate Array (FPGA) board. The FPGA board is programmed as a digital delay line with provisions for storing and generating signals. The interfacing between the digital FPGA and the analog circuit is done by using several 'PMOD'

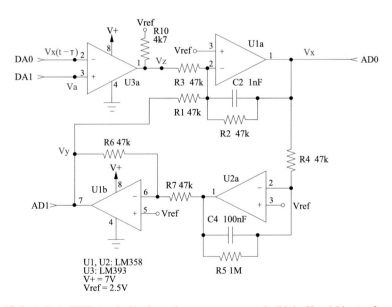

Fig. 17.6 A single FHN circuit. U_{3a} is used as a comparator to build the Heaviside step function. The delay is built using an FPGA equipped with AD and DA converters (not shown)

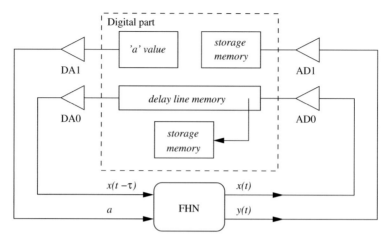

Fig. 17.7 Schematical view showing how the data flows in the experiment

analog-to-digital (AD) and digital-to-analog (DA) plug-in modules, also from Digilent. Figure 17.7 shows how the digital part is programmed and connected to the
analog FHN circuit. $x(t)$ is read by AD0 and a delayed version $x(t - \tau)$ is output by
DA1. In between is a patch of memory, programmed as a digital delay line. At the
same time another patch of memory is used to store a time trace of $x(t)$. DA1 is used
to output the value of a, which is fixed in each run of the experiment. Before each run
of the experiment, the initial function (17.3) is loaded in the delay line memory. $y(t)$
is read by AD1 and stored in another section of memory. A Python script running on
a host PC is responsible for controlling the experiment, initializing the FPGA board,
downloading the data, etc.

17.3.2 Voltage and Time Scaling

In order to compare experimental measures with theoretical results, we reformulate
Eqs. (17.22) and (17.23) in dimensionless form. Numerical simulations show the
dimensionless variables to be within the interval $[-1, +1]$. The micro-controller
uses voltages between 0 and 5 V as output and input on the DA and AD converters
respectively. Therefore, $V_{\text{ref}} = 2.5$ V was chosen as 'zero'. We introduce the new
variables x, y, and z as

$$x, y, z = \frac{V_{x,y,z} - V_{\text{ref}}}{V_{\text{ref}}}. \tag{17.24}$$

V_a is generated by a DA1 and its value is between 2.5 and 5 V. This enables us
to vary V_a in the Python script. The internal representation is signed sixteen bit,
having values between $-32{,}768$ and $32{,}767$ with 0 being V_{ref}. Comparison between

Table 17.3 Measured values of different components

Component	Nominal	Measured
R_4	$47\,k\Omega$	$46.8\,k\Omega$
C_4	$100\,nF$	$104.9\,nF$
R_2	$47\,k\Omega$	$46.9\,k\Omega$
C_2	$1\,nF$	$1.02\,nF$

Eqs. (17.22) and (17.23), and the system equations shows that real and dimensionless times are related as

$$t_{real} = R_4 C_4 t_{dimensionless}. \tag{17.25}$$

Inserting (17.24) and (17.25) into Eqs. (17.22) and (17.23) leads to

$$\varepsilon x' = -y - x + H[x(t - \tau) - a], \tag{17.26}$$

$$y' = x - y \left(\frac{R_4}{R_5}\right), \tag{17.27}$$

where

$$\varepsilon = \frac{R_2 C_2}{R_4 C_4}.$$

In order to properly obtain the Heaviside function, R_{10} must be much smaller than R_3 to assure that V_z goes to V_{ref} during the phase that the output of the comparator is not sinking current. On the other hand, R_{10} cannot be too small so that the comparator can adequately bring V_z low output during the other phase.

The ratio $R_4/R_5 \approx 0.006$ is sufficiently small to ignore the last term in Eq. (17.27). Equations (17.26) and (17.27) then have the same form as Eqs. (17.1) and (17.2). We choose $R_2 C_2 = 50\,\mu s$ and $R_4 C_4 = 5\,ms$ to fix ϵ to 0.01. The digital delay is programmed to sample at $T_s = 5\,\mu s$. Since $\tau = NT_s/(R_4 C_4)$, the length of the delay line is $N = 1000$ samples.

To compensate for tolerances, actual component values were measured (see Table 17.3) and the delay line length N set appropriately to yield $\varepsilon \approx 0.01$ and $\tau \approx 1$. R_1, R_2 and R_3 were hand selected to have values very close to each other. Since

$$\tau = \frac{NT_s}{R_4 C_4} = 1, \tag{17.28}$$

the delay line length N was set to

$$N = \tau \frac{R_4 C_4}{T_s} = \frac{1 \cdot 4.9093\,ms}{5\,\mu s} = 981.86. \tag{17.29}$$

Because N is an integer, we choose $N = 982$ which gives an actual $\tau = 1.0001$. The actual value for ε is

$$\varepsilon = \frac{R_2 C_2}{R_4 C_4} = \frac{47.738\,\mu s}{4.9093\,ms} = 0.00974 \approx 0.01. \tag{17.30}$$

17.3.3 Experimental Results

A scan over the parameter $a = 0 \ldots 0.5$ was made for $n = 1, 2, 3$ and 4. Each value for a and n was run for a time of 10 s, equivalent to 2000τ, to allow the circuit to stabilize. The reported extrema of y are the averages of the extrema over the last 10 delay line recordings. See Fig. 17.3b. For $n = 4$, the running time was increased by 1 min to conclude for stability, since even after 10 s, a jump could be observed. The critical values a_n^c for different n are listed in Table 17.1.

17.4 Discussion

In this chapter, we performed a theoretical and experimental study of a time-delayed FHN system with threshold nonlinearity. Different coexisting stable periodic regimes are observed for the same values of the parameters (multirhythmicity). The period of the oscillations is close to $T_n \simeq \tau/n$, where τ is the delay and n is a positive integer.

From numerical simulations of Eqs. (17.1) and (17.2), we found that these solutions exist from $a = 0$ to a critical value $a = a_n^c$. Beyond this point, the oscillations either jump to another oscillatory state with a larger period (smaller n) or to the stable steady state. We also noted that a_n^c decreases as n increases. In order to test their robustness with respect to noise, we built a system that is described mathematically by the same FHN equations. We obtained quantitative agreement between numerical and experimental bifurcation diagrams.

An analytical construction of the T_n-periodic solution is possible and leads to a nonlinear transcendental equation. Using the dichotomy method, we determined the bifurcation diagram and showed that the periodic solutions emerge from limit-points of limit-cycles. Simple analytic expressions for the location of the limit points are derived by considering the limit $\varepsilon \to 0$. Although numerical and experimental bifurcation diagrams are in good quantitative agreement, the point where a specific periodic solution disappears doesn't match the computed limit point. This difference is likely due to the weak stability of the solutions near the limit points. However, a possible instability near the limit point cannot be ruled out (see [29] for a simple example). We believe that this multirhythmicity is generic to a large class of slow-fast delay systems. Recently, coexistence of stable oscillations of period close to τ/n have been observed numerically and experimentally for an optoelectronic oscillator [30] as well as for a laser subject to polarization rotated feedback [31].

The stable periodic solutions however emerge from Hopf bifurcation points rather than limit points. The mathematical model for the optoelectronic oscillator displays quite similar equations as our FHN system. It is a two variables slow/fast system with an S-shaped slow manifold. The small parameter ε results from the large delay of the feedback loop.

For applications, the selection of a specific periodic solution is an important issue. In [30], a pattern generator is included in the electric part of the optoelectronic oscillator feedback loop. The system is then excited initially with a signal exhibiting the chosen frequency. In [31], a weak conventional optical feedback is added to the laser subject to a polarization rotated feedback. By controlling the delay of the feedback control with respect to the delay of the rotated feedback a specific square-wave with the desired period can be generated.

Acknowledgments L.W. acknowledges the Belgian F.R.I.A., the Conseil Régional de Lorraine, and the Agence Nationale de la Recherche (ANR) TINO project (ANR-12-JS03-005). G.F. acknowledges the Belgian F.R.I.A. T.E. acknowledges the support of the F.N.R.S. This work also benefited from the support of the Belgian Science Policy Office under Grant No IAP-7/35.

Appendix

A Connection at $t = t_1$ and $t = t_2$

We first require that the solutions (17.9)–(17.10) and (17.13)–(17.14) are equal at the critical times $t = t_1$ and $t = t_2$. This leads to the following four equations

$$C + D = Ae^{\lambda_+ t_1} + Be^{\lambda_- t_1}, \tag{17.31}$$

$$1 - (1 + \varepsilon\lambda_+)Ae^{\lambda_+ t_1} - (1 + \varepsilon\lambda_-)Be^{\lambda_- t_1} = -(1 + \varepsilon\lambda_+)C - (1 + \varepsilon\lambda_-)D, \tag{17.32}$$

$$Ce^{\lambda_+ t_{21}} + De^{\lambda_- t_{21}} = A + B, \tag{17.33}$$

$$-(1 + \varepsilon\lambda_+)Ce^{\lambda_+ t_{21}} - (1 + \varepsilon\lambda_-)De^{\lambda_- t_{21}} = 1 - (1 + \varepsilon\lambda_+)A - (1 + \varepsilon\lambda_-)B, \tag{17.34}$$

where $t_{21} \equiv t_2 - t_1$. We now determine the constants A, B, C, and D as functions of t_1 and t_2.

From Eqs. (17.31) and (17.33), we determine

$$Ae^{\lambda_+ t_1} = C + D - Be^{\lambda_- t_1}, \tag{17.35}$$

$$Ce^{\lambda_+ t_{21}} = A + B - De^{\lambda_- t_{21}}. \tag{17.36}$$

Inserting these expressions of $A \exp(\lambda_+ t_1)$ and $C \exp(\lambda_+ t_{21})$ into Eqs. (17.32) and (17.34), respectively, leads to two coupled equations for B and D

$$Be^{\lambda_- t_1} = D - \frac{1}{\varepsilon\,(\lambda_+ - \lambda_-)}, \tag{17.37}$$

$$De^{\lambda_- t_{21}} = \frac{1}{\varepsilon\,(\lambda_+ - \lambda_-)} + B. \tag{17.38}$$

By using (17.37), we eliminate D into Eq. (17.38) and find

$$B = \frac{1 - e^{\lambda_- t_{21}}}{\varepsilon\,(\lambda_+ - \lambda_-)\left[e^{\lambda_- t_2} - 1\right]}. \tag{17.39}$$

Introducing then B given by (17.39) into Eq. (17.37), we obtain D as

$$D = \frac{e^{\lambda_- t_1} - 1}{\varepsilon\,(\lambda_+ - \lambda_-)\left[e^{\lambda_- t_2} - 1\right]}. \tag{17.40}$$

Inserting (17.37) into (17.35) and (17.38) into (17.36) provides two coupled equations for A and C given by

$$Ae^{\lambda_+ t_1} = C + \frac{1}{\varepsilon\,(\lambda_+ - \lambda_-)}, \tag{17.41}$$

$$Ce^{\lambda_+ t_{21}} = A - \frac{1}{\varepsilon\,(\lambda_+ - \lambda_-)}. \tag{17.42}$$

Using (17.41), we eliminate C in Eq. (17.42) and find

$$A = \frac{e^{\lambda_+ t_{21}} - 1}{\varepsilon\,(\lambda_+ - \lambda_-)\left[e^{\lambda_+ t_2} - 1\right]}. \tag{17.43}$$

Finally, introducing A given by (17.43) into Eq. (17.41) provides C as

$$C = \frac{1 - e^{\lambda_+ t_1}}{\varepsilon\,(\lambda_+ - \lambda_-)\left[e^{\lambda_+ t_2} - 1\right]}. \tag{17.44}$$

From Fig. 17.4b, we note that x increases at time $t = 0$ when $x(t - \tau) = a$. At time $t = \delta$, it is the turn of x to equal a. From Fig. 17.4c, we note that $t = t_1$ and $t = t_1 + \delta$ mark the times where $x(t - \tau)$ and then x are equal to a. Using (17.9) with $x(\delta) = a$ and (17.13) with $x(t_1 + \delta) = a$, we obtain

$$Ae^{\lambda_+ \delta} + Be^{\lambda_- \delta} = a, \tag{17.45}$$

$$Ce^{\lambda_+ \delta} + De^{\lambda_- \delta} = a. \tag{17.46}$$

Equations (17.31)–(17.46) with

$$t_2 = T_n, \tag{17.47}$$

defined by (17.6), are six equations for seven unknowns, namely A, B, C, D, t_1, and δ. We introduce the expressions of A, B, C, and D into Eqs. (17.45) and (17.46), and find

$$a\varepsilon\,(\lambda_+ - \lambda_-) = \frac{1 - e^{\lambda_+ t_{21}}}{1 - e^{\lambda_+ t_2}} e^{\lambda_+ \delta} + \frac{e^{\lambda_- t_{21}} - 1}{1 - e^{\lambda_- t_2}} e^{\lambda_- \delta}, \tag{17.48}$$

$$a\varepsilon\,(\lambda_+ - \lambda_-) = \frac{e^{\lambda_+ t_1} - 1}{1 - e^{\lambda_+ t_2}} e^{\lambda_+ \delta} + \frac{1 - e^{\lambda_- t_1}}{1 - e^{\lambda_- t_2}} e^{\lambda_- \delta}. \tag{17.49}$$

Substracting side by side, we eliminate $a\varepsilon\,(\lambda_+ - \lambda_-)$. Multiplying then by $e^{-\lambda_+ \delta}$, we have

$$0 = \frac{2 - e^{\lambda_+ t_{21}} - e^{\lambda_+ t_1}}{1 - e^{\lambda_+ t_2}} + \left(\frac{e^{\lambda_- t_{21}} - 2 + e^{\lambda_- t_1}}{1 - e^{\lambda_- t_2}} \right) e^{(\lambda_- - \lambda_+)\delta}. \tag{17.50}$$

References

1. J. Keener, J. Sneyd, *Mathematical Physiology* (Springer, 1998)
2. C. Fall, *Computational Cell Biology* (Springer, 2002)
3. B. Ermentrout, D. Terman, *Mathematical Foundations of Neuroscience* (Springer, 2002)
4. A.L. Hodgkin, A.F. Huxley, J. Physiol. **117**(4), 500 (1952)
5. R. FitzHugh, Biophys. J. **1**(6), 445 (1961)
6. J. Nagumo, S. Arimoto, S. Yoshizawa, Proc IRE **50**(10), 2061 (1962)
7. E. Schöll, G. Hiller, P. Hövel, M. Dahlem, Philos. Trans. R. Soc. London Ser. A **367**(1891), 1079 (2009). doi:10.1098/rsta.2008.0258. http://rsta.royalsocietypublishing.org/content/367/1891/1079.abstract
8. W.J. Freeman, Int. J. Bifurcat. Chaos **10**(10), 2307 (2000)
9. S. Kim, S.H. Park, C.S. Ryu, Phys. Rev. Lett. **79**, 2911 (1997). doi:10.1103/PhysRevLett.79.2911
10. M.K.S. Yeung, S.H. Strogatz, Phys. Rev. Lett. **82**, 648 (1999). doi:10.1103/PhysRevLett.82.648
11. M.Y. Choi, H.J. Kim, D. Kim, H. Hong, Phys. Rev. E **61**, 371 (2000). doi:10.1103/PhysRevE.61.371
12. W.S. Lee, E. Ott, T.M. Antonsen, Phys. Rev. Lett. **103**, 044101 (2009). doi:10.1103/PhysRevLett.103.044101
13. G. Deco, V. Jirsa, A. McIntosh, O. Sporns, R. Kötter, Proc. Nat. Acad. Sci. **106**(25), 10302 (2009)
14. G. Deco, V.K. Jirsa, J. Neurosci. **32**(10), 3366 (2012)
15. R. Ton, G. Deco, A. Daffertshofer, PLoS Comput. Biol. **10**(7), e1003736 (2014)
16. C. Cakan, J. Lehnert, E. Schöll (2013). arXiv:1311.1919
17. N. Burić, D. Todorović, Phys. Rev. E **67**, 066222 (2003). doi:10.1103/PhysRevE.67.066222
18. M. Dahlem, G. Hiller, A. Panchuk, E. Schöll, Int. J. Bifurcat. Chaos **19**(02), 745 (2009). doi:10.1142/S0218127409023111
19. O. Vallès-Codina, R. Möbius, S. Rüdiger, L. Schimansky-Geier, Phys. Rev. E **83**, 036209 (2011). doi:10.1103/PhysRevE.83.036209

20. L. Weicker, T. Erneux, L. Keuninckx, J. Danckaert, Phys. Rev. E **89**, 012908 (2014). doi:10. 1103/PhysRevE.89.012908
21. P. Hövel, Control of complex nonlinear systems with delay. Ph.D. Thesis, Technischen Universität Berlin (2010)
22. T. Prager, H.P. Lerch, L. Schimansky-Geier, E. Schöll, J. Phys. A Math. Theor. **40**(36), 11045 (2007)
23. N.B. Janson, A.G. Balanov, E. Schöll, Phys. Rev. Lett. **93**, 010601 (2004). doi:10.1103/ PhysRevLett.93.010601
24. A. Balanov, N.B. Janson, E. Schöll, Physica D **199**(1), 1 (2004)
25. J. Rinzel, J.B. Keller, Biophys. J. **13**(12), 1313 (1973). doi:10.1016/S0006-3495(73)86065-5. http://www.sciencedirect.com/science/article/pii/S0006349573860655
26. S. Coombes, Physica D **160**, 173 (2011)
27. S. Coombes, C. Laing, Physica D **238**(3), 264 (2009). doi:10.1016/j.physd.2008.10.014. http:// www.sciencedirect.com/science/article/pii/S0167278908003692
28. T. Erneux, *Applied Delay Differential Equations* (Springer, New York, 2009)
29. S. Yanchuk, P. Perlikowski, Phys. Rev. E **79**, 046221 (2009). doi:10.1103/PhysRevE.79.046221
30. L. Weicker, T. Erneux, D.P. Rosin, D.J. Gauthier, Phys. Rev. E **91**, 012910 (2015). doi:10.1103/ PhysRevE.91.012910
31. G. Friart, G. Verschaffelt, J. Danckaert, T. Erneux, Opt. Lett. **39**(21), 6098 (2014). doi:10.1364/ OL.39.006098

Chapter 18
Exploiting Multistability to Stabilize Chimera States in All-to-All Coupled Laser Networks

Fabian Böhm and Kathy Lüdge

Abstract Large networks of optically coupled semiconductor lasers can be realized as on-chip solutions. They serve as general testbeds for delay-coupled networks but may also be candidates for new methods in signal processing. Our work focuses on all-to-all networks where the individual units are coupled to each other by a common mirror with very short delay times. Using the well-known Lang-Kobayashi-model for the local laser dynamics, we investigate the occurring bifurcation structure of the complex network in terms of numerical integration and path continuation techniques. We especially focus on the interrelation between material parameters of the laser and occurring synchronization patterns. In this respect we identify the time scale separation between photon and electron lifetimes T as well as the amplitude-phase coupling α to be the driving forces for multi-stability between different cluster solutions. As an example quantum-dot lasers with strongly damped relaxation oscillations are found to present less rich dynamics when coupled to a network. Depending on the initial conditions, one-color symmetric states (all lasers emit the same constant waves), inhomogeneous one-color symmetry-broken states (clusters form that emit at different constant wave intensities), and multi-color symmetry-broken states (clusters with different period pulsations) are found. Those solutions can be analytically understood by reducing the equations to two coupled lasers, where the dynamic bifurcation scenarios have been discussed (Clerkin et al. Phys Rev E 89:032919, 2014 [2]). Additionally we find chimera states, i.e. partially synchronized cluster solutions, where the desynchronized clusters are chaotic in phase, amplitude and carrier inversion (Böhm et al. Phys Rev E 91(4):040901(R), 2015 [1]). They form from random initial conditions within the regions of multistability for the case of large enough amplitude-phase coupling. These chimera states defy several of the previously established existence criteria. While chimera states in phase oscillators generally demand non-local coupling, large system sizes, and specially prepared ini-

K. Lüdge (✉) · F. Böhm
Institut für Theoretische Physik, Technische Universität Berlin,
Hardenbergstr. 36,
10623 Berlin, Germany
e-mail: kathy.luedge@tu-berlin.de

F. Böhm
e-mail: boehm@itp.tu-berlin.de

© Springer International Publishing Switzerland 2016   355
E. Schöll et al. (eds.), *Control of Self-Organizing Nonlinear Systems*,
Understanding Complex Systems, DOI 10.1007/978-3-319-28028-8_18

tial conditions, we find chimera states that are stable for global coupling in a network of only four coupled lasers for random initial conditions.

18.1 Introduction

With the advance of technology, semiconductor lasers have become increasingly important for numerous applications [3, 4]. Owning to their inexpensive fabrication, energy efficiency and good modulation characteristics, they are the backbone of modern communication technology. Optical interconnections with semiconductor lasers as light sources have long replaced conventional electrical interconnections in communication networks. However, they also find application in very different fields. If semiconductor lasers are subjected to time-delayed self-feedback, a variety of different rich dynamics can occur, i.e. periodic or chaotic dynamics, spiking behavior and multistability [3, 5–7]. These rich dynamics can be used for various applications. They have for example been successfully studied in regards to chaos encryption [8–12] and reservoir computing in the past [13, 14]. Furthermore, the analogy to spiking in neuronal networks allows for the construction of all-optical neurons [15, 16]. Semiconductor laser are thus a promising system for the advancement of the frontiers of optical signal processing and unconventional new computing paradigms such as all-optical computing. However, as the required complex dynamical behavior is induced by long feedback, experimental realization always requires long delay lines and thus large and complex setups which greatly hinders easy implementation. Another possibility to achieve the required richness in dynamics in semiconductor lasers is by mutual coupling in a network [17–22]. Small laser networks have shown that they can posses complex synchronization and dynamical behavior even for very short delay lines and thus small spatial dimensions [2, 23–26]. The advancement of epitaxial growth and lithographic techniques would allow for the construction of such networks as powerful on-chip solutions in photonic circuits and at the same time open the door for completely new ideas.

Networks of delay-coupled semiconductor lasers are also a promising platform for the study of complex networks [3, 25, 27]. In particular the study of complex partial synchronization patterns has recently become the focus of intense research [3, 17, 22, 26, 28–32]. One very prominent example are chimera states where an ensemble of identical elements self-organizes into spatially separated coexisting domains of coherent (synchronized) and incoherent (desynchronized) dynamics [33, 34]. Chimera states appear in many different fields of research and are believed to be linked to several natural phenomena, e.g. in the unihemispherical sleep of birds and dolphins [35], in neuronal bump states [36, 37], in power grids [38], or in social systems [39]. At this point, the appearance of chimera states has been reported in various systems in theoretical investigations [40–47] as well as in experiments: [48–56]. However, no universal mechanism for the formation of chimera states could yet be established. Many studies have found that there are three essential requirements for long living chimera states:

(i) a large number of coupled elements, (ii) non-local coupling, and (iii) specific initial conditions. Recent studies however have shown, that these paradigms can be broken and chimera states are observed also for for small system sizes [1, 57], global coupling [47, 56, 58–60] and random initial conditions [61]. With our work, we want to provide a bridge between laser networks and chimera patterns. While laser networks pose a type of system, that has not been previously investigated in relation with chimera states, they open up perspectives for application. In particular, laser networks raise the question how the specific features of the local dynamics of semiconductor lasers, i.e. three degrees of freedom and amplitude-phase-coupling influence the type of chimera states observable in these networks. By combining ideas from network science and laser dynamics, we want to gain new insights into these complex synchronization patterns and show that the dynamics of laser networks can give rise to interesting new types of chimera states that break all of the three previously mentioned existence criteria simultaneously.

As the laser network is envisioned to be realized in a small-scale photonic circuit, we choose a setup, where an array of Z identical oscillators is globally coupled into a single external cavity (see Fig. 18.1a). The lasers receive feedback from one common mirror at the end of the cavity. This particular setup has already been the focus of previous studies [1, 24–26]. In terms of network topology it is a completely connected network as sketched in the scheme in Fig. 18.1b. The advantages of such a setup lie within the easy experimental realization. Instead of using interconnections between all the units, coupling of all laser outputs into a single external cavity reduces the number of required interconnections and is thus better suited for up-scaling. The system incorporates an intrinsic time delay due to the finite speed of light. It is given by the external cavity roundtrip time τ. Since the external cavity in a photonic circuit is required to be very small (on the order of mm), the feedback-delay τ is much smaller than the intrinsic timescale of the internal dynamics (limit of short delay). Thus, no additional bifurcations are induced by the small delay [23] and the dynamics are not significantly altered.

The local dynamics of the system is modeled by dimensionless semi-classical rate equations, i.e., Lang-Kobayashi equations that govern the complex electric field

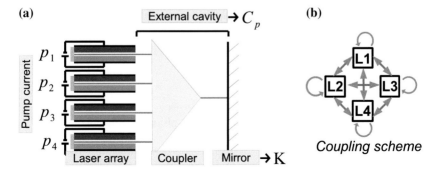

Fig. 18.1 **a** Setup of an all-to all coupled laser network. **b** Corresponding coupling scheme

amplitude E_n of the n-th laser in the array, and the carrier inversion $N_n, i = 1, ..., Z$.

$$\frac{dE_n}{dt} = (1 + i\alpha)E_n N_n + e^{-iC_p^n}\kappa \sum_{j=1}^{Z} e^{-iC_p^j} E_j(t - \tau) \tag{18.1}$$

$$\frac{dN_n}{dt} = \frac{1}{T}(p - N_n - (1 + 2N_n)|E_n|^2) \tag{18.2}$$

The lasers are pumped electrically with the excess pump rate p. Note that p is defined to yield a threshold pump current of $p = 0$. The feedback strength κ and the feedback phase C_p are the bifurcation parameters, which are used to tune the dynamics and the synchronization behavior of the system. They are determined from the reflectivity of the mirror and the length of the external cavity. We assume identical feedback phases for all lasers, i.e. $C_p = C_p^j$ for all $j = 1 ... Z$. The sum in Eq. (18.1) reflects the coupling. It includes self-coupling as long as $n = j$ is not excluded. In polar representation the dynamics of the complex electrical field amplitude $E_n(t) = A_n(t)e^{i\varphi_n(t)}$ can be visualized in the complex plane with amplitude A_n and phase φ_n (see Fig. 18.2a). The local dynamics of one laser thus has three degrees of freedom. In semiconductor lasers the amplitude and phase dynamics are coupled by the linewidth enhancement factor α, because the optical path-length of the laser cavity (defining the phase of E) is dynamically linked to changes in the charge carrier densities (defining the gain and therewith the amplitude of E). For quantum-well lasers, typical values for α are between 2 and 5 [62], while for more complex nanostructured lasers, e.g. quantum-dot lasers, the α-factor is much harder to define [63–65] and generally smaller than 1. Also outside the laser community, coupled amplitude-phase dynamics is a significant concept widely exploited in various fields of research. It refers to anisochronicity in nonlinear dynamics [66] and is known as shear in fluid dynamics [67].

The parameter T in Eq. (18.2) describes the ratio between electron and photon lifetime in the laser cavity. In semiconductor lasers this is a large quantity as there

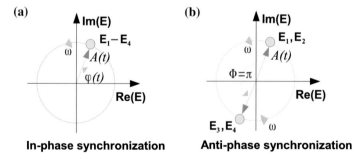

Fig. 18.2 Two possible compound laser modes (CLMs) in a 4-laser network visualized in the complex E-field plane (*filled circles* represent the lasers). **a** All lasers are in-phase synchronized. **b** The lasers form anti-phase synchronized pairs

the photons have short lifetimes on the order of *ps* compared to electron lifetimes on the order of *ns*. Using the parameter T, the dynamical behavior of lasers can be classified [6, 62, 68]: Class B lasers with large T show pronounced turn-on oscillations (relaxation oscillations (RO)) while the dynamics of class A lasers with very small T is sufficiently governed by the field dynamics and thus by only two degrees of freedom (the fast electronic inversion N can be adiabatically eliminated as it immediately follows the slow field dynamics [69, 70]).

18.2   Bifurcation Structure of All-to-All Laser Networks

We first want to consider the simplest case for a laser network which consists of two mutually coupled lasers (without self-feedback). This setup was already intensively studied (see review [3, 19]), however we shortly repeat some details needed for the subsequent analysis. At first we are interested in the stable solutions that exist in the C_p–κ-parameter space, i.e., solutions with constant wave(cw) emission. These basic solutions to the coupled Lang-Kobayashi equations are the compound laser modes (CLM). Their general form is characterized by constant inversion N_n and complex electric field vectors $E_n(t) = A_n e^{i\omega_n t} e^{i\Phi_n}$ with constant amplitudes A_n that are rotating with a frequency ω_n and an arbitrary phase offset Φ_n in the complex plane (Fig. 18.2). Note that ω_n is the frequency shift with respect to the rotating frame ω_0, i.e., with respect to the optical wavelength of the full electric field $\tilde{E}_n(t) = E_n(t) \cdot e^{i\omega_0 t}$. Depending on the choice for A_n, ω_n, and Φ_n, different kinds of synchronized and unsynchronized solutions can be realized. The simplest case of full (zero-lag) synchronization occurs when $\omega_n = \omega_j$, $\Phi_n - \Phi_j = 0$ and $A_n = A_j$ holds for all lasers $n, j = 1, ..., Z$ at all times t (see Fig. 18.2a). For 2 mutually coupled lasers the in-phase CLM solution for the case $\tau = 0$ is found to be:

$$\begin{pmatrix} A_n \\ N_n \\ \omega_n \\ \Phi_n \end{pmatrix} = \begin{pmatrix} \sqrt{2}[p + \kappa \cos(2C_p)]/[0.5 - \kappa \cos(2C_p)] \\ -\kappa \cos(2C_p) \\ -\kappa[\alpha \cos(2C_p) + \sin(2C_p)] \\ 0 \end{pmatrix}. \qquad (18.3)$$

The transverse stability of this synchronous solution can be obtained by a straightforward linear stability analysis for the case of vanishing delay ($\tau \to 0$) [23]. It yields that the synchronous solutions destabilize through a supercritical Hopf and a pitchfork bifurcation. Note that the constant wave emission within the synchronization manifold is globally stable. The Lang-Kobayashi equations Eqs. (18.1) and (18.2) are invariant under a change of sign in E, thus a solution shifted by an angle of π ($E e^{i\pi} = -E$) is also a solution. This relates to a CLM where not all lasers have the same phase offset Φ but where the phase difference of a pair of 2 lasers is π. This anti-phase CLM is defined by $\omega_n = \omega_j$, $\Phi_n - \Phi_j = \pi$ and $A_n = A_j$ (see Fig. 18.2b). For the case of 2 mutually coupled lasers it is again given by Eq. (18.3) but shifted in C_p by $\pi/2$. We will refer to these two cases of CLMs as in-phase

$(\Phi_n - \Phi_j = 0)$ and anti-phase $(\Phi_n - \Phi_j = \pi)$ synchronization in the following. Figure 18.3a illustrates the stability of theses CLMs in the C_p–κ parameter space by showing Hopf (black) and pitchfork (red) bifurcation lines that border the regions of stable in-phase and anti-phase synchronization, i.e., in-phase around $C_p = 0$ and anti-phase around $C_p = \pi/2$.

Besides the CLMs, stable periodic solutions can also be found in the mutually coupled 2-laser network. In this case, amplitude and inversion are no longer constant but instead show periodic oscillations. The existence and stability for these can for example be determined by numerical path continuation using a previously proposed decomposition of the Lang-Kobayashi equations [2]. At the supercritical Hopf bifurcation (black line in Fig. 18.3), a stable periodic solution forms and splits into two limit cycles upon decreasing the feedback phase and passing the nearby pitchfork bifurcation of limit cycles (not shown). Thereafter the lasers are unsynchronized with each laser being on a different limit cycle (indicated by the light grey region in Fig. 18.3a). This state is also referred two as a two-color periodic state. By further decreasing C_p the periodic solution looses its stability either by passing through a Torus bifurcation at low feedback strengths (labeled T in Fig. 18.3) or by a Hopf bifurcation at higher feedback strength. The dark grey region in Fig. 18.3a indicates a multi-stable region in parameter space, where the periodic two-color state and the CLM is stable. For details please see the extensive bifurcation analysis in [2].

The bifurcation structure of the CLMs for an all-to-all laser network of 4-lasers is shown in Fig. 18.3b and was also obtained by path continuation for vanishing delay $\tau \to 0$ to enable usage of the continuation software AUTO. Due to the increased complexity of different possible cluster states, only the CLM solutions are followed and not the periodic ones as done for the 2-laser case. The numeric results for the

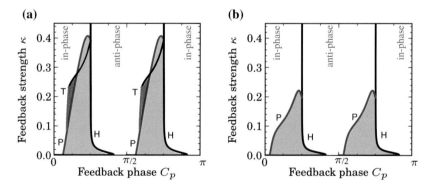

Fig. 18.3 Stability of the constant laser modes (CLM) for **a** 2 mutually coupled lasers (without self-feedback) and **b** 4 all-to-all coupled lasers (with optical self-feedback), obtained by path continuation. *Red, black* and *green lines* indicate pitchfork (*P*), Hopf (*H*) and Torus bifurcations (*T*) that border the stable in-phase and anti-phase CLMs (*white regions*). *Light grey* region indicate unstable CLMs. The *small dark grey* region in (**a**) mark multi stability between two-color periodic solution and CLM. Parameters as given in Table 18.1, with instantaneous coupling $\tau = 0$

Table 18.1 Parameters used for the simulations unless stated otherwise

Symbol	Name	Value
α	Linewidth enhancement factor	2.5
T	Ratio of electron versus photon lifetime	392
p	Pump parameter	0.23
τ	External cavity roundtrip time	0 or 1
κ	Feedback rate	$[0, 0.1]$
C_p	Feedback phase	$[0, \pi]$

cluster dynamics in the 4-laser case will be discussed later on in Sect. 18.3. We find that the general form of the bifurcation structure in Fig. 18.3b is similar to the mutually coupled 2-laser network in Fig. 18.3a. The regions of transversely stable in- and anti-phase synchronization are again bordered by Hopf and pitchfork bifurcations. Compared to the mutually coupled 2-laser network, the unsynchronized regions (light grey region in Fig. 18.3b) exist only for smaller feedback strengths while the lasers are completely synchronized for higher feedback strength. This behavior can be understood by considering the coupling term in Eq. (18.1). Inserting the CLM-ansatz $E_n(t) = A_n e^{i\omega_n t} e^{i\Phi_n}$ for in-phase synchronization into Eq. (18.1) yields solutions equivalent to the mutually coupled 2-laser network (Eq. 18.3) with an effective coupling strength of κZ [25]. The in-phase CLM of a mutually coupled 2-laser network with coupling strength κ (Eq. 18.3) is thus equivalent to the solution of an all-to-all coupled Z-laser network with feedback strength κZ. Since the 4-laser network investigated here is fully connected it also contains self-coupling which changes the symmetry and thus the solution for the anti-phase CLM, i.e., instead of Eq. (18.3) it is given by $A_n = \sqrt{p}$, $\omega_n = 0$ and $N_n = 0$ independent of C_p and K. Due to the scaling, the whole bifurcation structure is compressed by a factor Z, enabling to predict the general bifurcation structure of the CLMs of large globally coupled laser networks by rescaling the results of the 2-laser case (as done later on in Fig. 18.9). However, the shape of the Hopf-bifurcation line for small feedback strength strongly depends on the local dynamics (damping of the relaxation oscillations of a single laser) and thus does not change with Z.

18.3   Multi-stability in Laser Networks

While the global coupling topology of the laser network allows us to predict the stability of in- and anti-phase synchronization, a large number of nodes in a network also allows for more complex dynamics and synchronization phenomena as well as multistability between different solutions. Depending on the chosen parameters and the initial conditions of the single lasers, cluster, partial and complete desynchronization can occur. In the case of cluster synchronization, the different oscillators form groups with identical frequencies, amplitude and inversion. In a partially synchronized state the dynamical variables differ for one or more lasers while the rest stays synchronized. In a completely unsynchronized state there is no fixed phase relation between any given oscillators. For a given parameter set, many of those solutions might be stable, and care has to be taken performing numerical integration. By applying small perturbations perpendicular to the synchronization manifold, e.g. in the form of noise, small kicks or different initial conditions, it is possible to navigate between the solutions. In this section we apply perturbing kicks to the inversion N_n of the lasers in the fully connected 4-laser network. In experiments, these could be realized by pulses in the pump current.

The results of the dynamics in the C_p–κ parameter plane are shown in Fig. 18.4a, b for the cases of two and one perturbed laser, respectively. The two kicks (Fig. 18.4a) of $\Delta N_1 = 0.1$ and $\Delta N_2 = 0.15$ were administered to the first and second laser after the equilibrium state for each parameter set was reached. In Fig. 18.4b only the first laser was perturbed by a kick of $\Delta N_1 = 0.1$ leaving the remaining 3 lasers synchronized and thus forcing the system onto a 1–3–cluster solution. The number of maxima detected in the time series of one laser were extracted and plotted as color code in Fig. 18.4. The yellow regions indicate constant light output of the laser (cw) which can be either one of the CLM solution of in-phase or anti-phase synchronization discussed above, or a symmetry broken state, where all lasers emit cw light but with different intensities (region in the middle of Fig. 18.4b). The red and orange regions indicate periodic solutions, while blue and white indicates chaos found in the dynamics of laser no.1. In Fig. 18.4b, the dynamics of one laser is shown while

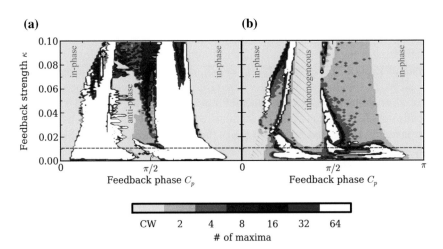

Fig. 18.4 Dynamics of a 4-laser network in C_p–κ parameter space with different perturbations applied to the inversion. In **a** for each parameter set two kicks of $\Delta N_1 = 0.15$ and $\Delta N_2 = 0.1$ were administered after transients died out. In **b** the system was perturbed by one kick of $\Delta N_1 = 0.1$. The *color code* indicates the number of maxima found in the time series of the first laser. *Yellow regions* with constant light output (*CW*) can be in-phase, anti-phase and inhomogeneous CW states of the network. The *dashed line* indicates the scans presented in Fig. 18.5 while the *star* shows the position for the attractor basins studied in Fig. 18.6. Parameters as given in Table 18.1 with $\tau = 0$

the 3 remaining unperturbed lasers form a synchronized cluster where all lasers show the same dynamics, however emit modulated light at different intensity than the perturbed laser. Only in the indicated in-phase regions complete synchronization is reached. For further analysis this cluster solutions can be analysed by reducing the system Eqs. (18.1) and (18.2) to a fully connected 2-laser network (with unequal coupling strength $\kappa_1 = \kappa$ and $\kappa_c = 3\kappa$) by summing up the three equal field equations and introducing a new variable $E_c = 3E_2$:

$$\frac{dE_c}{dt} = (1 + i\alpha)E_c N_2 + 3\kappa e^{-2iC_p}(E_c(t - \tau) + E_1(t - \tau)) \qquad (18.4)$$

$$\frac{dE_1}{dt} = (1 + i\alpha)E_1 N_1 + \kappa e^{-2iC_p}(E_1(t - \tau) + E_c(t - \tau)) \qquad (18.5)$$

To underline the different possible dynamic states of the 4-laser network, slices through the two parameter bifurcation diagrams of Fig. 18.4 (indicated by the dashed line) are discussed in Fig. 18.5. From the bifurcation scans different dynamic regimes can be seen, ranging from inhomogeneous steady states (1), self-pulsation (2), chaotic dynamics (3) and spiking (4), as indicated and marked in Fig. 18.5. The related time series of the electric field amplitude $A(t)$ of two lasers in the network (the perturbed

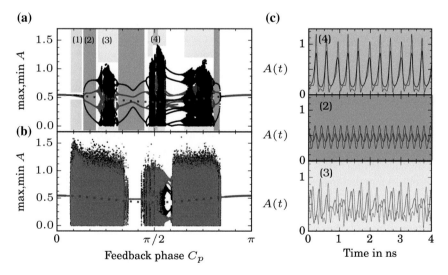

Fig. 18.5 One-parameter bifurcation diagrams for $\kappa = 0.01$, $\tau = 0$ with one kick (**a**) and two kicks (**b**) applied to the lasers (slices of Fig. 18.4a, b). *Black*, *red* and *green* indicating the maxima detected for laser no. 1, no. 2. and no. 3. *Solid* and *dashed blue line* stable and unstable CLM-solution obtained by path continuation. Indicated with *background color* are regimes of spiking (*4*), chaotic dynamics (*3*), period-2 oscillations (*2*) and symmetry broken 1-color solutions (*1*) with the related time series of laser no. 1 and 2 shown in (**c**) for the $C_p = 1.3$ (*2*), $C_p = 2.35$ (*3*) and $C_p = 1.78$ (*4*). *Densely packed areas* in (**b**) represent partially synchronized chaotic dynamics. Parameters as given in Table 18.1

laser and one of the lasers within the cluster) is shown in Fig. 18.5c. Both show similar dynamics but at different intensities as also visible by comparing the red and the black lines in the bifurcation scan of Fig. 18.5a.

The laser network dynamics for two perturbed lasers that are presented in Fig. 18.4a differ strongly from the 1–3–cluster case in Fig. 18.4b. Most prominent are the extended white chaotic regions that appear at the position where the CLMs are unstable (compare Fig. 18.3). Within these white regions a 1–1–2 cluster chaos synchronization is found, i.e., partial synchronization or in other words a chimera state is induced by the perturbation. One line scan for the case of two perturbing kicks is shown in Fig. 18.5b. It exemplarily visualizes the different solutions found numerically. Chaotic partially synchronized dynamics (green regions), anti-phase solutions with $A_n = \sqrt{p}$ and symmetry broken pulsations (periodic window within the densely packed areas) can be seen. In between the chaotic regions the anti-phase solution is stable but, remembering the results of the 1–3 cluster dynamics, it is multi-stable with other solutions, e.g. symmetry broken periodic pulsations (orange region seen in Fig. 18.4 around $C_p = \pi/2$).

By comparing Fig. 18.4a and b, it can be seen that the laser network is very sensitive to the number of the kicks. Large regions of multistability can be observed. This extend of rich dynamics and multi-stability cannot be found in a 2-laser network. To further investigate the level of multistability especially in the region where CLM and periodic solutions coexist, i.e., region suggested by the 2-laser bifurcation analysis in Fig. 18.3a, the influence of perturbations into different directions in phase space is considered at a fixed position in parameter space ($C_p = 0.75, \kappa = 0.09$ ($\kappa Z = 0.36$)) indicated by a star in Fig. 18.4a. This point is close to the pitchfork bifurcation and is still in the regime, where the CLM is transversely stable. The system was perturbed simultaneously by a single kick to one of the lasers and three identical ones to the rest of the array. The basin of attraction was mapped by varying the sizes of the kicks and mapping the stable dynamic state reached afterwards. The results are shown in Fig. 18.6a and three multi-stable dynamical solutions can be found: in-phase CLM (yellow), period-2 oscillation of the 1–3–cluster state (orange) and chaotic oscillations (blue). Thus, switching between these different attractors should be possible. A simulation of a complete back and forth switching event between the CLM and the period-2 oscillation is shown in Fig. 18.6b. The kicks are emulated by two different pulses in the pump current with an intensity of $p_1 = 0.7$ and $p_2 = 0.1$ and a duration of $t_1 = 2.56$ ns and $t_2 = 5.63$ ns (blue line in the upper panel of Fig. 18.6b). The transient of the resulting switching of the light output of the lasers is seen in Fig. 18.6b. It is on the order of several ns which allows for fast optical switching events, e.g., it can be utilized for fast optical memory.

So far we investigated the different solutions and the extend of multistability found by varying the initial conditions of the inversion. A different approach for our laser network is to investigate the dynamics with random initial conditions for the phases of the electric field. Figure 18.7a shows the resulting dynamics and Fig. 18.7b the synchronization properties for a four laser network, where the initial phases were chosen randomly while the initial inversions and field intensities are all set constant to $|E_n(0)|^2 = 1$ and $N_n(0) = 0$. The overall dynamics observed in this case

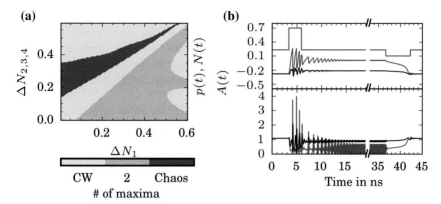

Fig. 18.6 **a** Attractor basin for kicks ΔN_1 (inversion of laser 1) and $\Delta N_{2,3,4}$ (inversion of laser 2, 3 and 4) in a 4-laser network for $\kappa = 0.09$ and $C_p = 0.75$ (position indicated by the star in Fig. 18.4a). Indicated by *colors* are the different dynamic regimes: *yellow* (stable CLM-solution), *orange* (symmetry broken period-2 oscillations), *blue* (chaotic dynamics). **b** Switching event between CLM-solution and symmetry broken period-2 oscillation by administering a pulse in the electric pump (*blue line*). *Red lines* show the dynamics of inversion and amplitude of the perturbed laser, while *black lines* show the remaining 3 unperturbed lasers. Parameters as given in Table 18.1 with $\tau = 0$

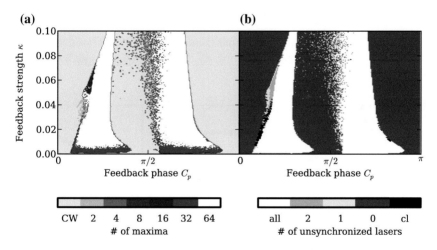

Fig. 18.7 **a** Dynamics of a 4-laser network in C_p–κ parameter space for random initial phases $\varphi_n(0)$ and constant initial field intensities $|E_n(0)|^2 = 1$. *Color coded* are the number of maxima found in the time series. **b** Corresponding synchronization map, where the number of synchronized lasers is indicated by the *color code*. Parameters as given in Table 18.1 with $\tau = 1$

resembles the case for two perturbed lasers in Fig. 18.4a, however strong differences appear in the synchronization pattern given by the color code in Fig. 18.7. The color indicates the number of unsynchronized lasers, with brown regions showing CLM solutions and white regions indicating complete desynchronization. The stable in- and anti-phase synchronization regions (yellow and brown regions in Fig. 18.7a and b, respectively) follow the bifurcation structure of the CLMs in Fig. 18.3. In between the bifurcation lines where the 1–3–cluster partial synchronization was found in Fig. 18.4, the lasers show completely chaotic unsynchronized dynamics (white regions). Within the anti-phase CLM region around $C_p = \pi/2$, we do not see the multi-stable periodic cluster solutions discussed before, which is due to the constant initial inversion chosen here.

Close to the pitchfork bifurcation, i.e., at the left border of the left white region in Fig. 18.7, we see regions where both partially and cluster synchronized regions exist (see green/yellow and black regions in Fig. 18.7b). A blowup of the region is shown in Fig. 18.9a. It reveals that the lasers gradually start to desynchronize as the bifurcation line is reached with increasing C_p, before complete desynchronization occurs.

To get more insights into the dynamics within those partially synchronized regions, we analyze the temporal dynamics of the involved lasers within the partially synchronized region at $C_p = 0.54$ and $\kappa = 0.01$ ($\kappa Z = 0.2$). A space-time plot of all 3 local laser variables, i.e. amplitude, phase, and inversion is shown in Fig. 18.8b (for the plot the dimensionless time is converted back to ps). A snapshot for one time moment is shown in Fig. 18.8a for the three local variables. Note that the results shown in the plot were obtained from a 20-laser network to better visualize the partial synchronization, although a similar behavior is found for the 4-laser network. We find that the unsynchronized and the synchronized lasers self-organize into domains of coherent and incoherent dynamics, thus chimera states [33, 34] are formed. Due to the global coupling no spatial ordering of the lasers is a priori defined and for

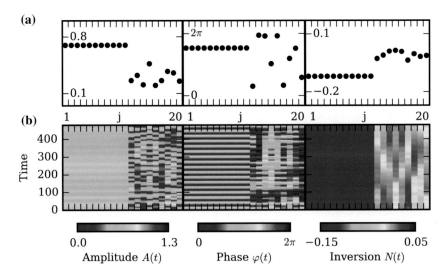

Fig. 18.8 a Snapshot and **b** space-time plots of the dynamics of amplitude, phase, and inversion (from *left* to *right*) in a chimera state for a 20-laser network after transients have died out (time is given in ps). *Parameters $C_p = 0.54, \kappa = 0.01$ ($\kappa Z = 0.2$), rest of parameters as in Table 18.1 with $\tau = 1$

better visualization, the lasers have been re-numbered such that coherent and incoherent domains are separated. Contrary to other examples of chimera states [47, 56, 58–61], e.g. amplitude chimeras, the coherence-incoherence pattern coexists simultaneously in the dynamics of the amplitude, the phase and the inversion of the laser with chaotic temporal dynamics. Furthermore, the chimera states have very long lifetimes. We found the partial synchronization to persist even for very long times in the range of microseconds, which is significantly longer than the timescales of transients in relaxation oscillations. Thus, this specific dynamic pattern seems to be unique to the laser network and further investigations are performed in the next section.

18.4   Tiny Chimera States in Laser Networks

Chimera states are a common phenomenon, that can be observed in a variety of different systems. They have also been previously investigated in optical systems, e.g. in optical combs [49], optical light modulators [48] or electro-optical feedback systems [53, 54]. However, no study of coupled lasers has been undertaken so far. The main difference of the system compared to other more common systems, e.g. Fitz-Hugh-Nagumo, phase oscillators or Ginzburg-Landau is that the local dynamics has three degrees of freedom. Furthermore, the laser network has a global coupling topology in contrast to the more common non-local coupling found in many studies. This is also interesting in regards to existence criteria related with the formation of chimera states, which generally demand non-local coupling, large network sizes and specific initial conditions in order to achieve long-living chimera states.

The chimera states found in laser networks posses chaotic temporal dynamics and coexistence of the coherence-incoherence pattern in amplitude, phase and inversion. These characteristics show similarities to amplitude-mediated chimera states that have been found for non-local [71] and also global coupling [47] in coupled Ginzburg-Landau-oscillators. These states typically show chaotic dynamics and coherence-incoherence patterns for both amplitude and phase, and emerge under random initial conditions. It is remarkable that in contrast to the amplitude-mediated chimeras in the Stuart-Landau system, the chimeras in laser networks also form in very small networks. The partially synchronized region in laser networks was found to exist for a minimum number of four lasers. Furthermore, it was found that the number of units Z has no influence on the chimera region. When we compare the partially synchronization for different system sizes, we can see that neither the shape nor the size of the region changes significantly with Z (see Fig. 18.9a–c). Comparing the chimera region to the bifurcation analysis of the 2-laser network in Fig. 18.9d, it appears to be correlated to the underlying multistability in region (2). While chimera states have also been found in small networks of coupled phase oscillators, non-local coupling is a mandatory requirement for their formation. Chimera states in laser networks on the other hand seem to break all of the existence criteria mentioned above simultaneously. This requires a critical reassessment of the question of necessary

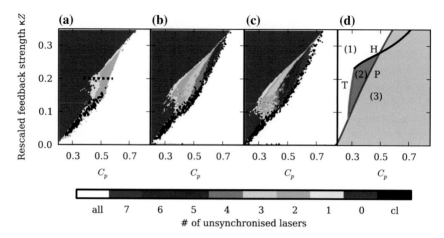

Fig. 18.9 Synchronization maps of chimera region for different system sizes **a** 4 laser, **b** 8 and **c** 12 laser with *color code* indicating the number of desynchronized lasers. **d** Bifurcation scenario of a mutually coupled 2-laser network. Indicated are the pitchfork (*P*), Hopf (*H*) and the Torus (*T*)-bifurcation lines as well as regions of stable CLM-solutions (*white labeled* (*1*)), stable 2-color-periodic solutions (*light grey* labeled (*3*)) and multistability between regions *1* and *3* (*dark grey* labeled (*2*)). Parameters as given in Table 18.1 a–c: $\tau = 1$ d: $\tau = 0$

criteria for the formation of chimera states in laser networks. Furthermore, it opens up the question of the behavior of chimera states in networks with higher dimensional local dynamics.

To gain a deeper understanding into the formation of the states, we address the influence of the specific features that semiconductor lasers posses in the following, namely the additional degree of freedom given by the carrier dynamics, the amplitude-phase coupling and the multi-stability. To start with the latter we investigate the basins of attraction of the different solutions by performing 2 dimensional sweeps of the initial phases $\varphi_n(0)$ at different positions in parameter space within the chimera region for the 4-laser network (Fig. 18.10a–c). In these figures the synchronization state is color coded after the initial phases of two lasers are varied in the interval $\varphi(0) = [0, 2\pi]$ while the initial phases of the other two lasers remained fixed. Three distinct basins of attraction can be found: If the initial phases are chosen close to each other, the synchronous in-phase CLM is reached (brown regions in the middle of Fig. 18.10a–c). For in-phase synchronization, it is sufficient if only 3 of the four lasers have similar phases in the beginning, while the remaining laser is dragged into the synchronous solution. If the initial phases are chosen further away from the in-phase state, first a state with one desynchronized laser is reached (yellow region), bordered by the attractor basin of a tiny chimera state where two lasers are desynchronized (green region). By choosing the appropriate initial phases, we can thus control the number of unsynchronized lasers. In larger networks, this can be used to control the size of the incoherent region in the chimera states. Similar to the multistable region for the 2-laser network in Fig. 18.9d, we find that the region

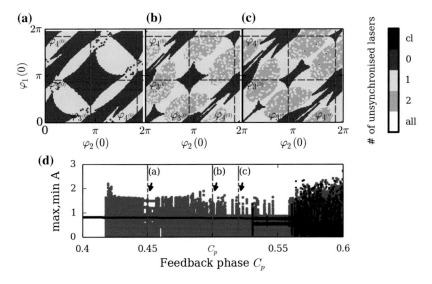

Fig. 18.10 Attractor basins for different initial phases in a 4-laser network for increasing feedback phase $C_p = 0.45$ (**a**), $C_p = 0.5$ (**b**), and $C_p = 0.52$ (**c**). The initial phases of two lasers are fixed at $\varphi_3(0) = 2.84$ and $\varphi_4(0) = 5.78$ (indicated by *blue dashed lines*) while the other initial phases are varied over the interval $\varphi(0) = [0, 2\pi]$. *Colors* indicate the number of unsynchronized lasers. **d** Bifurcation diagram across the chimera region (indicated by *dotted line* in Fig. 18.9a) for in-phase synchronized initial phases (*black*) and randomly distributed initial phases (*red*). Marked by the *vertical blue lines* are the positions of the attractor basins plotted in (*a*), (*b*), and (*c*). *Parameters* $\kappa = 0.05$, $\tau = 1$, other as in Table 18.1

of stable chimera states is again linked to multistability. This multistability is also visualized in Fig. 18.10d, where the bifurcation diagram of the dynamics of one laser across the chimera region is shown, both in the case where the initial phases are close to in-phase synchronization (black line) and for initial phases far away from in-phase synchronization (red line). The densely packed red regions are the chimera states with chaotic dynamics that form for much smaller C_p if random initial conditions are chosen while the simultaneously stable in-phase solution is reached for nearly synchronized initial values.

Looking at the basins of attraction, we can also explain the gradual desynchronization that was observed in Fig. 18.9. Figure 18.10a–c maps the basins of attraction for different values of C_p (indicated by arrows and horizontal blue lines in Fig. 18.10). We can observe that the size of the attractor basin for the in-phase state shrinks with increasing C_p and thus with approaching the desynchronized region. If the feedback phase is chosen close to the pitchfork bifurcation, the size of the in-phase region is small compared to the case when the position is chosen farther away from the pitchfork bifurcation. As the desynchronized region is approached, the basins for chimera states with larger incoherent regions thus grow and thereby increase the probability, that these states can be reached from the random initial phases.

So far we realized that the appearance of chimera states in laser networks is closely linked to an underlying multistability. Thus, next we aim to see how the laser parameters T and α influence the multi-stable region. While it is known that multistability can also be induced by time delay, we do not investigate this influence in the following. Our delay is very short compared to the intrinsic time scales, and we checked that no additional solutions are induced by the delay. In fact, we find chimera states also in the case without delay ($\tau = 0$) in the simulations.

When we consider the influence of the inversion N as an additional dynamic variable towards the formation of the multi-stable region, we can first elaborate the limit where we can eliminate the variable by adiabatic elimination. If the dynamics of N happens on a much faster timescale than the dynamics of the complex electrical field, the inversion can be taken to be in a quasi-steady state:

$$N_n^* = \frac{p - |E_n|}{1 + 2|E_n|} \qquad (18.6)$$

This quasi-steady state can be reached for very small lifetimes T of the electronic subsystem. Numeric simulations in the case of very small T show that the lasers tend to completely synchronize and that no multi-stable region can be found. The higher dimensionality is thus a necessary requirement in laser networks for the formation of complex unsynchronized dynamics and thus for the formation of chimera states.

This can also be corroborated by a bifurcation analysis with path continuation. Since the continuation of states with chaotic attractors is not possible with path continuation software, a simpler approach is needed and we consider again a minimal network of two coupled lasers, where the multi-stable region was identified in the previous Sect. 18.2. This region (see Fig. 18.9d) is bordered by a Hopf and a Torus bifurcation and coincides with the position of the chimera region in the parameter space. Now, changing T, the unsynchronized and the multi-stable region shrink with decreasing T as can be seen in Fig. 18.11a, where the change of the Torus (T), the Hopf (H) and the pitchfork (P) bifurcation with T are shown. While the position of the pitchfork bifurcation line remains unchanged with decreasing T, the Hopf bifurcation starts to close in on the pitchfork bifurcation as T approaches zero. The Torus bifurcation eventually crosses the pitchfork bifurcation at $T_{crit} \approx 90$ and thus multistability can no longer be found. For even smaller T, the periodic solution starts to destabilize through a set of different bifurcations, that is not shown here.

Another parameter that has a strong relation with multistability in semiconductor lasers is the linewidth enhancement factor α. It has been shown in theory and experiments that amplitude-phase coupling has an important influence upon the synchronization behavior. It is also well established that strong amplitude-phase coupling is able to induce multi-stability in single lasers with optical feedback. We thus want to understand the influence of α on the synchronization properties and on the formation of chimera states. Comparing numerically obtained synchronization maps for different values of α, we find that the unsynchronized regions (white regions in Fig. 18.7) shrink with decreasing α, so that the lasers are only unsynchronized for low feedback strengths (not shown). The unsynchronized region completely van-

(a) **(b)**

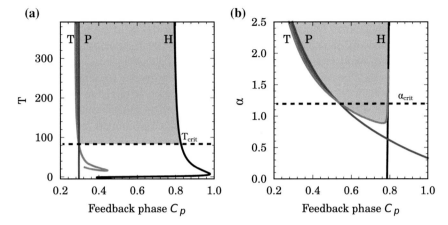

Fig. 18.11 Bifurcation lines found for the mutually coupled 2-laser network for fixed $\kappa = 0.1$ and $\tau = 0$. In **a** T is varied towards the case of adiabatic elimination. In **b** the amplitude phase coupling α is changed. *Red, black* and *green lines* indicate pitchfork (P), Hopf (H) and Torus (T) bifurcations, respectively. T_{crit} and α_{cit} indicate the values beneath which no multistability (*dark grey region*) and thus no chimera states are found

ishes for $\alpha = 0$. Amplitude-phase coupling is thus a requirement to find complex dynamics in our laser networks. Our numerical simulations of the chimera regions found for the fully connected 4-laser network close to the unsynchronized regions suggest that they indeed do not form above a minimal value of $\alpha_{crit} \approx 1.1$. To get a better understanding of the formation process, we again investigate the bifurcation scenario of the 2-laser network as presented in Fig. 18.11b. For a fixed value of $\kappa = 0.1$, we plot the Torus (T), Hopf (H) and pitchfork (P) bifurcation, that border the unsynchronized region and follow them over the feedback phase and the linewidth enhancement factor. We can observe, that the unsynchronized regions forms at a minimal value for $\alpha \approx 0.65$. The limit cycle stabilizes through the bifurcation at a critical value of $\alpha \approx 0.9$. It then starts to extend towards the pitchfork bifurcation, until it crosses the bifurcation line at critical value of $\alpha_{crit} \approx 1.2$. With this picture, it becomes clear how the multi-stability region forms in a 2-laser network. Comparing it to the numerical results of the dynamics of the 4-laser network, the interrelation between multi-stability and formation of chimera states is underlined.

18.5 Conclusions

To conclude, we investigated the emerging dynamics found in the light-output of all-to-all coupled laser networks of different network-size, ranging from minimal networks of two or four lasers to networks of up to 20 lasers. We characterized the occurring bifurcation scenarios and mapped regions of multistability as a function of the experimentally accessible parameters, i.e., strength and phase of the coupling via the external mirror. While the basic synchronized solutions of the laser dynamics within the network can be obtained from a simplified two laser system, we observe a high level of multistability to partially synchronized pulsating solutions that eventually evoke the formation of chimera states usually distinctive for larger networks. Concerning the local dynamics of the nodes, i.e. the lasers in our case, we identified the time scale separation between the lifetimes of electrons and photons as an important condition for the formation of long living chimera states chimera states. Without the additional degree of freedom of the carrier inversion within the laser the rich dynamics of the coupled network disappears. Further the existence of a coupling between the amplitude and the phase of the electric field (linewidth enhancement factor α) is crucial to observe the symmetry broken chimera state solutions.

The study of the dynamics of laser networks is done in the light of possible applications as all-optical devices. Having understood the bifurcation structure we showed that the multistability regions can be exploited for switching processes. By applying small perturbations to the pump current of one laser it is possible to switch the state of synchronization as well as the dynamic state of the lasers (chaotic vs. stable or regularly oscillating light emission).

References

1. F. Böhm, A. Zakharova, E. Schöll, K. Lüdge, Phys. Rev. E **91**(4), 040901 (R) (2015)
2. E. Clerkin, S. O'Brien, A. Amann, Phys. Rev. E **89**, 032919 (2014)
3. M.C. Soriano, J. García-Ojalvo, C.R. Mirasso, I. Fischer, Rev. Mod. Phys. **85**, 421 (2013)
4. K. Lüdge, *Nonlinear Laser Dynamics—From Quantum Dots to Cryptography* (Wiley-VCH, Weinheim, 2012)
5. C. Otto, B. Globisch, K. Lüdge, E. Schöll, T. Erneux, Int. J. Bifurcation Chaos **22**(10), 1250246 (2012). doi:10.1142/s021812741250246x
6. J. Ohtsubo, *Semiconductor Lasers: Stability, Instability and Chaos* (Springer, Berlin, 2005)
7. J. Mørk, B. Tromborg, J. Mark, IEEE J. Quantum Electron. **28**, 93 (1992)
8. A. Uchida, Optical Communication with Chaotic Lasers. Applications of Nonlinear Dynamics and Synchronization (Wiley, 2012)
9. L. Appeltant, M.C. Soriano, G. Van der Sande, J. Danckaert, S. Massar, J. Dambre, B. Schrauwen, C.R. Mirasso, I. Fischer, Nat. Commun. **2**, 468 (2011). doi:10.1038/ncomms1476
10. A. Argyris, D. Syvridis, L. Larger, V. Annovazzi-Lodi, P. Colet, I. Fischer, J. García-Ojalvo, C.R. Mirasso, L. Pesquera, K.A. Shore, Nature **438**, 343 (2005). doi:10.1038/nature04275
11. A. Argyris, M. Hamacher, K.E. Chlouverakis, A. Bogris, D. Syvridis, Phys. Rev. Lett. **100**(19), 194101 (2008). doi:10.1103/physrevlett.100.194101
12. T. Heil, J. Mulet, I. Fischer, C.R. Mirasso, M. Peil, P. Colet, W. Elsäßer, IEEE J. Quantum Electron. **38**(9), 1162 (2002). doi:10.1109/jqe.2002.801950
13. L. Larger, M.C. Soriano, D. Brunner, L. Appeltant, J.M. Gutierrez, L. Pesquera, C.R. Mirasso, I. Fischer, Opt. Express **20**(3), 3241 (2012). doi:10.1364/oe.20.003241
14. K. Vandoorne, P. Mechet, T. Van Vaerenbergh, M. Fiers, G. Morthier, D. Verstraeten, B. Schrauwen, J. Dambre, P. Bienstman, Nat. Photonics **5**(3541) (2014)
15. E.C. Mos, J.J.L. Hoppenbrouwers, M.T. Hill, M.W. Blum, J.J.H.B. Schleipen, H. de Waardt, I.E.E.E. Trans, Neural Netw. **11**(4), 988 (2000)
16. A. Hurtado, I.D. Henning, M.J. Adams, Opt. Express **18**(24), 25170 (2010)

17. T. Heil, I. Fischer, W. Elsäßer, J. Mulet, C.R. Mirasso, Phys. Rev. Lett. **86**, 795 (2001)
18. H. Erzgräber, D. Lenstra, B. Krauskopf, E. Wille, M. Peil, I. Fischer, W. Elsäßer, Opt. Commun. **255**(4–6), 286 (2005)
19. H. Erzgräber, B. Krauskopf, D. Lenstra, SIAM, J. Appl. Dyn. Syst. **5**(1), 30 (2006). doi:10.1137/040619958
20. N. Gross, W. Kinzel, I. Kanter, M. Rosenbluh, L. Khaykovich, Opt. Commun. **267**(2), 464 (2006)
21. V. Flunkert, O. D'Huys, J. Danckaert, I. Fischer, E. Schöll, Phys. Rev. E 79, 065201 (R) (2009). doi:10.1103/physreve.79.065201
22. V. Flunkert, E. Schöll, New. J. Phys. **14**, 033039 (2012). doi:10.1088/1367-2630/14/3/033039
23. S. Yanchuk, K.R. Schneider, L. Recke, Phys. Rev. E **69**(5), 056221 (2004). doi:10.1103/physreve.69.056221
24. G. Kozyreff, A.G. Vladimirov, P. Mandel, Phys. Rev. Lett. **85**(18), 3809 (2000)
25. G. Kozyreff, A.G. Vladimirov, P. Mandel, Phys. Rev. E **64**, 016613 (2001)
26. A.G. Vladimirov, G. Kozyreff, P. Mandel, EPL (Europhysics Letters) **61**(5), 613 (2003)
27. A. Amann, in *Nonlinear Laser Dynamics—From Quantum Dots to Cryptography*, ed. by K. Lüdge (WILEY-VCH Weinheim, 2011)
28. S. Yanchuk, Y. Maistrenko, E. Mosekilde, Math. Comp. Simul **54**, 491 (2001)
29. W. Poel, A. Zakharova, E. Schöll, Phys. Rev. E **91**, 022915 (2015). doi:10.1103/physreve.91.022915
30. H.J. Wünsche, S. Bauer, J. Kreissl, O. Ushakov, N. Korneyev, F. Henneberger, E. Wille, H. Erzgräber, M. Peil, W. Elsäßer, I. Fischer, Phys. Rev. Lett. **94**, 163901 (2005)
31. K. Hicke, O. D'Huys, V. Flunkert, E. Schöll, J. Danckaert, I. Fischer, Phys. Rev. E **83**, 056211 (2011). doi:10.1103/physreve.83.056211
32. M.C. Soriano, G. Van der Sande, I. Fischer, C.R. Mirasso, Phys. Rev. Lett. **108**, 134101 (2012). doi:10.1103/physrevlett.108.134101
33. Y. Kuramoto, D. Battogtokh, Nonlinear Phen. Complex Syst. **5**(4), 380 (2002)
34. D.M. Abrams, S.H. Strogatz, Phys. Rev. Lett. **93**(17), 174102 (2004). doi:10.1103/physrevlett.93.174102
35. N.C. Rattenborg, C.J. Amlaner, S.L. Lima, Neurosci. Biobehav. Rev. **24**, 817 (2000)
36. C.R. Laing, C.C. Chow, Neural Comput. **13**(7), 1473 (2001)
37. H. Sakaguchi, Phys. Rev. E **73**(3), 031907 (2006). doi:10.1103/physreve.73.031907
38. A.E. Motter, S.A. Myers, M. Anghel, T. Nishikawa, Nature Physics **9**, 191 (2013). doi:10.1038/nphys2535
39. J.C. Gonzalez-Avella, M.G. Cosenza, M.S. Miguel, Physica A **399**, 24 (2014). doi:10.1016/j.physa.2013.12.035
40. O.E. Omel'chenko, M. Wolfrum, S. Yanchuk, Y. Maistrenko, O. Sudakov, Phys. Rev. E **85**, 036210 (2012). doi:10.1103/physreve.85.036210
41. M.J. Panaggio, D.M. Abrams, Phys. Rev. Lett. **110**, 094102 (2013)
42. M.J. Panaggio, D.M. Abrams, Nonlinearity **28**, R67 (2015). doi:10.1088/0951-7715/28/3/r67
43. S.I. Shima, Y. Kuramoto, Phys. Rev. E **69**(3), 036213 (2004). doi:10.1103/physreve.69.036213
44. C. Gu, G. St-Yves, J. Davidsen, Phys. Rev. Lett. **111**, 134101 (2013)
45. C.R. Laing, Phys. Rev. E **81**(6), 066221 (2010). doi:10.1103/physreve.81.066221
46. G. Bordyugov, A. Pikovsky, M. Rosenblum, Phys. Rev. E **82**(3), 035205 (2010). doi:10.1103/physreve.82.035205
47. G.C. Sethia, A. Sen, Phys. Rev. Lett. **112**, 144101 (2014). doi:10.1103/physrevlett.112.144101
48. A.M. Hagerstrom, T.E. Murphy, R. Roy, P. Hövel, I. Omelchenko, E. Schöll, Nat. Phys. **8**, 658 (2012). doi:10.1038/nphys2372
49. E.A. Viktorov, T. Habruseva, S.P. Hegarty, G. Huyet, B. Kelleher, Phys. Rev. Lett. **112**, 224101 (2014)
50. M.R. Tinsley, S. Nkomo, K. Showalter, Nat. Phys. **8**, 662 (2012). doi:10.1038/nphys2371
51. E.A. Martens, S. Thutupalli, A. Fourriere, O. Hallatschek, Proc. Nat. Acad. Sci. **110**, 10563 (2013). doi:10.1073/pnas.1302880110
52. T. Kapitaniak, P. Kuzma, J. Wojewoda, K. Czolczynski, Y. Maistrenko, Sci. Rep. **4**, 6379 (2014)

53. L. Larger, B. Penkovsky, Y. Maistrenko, Phys. Rev. Lett. **111**, 054103 (2013). doi:10.1103/physrevlett.111.054103
54. L. Larger, B. Penkovsky, Y. Maistrenko, Nat. Comm. **6**, 7752 (2015)
55. M. Wickramasinghe, I.Z. Kiss, PLoS ONE **8**(11), e80586 (2013). doi:10.1371/journal.pone.0080586
56. L. Schmidt, K. Schonleber, K. Krischer, V. Garcia-Morales, Chaos **24**(1), 013102 (2014). doi:10.1063/1.4858996
57. P. Ashwin, O. Burylko, Chaos **25**, 013106 (2015). doi:10.1063/1.4905197
58. L. Schmidt, K. Krischer, Phys. Rev. Lett. **114**, 034101 (2015). doi:10.1103/physrevlett.114.034101
59. T. Banerjee, Europhys. Lett. **110**, 60003 (2015). arXiv:1409.7895v1
60. A. Yeldesbay, A. Pikovsky, M. Rosenblum, Phys. Rev. Lett. **112**, 144103 (2014). doi:10.1103/physrevlett.112.144103
61. J. Xie, E. Knobloch, H.C. Kao, Phys. Rev. E **90**, 022919 (2014). doi:10.1103/physreve.90.022919
62. T. Erneux, P. Glorieux, *Laser Dynamics* (Cambridge University Press, UK, 2010)
63. S. Melnik, G. Huyet, A.V. Uskov, Opt. Express **14**(7), 2950 (2006)
64. B. Lingnau, K. Lüdge, W.W. Chow, E. Schöll, Phys. Rev. E **86**(6), 065201(R) (2012). doi:10.1103/physreve.86.065201
65. B. Lingnau, W.W. Chow, E. Schöll, K. Lüdge, New J. Phys. **15**, 093031 (2013)
66. A. Zakharova, T. Vadivasova, V. Anishchenko, A. Koseska, J. Kurths, Phys. Rev. E **81**, 011106 (2010). doi:10.1103/physreve.81.011106
67. S.H.L. Klapp, S. Hess, Phys. Rev. E **81**, 051711 (2010). doi:10.1103/physreve.81.051711
68. F.T. Arecchi, G.L. Lippi, G.P. Puccioni, J.R. Tredicce, Opt. Commun. **51**(5), 308 (1984)
69. C.Z. Ning, H. Haken, Appl. Phys. B **55**(2), 117 (1992). doi:10.1007/bf00324060
70. H. Haken, *Light*, vol. 2 (North-Holland, Amsterdam, 1985)
71. G.C. Sethia, A. Sen, G.L. Johnston, Phys. Rev. E **88**(4), 042917 (2013)

Chapter 19
Feedback Control of Colloidal Transport

Robert Gernert, Sarah A.M. Loos, Ken Lichtner and Sabine H.L. Klapp

Abstract We review recent work on feedback control of one-dimensional colloidal systems, both with instantaneous feedback and with time delay. The feedback schemes are based on measurement of the average particle position, a natural control target for an ensemble of colloidal particles, and the systems are investigated via the Fokker-Planck equation for overdamped Brownian particles. Topics include the reversal of current and the emergence of current oscillations, transport in ratchet systems, and the enhancement of mobility by a co-moving trap. Beyond the commonly considered case of non-interacting systems, we also discuss the treatment of colloidal interactions via (dynamical) density functional theory and provide new results for systems with attractive interactions.

19.1 Background

Within the last years, feedback control [1] of colloidal systems, that is, nano- to micron-sized particles in a thermally fluctuating bath of solvent particles, has become a focus of growing interest. Research in that area is stimulated, on the one hand, by the fact that colloidal systems have established their role as theoretically and experimentally accessible model systems for equilibrium and nonequilibrium phenomena [2–4] in statistical physics. Thus, colloidal systems are prime candidates to explore *concepts* of feedback control and its consequences. On the other hand, feedback control of colloidal particles has nowadays found its way into experimental applications. Recent examples include control of colloids, bacteria and artificial motors in microfluidic set-ups [5–7], biomedical engineering [8], and the manipulation of colloids by feedback traps [9–11]. Further, a series of recent experiments involving feedback control aims at exploring fundamental concepts of thermodynamics and information exchange in small stochastic systems [11–13]. As a consequence of these developments, feedback control of colloids is now an emerging field with

R. Gernert · S.A.M. Loos · K. Lichtner · S.H.L. Klapp (✉)
Institut für Theoretische Physik, Technische Universität Berlin,
Hardenbergstraße 36, 10623 Berlin, Germany
e-mail: klapp@physik.tu-berlin.de

© Springer International Publishing Switzerland 2016
E. Schöll et al. (eds.), *Control of Self-Organizing Nonlinear Systems*,
Understanding Complex Systems, DOI 10.1007/978-3-319-28028-8_19

relevance in diverse contexts, including optimization of self-asssembly processes [14], and the manipulation of flow-induced behavior [15, 16] and rheology [17, 18].

Within this area of research, the present article focuses on feedback control of *one-dimensional* (1D) colloidal transport. Transport in 1D systems without feedback control has been extensively studied in the past decades, yielding a multitude of analytical and numerical results (see, e.g., [19–21]). These have played a major role in understanding fundamentals of diffusion through complex landscapes and the role of noise. Paradigm examples of such 1D systems are Brownian particles driven through a periodic 1D "washboard" potential, or ratchet systems (Brownian motors) operating by a combination of asymmetric static potentials and time-periodic forces. It is therefore not surprising that the first applications of feedback control of colloids involve just these kinds of systems, pioneering studies being theoretical [22–24] and experimental [25] investigations of a feedback-controlled 1D "flashing ratchet". Here it has been shown that the fluctuation-induced directed transport in the ratchet system can be strongly enhanced by switching not under an externally defined, "open-loop" protocol, but with a "closed-loop" feedback scheme.

From the theoretical side, most studies focus on manipulating *single* colloidal particles (or an ensemble of non-interacting particles) in a 1D set-up, the basis being an overdamped or underdamped Langevin equation. The natural control target is then the position or velocity of the colloidal particle at hand. Within this class, many earlier studies assume *instantaneous* feedback, i.e., no time lag between measurement and control action [22]. However, there is now increasing interest in exploring systems with time delay [23, 24, 26–28]. The latter typically arises from a time lag between the detection of a signal and the control action, an essentially omnipresent situation in experimental setups. Traditionally, time delay was often considered as a perturbation; for example, in some ratchet systems it reduces the efficiency of transport [23]. However, it is known from other areas that time delay can also have significant positive effects. For example, it can stabilize desired stationary states in sheared liquid crystals [16], it can be used to probe coherent effects in electron transport in quantum-dot nanostructures [29], and it can generate new effects such as current reversal [30, 31] and spatiotemporal oscillations in extended systems [32, 33]. Moreover, time delay can have a *stabilizing* effect on chaotic orbits, a prime example being Pyragas' control scheme [34] of time-delayed feedback control [35]. Apart from the effects of time delay on the dynamical behavior, a further issue attracting increasing attention is the theoretical treatment of time-delayed, feedback-controlled (single-particle) systems via stochastic thermodynamics [28, 36–38].

Finally, yet another major question concerns the role of particle interactions. We note that, even in the idealized situation of a (dilute) suspension of non-interacting particles, feedback can induce *effective* interactions if the protocol involves system-averaged quantities [22]. For many real colloidal systems, however, direct interactions between the colloids stemming e.g., from excluded volume effects, charges on the particles' surfaces, or (solvent-induced) depletion effects cannot be neglected. Within the area of transport under feedback, investigations of the role of interactions have started only very recently. Understanding the impact of interactions clearly becomes particularly important when one aims at feedback-controlling systems with

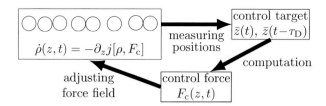

Fig. 19.1 Concept of feedback control for a system of (interacting) colloids. The control target is the average particle position $\bar{z}$ measured either at time t or at a delayed time $t - \tau_D$. This average position determines the control force $F_c(z, t)$. The system is investigated based on the Fokker-Planck equation where ρ is the probability density and j is the current

phase transitions, pattern (or cluster-)forming systems, and systems with collective dynamic phenomena such as synchronization.

For an interacting, 1D colloidal system, one natural control variable is the *average* particle position, which is experimentally accessible e.g. by video microscopy. Theoretically, the average position can be calculated from the time-dependent probability distribution $\rho(z, t)$, whose dynamics is determined by the Fokker-Planck (FP) equation [39] (for overdamped particles often called Smoluchowski equation). About two decades ago, Marconi and Tarazona [40, 41] have proposed a special type of FP equation, the so-called dynamical density functional theory, which is suitable for an interacting, overdamped system of colloidal particles. Within this framework, dynamical correlations are approximated adiabatically, and correlation effects enter via a free energy functional.

In this spirit, we have recently started to investigate a number of feedback-controlled 1D systems based on the FP formalism [31, 32, 42, 43]. The general scheme of feedback control used in these studies is sketched in Fig. 19.1. The purpose of the present article is to summarize main results of these investigations. We cover both, non-interacting systems and interacting systems, including new results for systems with attractive interactions. Also, we discuss examples with instantaneous feedback and with time delay. We note that, in presence of time delay, the connection between the FP equation and the underlying Langevin equation is not straightforward (see, e.g., Refs. [28, 36, 44–46]), and this holds particularly for control schemes involving *individual* particle positions. However, here we consider the *mean* particle position as control target. For this situation, the results become consistent with those from a corresponding Langevin equation (with delayed force), if the number of realizations goes to infinity [42].

19.2 Theory

We consider the motion of a system of N overdamped colloidal particles at temperature T in an external, one-dimensional, periodic potential $V_{\text{ext}}(z)$ supplemented by a constant driving force F_{ext}, where z is the space coordinate. The particles are assumed to be spherical, with the size being characterized by the diameter σ. In addition to thermal fluctuations, each particle experiences a time-dependent force $F_c(z, t)$ which we will later relate to feedback control. We also allow for direct particle interactions which are represented by an interaction field $V_{int}(z)$ to be specified later. The dynamics is investigated via the FP equation [39] for the space- and time-dependent one-particle density $\rho(z, t) = \langle \sum_{i=1}^{N} \delta(z - z_i(t)) \rangle$ (where $\langle \ldots \rangle$ denotes a noise average), yielding

$$\partial_t \rho(z, t) = \partial_z \left[\gamma^{-1} \left(V'_{ext}(z) - F_{ext} - F_c(z, t) + \partial_z V_{int}(z, \rho) \right) \rho(z, t) + D_0 \partial_z \rho(z, t) \right]$$
$$= -\partial_z j(z, t), \tag{19.1}$$

where D_0 is the short-time diffusion coefficient, satisfying the fluctuation-dissipation theorem [39] $D_0 = k_B T / \gamma$ (with k_B and γ being the Boltzmann and the friction constant, respectively), and $j(z, t)$ is the probability current. Throughout the paper, we measure the time t in units of the "Brownian" time scale, $\tau_B = \sigma^2 / D_0$. For typical, μm-sized particles τ_B is about 1 s [25, 47] or larger [48].

Feedback control is implemented through the time-dependent force $F_c(z, t)$. Specifically, we assume this force to depend on the (time-dependent) average position

$$\bar{z}(t) = \frac{1}{N} \int dz \, \rho(z, t) \, z, \tag{19.2}$$

where we have used that $N = \int dz \, \rho(z, t)$. The density is calculated with periodic boundary conditions, that is, $\rho(z + L_{sys}, t) = \rho(z, t)$ with L_{sys} being the system size. Thus, the time dependency of $F_c(z, t)$ arises through the internal state of the system.

Our reasoning behind choosing the *mean* particle position rather than the individual position as control target is twofold: First, within the FP equation treatment we have no access to the particle's position for a given realization of noise, because the latter has already been averaged out. This is in contrast to previous studies using Langevin equations [25, 26, 49] where the dynamical variable is the particle position itself. Second, the mean position is an experimentally accessible quantity, which can be monitored, e.g., by video microscopy [26].

19.3   Non-interacting Systems Under Feedback Control

19.3.1   Particle in a Co-moving trap

As a starting point [43], we consider a single particle (or non-interacting colloids in a dilute suspension) under the combined influence of a static, "washboard" potential,

$$V_{ext}(z) = u(z) = u_0 \sin^2(\pi z / a) \tag{19.3}$$

supplemented by a constant tilting force F_{ext} and the feedback force

$$F_c(z, t) = -\partial_z V_{DF}(z, t) \tag{19.4}$$

derived from the potential

$$V_{DF}(z, t) = \eta(z - \bar{z}(t))^2 . \tag{19.5}$$

Physically speaking, Eq. (19.5) describes a parabolic confinement, which moves instantaneously with the mean position, thus resembling the potential seen by particles in moving optical traps [50, 51]. The strength of the harmonic confinement, η, is set to constant.

In the absence of the potential barriers ($u_0 = 0$) the problem can be solved analytically. Starting from the initial condition $\rho(z, t=0) = \delta(z - z_0)$ one finds $\bar{z}(t) = (F_{ext}/\gamma) t + z_0$, yielding the mobility

$$\mu := \lim_{t \to \infty} \frac{\partial_t \bar{z}}{F_{ext}} = \frac{1}{\gamma}. \tag{19.6}$$

Moreover, the mean-squared displacement describing the width of the distribution,

$$w(t) = \langle (z - \bar{z}(t))^2 \rangle \tag{19.7}$$

becomes

$$w(t) = \frac{k_B T}{2\eta} \left(1 - e^{-4\eta t/\gamma}\right), \tag{19.8}$$

showing that density fluctuations *freeze* in the long-time limit. Interestingly, the same type of behavior of $w(t)$ occurs in a model of quantum feedback control [52].

For non-vanishing potential barriers and in presence of feedback, Eq. (19.1) has to be solved numerically. Figure 19.2a, b shows representative results for the average position and the width.

Upon increase of η the slope of $\bar{z}(t)$ first increases but then decreases again. A further characteristic feature is the emergence of oscillations in $\bar{z}(t)$, the velocity $v(t) = \partial \bar{z}/\partial t$ and the width $w(t)$. These oscillations can be traced back to the periodic reconstruction of the effective energy landscape, $V_{DF}(z, t) + u(z)$, which consists of a periodic increase and decrease of the energy barriers [43]. The period $\mathcal{T}$ of oscillations roughly coincides with the inverse Kramers rate [21, 39], which is the relevant time scale for the slow barrier-crossing. Also, the regime of pronounced oscillations partly coincides with the regime where a "speed up" of the motion occurs. We quantify this "speed up" via an average mobility $\mu = \bar{v}/F_{ext}$ based on the time-averaged velocity $\bar{v} = \mathcal{T}^{-1} \int_{t_1}^{t_1+\mathcal{T}} dt\, v(t)$. Figure 19.2c shows μ/μ_0 depending on η, where $\mu_0 \approx 1.2 \cdot 10^{-4}/\gamma$ is the mobility of the uncontrolled system ($\eta=0$) with the same external potential [39, 53]. For small η, we find $\mu \approx \mu_0$. At intermediate values of η the mobility shows a global maximum. This maximum occurs in the range of η

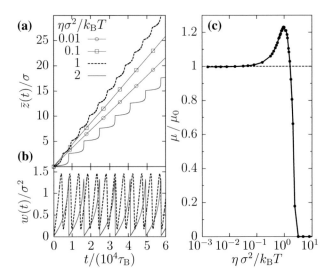

Fig. 19.2 Single particle in a trap [43]. **a** Average position and **b** width of the density distribution as functions of time. **c** Mobility (normalized by the mobility of the uncontrolled system) as function of control strength

where the oscillation periods of $v(t)$ are about (in fact, somewhat smaller than) the inverse Kramers rate. For even larger values of η one observes a sharp decrease of the mobility to zero. Here, the confinement induced by the trap becomes so strong that barrier diffusion is prohibited (note that this effect would be absent if the trap was moved by an externally imposed velocity). Overall, the increase of mobility by the co-moving trap is about twenty percent. As we will see in Sect. 19.4, a much more significant enhancement of mobility occurs when the particles interact.

19.3.2 Feedback Controlled ratchet

In the second example [42], $V_{\text{ext}}(z)$ is a periodic, piecewise linear, "sawtooth" potential [25, 54, 55] defined by $V_{\text{ext}}(z + a) = V_{\text{ext}}(z)$ with

$$V_{\text{ext}}(z) = \begin{cases} u_0 z/(\alpha a), & 0 < z \le \alpha a, \\ u_0 z/((\alpha - 1)a), & (\alpha - 1)a < z \le 0, \end{cases} \tag{19.9}$$

where u_0 and a are again the potential height and the period, respectively, and $\alpha \in [0, 1]$ is the asymmetry parameter. The potential minimum within the central interval $S = [(\alpha - 1)a, \alpha a]$ is at $z = z_{\min} = 0$. We further assume periodic boundary conditions such that $\rho(z + a, t) = \rho(z, t)$ (i.e., $a = L_{\text{sys}}$), and we calculate the mean position from Eq. (19.2) with the integral restricted to the interval S.

In the absence of any further force ($F_{\text{ext}} = 0$ and $F_c = 0$) beyond that arising from $V_{\text{ext}}(z)$, the system approaches for $t \to \infty$ an equilibrium state and thus there is no transport (i.e., no net particle current). It is well established, however, that by supplementing $V_{\text{ext}}(z)$ by a time-dependent oscillatory force (yielding a "rocking ratchet"), the system is permanently out of equilibrium and macroscopic transport can be achieved [20, 56, 57].

Here we propose an alternative driving mechanism which is based on a *time-delayed* feedback force $F_c(z, t)$ depending on the average particle position at an earlier time. Specifically,

$$F_c(t) = -F \cdot \text{sign}(\bar{z}(t - \tau_D) - z_0), \qquad (19.10)$$

where τ_D is the delay time, F is the amplitude (chosen to be positive), z_0 is a fixed position within the range $[0, \alpha a]$ (where V_{ext} increases with z), and the sign function is defined by $\text{sign}(x) = +1 \, (-1)$ for $x > 0 \, (x < 0)$. From Eq. (19.10) one sees that the feedback force changes its sign whenever the delayed mean particle position $\bar{z}(t - \tau_D)$ becomes smaller or larger than z_0; we therefore call z_0 the "switching" position.

In the limit $\tau_D \to 0$ any transport vanishes since the feedback force leads to a trapping of the particle at z_0. This changes at $\tau_D > 0$. Consider a situation where the mean particle position at time t is at the right side of z_0, while it has been on the left side at time $t - \tau_D$. In this situation the force $F_c(t)$ points *away* from z_0 (i.e., $F_c > 0$), contrary to the case $\tau_D = 0$. Thus, the particle experiences a driving force towards the next potential valley, which changes only when the delayed position becomes larger than z_0. The force then points to the left until the delayed position crosses z_0 again. This oscillation of the force, together with the asymmetry of $V_{\text{ext}}(z)$, creates a ratchet effect.

To illustrate the effect, we present in Fig. 19.3a exemplary data for the time evolution of the mean particle position, $\bar{z}(t)$, which determines the control force. It is seen that $\bar{z}(t)$ displays regular oscillations between values above and below z_0 for both force amplitudes considered. The period of these oscillations, T, is roughly twice the delay time. We note that the precise value of the period as well as the shape of the oscillations depend on the values of F and z_0 [42]. Due to the oscillatory behavior of $\bar{z}(t)$ the delayed position $\bar{z}(t - \tau_D)$ oscillates around z_0 as well, yielding a periodic switching of the feedback force between $+F$ and $-F$ with the same period as that observed in $\bar{z}(t)$ (see Fig. 19.3b). The oscillatory behavior of the feedback force then induces a net current defined as

$$J = \frac{1}{T} \int_{t_1}^{t_1 + T} dt' \, v(t') \qquad (19.11)$$

where t_1 is an arbitrary time after the "equilibration" period, $v(t) = \int_S dz \, j(z, t)$ is the velocity, and $j(z, t)$ is calculated from the FP equation (19.1) with periodic boundary conditions. Numerical results for J in dependence of the delay time τ_D and the force amplitude are plotted in Fig. 19.3c.

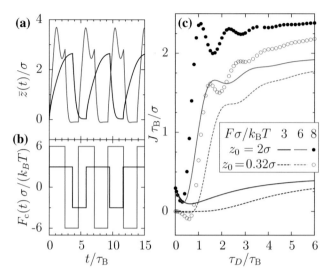

Fig. 19.3 Feedback-controlled ratchet [42]. **a** Average position and **b** feedback force at two values of the force amplitude for delay time $\tau_D = 2\tau_B$ and $z_0 = 2\sigma$. **c** Average current (for different amplitudes and positions z_0 of the control force) as function of delay time. Parameters are $L = 8\sigma$ and $a = 0.8$

The results clearly show that the time delay involved in the feedback protocol is *essential* for the creation of a ratchet effect and, thus, for a nonzero net current. For finite delay times ($\tau_D \gtrsim 3\tau_B$), the current generally increases with τ_D. Also, for a fixed τ_D, J increases with increasing force amplitude (or with larger z_0). At small delay times ($\tau_D \lesssim 3\tau_B$) the behavior of the function $J(\tau_D)$ is sensitive (in fact, behaves non-monotonous) with respect to both, F and z_0 [42].

Given the feedback-induced transport, it is interesting to compare the resulting current with that generated by a conventional rocking ratchet. The latter is defined by replacing the force $F_c(t)$ in Eq. (19.1) with a time-periodic (rectangular) force $F_{osc}(t) = -F \cdot \text{sign}\left[\cos\left((2\pi/T')\,t\right)\right]$, where the period T' is set to the resulting period $\mathcal{T}$ in the feedback-controlled case. While the general behavior of the current (that is, small values of J for small periods, saturation at large values for large periods) is similar for both, open-loop and closed-loop systems [42], the actual values of J for a given period strongly depend on the type of control. It turns out that, for a certain range of switching positions (and not too large delay times), the net current in the feedback-controlled system is actually *enhanced* relative to the open-loop system.

A somewhat subtle aspect of the present model is that we introduce feedback on the level of the Fokker-Planck equation describing the evolution of the probability density. This is different from earlier studies based on the Langevin equation (see, e.g., [25, 26, 49]), where the feedback is applied directly to the position of one particle, $\chi_i(t)$, or to the average of N particle positions $N^{-1}\sum_{i=1}^{N}\chi_i(t)$. Introducing feedback control in such systems implies to introduce effective *interactions* between the particles. As a consequence, the transport properties in these particle-based models depend explicitly on the number of particles, N. From the perspective of these Langevin-based models, the present model corresponds to the "mean field" limit $N \to \infty$ (for a more detailed discussion, see [42]).

19.4 Impact of Particle Interactions

We now turn to (one-dimensional) transport in systems of interacting colloids. To construct the corresponding contribution $V_{int}(z)$ in the FP equation (19.1), we employ concepts from dynamical density functional theory (DDFT) [40, 41, 58]. Within the DDFT, the exact FP equation for an overdamped system with (two-particle) interactions is approximated such that non-equilibrium two-particle correlations at time t are set to those of an equilibrium system with density $\rho(z, t)$. This *adiabatic approximation* allows to formally relate the interaction contribution to the FP equation to the excess free energy of an equilibrium system [whose density profile $\rho_{eq}(z)$ coincides with the instantaneous density profile $\rho(z, t)$]. It follows that

$$V_{int}(z) = \frac{\delta \mathcal{F}^{int}[\rho]}{\delta \rho(z, t)} \tag{19.12}$$

where $\mathcal{F}^{int}[\rho]$ is the excess (interaction) part of the *equilibrium* free energy functional. Thus, one can use well-established equilibrium approaches as an input into the (approximate) dynamical equations of motion.

19.4.1 Current reversal

Our first example involves "ultra-soft" particles interacting via the Gaussian core potential (GCM)

$$v_{GCM}(z, z') = \varepsilon \exp\left(-\frac{(z - z')^2}{\sigma^2}\right), \tag{19.13}$$

(with $\varepsilon > 0$), a typical coarse-grained potential modeling a wide class of soft, partially penetrable macroparticles (e.g., polymer coils) with effective (gyration) radius σ [59, 60]. Due to the penetrable nature of the Gaussian potential which allows an, in principle, infinite number of neighbors, the equilibrium structure of the GCM model can be reasonably calculated within the mean field (MF) approximation

$$\mathcal{F}^{int}[\rho] = \frac{1}{2} \int dz \int dz' \rho(z, t) v^{GCM}(z, z') \rho(z', t). \tag{19.14}$$

The MF approximation is known to become quasi-exact in the high-density limit and yields reliable results even at low and moderate densities [60].

The particles are subject to an external washboard potential of the form defined in Eq. (19.3) plus a constant external force $F_{\text{ext}} = 3k_B T/\sigma$. To implement feedback control we use the time-delayed force

$$F_c(z, t) = F_c(t) = -K_0 \left(1 - \tanh \left[\frac{N}{\sigma} (\bar{z}(t) - \bar{z}(t - \tau_D)) \right] \right), \qquad (19.15)$$

which involves the difference between the average position at times t and $t - \tau_D$. By construction, $F_c(t)$ vanishes in the absence of time delay ($\tau_D = 0$). The idea to use a feedback force depending on the difference of the control target at two times is inspired by the time-delayed feedback control method suggested by Pyragas [34] in the context of chaos control. Indeed, the original idea put foward by Pyragas was to stabilize certain unstable *periodic* states in a non-invasive way (notice that $F_c(t)$ vanishes if $\bar{z}(t)$ performs periodic motion with period τ_D). Later, Pyragas control has also been used to stabilize steady states (for a recent application in driven soft systems, see [16]). We also note that a similar strategy has been used on the level of an (underdamped) Langevin equation by Hennig et al. [61].

The impact of the control force $F_c(t)$ on the average particle position $\bar{z}(t)$ is illustrated in Fig. 19.4, where we have chosen a moderate value of the driving force (yielding rightward motion in the uncontrolled system) and a delay time equal to the "Brownian" time, $\tau_D = \tau_B$. In the absence of control ($K = K_0 \sigma/k_B T = 0$) the average position just increases with t reflecting rightward motion, as expected. The slope of the function $\bar{z}(t)$ at large t may be interpreted as the long-time velocity $v_\infty = \lim_{t \to \infty} d\bar{z}(t)/dt$. Increasing K from zero, the velocity first decreases until the motion stops (i.e., the time-average of $\bar{z}(t)$ becomes constant) at $K = 3$. This value corresponds to a balance between control force and biasing driving force. Here, the average position $\bar{z}(t)$ displays an *oscillating* behavior changing between small backward motion and forward motion, with a period of about $5\tau_B$ (that is, much larger than the delay time). These oscillations are accompanied by oscillations of the *effective* force $F_{\text{eff}} = F_c(t) + F_{\text{ext}}$ around zero (notice the restriction $-2K_0 \leq F_c(t) \leq 0$). Consistent with this observation, there is no directed net motion. A more detailed discussion of the onset of oscillations is given in Ref. [32], where we have focussed on a non-interacting system ($\epsilon = 0$). Indeed, for the present situation we have found that a non-interacting ensemble subject to the Pyragas control (19.15) behaves qualitatively similar to its interacting counterpart. Moreover, for the non-interacting case, we have identified the onset of oscillations as supercritical Hopf bifurcation.

Turning back to Fig. 19.4a we see that even larger control amplitudes ($K > 3$) result in a significant backward motion, i.e., $\bar{z}(t)$ and v_∞ become negative. Thus, the feedback control induces *current reversal*.

To complete the picture, we plot in Fig. 19.4b the long-time velocity v_∞ (averaged over the oscillations of $\bar{z}(t)$, if present) as function of the control amplitude. We have included data for different delay times τ_D and different interaction (i.e., repulsion) strengths ε. All systems considered display a clear current reversal at $K = 3$ (balance

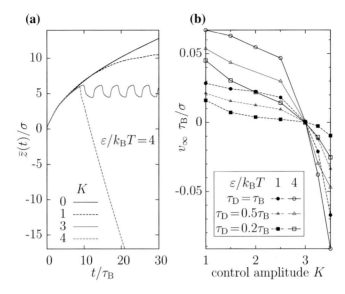

Fig. 19.4 Feedback-controlled particle in a washboard potential with tilt [31]. **a** Average particle position as a function of time for various control amplitudes K. **b** Long-time velocity as function of control amplitude for various coupling strengths and time delays

between feedback and bias), where the velocity v_∞ changes from positive to negative values irrespective of ε and τ_D. Regarding the role of the delay we find that, at fixed coupling strength ε, v_∞ decreases in magnitude when the delay time decreases from $\tau_D = \tau_B$ towards $\tau_D = 0.2\tau_B$. In other words, the time delay *supports* the current reversal in the parameter range considered. Regarding the interactions, Fig. 19.4b shows that reduction of ε (at fixed τ_D) yields a decrease of the magnitude of v_∞ as compared to the case $\varepsilon/k_B T = 4$. Thus, repulsive interactions between the particles yield a "speed up" of motion.

19.4.2 Interacting Particles in a Trap

As a second example illustrating the impact of particle interactions we turn back to the feedback setup discussed in Sect. 19.3.1, that is, feedback via a co-moving harmonic trap. In Sect. 19.3.1 we have discussed this situation for a single colloidal particle driven through a washboard potential. In that case, feedback leads to a slight, yet no dramatic increase of the transport efficiency as measured by the mobility.

This changes dramatically when the particles interact. In [43] we have explored the effect of two types of repulsive particle interactions, one of them being the Gaussian core potential introduced in Eq. (19.13). Here we focus on results for hard particles described by the interaction potential

$$v_{\text{hard}}(z, z') = \begin{cases} 0 & , \text{ for } |z - z'| \geq \sigma \\ \infty & , \text{ for } |z - z'| < \sigma \end{cases} . \tag{19.16}$$

For one-dimensional systems of hard spheres there exists an exact free energy functional [62] derived by Percus, which corresponds to the one-dimensional limit of fundamental measure theory [63]. This functional is given by

$$\mathcal{F}^{\text{int}}[\rho] = -\frac{1}{2} \int dz \, \ln\left(1 - \ell[\rho, z, t]\right) [\rho(z + \frac{\sigma}{2}, t) + \rho(z - \frac{\sigma}{2}, t)], \tag{19.17}$$

where

$$\ell[\rho, z, t] = \int_{z-\sigma/2}^{z+\sigma/2} dz' \, \rho(z', t) \tag{19.18}$$

is the local packing fraction. Corresponding results for the mobility are shown in Fig. 19.5a. For appropriately chosen lattice constants ($a > \sigma$), we observe a dramatic increase of μ with η and N over several orders of magnitude. This is in striking contrast to the corresponding single-particle result (see dotted line in Fig. 19.5a), and similar behavior occurs for ultra-soft particles [43]. In fact, for specific values η and N, the mobility increases up to the maximal possible value $\mu = 1/\gamma$, the mobility of free (overdamped) motion.

The dramatic enhancement of transport can be understood by considering the (time-dependent) energy landscape formed by the combination of external potential $u(z)$, feedback potential $V_{\text{DF}}(z, t)$ (see Eq. (19.5)) and interaction contribution $V_{\text{int}}(z)$ [43]. It turns out that the $V_{\text{int}}(z, t)$ develops peaks at the minima of the potential $V_{\text{DF}} + u$. The interaction contribution thus tends to "fill" the valleys, implying that the energy barriers between the minima decrease. This results in an enhancement of diffusion over the barriers and thus, to faster transport. In other words, interacting particles "help each other" to overcome the external barriers.

Delayed trap

Given that any experimental setup of our feedback control involves a finite time to measure the control target (i.e., the mean position), we briefly consider the impact of time delay. To this end we change the control potential defined in Eq. (19.5) into the expression

$$V_{\text{DF}}^{\text{delay}}(z, \rho) = \eta \left(z - \bar{z}(t - \tau_{\text{D}})\right)^2 . \tag{19.19}$$

We now consider two special cases involving hard particles, where the non-delayed feedback control leads to a particularly high mobility. Numerical results are plotted in Fig. 19.5b, showing that the delay causes a pronounced decrease of mobility. To estimate the consequences for a realistic colloidal system, we note that feedback mechanisms can be implemented at the time scale of 10 ms [6, 25, 64] where τ_B (the timescale of Brownian motion) is for μm sized particles of the order of 1 s [25, 47] or

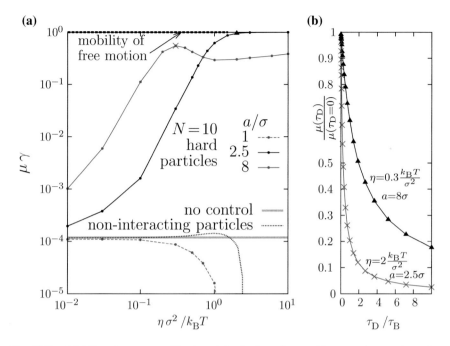

Fig. 19.5 a Mobility of a system of hard particles in dependence of the control strength, η, for various values of the lattice constant. The mobility can rise up to $1/\gamma$, the mobility of free motion. The *thick line* indicates the mobility in the uncontrolled case. **b** Impact of time delay at the parameters indicated by a triangle and cross in part **a**

larger [48]. Hence, we expect that the ratio τ_D/τ_B is rather small, that is, of the order 10^{-1}. For such situations, our results in Fig. 19.5b predict only a small decrease of μ relative to the non-delayed case.

Attractive interactions

Given the strong enhancement of mobility it clearly is an interesting question to which extent these observations depend on the type of the interactions. In [43] we have observed very similar behavior for two, quite different types of *repulsive* interactions. What would happen in presence of additional attractive interactions?

Indeed, in colloidal systems attractive forces quite naturally arise through the so-called depletion effect, which originates from the large size ratio between the colloidal and the solvent particles: when two colloids get so close that solvent particles do not fit into the remaining space, the accessible volume of the colloids effectively increases, yielding a short-range "entropic" attraction with a range determined by the solvent particles' diameter. Other sources of attraction are van-der-Waals forces [65], or the screened Coloumbic forces between oppositely charged colloids [66]. A generic model to investigate the impact of attractive forces between colloids is the hard-core attractive Yukawa (HCAY) model [67] defined by

$$v_{\text{hcay}}(z, z') = v_{\text{hard}}(z, z') - Y \frac{\exp(\kappa(\sigma - |z - z'|))}{|z - z'|/\sigma}, \qquad (19.20)$$

where $v_{\text{hard}}(z, z')$ has been defined in Eq. (19.16), and the parameters Y and κ determine the strength and range of the attractive part, respectively. Here we set $Y/k_B T = 10$ and consider the range parameters $\kappa\sigma = 7$ and $\kappa\sigma = 1$. The former case refers to a typical depletion interactions (whose range is typically much smaller than the particle diameter) [68–70], whereas the second case rather relates to screened Coloumb interactions. In both cases, the *three-dimensional* HCAY system at $Y/k_B T = 10$ would be phase-separated (gas-solid coexistence) [69]. In other words, our choice of Y corresponds to a strongly correlated situation. To treat the HCAY interaction within our theory, we construct a corresponding potential (see Eq. (19.12)) from the derivative of the (exact) hard-sphere functional given in Eq. (19.17) combined with the mean field functional (19.14) for the Yukawa attraction.

Numerical results for the mobility of the (one-dimensional) HCAY system under feedback control are plotted in Fig. 19.6a together with corresponding results for the (purely repulsive) hard sphere system. The general dependence of the mobility on the feedback strength seems to be quite insensitive to the detail of interactions: In all three cases we find an enhancement of μ towards the value characterizing a freely (without barriers) diffusing particle. Quantitatively, the results in the range $\eta\sigma^2 \lesssim 0.7k_B T$ depend on the range parameter κ. In particular, the system with the longer range of attraction ($\kappa\sigma = 1$) has a higher mobility than the one at $\kappa\sigma = 7$, with the mobility of the second one being even smaller than that in the hard-sphere system. However, at $\eta\sigma^2 \geq 0.7k_B T$ both HCAY mobilities exceed the hard-sphere mobility. The physical picture is that of a moving "train" of particles, where each particle not only pushes its neighbors (such as in the repulsive case) but also drags them during motion.

Finally, we consider in Fig. 19.6b the dependence of the mobility on the total number of particles, N (at fixed feedback strength η). This dependence arises from the fact that the length of the particle "train", $N\sigma$, competes with the two other relevant length scales, that is, the effective size of the trap (controlled by η), and the wavelength a. Thus, increasing N in the presence of particle interactions means to "compress" the train. For all systems considered in Fig. 19.6b this compression leads to an increase of mobility since, as shown explicitly in [43] for hard-sphere systems, the barriers in the *effective* potential landscape become successively smaller. From Fig. 19.6b we see that the increase of μ with N is even more pronounced in presence of colloidal attraction, suggesting that attractive forces enhance the rigidity of the train.

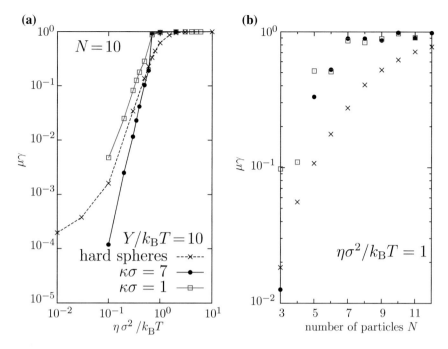

Fig. 19.6   a Mobility of a system of hard particles with additional short-ranged attractive interactions as function of the control strength, η, and two range constants. **b** Impact of particle number

19.5   Conclusions and Outlook

In this article we have summarized recent research on feedback control in 1D colloidal transport. We close with pointing out some open questions and possible directions for future research.

A first notion concerns the role of the control target and the theoretical formalism employed. The (Fokker-Planck based) approach described in Sects. 19.2–19.4 assumes control schemes targeting the *average* particle position, which seems to be the natural, i.e., experimentally accessible, choice for a realistic system of (interacting) colloids. Moreover, the FP approach allows for a convenient treatment of colloidal interactions via the DDFT approach, which have been typically neglected in earlier, (Langevin-based) investigations. However, it remains to be clarified how the FP results relate to findings from Langevin-based investigations targeting the *individual* positions (or other degrees of freedom), which is the straightfoward way to control a *single* colloidal particle. In other words, in which respect does an ensemble of colloids behave differently from a single one under feedback control? These issues become particularly dramatic in the case of time-delayed feedback control, where the Langevin equation is non-Markovian and the FP description consists, in principle, of an infinite hierarchy of integro-differential equations (see, e.g., [36]).

We note that even if one takes the average position as control target on the Langevin level, the results become consistent with those from our FP approach only in the limit $N \to \infty$ [42].

Conceptual questions of this type are also of importance in the context of stochastic thermodynamics. As pointed out already in Sect. 19.2 there is currently a strong interest (both in the classical and in the quantum systems community) to explore the role of feedback for the exchange of heat, work and entropy of a system with its environment [28, 36–38]. This is usually done by considering the entropy production, second-law like inequalities and fluctuation relations. In [42], we have presented some numerical results for the entropy production in the time-delayed feedback controlled rocking ratchet described in Sect. 19.3, the goal being to evaluate the efficiency of feedback control versus open-loop control. However, systematic investigations of feedback systems with time delay are just in their beginnings. This is even more true for systems with direct (pair) interactions.

A further interesting question from a physical point of view concerns the role of spatial dimension. In the present article we have focused (as it is mostly done) on 1D systems. Clearly, it would be very interesting to develop feedback control concepts for two-dimensional, interacting colloidal systems where, in addition to particle chain and cluster formation, anisotropic collective transport mechanisms [71], phase transitions [72], spinodal decomposition, and more complex pattern formation such as stripe formation [73] can occur. From the perspective of the present theoretical approach, which is based on the FP equation, a main challenge for the 2D case arises through the fact that we handle interaction effects on the basis of dynamical density functional theory (DDFT). For example, contrary to the 1D case there is no exact functional for hard spheres in two dimensions, making the entire approach less accurate. Thus, it will become even more important to test any FP-DDFT results against particle-resolved (Brownian Dynamics) simulations. One distinct advantage of the FP-DDFT approach, however, is that one can perform further approximations such as gradient expansions. This would allow to establish a relation to the large amount of work on feedback-controlled pattern forming systems based on (continuum) partial differential equations (see, e.g., [33, 35]).

Finally, we want to comment on the experimental feasibility of our feedback protocols. To this end we first note that state-of-the-art video microscopy techniques allow to monitor particles as small as 20 nm [74]. This justifies the use of (average) particle positions as control targets for colloids with a broad range of sizes from the nanometer to the micron scale. Typical experimental delay times (arising from the finite time required for particle localisation) are about 5–10 ms for single particles (see, e.g., [6, 13]). These values are substantially smaller than typical diffusion ("Brownian") time scales ($\approx$500 ms–1 μs), which underlines the idea that the relative time delay in colloidal transport is typically small. Naturally, somewhat larger delay times are expected to arise in feedback control of several (interacting) particles. Still, we think that our feedback protocols for many-particle systems are feasible, last but not least because many-particle monitoring techniques are being continuously improved [75]. We thus hope that the recent theoretical advancements reported in this article and in related theoretical studies will stimulate further experimental work.

Acknowledgments This work was supported by the Deutsche Forschungsgemeinschaft through SFB 910 (project B2).

References

1. J. Bechhoefer, Rev. Mod. Phys. **77**, 783 (2005)
2. H. Löwen, Eur. Phys. J. Spec. Topics **222**, 2727 (2013)
3. U. Seifert, Rep. Prog. Phys. **75**, 126001 (2012)
4. T. Sagawa, M. Ueda, Phys. Rev. E **85**, 021104 (2012)
5. B. Qian, D. Montiel, A. Bregulla, F. Cichos, H. Yang, Chem. Sci. **4**, 1420 (2013)
6. A.P. Bregulla, H. Yang, F. Cichos, Acs Nano **8**, 6542 (2014)
7. M. Braun, A. Würger, F. Cichos, Phys. Chem. Chem. Phys. **16**, 15207 (2014)
8. J. Fisher, J. Cummings, K. Desai, L. Vicci, B. Wilde, K. Keller, C. Weigle, G. Bishop, R. Taylor, C. Davis, Rev. Sci. Instrum. **76**, 053711 (2005)
9. A.E. Cohen, Phys. Rev. Lett. **94**, 118102 (2005)
10. Y. Jun, J. Bechhoefer, Phys. Rev. E **86**, 061106 (2012)
11. J. Gieseler, L. Novotny, C. Moritz, C. Dellago, New. J. Phys. **17**, 045011 (2015)
12. S. Toyabe, T. Sagawa, M. Ueda, E. Muneyuki, M. Sano, Nat. Phys. **6**, 988 (2010)
13. Y. Jun, M. Gavrilov, J. Bechhoefer, Phys. Rev. Lett. **113**, 190601 (2014)
14. J.J. Juarez, M.A. Bevan, Adv. Funct. Mat. **22**, 3833 (2012)
15. C. Prohm, H. Stark, Lab Chip **14**, 2115 (2014)
16. D.A. Strehober, E. Schöll, S.H.L. Klapp, Phys. Rev. E **88**, 062509 (2013)
17. S.H.L. Klapp, S. Hess, Phys. Rev E **81**, 051711 (2010)
18. T.A. Vezirov, S. Gerloff, S.H.L. Klapp, Soft Matter **11**, 406 (2015)
19. P. Hänggi, F. Marchesoni, Rev. Mod. Phys. **81**, 387 (2009)
20. P. Reimann, Phys. Rep. 361 (2002)
21. R. Gernert, C. Emary, S.H.L. Klapp, Phys. Rev. E **90**, 062115 (2014)
22. F.J. Cao, L. Dinis, J.M.R. Parrondo, Phys. Rev. Lett. **93**, 040603 (2004)
23. M. Feito, F.J. Cao, Phys. Rev. E **76**, 061113 (2007)
24. E.M. Craig, B.R. Long, J.M.R. Parrondo, H. Linke, EPL **81**, 10002 (2008)
25. B.J. Lopez, N.J. Kuwada, E.M. Craig, B.R. Long, H. Linke, Phys. Rev. Lett. **101**, 220601 (2008)
26. E.M. Craig, N.J. Kuwada, B.J. Lopez, H. Linke, Ann. Phys. (Berlin) **17**, 115 (2008)
27. F.J. Cao, M. Feito, Entropy **14**, 834 (2012)
28. T. Munakata, M.L. Rosinberg, Phys. Rev. Lett. **112**, 180601 (2014)
29. C. Emary, Phil. Trans. R. Soc. A **371**, 20120468 (2013)
30. D. Hennig, L. Schimansky-Geier, and P. Hnggi, Phys. Rev. E **79**, 041117 (2009)
31. K. Lichtner, S.H.L. Klapp, EPL **92**, 40007 (2010)
32. K. Lichtner, A. Pototsky, S.H.L. Klapp, Phys. Rev. E **86**, 051405 (2012)
33. S.V. Gurevich, R. Friedrich, Phys. Rev. Lett. **110**, 014101 (2013)
34. K. Pyragas, Phys. Lett. A **170**, 421 (1992)
35. E. Schöll, H.G. Schuster (eds.), *Handbook of Chaos Control* (Wiley, 2007)
36. M.L. Rosinberg, T. Munakata, G. Tarjus, Phys. Rev. E **91**, 042114 (2015)
37. T. Munakata, S. Iwama, M. Kimizuka, Phys. Rev. E **79**, 031104 (2009)
38. H. Jiang, T. Xiao, Z. Hou, Phys. Rev. E **83**, 061144 (2011)
39. H. Risken, *The Fokker-Planck Equation* (Springer, 1984)
40. U.M.B. Marconi, P. Tarazona, J. Chem. Phys. **110**, 8032 (1999)
41. U.M.B. Marconi, P. Tarazona, J. Phys.: Condens. Matter **12**, 413 (2000)
42. S.A.M. Loos, R. Gernert, S.H.L. Klapp, Phys. Rev. E **89**, 052136 (2014)
43. R. Gernert, S.H.L. Klapp, Phys. Rev. E **92**, 022132 (2015)
44. S. Guillouzic, I. L'Heureux, A. Longtin, Phys. Rev. E **59**, 3970 (1999)

45. T.D. Frank, P.J. Beek, R. Friedrich, Phys. Rev. E **68**, 021912 (2003)
46. C. Zeng, H. Wang, Chem. Phys. **402**, 1 (2012)
47. C. Dalle-Ferrier, M. Krüger, R.D.L. Hanes, S. Walta, M.C. Jenkins, S.U. Egelhaaf, Soft Matter **7**, 2064 (2011)
48. S.-H. Lee, D.G. Grier, Phys. Rev. Lett. **96**, 190601 (2006)
49. M. Feito, J.P. Baltanás, F.J. Cao, Phys. Rev. E **80**, 031128 (2009)
50. E.L. Florin, A. Pralle, E.H.K. Stelzer, J.K.H. Hörber, Appl. Phys. A **66**, 75 (1998)
51. D.G. Cole, J.G. Pickel, J. Dyn. Syst. Meas. Control **134**, 011020 (2012)
52. T. Brandes, Phys. Rev. Lett. **105**, 060602 (2010)
53. R.L. Stratonovich, Radiotekh. Elektron. **3**, 497 (1958)
54. H. Kamegawa, T. Hondou, F. Takagi, Phys. Rev. Lett. **80**, 5251 (1998)
55. C. Marquet, A. Begun, L. Talini, P. Silberzan, Phys. Rev. Lett. **88**, 168301 (2002)
56. R. Bartussek, P. Hänggi, J.G. Kissner, EPL **28**, 459 (1994)
57. M.O. Magnasco, Phys. Rev. Lett. **71**, 1477 (1993)
58. A.J. Archer, R. Evans, J. Chem. Phys. **121**, 4246 (2004)
59. C.N. Likos, Phys. Rep. **348**, 267 (2001)
60. A.A. Louis, P.G. Bolhuis, J.P. Hansen, Phys. Rev. E **62**, 7961 (2000)
61. D. Hennig, Phys. Rev. E **79**, 041114 (2009)
62. J.K. Percus, J. Stat. Phys. **15**, 505 (1976)
63. R. Roth, J. Phys.: Cond. Mat. 22, 063102 (2010)
64. A.E. Cohen, W.E. Moerner, PNAS **103**, 4362 (2006)
65. K.J.M. Bishop, C.E. Wilmer, S. Soh, B.A. Grzybowski, Small **5**, 1600 (2009)
66. M.E. Leunissen, C.G. Christova, A.-P. Hynninen, C.P. Royall, A.I. Campbell, A. Imhof, M. Dijkstra, R. van Roij, A. van Blaaderen, Nature **437**, 235 (2005)
67. C. Caccamo, Phys. Rep. **274**, 1 (1996)
68. M.H.J. Hagen, D. Frenkel, J. Chem. Phys. **101**, 4093 (1994)
69. M. Dijkstra, Phys. Rev. E **66**, 021402 (2002)
70. L. Mederos, G. Navascues, J. Chem. Phys. **101**, 9841 (1994)
71. F. Martinez-Pedrero, A. Straube, T.H. Johansen, P. Tierno, Lab Chip **15**, 1765 (2015)
72. A.J. Archer, A.M. Rucklidge, E. Knobloch, Phys. Rev. Lett. **111**, 165501 (2013)
73. K. Lichtner, S.H.L. Klapp, Europhys. Lett. **106**, 56004 (2014)
74. M. Selmke, A. Heber, M. Braun, F. Cichos, Appl. Phys. Lett. **105**, 013511 (2014)
75. J.R. Gomez-Solano, C. July, J. Mehl, C. Bechinger, New. J. Phys. **17**, 045026 (2015)

Chapter 20
Swarming of Self-propelled Particles on the Surface of a Thin Liquid Film

Andrey Pototsky, Uwe Thiele and Holger Stark

Abstract We consider a colony of self-propelled particles (swimmers) in a thin liquid film resting on a solid plate with deformable liquid-gas interface. Individual particles swim along the surface of the film predominantly in circles and interact via a short range alignment and longer-range anti-alignment. The local surface tension of the liquid-gas interface is altered by the local density of swimmers due to the soluto-Marangoni effect. Without the addition of swimmers, the flat film surface is linearly stable. We show that a finite wave length instability of the homogeneous and isotropic state can be induced by the carrier film for certain values of the rotational diffusivity and a nonzero rotation frequency of the circular motion of swimmers. In the nonlinear regime we find square arrays of vortices, stripe-like density states and holes developing in the film.

20.1 Introduction

Emerging spatio-temporal density and velocity patterns in suspensions of motile living cells became the focus of many experimental and theoretical studies over the last decade. With the typical body size of several µm, the colonies of swimmers exhibit a wide range of meso-scale and large-scale coherent structures such as circular vortices, swirls and meso-scale turbulence with the correlation length of the collective motion ranging between $\sim$10 and $\sim$100 µm [6, 10, 12, 19–23, 27].

A. Pototsky (✉)
Department of Mathematics, Faculty of Science Engineering and Technology,
Swinburne University of Technology, Hawthorn, VIC 3122, Australia
e-mail: apototskyy@swin.edu.au

U. Thiele
Institut für Theoretische Physik, Westfälische Wilhelms-Universität Münster,
48149 Münster, Germany

H. Stark
Institut für Theoretische Physik, Technische Universität Berlin,
Hardenbergstraße 36, 10623 Berlin, Germany

© Springer International Publishing Switzerland 2016
E. Schöll et al. (eds.), *Control of Self-Organizing Nonlinear Systems*,
Understanding Complex Systems, DOI 10.1007/978-3-319-28028-8_20

It was recognized that the onset of the observed large-scale patterns is associated with a finite wave length instability of the homogeneous and isotropic distribution of swimmers [2]. Recently, a minimal phenomenological model of the spatio-temporal pattern formation of living matter was developed on the basis of a Swift-Hohenberg type scalar field theory [7]. The physical mechanism, underlying the instability was traced down to the short-range aligning and longer-range anti-aligning interaction between the orientation of swimmers [9]. On the microscopic level, the short-range alignment can be explained by the collisions between swimmers with elongated bodies, or by flagellar bundling [27]. The longer-range anti-alignment is linked to hydrodynamic interactions [9] that are known to destabilize the polar order at high densities.

In experiments with bacterial suspensions confined between solid boundaries, the role of the solvent fluid is seen as a passive carrier that gives rise to hydrodynamic interactions between individual swimmers. However, in the case of freely suspended soaplike liquid films loaded with bacteria, the deformations of the liquid-gas interface and the resulting motion of the carrier fluid can no longer be neglected. Thus, in the early experiments with *E. coli* bacteria [29], a droplet of bacterial suspension was stretched between $10\,\mu$m thin fibers to form a soaplike film. In order to delay the film rupture, a stabilizing chemical surfactant had to be added. In later studies with $1\,\mu$m soaplike films, metabolic products, secreted by the *B. subtilis* bacteria, played the role of the stabilizing surfactant [20–22]. Without the addition of a stabilizing surfactant, the life time of the film is determined by the film thickness and the surface tension. In most recent experiments with *E. coli* bacteria, the rupture of a $20\,\mu$m film was detected after several minutes [12].

Motivated by these recent experiments, we address here the question of how a suspension of swimmers, confined to move in a thin liquid film on a solid substrate, is affected by the presence of a deformable liquid-gas interface. To this end, we consider a non-evaporating 10–$100\,\mu$m thin liquid film with a deformable liquid-gas interface, resting on a solid plate. Lubrication theory [14] predicts that without the addition of swimmers, the flat film is linearly stable, with respect to small amplitude variations of its thickness. The film is loaded with surface swimmers that interact with each other via a short-range alignment and longer-range anti-alignment, as described in [9]. In case of the resting fluid, a homogeneous and isotropic distribution of swimmers is linearly unstable with respect to a finite wave length instability.

We extend the model of self-propelled particles, used in Ref. [9], by additionally taking into account a deterministic rotation of the bodies of individual swimmers that gives rise to their circular motion. Thus, it is known that bacteria with helical flagellas swim in circles, predominantly clockwise near a solid-liquid interface and anticlockwise near a liquid-gas interface [5, 8]. We neglect steric repulsion between the swimmers and introduce their translational surface diffusion.

As suggested by earlier studies, the coupling between the swimmers and the liquid film occurs through the soluto-Marangoni effect [1, 18]. Indeed, some living cells, such as *B. subtilis* bacteria, excrete metabolic products [21] that act as a surfactant and change the local surface tension of the liquid film. Consequently, the local concentration of the surfactant particles is proportional to the local concentration of swimmers. It should be emphasized that the surface tension decreases with the surfactant concentration, implying that soluto-Marangoni effect stabilizes a flat film [14].

Our resulting model consists of the thin film equation for the local film thickness, coupled to the Smoluchowski equation for the swimmer density distribution function. In the coupled system, the emergence of a density patterns always occurs in conjunction with film surface deformations. We find that a seemingly passive liquid film has a profound effect on the linear stability of a homogeneous isotropic distribution of swimmers on the surface of a flat film. In particular, there exists a window of parameters, for which the isotropic state is linearly stable in the absence of the liquid film and is linearly unstable when the liquid film is included.

By numerically solving the evolution equations for film height and swimmer density, we find square arrays of vortices for parameter combinations close to the instability threshold. Deep in the unstable region, we find stripes in the density distribution for small values of the self-propulsion velocity. These long-lasting states are accompanied by stripe-like small amplitude deformations of the film surface. Typically the stripes on the film surface are in antiphase with the density stripes. For large self-propulsion velocities, we demonstrate the development of a depression region on the film surface that has a lateral size comparable to the system size. The depth of the depression gradually increases with time, thus, increasing the probability of film rupture. In our numerical simulations we have observed film rupture at finite times.

20.2   Model Equations

Consider a colony of active Brownian particles that swim along the liquid-gas interface of a thin liquid film. The swimming direction of the ith particle is given by the unit vector p_i, which is tangential to the liquid-gas interface at all times. We only take into account long wavelength deformations of the liquid-gas interface at height $h(x, y)$, whereby the gradient ∇h is small at all times. In this case, the orientation vector p_i is approximately two-dimensional $p_i = (\cos(\phi), \sin(\phi))$, where ϕ denotes the polar angle.

The interaction between the swimmers is characterized by pair-wise alignment at short distances and anti-alignment at large distances. The interaction strength is given by a certain coupling function $\mu(|r_i - r_j|)$ of the separation distance $|r_i - r_j|$ between the ith and the jth swimmer. Positive (negative) values of $\mu(r)$ correspond to pair-wise alignment (anti-alignment) [9].

The stochastic equations of motion for the ith swimmer can be written as

$$\dot{r}_i = v_0 p_i + U_i + \xi_i,$$
$$\dot{\phi}_i = \chi_i + \omega_0 + \frac{1}{2}\Omega_z - \sum_{k \neq i} \mu(|r_i - r_k|)\sin(\phi_i - \phi_k), \qquad (20.1)$$

where $\omega_0 > 0$ is an intrinsic rotation frequency that gives rise to the circular clock-wise motion of the swimmer at the liquid-gas interface. $\vec{U} = (U_x, U_y)$ is the surface velocity field of the fluid, $\Omega_z = (\nabla \times U)_z = \partial_x U_y - \partial_y U_x$ is the z–component of the curl of the surface velocity field. Random rotation of the vector p_i is characterized by a Wiener process $\chi_i(t)$ with $\langle \chi_i(t)\chi_k(t')\rangle = 2D_r\delta(t - t')\delta_{ik}$, where D_r is the rotational diffusivity. The two-dimensional vector $\xi_i = (\xi_x, \xi_y)_i$ with $\langle \xi_i(t)\xi_k(t')\rangle = 1_{2\times2}2Mk_BT\delta(t - t')\delta_{ik}$ represents the translational noise, where $1_{2\times2}$ is a unit 2×2 matrix, M is the mobility of a single swimmer and T is the absolute temperature.

It is worthwhile to mention the relation of Eq. (20.1) to the previously studied models of self-propelled particles. Thus in case of $\mu = 0$, $\chi = 0$, $\xi = 0$ and $\omega_0 = 0$, the Eq. (20.1) describe a deterministic self-propelled particle that moves in a fluid with a given flow velocity U, as considered in Ref. [30]. In the absence of the fluid and without the deterministic rotation, i.e. $U = 0$, $\Omega_z = 0$, $\omega_0 = 0$ and $\xi = 0$ the system Eq. (20.1) reduces to a swarming model, studied in Ref. [9]. For non-interacting and non-rotating swimmers, i.e. for $\mu = 0$ and $\omega_0 = 0$, Eq. (20.1) in conjunction with the thin film equation, have been studied in Ref. [18].

The Smoluchowski equation, derived from Eq. (20.1), for the surface density of swimmers $\rho(r, \phi, t)$ is then given by

$$\partial_t \rho + \nabla \cdot J_t + \partial_\phi J_\phi = 0, \tag{20.2}$$

with translational and rotational currents J_t and J_ϕ, given respectively by

$$J_t = v_0 p \rho + U\rho - Mk_BT\nabla\rho, \tag{20.3}$$

$$J_\phi = \left(\omega_0 + \frac{1}{2}\Omega_z\right)\rho - D_r\partial_\phi\rho - \int\int d\phi'dr'\rho_2(r, \phi, r + r', \phi')\mu(r')\sin(\phi - \phi'),$$

where ρ_2 is the two-particle density function.

Following [9], we employ a mean-field approximation and replace the two-body density $\rho_2(r, \phi, r + r', \phi')$ in Eq. (20.3) by $\rho(r, \phi)\rho(r + r', \phi')$. Next, we recall that the coupling strength $\mu(r')$ rapidly decays with the distance r' between the swimmers. This allows us to expand $\rho(r + r', \phi')$ about r and truncate the expansion after a certain number of leading terms. As shown in Ref. [9], in order to recover a finite wave length instability, one should retain quartic terms $\sim(r')^4$ in the density expansion.

The resulting rotational current can be written as

$$J_\phi = \left(\omega_0 + \frac{1}{2}\Omega_z\right)\rho - D_r\partial_\phi\rho - \rho(r, \phi)\left[\sin\phi\,\hat{\mu}C(r) - \cos\phi\,\hat{\mu}S(r)\right], \tag{20.4}$$

where

$$C(r) = \int_0^{2\pi} \rho(r, \phi')\cos\phi'd\phi', \quad S(r) = \int_0^{2\pi} \rho(r, \phi')\sin\phi'd\phi' \tag{20.5}$$

and the operator $\hat{\mu}$ is given by

$$\hat{\mu} = \mu_0 + \mu_2 \Delta + \mu_4 \Delta^2, \qquad (20.6)$$

with $\Delta = \nabla^2$. The coefficients μ_0, μ_2 and μ_4 in Eq. (20.6) can be explicitly written as functionals of the coupling strength $\mu(r)$ [9].

Swimmers at the liquid-gas interface may excrete metabolic products that act as a surfactant and change the local surface tension $\sigma(r)$ of the liquid film. This phenomenon, called the soluto-Marangoni effect, typically implies a linear decrease of $\sigma(r)$ due to the local surface concentration of swimmers $\langle \rho \rangle (r) = \int_0^{2\pi} \rho(r, \phi) \, d\phi$ (cf. Ref. [24]):

$$\sigma(r) = \sigma_0 - \Gamma \langle \rho \rangle (r), \qquad (20.7)$$

where $\Gamma > 0$ and Σ_0 is the reference surface tension in the absence of swimmers.

The thin film equation for the local film thickness $h(r, t)$, derived in the lubrication approximation [14], is then coupled to the average concentration $\langle \rho \rangle$ [1, 18]

$$\partial_t h + \nabla \cdot \left(\frac{h^3}{3\eta} \nabla \left[\sigma_0 \Delta h \right] \right) - \Gamma \nabla \cdot \left(\frac{h^2}{2\eta} \nabla \langle \rho \rangle \right) = 0, \qquad (20.8)$$

where η is the dynamic viscosity. The surface fluid velocity $U(r)$ is found as a function of the local film thickness h [14]

$$U = -\Gamma \frac{h}{\eta} \nabla \langle \rho \rangle + \frac{h^2}{2\eta} \nabla \left(\sigma_0 \Delta h \right). \qquad (20.9)$$

Equations (20.2), (20.8) and (20.9) form a closed system of integro-differential equations for the density $\rho(r, \phi, t)$ and the film height $h(r, t)$.

20.3 Linear Stability of the Homogeneous and Isotropic State

In this section we discuss the linear stability of a spatially homogeneous and isotropic stationary solution of Eqs. (20.2), (20.8) and (20.9), given by $\rho(r, \phi) = \rho_0/(2\pi)$ and $h(r) = h_0$, where ρ_0 is the stationary total swimmer density. Using the ansatz $h = h_0 + \delta h$ and $\rho = \rho_0/(2\pi) + \delta \rho$, we expand the perturbation functions δh and $\delta \rho$ according to

$$\delta h(r, t) = h_0 \int \hat{h}(k) e^{\gamma(k)t} e^{ikr} \, dk, \qquad (20.10)$$

$$\delta \rho(r, \phi, t) = \lim_{N \to \infty} \frac{\rho_0}{2\pi} \sum_{n=-N}^{N} e^{in\phi} \int W_n(k) e^{\gamma(k)t} e^{ikr} \, dk, \qquad (20.11)$$

with the Fourier amplitudes $\hat{h}(k)$ and $W_n(k)$, the wave vector of the perturbation $k = (k_x, k_y)$, and the growth rate $\gamma(k)$.

Substituting the expansions Eq. (20.10) into Eqs. (20.2) and (20.8) and linearizing about the steady state, we obtain the eigenvalue problem

$$\gamma(k)H = \mathcal{J}(k)H, \tag{20.12}$$

with the eigenvector H

$$H(k) = (\hat{h}, W_0, W_1, W_{-1}, W_2, W_{-2}, \dots), \tag{20.13}$$

and the Jacobi matrix $\mathcal{J}$, which corresponds to a banded matrix of the structure

$$
- \mathcal{J}(k) =
\begin{pmatrix}
T_{11} & T_{12} & 0 & & 0 & & 0 & 0 & 0 & \dots \\
T_{21} & T_{22} & V^- & & V^+ & & 0 & 0 & 0 & \dots \\
0 & V^+ & d_1 - \frac{\hat{\mu}(k)}{2} & 0 & & & V^- & 0 & 0 & \dots \\
0 & V^- & 0 & & d_{-1} - \frac{\hat{\mu}(k)}{2} & 0 & & V^+ & 0 & \dots \\
0 & 0 & V^+ & & 0 & & d_2 & 0 & V^- & \dots \\
0 & 0 & 0 & & V^- & & 0 & d_{-2} & 0 & \dots \\
0 & 0 & 0 & & 0 & & V^+ & 0 & d_3 & \dots \\
\dots & & & & & & & & &
\end{pmatrix}. \tag{20.14}
$$

Here $\hat{\mu}(k) = \mu_0 - \mu_2 k^2 + \mu_4 k^4$, $k^2 = k_x^2 + k_y^2$, $V^+ = v_0 \left(\frac{k_y}{2} + \frac{ik_x}{2} \right)$, $V^- = v_0 \left(-\frac{k_y}{2} + \frac{ik_x}{2} \right)$ and $d_m = im\omega_0 + m^2 D_r + Mk_B Tk^2$ with $m = \pm 1, \pm 2, \pm 3, \dots$ The (2×2) matrix T in the upper left corner of $\mathcal{J}$ is given by

$$
T(k) =
\begin{pmatrix}
\frac{h_0^3}{3\eta} \sigma_0 k^4 & \frac{\Gamma h_0^2}{2\eta} k^2 \\
\frac{h_0^2}{2\eta} \sigma_0 k^4 & \left(\Gamma \frac{h_0}{\eta} + Mk_B T \right) k^2
\end{pmatrix}. \tag{20.15}
$$

In practice, we truncate the expansion in the angle ϕ and only take a certain number of the first N Fourier modes into account. Then, the Jacobi matrix $\mathcal{J}$ is an $(2N+2) \times (2N+2)$ matrix and the truncated eigenvector $\mathcal{H} = (\hat{h}, W_0, W_1, W_{-1}, \dots, W_N, W_{-N})$ is $(2N+2)$ dimensional.

Because the perturbations δh and $\delta \rho$ are both real, the eigenvectors of H satisfy the following symmetry conditions

$$
\begin{aligned}
\hat{h}(k)_\gamma^* &= \hat{h}(-k)_{\gamma^*} \\
W_n(k)_\gamma^* &= W_{-n}(-k)_{\gamma^*},
\end{aligned} \tag{20.16}
$$

where the asterisk denotes complex conjugation and the subscript γ indicates that the eigenvector $(\hat{h}(k)_\gamma, W_0(k)_\gamma, W_1(k)_\gamma, W_{-1}(k)_\gamma, \dots)$ corresponds to the eigenvalue γ.

In what follows, we non-dimensionalise the evolution equations employing the scaling as in Ref. [18]. Thus, we use h_0 as the vertical length scale, $h_0\sqrt{\sigma_0/\Gamma\rho_0}$ as the horizontal length scale, $\eta h_0\sigma_0/(\Gamma^2\rho_0^2)$ as the time scale and the direction-averaged density of swimmers in the homogeneous state ρ_0 as the density scale.

The complete set of the dimensionless system parameters consists of: the self-propulsion velocity $V = v_0\eta\sigma_0^{1/2}/(\Gamma\rho_0)^{3/2}$, the dimensionless rotational diffusivity $D = D_r h_0\eta\sigma_0/(\Gamma\rho_0)^2$, the translational surface diffusivity $d = k_B T M\eta/(h_0\rho_0\Gamma)$, the rotation frequency $\Omega_0 = \omega_0 h_0\eta\sigma_0/(\Gamma\rho_0)^2$ and the alignment/anti-alignment interaction parameters $\tilde{\mu}_i = \mu_i\eta h_0\sigma_0/(\Gamma^2\rho_0)$. For simplicity we drop the tildes in the dimensionless interaction parameters. The dimensionless evolution equations are summarized in Appendix 1.

In what follows, we focus on the effect of the parameter triplet (V, D, Ω_0) on the linear stability of the homogeneous isotropic state. From here on we fix the interaction parameters at $\mu_0 = 1$, $\mu_2 = -1$, $\mu_4 = -10^{-2}$ that can be achieved by the appropriate choice of the coupling function $\mu(r)$. This choice of μ_i corresponds to the finite wave length instability of the homogeneous distribution of non-rotating swimmers, i.e. $\Omega_0 = 0$, in the absence of the liquid film, as studied in Ref. [9].

20.3.1 Singularity of the Instability at $V = 0$

Linear stability analysis reveals remarkable behaviour of the system at vanishingly small swimming velocity $V \approx 0$. By setting $V = 0$ in Eq. (20.14), the eigenvalue with the largest real part can be found analytically

$$\gamma_{\max}(k) = \frac{\hat{\mu}(k)}{2} - d_{\pm 1} = \pm i\Omega_0 - D - dk^2 + \frac{1}{2}\left(\mu_0 - \mu_2 k^2 + \mu_4 k^4\right). \qquad (20.17)$$

The fastest growing wave number $k_{\max}$ and the corresponding growth rate $\mathrm{Re}[\gamma(k_{\max})]$ are

$$(k_{\max})^2 = \frac{2d + \mu_2}{2\mu_4}, \quad \mathrm{Re}[\gamma(k_{\max})] = -D + \frac{\mu_0}{2} - \frac{(2d + \mu_2)^2}{8\mu_4}. \qquad (20.18)$$

At $V = 0$, the matrix in the lower right corner of Eq. (20.14) is diagonal. The eigenvectors that corresponds to each of the two complex-conjugate eigenvalues Eq. (20.17), have only one non-zero component: either $W_1 \neq 0$, or $W_{-1} \neq 0$. Indeed, the eigenvector $\boldsymbol{H}^+$ that corresponds to $\gamma(k) = \hat{\mu}(k)/2 - d_1$ is given by $\boldsymbol{H}^+ = (0, 0, W_1, 0, 0, \dots)$. Similarly, the eigenvector $\boldsymbol{H}^-$ that corresponds to $\gamma(k) = \hat{\mu}(k)/2 - d_{-1}$ is given by $\boldsymbol{H}^- = (0, 0, 0, W_{-1}, 0, \dots)$. From the physical point of view this means that the colony of swimmers is unstable for a certain interval of k, however this is a purely orientational instability that is reflected in the first Fourier mode, i.e. $W_{\pm 1} \neq 0$. This orientational instability *does not* translate into the instability of h and W_0, as the orientation averaged density $\int_0^{2\pi} \rho(r\phi)\, d\phi$ is insensitive w.r.t. the orientational order of swimmers.

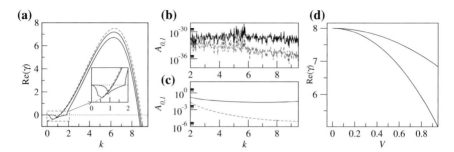

Fig. 20.1 **a** *Solid lines* show the real part of the two largest eigenvalues versus wave number $k = \sqrt{k_x^2 + k_y^2}$ for $V = 0.7$, $D = 1$, $\Omega_0 = 0$ and $d = 0.1$. The *dashed line* corresponds to the dispersion curve at $V = 0$ from Eq. (20.17). The *inset* shows a zoom of the highlighted area near the origin. **b, c** Amplitudes $A_0 = |\hat{h}(k)|^2$ (*dashed line*) and $A_1 = |W_0(k)|^2$ (*solid line*) of the eigenvectors of the **b** largest and the **c** second largest eigenvalue for $2 < k < 10$ where at least one mode is unstable. **d** Two most unstable eigenvalues $Re(\gamma)(k_{max})$ at the fastest growing wave number k_{max} versus V

Thus, any non-zero velocity $V \neq 0$, no matter how small, introduces the coupling between the first Fourier mode $W_{\pm 1}$ and all other modes W_n, $n = 0, \pm 2, \pm 3, \ldots$, including the amplitude of the film surface deformation $\hat{h}$. As the consequence, the emerging orientational order translates into an instability of the uniform swimmer density and the plane film surface. In order to examine the coupling between the orientational instability of swimmers and the film surface deformations, we set $V = 0.7$, $D = 1$, $\Omega_0 = 0$ and $d = 0.1$ and numerically determine the eigenvalues and the corresponding eigenvectors of the truncated Jacobi matrix Eq. (20.14) with the total number of Fourier modes $N = 50$.

The real parts of the first and the second most unstable eigenvalues are given in Fig. 20.1a as solid lines. The analytic eigenvalue, corresponding to $V = 0$ from Eq. (20.17) is shown by the dashed line. Approximately, for $k > 2$, the first two most unstable eigenvalues for $V = 0.7$ have positive real parts, implying an instability. The fastest growing wave number for $V = 0.7$ is approximately the same as for $V = 0$, i.e. $k_{max} = \sqrt{40}$ from Eq. (20.18).

By examining the eigenvector that corresponds to the most unstable eigenvalue, we find that the first two components of this eigenvector are given by a numerical zero, i.e. $|\hat{h}(k)|^2 \sim 10^{-32}$ and $|W_0(k)|^2 \sim 10^{-32}$, as shown in Fig. 20.1b. This means that the corresponding perturbation is purely orientational and does not couple to the thin film instability. However, the eigenvector of the second most unstable eigenvalue has $|\hat{h}(k)|^2 \neq 0$ and $|W_0(k)|^2 \neq 0$, as shown in Fig. 20.1c. This mode corresponds to a simultaneous instability of the film thickness and the average swimmer density.

The singularity of the instability at vanishingly small V is visualized in Fig. 20.1d, where the real parts of the first two most unstable eigenvalues, computed at the fastest growing wave number k_{max}, are plotted against V. We are interested in the second most unstable eigenvalue that corresponds to the coupling between the orientational instability and the film surface deformation. Thus, at any nonzero $V \neq 0$, no matter how small, the colony of swimmers is linearly unstable with the finite growth rate $\mathrm{Re}[\gamma(k_{\max})] = -D + \mu_0/2 - [(2d + \mu_2)^2]/[8\mu_4]$, as given by Eq. (20.18).

It is important to remark that the role of the thin film in the onset of the instability at vanishingly small V is purely passive. The above described coupling between the orientational instability and the instability of the average density of swimmers $\langle \rho \rangle$ occurs with or without the liquid film, which is linearly stable without the colony of swimmers. However, the situation changes dramatically, if V is finite and if one takes into account the rotation frequency Ω_0, as discussed in the next section.

20.3.2 Effect of the Liquid Film on the System Stability

In order to study the effect of the liquid film on the system stability, we distinguish between the film loaded with swimmers and the bare colony of swimmers without the liquid film. Technically, the latter case corresponds to the matrix T in Eq. (20.14), replaced by $T_{11} = T_{12} = T_{21} = 0$ and $T_{22} = Mk_BTk^2$. We numerically solve the eigenvalue problem Eq. (20.14) for the swimmers with and without the liquid film. In the presence of the liquid film, we determine the largest eigenvalue that corresponds to the coupling mode between the orientational instability and the film surface deformation.

We fix $V = 1, d = 0.1$ and determine the stability threshold in the plane of parameters (D, Ω_0). In Fig. 20.2a the shaded region marks the values of (D, Ω_0), where

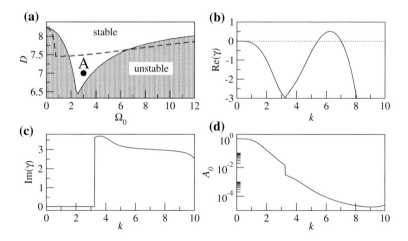

Fig. 20.2 **a** Stability diagram for $V = 1$ and $d = 0.1$. *Shaded area* corresponds to linearly unstable colony of swimmers in the absence of the liquid film. *Dashed line* is the stability threshold for the flat liquid film with swimmers: the system is stable above and unstable below the *dashed line*. In **b, c, d** we set $D = 7$ and $\Omega_0 = 3$ for a flat liquid film with swimmers (point A in (**a**)): **b** real part, **c** imaginary part and **d** the amplitude $A_0 = |\hat{h}(k)|^2$ of the most unstable eigenvalue versus k

the colony of swimmers is unstable in the absence of the liquid film. For the values of the interaction parameters μ_i chosen here, the homogeneous distribution of swimmers becomes linearly unstable along the border of the shaded region via an oscillatory instability at a finite wave number.

Remarkably, the addition of a stable liquid film, changes the stability threshold dramatically, as shown by the dashed line in Fig. 20.2a. Thus, we find a window of the rotation frequency $2 \lesssim \Omega_0 \lesssim 4$ and of the rotational diffusivity $6.5 \lesssim D \lesssim 7.5$, where the inclusion of a seemingly passive liquid film destabilizes the system. The dispersion curve $\mathrm{Re}(\gamma)(k)$ computed for $D = 7$ and $\Omega_0 = 3$ (point A in Fig. 20.2a) is shown in Fig. 20.2b. In Fig. 20.2c, d we plot the imaginary part of the most unstable eigenvalue and the amplitude $A_0 = |\hat{h}(k)|^2$ versus k, respectively. The finite values of $|\hat{h}(k)|^2$ confirm the coupling between the orientational instability and the deformation of the film surface.

Our results show that a colony of swimmers that move on top of a deformable liquid film is less stable than the bare system of swimmers in the case when the film is absent. Such effects are also known for passive systems where e.g. for a film of a binary mixture, the decomposition process couples to the dewetting process in such a way that the flat film becomes linearly unstable. However, the flat film remains linearly stable w.r.t. the each process separately: i.e. it is stable w.r.t. the dewetting and stable w.r.t. the decomposition process [25]. Another example of a coupled system that is less stable than each of its components when decoupled, is a two-layer liquid film on a solid substrate [16, 17]. Thus, for certain immiscible polymer films of different film thickness, placed on top of each other, a two-layer film can be linearly unstable due to weak van der Waals forces that exists between apolar molecules. However, when separated, each of the two layers supported by the same substrate may be linearly stable.

20.4 Nonlinear Evolution from a Homogeneous Isotropic State

The nondimensional evolution equations for the swimmer density $\rho(\boldsymbol{r}, \phi, t)$ and the local film thickness $h(\boldsymbol{r}, t)$ are summarized in Appendix 1. In order to numerically solve the system of Eq. (20.22), we discretize the film thickness $h(x, y, t)$ in a square box $-L \times L$ and the density $\rho(x, y, \phi, t)$ in a rectangle ($x \in [-L/2, L/2]$) × ($y \in [-L/2, L/2]$) × ($\phi \in [0, 2\pi]$) with periodic boundaries. We use 100×100 mesh points for the discretisation in space and 20 Fourier modes for the decomposition of the ϕ-dependency. We adopt a semi-implicit pseudo-spectral method for the time integration, as outlined in Appendix 2. In order to quantify the patterns, we introduce three global measures: the space-averaged mode type M

$$M = L^{-2} \int \int (h(x, y, t) - 1)(\langle \rho \rangle (x, y, t) - 1) \, dx dy, \qquad (20.19)$$

the space-averaged flux of the fluid $\bar{\boldsymbol{J}}_h$, determined by

$$\bar{\boldsymbol{J}}_h = L^{-2} \int \int \left[(h^3/3) \nabla (\Delta h) - (1/2) \left(h^2 \nabla \langle \rho \rangle \right) \right] dx dy \qquad (20.20)$$

and the space-averaged orientation field from Eq. (20.5)

$$(\langle C \rangle (t), \langle S \rangle (t)) = L^{-2} \int \int dx dy \, (C(x, y, t), S(x, y, t)) \qquad (20.21)$$

The mode type M can be used to analyse the phase shift between the patterns of h and $\langle \rho \rangle$. Non-zero values of the space-averaged fluid flux $\bar{\boldsymbol{J}}_h$ indicates global propagation of patterns. The active part of the space-averaged translational flux of swimmers in Eq. (20.3) is given by $v_0(\langle C \rangle (t), \langle S \rangle (t))$.

20.4.1 Square Array of Vortices

We demonstrate the destabilizing action of the liquid film by choosing the parameters as in point A in Fig. 20.2a.

The time evolution of the mode type M and of the film height h at arbitrarily chosen point on the film surface (x^*, y^*) are shown in Fig. 20.3a, b, respectively. The amplitudes of the average density $\langle \rho \rangle_{\min}$ and $\langle \rho \rangle_{\max}$ are shown in Fig. 20.3c. After passing a certain relaxation time of approximately ≈ 100 time units, the system reaches a stable time-periodic solution that can be characterized as a standing square wave with a well defined spatial period. A typical snapshot of the average density $\langle \rho \rangle (x, y)$ and of the film thickness $h(x, y)$ taken at $t = 150$ is shown in the lower panels in Fig. 20.3. The spatial period l of the square pattern is $l \approx 2\pi / k_{\max}$, where $k_{\max} = 6.2$ is the fastest growing wave number, as extracted from the dispersion curve in Fig. 20.2b. During the entire time evolution, the space-averaged flux of the fluid $\bar{\boldsymbol{J}}_h$ (not shown here) from Eq. (20.20) is of order of 10^{-6}, dropping to a numerical zero for $t > 100$. As the standing square wave regime is established, the space-averaged orientation Eq. (20.21) is numerically zero (not shown). The temporal oscillation period of the standing wave, $T = 2.1$, is extracted from the evolution of $h(x^*, y^*)$ is shown in the inset of Fig. 20.3b. Interestingly, the mode type M oscillates with only a half of the period, $T/2 = 1.05$, indicating that the pattern oscillates between two identical states that are shifted in space.

In order to gain a better understanding of the different phases of the temporal oscillations of the vortex state, we show in Fig. 20.4 three snapshots of the average density $\langle \rho \rangle (x, y)$ and the film thickness $h(x, y)$ from the zoomed area around the bottom left corner of the domain. In addition, we overlay the density snapshot with the vector field of the average orientation of swimmers $\alpha(C(x, y), S(x, y))$, with a conveniently chosen scaling factor α. The three snapshots are taken over one half of the temporal period, between $t = 150.4$ and $t = 151.2$.

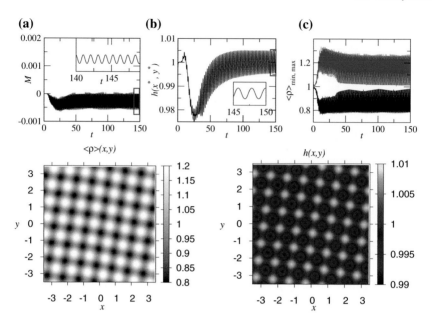

Fig. 20.3 Nonlinear evolution from a homogeneous isotropic state for parameters as in point A in Fig. 20.2a. **a** The mode type M from Eq. (20.19), **b** time evolution of the film height h at an arbitrarily chosen point on the film surface (x^*, y^*) and **c** $\langle \rho \rangle_{min}$ and $\langle \rho \rangle_{max}$ as a function of time. *Lower panel* snapshot of the average density $\langle \rho \rangle(x, y)$ and of the film thickness $h(x, y)$ at $t = 150$

At $t = 150.4$ the orientation field is represented by vortices arranged in a square lattice. Vortices located over the depression (elevation) regions of the film thickness profile have an anticlockwise (clockwise) polarity. The average density is in anti-phase with the film profile, implying that the mode type is negative. The dynamics of the vortex polarity can be appreciated from the snapshot taken at $t = 151.0$, where the orientation field is almost radially symmetric. Note that at $t = 151.0$ the depression regions in the average density and in the film thickness profile have turned into the elevation regions and vise-versa. At $t = 151.2$, the polarity of the vortices has reversed as compared with the snapshot taken at $t = 150.4$.

20.4.2 Stripe-Like Density Patterns

Next, we explore the temporal evolution of the system deep in the unstable region. By setting $\Omega_0 = 0$ and $D = 1$ we vary the self-propulsion velocity V and compare the evolution of the swimmers in the absence of the liquid film with the dynamics of the coupled system. Numerically, the Smoluchowski equation is decoupled from the thin film equation by setting $U = 0$ and $\Omega_z = 0$. In the absence of the liquid film we find stripe-like density patterns at small velocities. In Fig. 20.5a the evolution of $\langle \rho \rangle_{min}$ and $\langle \rho \rangle_{max}$ is shown by the solid (the dashed) line in the presence (in the

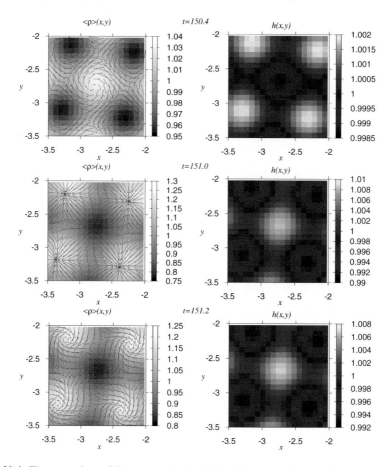

Fig. 20.4 Three snapshots of the average density $\langle\rho\rangle(x, y)$ in grey scale and of the film thickness $h(x, y)$, taken over one half of the oscillation period of the vortex state. The density is overlayed by the vector field of the average orientation of swimmers $\sim(C(x, y), S(x, y))$

absence) of the liquid film. Figure 20.5e shows a snapshot of the density patterns at $t = 100$ in the absence of the film.

The inclusion of the liquid film leads to a significantly smaller amplitude of the average density fluctuations $\langle\rho\rangle_{max} - \langle\rho\rangle_{min}$, as can be seen from Fig. 20.5a. In the long time limit, almost parallel stripe-like density patterns are found in the presence of the film as given in Fig. 20.5f. The surface of the film is covered with similar stripe-like patterns that are oriented parallel to the density stripes (Fig. 20.5g). The amplitude of the film surface deformation is of the order of 0.5 % of the average film thickness $h = 1$ (Fig. 20.5b). Stripes on the film surface are in anti-phase with the density stripes, so that the mode type M is negative (Fig. 20.5c). The fluid flux $\bar{J}_h$ is zero in the long time limit (Fig. 20.5d). The space-averaged orientation Eq. (20.21) in the long time limit is a certain non-zero constant (not shown).

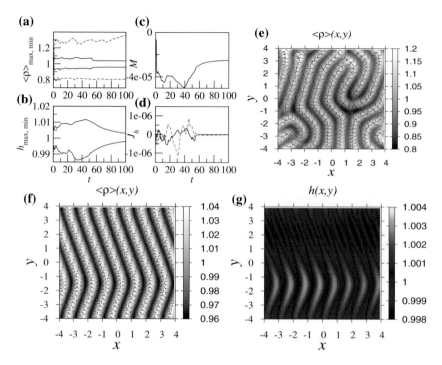

Fig. 20.5 Evolution with and without liquid film at $V = 0.1$, $D = 1$, $\Omega_0 = 0$ and $d = 0.1$. **a, b, c, d** Time evolution of local and global measures. *Solid (dashed) lines* in (**a**) show $\langle \rho \rangle_{\min}$ and $\langle \rho \rangle_{\max}$ in the presence (in the absence) of the liquid film. *Dashed* and *solid lines* in (**d**) correspond to $(J_h)_x$ and $(J_h)_y$, respectively. **e** Snapshot of the density patterns at $t = 100$ in the absence of the film. **f, g** Snapshots of the density patterns and the film surface patterns at $t = 100$ in the presence of the film

20.4.3 Large-Scale Holes in the Film and Film Rupture

When the self-propulsion velocity is increased to $V = 1$, the stripe-like density patterns in the absence of the film are no longer found in the long time limit, as shown in Fig. 20.6f, where we plot the average density field $\langle \rho \rangle(x, y)$ taken at $t = 60$. Instead, the density field corresponds to an irregular time-varying array of high- and low-density spots that have the size of the fastest growing wave length.

In the case, when the film is present, the density field shows maze-like patterns with a typical size comparable to the fastest growing wave length, as given in Fig. 20.6g. These maze-like patterns are overlayed with large-scale modulations with the typical length approximately equal to the domain size. Thus, an elevation region, resembling a droplet, can be seen in the density field in Fig. 20.6g concentrated around $x = -3$, $y = -1$. The film height h is nearly zero in this point, as seen in Fig. 20.6h. The amplitude of density modulations $\langle \rho \rangle_{\max} - \langle \rho \rangle_{\min}$ remains largely unaffected by the liquid film (Fig. 20.6a), fluctuating around the value of $\langle \rho \rangle_{\max} - \langle \rho \rangle_{\min} \approx 1.5$.

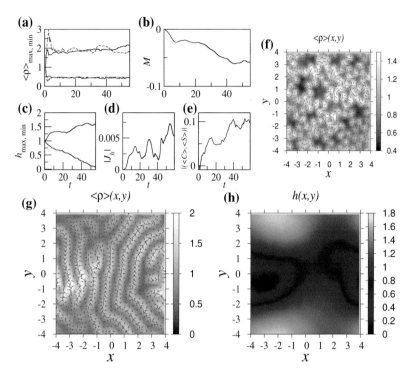

Fig. 20.6 Evolution with and without liquid film at $V = 1$, $D = 1$, $\Omega_0 = 0$ and $d = 0.1$. Line styles in (**a**) as in Fig. 20.5a. Solution measures: **b** the mode type, **c** $h_{\min}$, $h_{\max}$, **d** $| \bar{J}_h |$ and **e** $|(\langle C \rangle, \langle S \rangle)|$. **f** Snapshot of the density patterns at $t = 60$ in the absence of the film. **g**, **h** Snapshots of the density patterns and the film surface patterns at $t = 60$ in the presence of the film

Remarkably, we find a large-scale hole in the film that develops at a late stage of the time evolution, as shown in Fig. 20.6h. The lateral hole size is of the order of the domain length. The amplitude of the film surface deformations increases with time and, eventually, reaches the point, where $h_{\min} \approx 0$ and the film rupture occurs ($t = 60$ in Fig. 20.6c). The hole in the film is in anti-phase with the elevation region in the average density field, implying negative mode type M in Fig. 20.6b. The magnitudes of the fluid flux $| \bar{J}_h |$ and the space-averaged orientation $| (\langle C \rangle, \langle S \rangle) |$ fluctuate randomly with time, as shown in Fig. 20.6d, e, respectively.

From our numerical results we can not definitely decide whether the observed film rupture corresponds to a true finite time singularity or whether it is due to the limited numerical resolution. There exist extensive studies on the rupture of films of simple liquids [4]. In the case of the long-wave thermocapillary instability [3, 13], the film rupture does not seem to occur after finite time. In the case of the destabilising van der Waals interactions [15, 26, 28], the rupture clearly occurs after finite time and close to rupture self-similar solutions can be given. In the model studied here, there are no van der Waals or indeed any other destabilizing (or stabilizing) film surface—substrate interactions, what makes it unlikely that finite time rupture occurs. However, the issue should be studied in detail in the future, in particular, the role of the interaction of the swimmer density and the film height close to rupture. Average density contains two distinct dominant wave lengths: one is of the order of $2\pi / k_{\max} \approx 1$, and the other one is of the order of the domain size.

20.5   Conclusion

In summary we have considered the dynamics of a colony of swimmers that interact with each other via a short-range alignment and longer-range anti-alignment mechanism and move along the surface of a thin liquid film with deformable liquid-gas interface. We have derived a dynamical model that consists of a thin film equation in the long-wave approximation for the evolution of the local film thickness coupled to the Somluchowski equation for the evolution of the swimmer density function. In contrast to previously used models [1, 9, 18], we have included a deterministic rotation of the swimmers bodies that gives rise to their circular motion along the film surface.

   We have focused on the effect of the liquid film on the linear stability of the homogeneous isotropic distribution of swimmers and on its role in the nonlinear time evolution of the system. To this end, we have compared the coupled system of swimmers on the deformable film surface with the bare system of swimmers without a liquid film.

   Our results show that the inclusion of the flat film, which is linearly stable without the colony of swimmers on its surface, can induce a finite wave length instability of the isotropic density distribution. This effect is only found for a certain combination of the rotational diffusivity, the self-propulsion velocity and the rotation frequency that gives rise to the circular motion of swimmers. It is not surprising that the coupled system of swimmers on top of a deformable liquid film appears to be less stable than the bare system of swimmers. Generally, a higher degree of complexity of a system implies less stability. Thus, a similar effect was observed earlier for some passive systems that do not contain any active matter [16, 17, 25].

   By numerically solving the equations of motion we investigated the nonlinear dynamics of the system from the isotropic state for parameters close to the stability threshold and deep in the unstable region. Close to the stability threshold we found square array of vortices in the density distribution, accompanied by small amplitude deformations of the film surface. Deep in the unstable region, for small values of the self-propulsion velocity, small amplitude stripes in the density field emerge. The film surface remains almost flat with the maximal deformation amplitude reaching as less as 0.5 % of the average film thickness.

   For larger values of the self-propulsion velocity, large-scale deep depression forms in the film. The size of the depression is of the order of the domain size and its depth gradually increases with time. The depth of the emerging depression may eventually reach the value of the average film thickness, thus inducing the film rupture.

On a qualitative level, our results can be used to explain the rupture of soaplike liquid films loaded with bacteria, as observed in a series of experiments [20–22]. Without bacteria, any soaplike liquid film, regardless of the film thickness, is linearly unstable due to long-range van der Waals forces that act between apolar molecules that make up the ambient gas layer separated by the liquid film [11]. This implies that even in the ideal case when the evaporation of the liquid can be neglected and the liquid is not drained due to gravity, the flat film is linearly unstable w.r.t. the long-wave deformations of the two liquid-gas interfaces. This instability eventually leads to film rupture after a certain interval of time. The life time of the film is determined by the Hamaker constant that characterizes the strength of the van der Waals interaction, the film thickness h, the surface tension σ and the viscosity of the liquid η. In fact, by using the lubrication approximation [14], it can be shown that the typical life time of the film scales as $\sim h^5 \sigma \eta$.

However, the situation changes dramatically, when the film is loaded with swimmers, whose motion couples to the film deformations via the soluto-Marangoni effect. The orientational instability of the colony of swimmers couples to the instability of the flat film. For films thicker than several μm, the destabilizing action of the van der Waals forces can be neglected as compared with the strength of the orientational instability. In this case, the life time of the film is determined by the typical time scale of the orientational instability, which does not depend on the film thickness. This may explain that surfactant-covered films loaded with bacteria break down earlier than expected.

Appendix 1: Non-dimensional Thin Film Equation and the Smoluchowski Equation

In the here employed dimensionless quantities, the resulting coupled system consists of the reduced Smoluchowski equation for the swimmers density $\rho(\boldsymbol{r}, \phi, t)$ and the thin film equation for the local film thickness $h(\boldsymbol{r}, t)$

$$\partial_t h + \nabla \cdot \boldsymbol{J}_h = 0, \quad \partial_t \rho + \nabla \cdot \boldsymbol{J}_t + \partial_\phi J_\phi = 0, \tag{20.22}$$

with the vorticity of the fluid flow $\Omega_z = \partial_x U_y - \partial_y U_x$, the fluid flux $\boldsymbol{J}_h$, the translational and rotational probability currents $\boldsymbol{J}_t$ and J_ϕ, the surface fluid velocity $\boldsymbol{U}_\parallel$ given by

$$\boldsymbol{J}_h = (h^3/3)\nabla(\Delta h) - (h^2/2)\nabla\langle\rho\rangle, \quad \boldsymbol{J}_t = (V\boldsymbol{q} + \boldsymbol{U} - d\nabla)\rho,$$
$$J_\phi = (\Omega_0 + \Omega_z/2)\rho - D\partial_\phi\rho - \rho\left[\sin\phi\,\hat{\mu}C(\boldsymbol{r}, t) - \cos\phi\,\hat{\mu}S(\boldsymbol{r}, t)\right],$$
$$C(\boldsymbol{r}, t) = \int_0^{2\pi} \rho(\boldsymbol{r}, \phi, t)\cos\phi\,d\phi, \quad S(\boldsymbol{r}, t) = \int_0^{2\pi} \rho(\boldsymbol{r}, \phi, t)\sin\phi\,d\phi,$$
$$\hat{\mu} = \mu_0 + \mu_2\Delta + \mu_4\Delta^2, \quad \boldsymbol{U} = -h\nabla\langle\rho\rangle + h^2/2\nabla(\Delta h). \tag{20.23}$$

Appendix 2: Semi-implicit Numerical Scheme
for Eqs. (20.22)

The coupled evolution equations Eq. (20.22) are solved numerically using the following version of the semi-implicit spectral method. First, we average the density equation over the orientation angle ϕ. This yields

$$\partial_t \langle \rho \rangle + \nabla \cdot \langle J_{\text{trans}} \rangle = 0, \qquad (20.24)$$

with the average translational current $\langle J_{\text{trans}} \rangle = V \langle q\rho \rangle + (U - d\nabla)\langle \rho \rangle$ and $q = (\cos\phi, \sin\phi)$. It is worthwhile noticing that the only term in Eq. (20.24) that depends on the three-dimensional density $\rho(x, y, \phi, t)$ is the average orientation vector $\langle q\rho \rangle$. All other terms in Eq. (20.24), including the surface fluid velocity U explicitly depend on the average density $\langle \rho \rangle$.

Next, we group the thin film equation together with Eq. (20.24)

$$\partial_t h + \nabla \cdot J_h = 0, \quad \partial_t \langle \rho \rangle + \nabla \cdot \langle J_{\text{trans}} \rangle = 0, \qquad (20.25)$$

with the fluid flux $J_h = \frac{h^3}{3}\nabla(\Delta h) - \frac{1}{2}\nabla(h^2\nabla\langle\rho\rangle)$.

At the next step, we single out the linear parts in all the terms in Eq. (20.25) that explicitly depend on the average density $\langle \rho \rangle$. This is done by linearising the current J_t and the fluid flux J_h about the trivial steady state given by $h = 1$ and $\langle \rho \rangle = 1$.

Next, following the standard implicit time-integration scheme, we replace $\partial_t h$ and $\partial_t \langle \rho \rangle$ by $(h^{t+dt} - h^t)/dt$ and by $(\langle \rho \rangle^{t+dt} - \langle \rho \rangle^t)/dt$, respectively and take all linear terms at time $t + dt$ and all nonlinear terms, including the term $V\langle q\rho \rangle$, at time t. Upon these transformations Eq. (20.25) become

$$h^{t+dt} - h^t + (dt/3)\Delta^2 h^{t+dt} - (dt/2)\Delta\langle\rho\rangle^{t+dt} + dt\nabla \cdot (NL_h)^t = 0, \qquad (20.26)$$
$$\langle\rho\rangle^{t+dt} - \langle\rho\rangle^t + (dt/2)\Delta^2 h^{t+dt} - dt(1+d)\Delta\langle\rho\rangle^{t+dt} + dt\nabla \cdot (\langle NL_{\text{trans}}\rangle)^t = 0.$$

where NL denotes the nonlinear parts. After taking the discrete Fourier transforms of Eq. (20.26), we find the updated fields h^{t+dt} and $\langle\rho\rangle^{t+dt}$ at the time step $t + dt$.

With the update average density $\langle\rho\rangle^{t+dt}$ and the film thickness h^{t+dt} at hand, we find the updated surface fluid velocity U^{t+dt} and the updated vorticity Ω_z^{t+dt}.

At the next step, we decompose the currents in the second equation in Eq. (20.22) into a linear and a non-linear parts and make use of the semi-implicit integration scheme

$$\rho^{t+dt} - \rho^t + dt\hat{L}\rho^{t+dt} + dt\text{NL}(\rho^t) = 0. \qquad (20.27)$$

After taking the Fourier transform of Eq. (20.27) both, in space as well as in the angle ϕ, the operator $\hat{L}_F$ can be written as

$$\hat{L}_F = dk^2 + Dn^2 + in\Omega_0 - 0.5(\delta_{n,1} + \delta_{n,-1})\left(\mu_0 - \mu_0 k^2 + \mu_4 k^4\right). \qquad (20.28)$$

The nonlinear part NL is given by

$$NL(\rho^t) = 0.5\Omega_z^{t+dt}\partial_\phi\rho^t + \boldsymbol{\nabla} \cdot (V\boldsymbol{q}\rho^t + \boldsymbol{U}^{t+dt}\rho^t)$$
$$- \partial_\phi \left(\rho^t - (2\pi)^{-1}\right)\left[\sin\phi(\tilde{\tilde{\mu}}C^t) - \cos\phi(\tilde{\tilde{\mu}}S^t)\right]. \qquad (20.29)$$

By solving Eq. (20.27) w.r.t. ρ^{t+dt} in the Fourier space, we apply the backward Fourier transform and find the updated three-dimensional density $\rho^{t+dt}(x, y)$ in the real space.

References

1. S. Alonso, A.S. Mikhailov, Towards active microfluidics: interface turbulence in thin liquid films with floating molecular machines. Phys. Rev. E **79**, 061906 (2009)
2. I.S. Aranson, A. Sokolov, J.O. Kessler, R.E. Goldstein, Model for dynamical coherence in thin films of self-propelled microorganisms. Phys. Rev. E **75**, 040901 (2007)
3. W. Boos, A. Thess, Cascade of structures in long-wavelength Marangoni instability. Phys. Fluids **11**, 1484–1494 (1999)
4. R.V. Craster, O.K. Matar, Dynamics and stability of thin liquid films. Rev. Mod. Phys. **81**, 1131–1198 (2009)
5. R. Di Leonardo, D. Dell'Arciprete, L. Angelani, V. Iebba, Swimming with an image. Phys. Rev. Lett. **106**, 038101 (2011)
6. Ch. Dombrowski, L. Cisneros, S. Chatkaew, R.E. Goldstein, J.O. Kessler, Self-concentration and large-scale coherence in bacterial dynamics. Phys. Rev. Lett. **93**, 098103 (2004)
7. J. Dunkel, S. Heidenreich, K. Drescher, H.H. Wensink, M. Bär, R.E. Goldstein, Fluid dynamics of bacterial turbulence. Phys. Rev. Lett. **110**, 228102 (2013)
8. P.D. Frymier, R.M. Ford, H.C. Berg, P.T. Cummings, Three-dimensional tracking of motile bacteria near a solid planar surface. Proc. National Acad. Sci. **92**(13), 6195–6199 (1995)
9. R. Großmann, P. Romanczuk, M. Bär, L. Schimansky-Geier, Vortex arrays and mesoscale turbulence of self-propelled particles. Phys. Rev. Lett. **113**, 258104 (2014)
10. T. Ishikawa, N. Yoshida, H. Ueno, M. Wiedeman, Y. Imai, T. Yamaguchi, Energy transport in a concentrated suspension of bacteria. Phys. Rev. Lett. **107**, 028102 (2011)
11. J.N. Israelachvili, *Intermolecular and Surface Forces* (Academic, London, 1992)
12. K.-A. Liu, I. Lin, Multifractal dynamics of turbulent flows in swimming bacterial suspensions. Phys. Rev. E **86**, 011924 (2012)
13. A. Oron, Nonlinear dynamics of three-dimensional long-wave Marangoni instability in thin liquid films. Phys. Fluids **12**, 1633–1645 (2000)
14. A. Oron, S.H. Davis, S.G. Bankoff, Long-scale evolution of thin liquid films. Rev. Mod. Phys. **69**, 931 (1997)
15. D. Peschka, A. Münch, B. Niethammer, Thin-film rupture for large slip. J. Eng. Math. **66**, 33–51 (2010)
16. A. Pototsky, M. Bestehorn, D. Merkt, U. Thiele, Alternative pathways of dewetting for a thin liquid two-layer film. Phys. Rev. E **70**, 025201 (2004)
17. A. Pototsky, M. Bestehorn, D. Merkt, U. Thiele, Morphology changes in the evolution of liquid two-layer films. Phys. Rev. E **122**, 224711 (2005)
18. A. Pototsky, U. Thiele, H. Stark, Stability of liquid films covered by a carpet of self-propelled surfactant particles. Phys. Rev. E (R), **90**, 030401 (2014)
19. I.H. Riedel, K. Kruse, J. Howard, A self-organized vortexarray of hydradynamically entrained sperm cells. Science **309**, 300 (2005)

20. A. Sokolov, I.S. Aranson, Physical properties of collective motion in suspensions of bacteria. Phys. Rev. Lett. **109**, 248109 (2012)
21. A. Sokolov, I.S. Aranson, J.O. Kessler, R.E. Goldstein, Concentration dependence of the collective dynamics of swimming bacteria. Phys. Rev. Lett. **98**, 158102 (2007)
22. A. Sokolov, R.E. Goldstein, F.I. Feldchtein, I.S. Aranson, Enhanced mixing and spatial instability in concentrated bacterial suspensions. Phys. Rev. E **80**, 031903 (2009)
23. Y. Sumino, K.H. Nagai, Y. Shitaka, D. Tanaka, K. Yoshikawa, H. Chate, K. Oiwa, Large-scale vortex lattice emerging from collectively moving microtubules. Nature **483**, 448–452 (2012)
24. U. Thiele, A.J. Archer, M. Plapp, Thermodynamically consistent description of the hydrodynamics of free surfaces covered by insoluble surfactants of high concentration. Phys. Fluids **24**, 102107 (2012)
25. U. Thiele, D.V. Todorova, H. Lopez, Gradient dynamics description for films of mixtures and suspensions: dewetting triggered by coupled film height and concentration fluctuations. Phys. Rev. Lett. **111**, 117801 (2013)
26. D. Tseluiko, J. Baxter, U. Thiele, A homotopy continuation approach for analysing finite-time singularities in thin liquid films. IMA J. Appl. Math. **78**, 762–776 (2013)
27. H.H. Wensink, J. Dunkel, S. Heidenreich, K. Drescher, R.E. Goldstein, H. Lwen, J.M. Yeomans. Meso-scale turbulence in living fluids. Proc. Natl. Acad. Sci. **109**(36), 14308–14313 (2012)
28. T.P. Witelski, A.J. Bernoff, Dynamics of three-dimensional thin film rupture. Phys. D **147**, 155–176 (2000)
29. X.-L. Wu, A. Libchaber, Particle diffusion in a quasi-two-dimensional bacterial bath. Phys. Rev. Lett. **84**, 3017–3020 (2000)
30. A. Zöttl, H. Stark, Nonlinear dynamics of a microswimmer in poiseuille flow. Phys. Rev. Lett. **108**, 218104 (2012)

Chapter 21
Time-Delayed Feedback Control
of Spatio-Temporal Self-Organized Patterns
in Dissipative Systems

Alexander Kraft and Svetlana V. Gurevich

Abstract We are interested in the dynamical properties of spatio-temporal self-organized patterns in a Swift-Hohenberg equation subjected to time delayed feedback. We show that variation in the delay time and the feedback strength can lead to complex dynamical behavior of the system in question including formation of traveling hexagons, traveling zigzag patterns, or intricate oscillatory structures. Furthermore, we provide a bifurcation analysis of the system and derive a set of order parameter equations which allow us to analytically demonstrate how the time delayed feedback can change the stability of the homogeneous steady state as well as of periodic patterns. Direct numerical simulations are carried out, showing good agreement with analytical predictions based on linear stability analysis and bifurcation theory. The presented results are derived in general form and can be applied to a wide class of spatially extended systems.

21.1 Introduction

Inspired by the work of Ott et al. [1], a variety of different techniques for controlling unstable or chaotic states in complex systems have been developed within the past decade (see e.g., [2, 3] and references therein). Among other control techniques, global feedback methods have attracted much attention due to a rather simple and easy experimental implementation and mathematical treatment. However, local feedback methods also have gained interest in recent years. A particularly quite simple and efficient scheme is time-delayed feedback (TDF), also referred to as Pyragas control or time-delay autosynchronization, which was first proposed by Pyragas [4].

A. Kraft (✉)
Institut für Theoretische Physik, Technische Universität Berlin,
Hardenbergstr. 36, 10623 Berlin, Germany
e-mail: alexander.kraft@tu-berlin.de

S.V. Gurevich
Institut für Theoretische Physik, Westfälische Wilhelms-Universität Münster,
Wilhelm-Klemm-Str. 9, 48149 Münster, Germany
e-mail: gurevics@uni-muenster.de

© Springer International Publishing Switzerland 2016   413
E. Schöll et al. (eds.), *Control of Self-Organizing Nonlinear Systems*,
Understanding Complex Systems, DOI 10.1007/978-3-319-28028-8_21

According to this scheme, a control force is constructed as a difference of the output signal $s(t)$ of some dynamical system at a time t and a delayed time $t - \tau$, so that the controlled system in question reads

$$\partial_t \psi = F[\psi] + K \left(s(t) - s(t - \tau) \right), \tag{21.1}$$

where $F[\psi]$ describes the intrinsic dynamics of the dynamical system and τ denotes the delay time. Thereby, the coupling matrix K allows to couple different components of the state vector ψ within the feedback loop.

This method allows a noninvasive stabilization of unstable periodic orbits of dynamical systems (see [3] and references therein) and has been also successfully applied to a number of both theoretical and experimental high-dimensional spatially extended systems including e.g., semiconductor systems [5, 6], plasma physics [7, 8], nonlinear optics [9–15] as well as electrochemical [16] and neural systems [17, 18].

In particular, dynamics of spatio-temporal patterns in dissipative systems under influence of TDF control has been intensively studied in recent years. We mention only the control of a kink solution in a reaction-diffusion system subjected to time-delayed feedback [19], biological range expansions problems [20, 21], control of turbulent structures in a diffusive Hutchinson equation [22], spatial and temporal feedback control of standing and traveling waves of the complex Ginzburg-Landau equation [23–26], as well as TDF control of spatio-temporal periodic and localized patterns in a predator-prey plankton system [27], a Gray-Scott [28], a Brusselator [29], a Lengyel-Epstein [30] or FitzHugh-Nagumo like reaction-diffusion systems [31–33].

In this manuscript, we are interested in the influence of the delayed feedback on the stability properties of the homogeneous steady state as well as of periodic patterns in a Swift-Hohenberg equation subjected to TDF:

$$\partial_t \psi(r, t) = \left[\epsilon - \left(k_c^2 + \nabla^2 \right)^2 \right] \psi(r, t) + \delta \cdot \psi(r, t)^2 - \psi^3(r, t) \tag{21.2}$$
$$+ \alpha \left[\psi(r, t) - \psi(r, t - \tau) \right],$$

where α denotes the feedback strength and τ is the delay time. In the absence of the TDF term, Eq. (21.2) reduces to the classical version of the Swift-Hohenberg equation for the real distributed order parameter $\psi(r, t)$, $r \in \mathbb{R}^2$. Here, ϵ is the bifurcation parameter, measuring the distance to the supercritical bifurcation with the most unstable wave number k_c. Parameter δ breaks the inversion symmetry and has in general an arbitrary sign. The Swift-Hohenberg equation often serves as a paradigm for general pattern forming systems and has been applied in various fields of nonlinear science such as hydrodynamics [34], chemical [35], ecological [36], and optical [37, 38] systems, or elastic materials [39].

Among other patterns, the control of localized solutions of the Swift-Hohenberg equation has been of increasing interest in recent years. In particular, properties of two-dimensional cavity solitons in the Swift-Hohenberg equation subjected to time-delayed feedback were studied in [11, 40]. It was shown that when the value of the product of the delay time τ and the feedback strength α exceeds some critical value, a single cavity soliton starts to move in an arbitrary direction. Moreover, an analytical formula for its velocity was derived. Recently, the influence of TDF on the stability properties of a single localized structure in the Swift-Hohenberg equation was investigated in detail [41]. It was demonstrated that the variation of the product of α and τ can lead to the formation of oscillons, soliton rings, or labyrinth patterns. Moreover, a bifurcation analysis of the delayed system was provided and a system of order parameter equations for the position of the localized structure as well as for its shape was derived. Note that in the context of nonlinear optics [37], Eq. (21.2) includes a free constant in the nonlinear function rather than the quadratic term in ψ. However, Eq. (21.2) can be rewritten by an offset transformation $\psi \to \psi - \psi_0$ with $\psi_0 = -\delta/3$ in such a way that the quadratic nonlinearity is removed and a free constant $Y = \dfrac{\delta}{3}\left(\epsilon - k_c^4 + \dfrac{2\delta^2}{9}\right)$ appears in (21.2).

In this chapter, we show that the variation of the delay time and the feedback strength can lead to a traveling wave bifurcation of the homogeneous solution of Eq. (21.2), whereas a drift bifurcation sets in for periodic structures, giving rise to complex dynamical behavior of the system including formation of traveling waves and hexagons as well as traveling zigzag patterns. We provide a bifurcation analysis of the system (21.2) and derive a set of order parameter equations which allow us to analytically demonstrate how the time delayed feedback can change the stability of the homogeneous steady state as well as of periodic patterns. In particular, we show that the aforementioned delay-induced drift bifurcation of the stationary solution always takes place if the system in question features translational invariance with respect to its spatial coordinates. The presented results are derived in general form and can be applied to a wide class of spatially extended systems, assuming the corresponding linear stability problem can be diagonalized.

21.2 Linear Stability Analysis

Notice that the function $\psi(\mathbf{r}, t)$ given by the delayed Swift-Hohenberg equation (DSHE) (21.2) is a scalar quantity. Nevertheless, for the sake of generality of the following analysis we rewrite Eq. (21.2) in terms of an n-dimensional vector function $\psi = \psi(\mathbf{r}, t)$, $\mathbf{r} \in \mathbb{R}^n$ and use the general form of the evolution Eq. (21.1)

$$\partial_t \psi(\mathbf{r}, t) = \mathbf{F}[\psi(\mathbf{r}, t)] + \mathbf{K}\left(\psi(\mathbf{r}, t) - \psi(\mathbf{r}, t - \tau)\right), \qquad (21.3)$$

where the output function $\mathbf{s}$ is chosen as the state variable ψ and $\mathbf{F}$ is a nonlinear operator. The coupling matrix $\mathbf{K}$ is in general a constant matrix of low rank, e.g., a multiple of the identity $\alpha \mathbf{E}$ with feedback strength α for the scalar DSHE (21.2).

Let ψ_0 be a stationary solution of (21.3) in the absence of the TDF term. Note that although this solution is not affected by the TDF, its stability can change. The linear stability of ψ_0 is characterized by the linear eigenvalue problem

$$\lambda \varphi = \left[F'[\psi_0] + K \left(1 - e^{-\lambda \tau}\right)\right] \varphi \qquad (21.4)$$

for eigenvalues λ and eigenfunctions φ. Here, $F'[\psi_0]$ is the linearization of the operator F evaluated at ψ_0. Equation (21.4) is a transcendental equation and its analytical solution is in general involved. However Eq. (21.4) can be simplified if, e.g., the linearization operator $F'[\psi_0]$ and the coupling matrix K commutate. In this case, there exists a common basis of eigenfunctions with corresponding eigenvalues μ and μ_K of $F'[\psi_0]$ and K, respectively. This leads to the following characteristic equation for the yet unknown set of eigenvalues λ:

$$\lambda = \mu + \mu_K \left(1 - e^{-\lambda \tau}\right).$$

For the scalar DSHE (21.2), where $K = \alpha E$, the linearization operator and coupling matrix commutate for trivial reasons. In this case, the above characteristic equation reads [11, 32, 33, 41, 42]

$$\lambda = \mu + \alpha \left(1 - e^{-\lambda \tau}\right). \qquad (21.5)$$

Notice that, from a practical perspective, the requirement/usage of simultaneously diagonalizable matrices $F'[\psi_0]$, K may be quite restrictive. However, a close analytical treatment of the simplest case of the control force enables one to gain deeper insights into the impact of the time-delayed feedback on the dynamical properties of the complex system in question.

The transcendental characteristic equation (21.5) links the eigenvalues λ of the stability problem with TDF, to the eigenvalues μ of the linear stability problem in the absence of TDF. The eigenvalues λ of the eigenvalue problem 21.4 can be phrased in terms of the Lambert-W function [43], which is defined as the multi-valued inverse of the function $z\, e^z$ with complex z:

$$\lambda = \mu + \alpha + \frac{1}{\tau} W_n \left(-\alpha \tau e^{-(\mu+\alpha)\tau}\right), \quad n \in \mathbb{Z}, \qquad (21.6)$$

where n is the branch index of the Lambert-W function. Note that the branches of the Lambert-W function have near-conjugate symmetry. In addition, except for $n = 0$ and $n = -1$, all branches are complex, whereas W_0 and W_{-1} can be real-valued only for certain ranges of z. This implies that the eigenvalues λ are in general complex, even if μ is real-valued. Furthermore, every eigenvalue μ of the linear stability problem without TDF induces an infinite number of eigenvalues $\lambda = \lambda(n)$, $n \in \mathbb{Z}$ of the stability problem with TDF. One of these branches, namely $n = 0$, starts at μ, while the other infinite number of branches $n \in \mathbb{Z} \setminus \{0\}$ start at $-\infty$, all belonging to the same eigenvalue μ.

21.2.1 Neutral Stability Curves

Our goal now is to solve the characteristic equation (21.5) for the neutral stability curves, at which the stability of the stationary solution ψ_0 changes, i.e., $\mathrm{Re}(\lambda)$ changes the sign. To this end, we consider the feedback strength α as a parameter and solve Eq. (21.5) for the critical delay time τ_c at which the corresponding eigenvalue λ crosses the imaginary axis. Notice that since there is an infinite number of branches of eigenvalues $\lambda(n)$, $n \in \mathbb{Z}$, one obtains neutral stability curves depending on the branch index.

Separating the real and imaginary parts of the Eq. (21.5) and solving the obtained system for $\mathrm{Re}(\lambda) = 0$, the following relation for the neutral stability curve $\tau = \tau_c(\alpha)$ can be derived:

$$\tau_c(m, \alpha) = \frac{\arccos\left(1 + \frac{\mathrm{Re}(\mu)}{\alpha}\right) \pm 2\pi m}{\alpha\sqrt{1 - \left(1 + \frac{\mathrm{Re}(\mu)}{\alpha}\right)^2} \pm \mathrm{Im}(\mu)}, \quad m \in \mathbb{Z}. \tag{21.7}$$

In addition, at $\tau = \tau_c$, the critical value of $\omega_c = \mathrm{Im}(\lambda)$ is given by

$$\omega_c = \pm\alpha\sqrt{1 - \left(1 + \frac{\mathrm{Re}(\mu)}{\alpha}\right)^2} + \mathrm{Im}(\mu). \tag{21.8}$$

Due to causality, the delay time τ and $\tau_c(\alpha)$ have to be positive, and therefore the nominator and the denominator of Eq. (21.7) must have the same sign. For the indices m for which this physical requirement can be satisfied, a neutral stability curve exists and the stability of the corresponding branch changes when crossing this curve in control parameter space. It is evident that the solution only exists if $\left|1 + \frac{\mathrm{Re}(\mu)}{\alpha}\right| \leq 1$. This is equivalent to the requirement that $\mathrm{Re}(\mu)$ and the feedback strength α must have opposite signs, i.e. the feedback strength α should be chosen positive in order to destabilize a stable solution ψ_0 ($\mathrm{Re}(\mu) < 0$) and chosen negative to stabilize an unstable solution ($\mathrm{Re}(\mu) > 0$). Furthermore, it states that the minimum feedback strength α to change the stability of the stationary solution ψ_0 is given by $|\alpha| \geq \frac{|\mathrm{Re}(\mu)|}{2}$.

21.2.1.1 Implications for $\mu \in \mathbb{R}$

Note that the index m of the neutral stability curve and the branch index n of our eigenvalues $\lambda(n)$ are connected. In order to establish this relation for the case of real eigenvalues μ, a new index $\tilde{m}$ can be defined by $m = \pm\mathrm{sign}(\alpha)\tilde{m}$. It turns out that the index $\tilde{m}$ is the index of conjugate branch pairs of the Lambert-W function, which can be understood as follows: Consider the argument $z = -\alpha\tau e^{-(\mu+\alpha)\tau}$ which is supplied to the Lambert-W function $W_n(z)$ in the Eq. (21.6) for eigenvalues λ. One can see that for $\mu \in \mathbb{R}$, z has the opposite sign of α, i.e., for $\alpha < 0$ and $z > 0$ the pairing of conjugate branches is $\tilde{m} = \pm n$, whereas for the opposite case ($\alpha > 0$ and $z < 0$) the pairing is $\tilde{m} = 0 \leftrightarrow n = \{0, -1\}, \tilde{m} = 1 \leftrightarrow n = \{1, -2\}$, etc. The above asymmetry arises since the pairing of corresponding near-conjugate branch pairs of the Lambert-W function shifts by one element when crossing $W_n(0)$ [43].

For $\mu \in \mathbb{R}$, we can rewrite Eq. (21.7) by factoring out the sign(α):

$$\tau_c(m, \mu, \alpha) = \frac{\text{sign}(\alpha)\left[\arccos\left(1 + \frac{\text{Re}(\mu)}{\alpha}\right) \pm 2\pi m\right]}{|\alpha|\sqrt{1 - \left(1 + \frac{\text{Re}(\mu)}{\alpha}\right)^2}}.$$

Note that $\tau_c > 0$ must hold due to causality and that the range of arccos(z) $\in [0, \pi]$. Since the denominator is positive, the nominator must be positive as well, which yields: (i) For the destabilization scenario ($\mu < 0$, $\alpha > 0$), the nontrivial solution of this equation exists for all m by choosing the correct sign out of $\pm$; Here, all branches start on the left half-plane of the complex plane and change half-plane when $\tau > \tau_c$ whereas (ii) for the opposite case ($\mu > 0$, $\alpha < 0$), the solution only exists for $m \neq 0$. That is, the branches starting from $-\infty$ on the left half-plane can change to the positive half-plane, which would lead to a destabilization of the solution in question, but the branch with $m = 0$, which starts at $\mu > 0$, never changes the half-plane, i.e., the eigenvalue λ approaches the imaginary axis, but it never crosses it. Therefore, a stabilization of solutions with eigenvalue $\mu \in \mathbb{R}_+$ is in general impossible with the chosen type of simple scalar TDF control in Eq. (21.3) with $K = \alpha E$, whereas complex eigenvalues $\mu \in \mathbb{C}$ with positive real part can be stabilized, as seen in Ref. [33].

Figure 21.1 shows a representative example of ten leading branches of neutral stability curves τ_c calculated from Eq. (21.7), illustrating the destabilization of ψ_0 with negative eigenvalue $\mu = -1$. One can see that τ_c curves approach the asymptote at $\alpha_{\min} = \frac{|\text{Re}(\mu)|}{2}$ for all indices $-5 \leq m \leq 5$. The critical delay time τ_u at which the first eigenmode becomes unstable is given by the first curve which is crossed in control parameter space when varying α and τ, which is reflected by the condition

$$\tau_u(\alpha) = \min_{m,\mu} \tau_c(m, \mu, \alpha). \tag{21.9}$$

21.2.1.2 Neutral Stability Curves for $\mu = 0$

Notice that if the system (21.3) in the absence of TDF features translational invariance with respect to its spatial coordinates, $\mu = 0$ is an eigenvalue of the operator $F'(\psi_0)$. For $\mu = 0$, the characteristic equation (21.5) can be simplified to

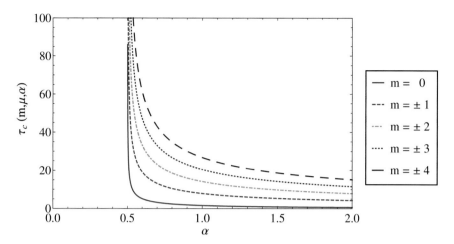

Fig. 21.1 Neutral stability curves $\tau_c(m, \mu, \alpha)$ calculated from Eq. (21.7) as function of the feedback strength α for fixed value of the eigenvalue $\mu = -1$ and branches $-5 \leq m \leq 5$. One can see that τ_c curves approach the asymptote at $\alpha_{min} = \frac{|Re(\mu)|}{2}$ for all indices m

$$\lambda = \alpha\left(1 - e^{-\lambda\tau}\right),$$

yielding two real-valued solutions $\lambda_{1,2}$ [11, 41]:

$$\lambda_1 = 0, \quad \lambda_2 \approx \frac{2(\alpha\tau - 1)}{\alpha\tau^2}.$$

The eigenvalue λ_1 remains zero, whereas the eigenvalue λ_2 is negative only for $\tau < 1/\alpha$ and coincides with λ_1 at $\tau_c = 1/\alpha$, which corresponds to the onset of spontaneous motion, first observed for localized states in the Swift-Hohenberg equation subjected to time-delayed feedback [11]. Notice that since the real-valued $\lambda_{1,2}$ correspond to the $n = 0$ and $n = -1$ branches of the Lambert-W function, the change of stability for ψ_0 at $\tau_c = 1/\alpha$ happens exactly at the branch point $(z, W_n(z)) = (-1, -e^{-1})$, where both $n = 0$ and $n = -1$ branches intersect.

Note that the aforementioned neutral stability curve $\tau_c = 1/\alpha$ can also be derived within our formal approach, when taking the limit $\mu \to 0$ of Eq. (21.7). From the point of view of bifurcation theory, the TDF leads to a spontaneous symmetry breaking of translational invariance, which causes a drift bifurcation of the solution ψ_0.

21.3 The Swift-Hohenberg Equation with Time-Delayed Feedback Control

In this section, we illustrate how the results obtained for general systems in Sect. 21.2 can be applied for the Swift-Hohenberg equation with time delayed feedback (21.2) introduced in Sect. 21.1. In the absence of TDF control, i.e., for $\alpha = 0$ or $\tau = 0$, the Swift-Hohenberg equation (21.2) possesses a Lyapunov functional. This implies that any given system state evolves monotonously towards a stable stationary state, represented by local minima of the functional. However, in the presence of TDF, the system loses this gradient structure and can show non-monotonous behavior. As discussed in Sect. 21.2, in the presence of TDF, the set of steady state solutions persist, but their stability may change.

Here, we focus on the stability of the trivial homogeneous solution, stripes and hexagons. Notice that more complex nontrivial solutions like localized states, rings, localized hexagon patches, or stripes can also be found [11, 37, 41, 44, 45], but they are out of the scope of this manuscript.

21.3.1 Stability of the Homogeneous Solution

Suppose that the homogeneous solution $\psi_0 = 0$ of Eq. (21.2) is stable for $\tau = 0$, i.e., the dispersion relation satisfies

$$\mu(\epsilon, k) = \epsilon - (k_c^2 - k^2)^2 < 0$$

for all values of k, where one can set $k_c = 1$ without loss of generality. For $\tau > 0$ the linear stability analysis with respect to perturbations $\sim \exp(i\,k\,x + \lambda t)$ yields eigenvalues of the form (cf. Eq. (21.6))

$$\lambda\left(n, \mu(\epsilon, k), \alpha, \tau\right) = \mu(\epsilon, k) + \alpha + \frac{1}{\tau} W_n\left(-\alpha\tau e^{-(\mu(\epsilon, k)+\alpha)\tau}\right), \quad n \in \mathbb{Z}. \quad (21.10)$$

By assumption ψ_0 is stable for $\tau = 0$ and we fix $\epsilon = -1$ and choose $\alpha > 0$ in order to be able to achieve a destabilization (see Sect. 21.2.1.2). To get a first insight, it is instructive to map a reasonable choice of sets $\{n, k, \alpha, \tau\}$ to $\lambda(n, \mu, \alpha, \tau)$, in order to get an overview of the whole spectrum (see Fig. 21.2). Thereby, it can be observed that the branches n build near-conjugate branch pairs, as expected from the properties of the Lambert-W function. In addition, we can identify the branches n with the largest growth rate $\mathrm{Re}(\lambda)$ as $n = 0$ and $n = -1$. Furthermore, from Eq. (21.8) one can see that all branches n have a non-vanishing oscillation frequency $\omega_c = \mathrm{Im}(\lambda) \neq 0$, when crossing the imaginary axis.

First, we investigate which branch becomes unstable first, when varying TDF control parameters (α, τ). To this end, we consider the growth rate $\max_k \mathrm{Re}\,\lambda(n, \mu(\epsilon, k), \alpha, \tau)$ as function of α and τ for fixed ϵ and at the wave number k_{max} which maximizes it (see Fig. 21.3).

Figure 21.3 shows the first unstable branches $n = 0$ and $n = -1$, which we denote as the *leading branches* as they govern the primary destabilization of ψ_0. Since Fig. 21.3 shows only values with $\mathrm{Re}(\lambda) \geq 0$, the boundary of these regions satisfy $\mathrm{Re}(\lambda) = 0$ and is given by the obtained neutral stability curves for $\mu \in \mathbb{R}$ from Eq. (21.7) (green solid line on Fig. 21.3), provided that their construction was appropriate and their indices m match the branch indices n of the Lambert-W function.

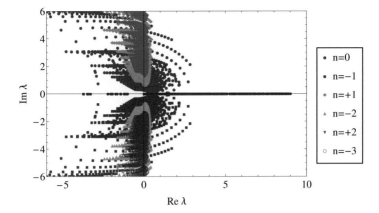

Fig. 21.2 Spectrum $\lambda(n, \mu, \alpha, \tau)$ for a set of $\{n, k, \alpha, \tau\}$ calculated for $\epsilon = -1$. Parameter ranges are $k \in [0, 3]$, $\alpha \in [0, 10]$, $\tau \in [0, 10]$ for branches $-3 \le n \le 2$. Branches appear in the legend in order of grouping of near-conjugate branch pairs. Branches with largest growth rate Re λ are $n = 0$ and $n = -1$

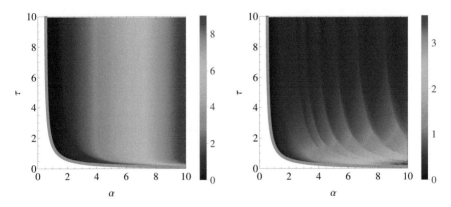

Fig. 21.3 Re $\lambda(n, \mu(\epsilon, k_{max}), \alpha, \tau)$ *versus* (α, τ) for leading branches $n = 0$ (*left*), $n = -1$ (*right*) and fixed $\epsilon = -1$ at fastest growing wave number k_{max}, which was found by numerical maximization with respect to k. Contour colors encode the size of the growth rate Re(λ). Only values with Re$(\lambda) \ge 0$ are shown. The *green solid line* represents the neutral stability curve $\tau_c(m, \mu, \alpha)$ calculated for $m = 0$

One can easily see [40, 41] that the fastest growing wave number could depend on α and τ and on the particular branch n, whereas the dominant wave number for the leading branch $n = 0$ is $k_{max} = k_c = 1$ with overall fastest growth rate of all branches n. As was mentioned above, all branches n have a non-vanishing oscillation frequency, when crossing the imaginary axis. This implies that the destabilization of the homogeneous steady state ψ_0 occurs via a *traveling wave bifurcation*, since predominantly the fastest growing wave number is $k_{max} = 1$ and its corresponding oscillation frequency Im(λ) is non-zero.

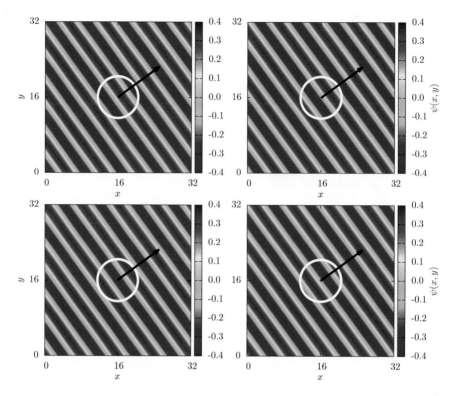

Fig. 21.4 Four snapshots of the time evolution of Eq. (21.2) showing a traveling wave solution as a result of destabilizing the homogeneous steady state by TDF. The *white circle* is a reference point, and the *black arrow* shows that the traveling wave moves towards the *top right corner*. Parameters are: $\epsilon = -1$, $\delta = 0$, $\alpha = 1.9$, $\tau = 0.8$, TDF turned on at $t = 0$, time interval between successive snapshots is $\Delta t = 0.8$

Figure 21.4 shows four snapshots of a traveling wave solution, calculated from direct numerical simulation of Eq. (21.2). The white circle serves a reference point and the black arrow illustrates the movement towards the top right corner. However, for some TDF control parameter values (α, τ) further away from the bifurcation curve, numerical solutions were found in which an oscillation between two opposite states was observed for larger times (see Fig. 21.5).

21.3.2 Stability of Stripes and Hexagons

We now analyze the influence of TDF control on the stability of nontrivial periodic solutions of Eq. (21.2) like stripes and hexagons. Since the TDF term vanishes at the steady states, the stripe and the hexagon steady state solution persist, but their stability may change due to TDF. The derivation of the corresponding linear stability

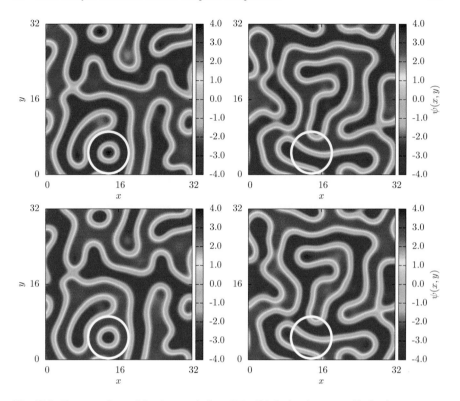

Fig. 21.5 Four snapshots of the time evolution of Eq. (21.2) showing an oscillation between two opposite states, as a result of destabilizing the homogeneous steady state by TDF. Parameters are: $\epsilon = -1$, $\delta = 0$, $\alpha = 4$, $\tau = 4$, TDF turned on at $t = 0$. Again, the *white circle* provides a reference point

problem for both solutions is straightforward and can be made by means of the general ansatz which incorporates both patterns as a special case [46, 47]:

$$\psi(\mathbf{r}, t) = \sum_{j=1}^{3} \xi_j e^{i\mathbf{k}_j \mathbf{r}} + c.c. \tag{21.11}$$

$$= \sum_{j=1}^{3} |\xi_j| \cos(\mathbf{k}_j \mathbf{r} + \varphi_j),$$

with complex amplitudes $\xi_j(t) \in \mathbb{C}$ and wave vectors $\mathbf{k}_j$, $j = 1, 2, 3$:

$$\mathbf{k}_1 = k_c \begin{pmatrix} 1 \\ 0 \end{pmatrix}, \quad \mathbf{k}_2 = \frac{k_c}{2} \begin{pmatrix} -1 \\ \sqrt{3} \end{pmatrix}, \quad \mathbf{k}_3 = \frac{k_c}{2} \begin{pmatrix} -1 \\ -\sqrt{3} \end{pmatrix}. \tag{21.12}$$

As stated earlier, this ansatz contains both stripes and hexagons: One obtains hexagons by considering $|\xi_1| = |\xi_2| = |\xi_3|$ and stripes by setting $|\xi_1| \neq 0$, $|\xi_2| = |\xi_3| = 0$.

When substituting the ansatz (21.11) in the delayed Swift-Hohenberg equation (21.2) and projecting on the modes $e^{ik_j r}$ of the ansatz, one obtains the amplitude equations for the complex amplitudes ξ_j in the presence of TDF:

$$\dot{\xi_1} = \epsilon\xi_1 + 2\delta\xi_2^*\xi_3^* - 3\xi_1\left[|\xi_1|^2 + 2|\xi_2|^2 + 2|\xi_3|^2\right] + \alpha\left(\xi_1(t) - \xi_1(t-\tau)\right),$$
$$\text{(21.13a)}$$

$$\dot{\xi_2} = \epsilon\xi_2 + 2\delta\xi_3^*\xi_1^* - 3\xi_2\left[2|\xi_1|^2 + |\xi_2|^2 + 2|\xi_3|^2\right] + \alpha\left(\xi_2(t) - \xi_2(t-\tau)\right),$$
$$\text{(21.13b)}$$

$$\dot{\xi_3} = \epsilon\xi_3 + 2\delta\xi_1^*\xi_2^* - 3\xi_3\left[2|\xi_1|^2 + 2|\xi_2|^2 + |\xi_3|^2\right] + \alpha\left(\xi_3(t) - \xi_3(t-\tau)\right).$$
$$\text{(21.13c)}$$

For vanishing TDF control ($\alpha = 0$ or $\tau = 0$), the amplitude equations (21.13) reduce to the amplitude equations derived for the classical Swift-Hohenberg equation [46, 47]. It is remarkable to note that the new terms induced by TDF enter linearly to the amplitude equations, since the control mechanism is linear and additive, and the ansatz consist of a linear superposition of plane waves. Notice that the system (21.13) was discussed in [30] for the special case of a stripe solution in the Lengyel-Epstein model subjected to TDF.

21.3.2.1 Linear Stability of Stripes

We start to analyze the amplitude equations (21.13) for the case of a stripe solution. To this aim, we split the complex amplitudes $\xi_j(t) = R_j(t)e^{i\varphi_j(t)}$ in Eq. (21.13) into their real amplitudes $R_j(t)$ and phases $\varphi_j(t)$. For the small perturbations $\delta R_j(t)$ of the real stationary stripe amplitudes $R_j^{(0)} = \left(\frac{\sqrt{\epsilon}}{3}, 0, 0\right)^T$, one obtains

$$\frac{d}{dt}\begin{pmatrix} \delta R_1(t) \\ \delta R_2(t) \\ \delta R_3(t) \end{pmatrix} = \underbrace{\begin{pmatrix} -2\epsilon & 0 & 0 \\ 0 & -\epsilon & 2\delta \cdot R_1^{(0)} \\ 0 & 2\delta \cdot R_1^{(0)} & -\epsilon \end{pmatrix}}_{=:S^R} \cdot \begin{pmatrix} \delta R_1(t) \\ \delta R_2(t) \\ \delta R_3(t) \end{pmatrix}$$

$$+ \alpha \cdot E \begin{pmatrix} \delta R_1(t) - \delta R_1(t-\tau) \\ \delta R_2(t) - \delta R_2(t-\tau) \\ \delta R_3(t) - \delta R_3(t-\tau) \end{pmatrix}.$$

One can show that the matrix S^R is the same as in the linear stability problem of amplitude perturbations δR_j without TDF with eigenvalues $\mu_1^R = -2\epsilon$, $\mu_2 = -\epsilon - 2\delta \cdot R_1^{(0)}$ and $\mu_3^R = -\epsilon + 2\delta \cdot R_1^{(0)}$. The stability for the small perturbation $\delta\varphi_1(t)$ of the phase $\varphi_1^{(0)}$ yields:

$$\frac{d}{dt}\delta\varphi_1(t) = 0 \cdot \delta\varphi_1(t) + \alpha(\delta\varphi_1(t) - \delta\varphi_1(t-\tau)).$$

Since $\mu_j^R < 0\,\forall j$ holds for stable stripes, we know that the occurring instability due to the presence of TDF is of oscillatory type and that the fastest growing wave number is $k = k_c = 1$. Therefore, a destabilization of an eigenvalue $\mu_j^R < 0$ via TDF results in a traveling wave bifurcation. Furthermore, one can see that the eigenvalue of the linear stability problem for the phase perturbation is $\mu^\varphi = 0$. This involves *a drift bifurcation* at $\alpha\tau = 1$ of the stripe solution of the DSHE.

In order to confirm this prediction, direct numerical simulations for different values of α and τ were performed (see Fig. 21.6). Here, green (red) points indicate that the stripes steady state is stable (unstable), when subjected to TDF. The change of stability is well described by the obtained neutral stability curve $\tau_c = 1/\alpha$ (solid blue line).

However, even more complex solutions are found in numerical simulations of Eq. (21.2): For instance, starting with stable stripes with wave number $k = k_c$, a traveling zigzag pattern can be obtained (see Fig. 21.7). Indeed, it is known that the Swift-Hohenberg equation without TDF can become unstable with respect to transversal or longitudinal modulations, the so-called ZigZag and Eckhaus instability, respectively. These modulation instabilities of stripes set in when the mismatch of wave numbers $k \neq k_c$ exceeds certain boundaries [47]. That is, since the parameter setting is chosen so that the stripe solution of the Swift-Hohenberg equation

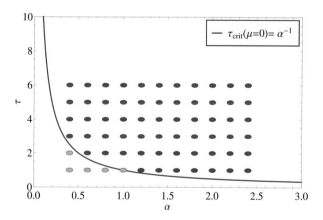

Fig. 21.6 Comparison of analytical predictions with a direct numerical simulation of a destabilization scenario of stripes. *Green (red)* points indicate that the stripes steady state is stable (unstable), when subjected to TDF with feedback strength α and delay time τ. The change of stability is well described by the obtained neutral stability curve $\tau_c(\alpha)$ from our analytical considerations. The first eigenvalue that becomes unstable is $\mu = 0$, which creates a drift bifurcation. Simulation parameters are: $\epsilon = 1$, $\delta = 0$, TDF turned on at time $T = 500$ after the stripes have fully developed

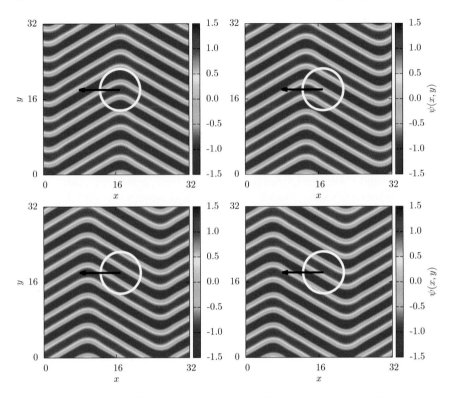

Fig. 21.7 Four snapshots of the time evolution of Eq. (21.2) showing a drift bifurcation of a zigzag pattern. The *white circle* is a reference point, whereas the *black arrow* allows to identify the movement to the *left*. Parameters are: $\epsilon = 1$, $\delta = 0$, $\alpha = 0.5$, $\tau = 2.2$, start with $\sin(k_c y)$, where $k_c = 1$ is the critical wave number. TDF was turned on after the relaxation into stripes

without TDF is stable with respect to transversal modulations, it can be concluded that the threshold of the modulation instability can also be influenced by TDF and its instability region becomes larger in parameter space.

21.3.2.2 Linear Stability of Hexagons

We consider the corresponding linear stability problem for hexagons. For the small perturbations $\delta R_j(t)$ to the real stationary amplitudes $R_j^{(0)} = R_h (1, 1, 1)^T$, $R_h = \frac{1}{15}\left((-1)^n \delta \pm \sqrt{\delta^2 + 15\epsilon}\right)$, one obtains

$$\frac{d}{dt}\begin{pmatrix}\delta R_1(t)\\\delta R_2(t)\\\delta R_3(t)\end{pmatrix}=\underbrace{\begin{pmatrix}A & B & B\\B & A & B\\B & B & A\end{pmatrix}}_{=:S^R}\cdot\begin{pmatrix}\delta R_1(t)\\\delta R_2(t)\\\delta R_3(t)\end{pmatrix}+\alpha\cdot E\begin{pmatrix}\delta R_1(t)-\delta R_1(t-\tau)\\\delta R_2(t)-\delta R_2(t-\tau)\\\delta R_3(t)-\delta R_3(t-\tau)\end{pmatrix},$$

with the abbreviations $A=\epsilon-21R_h^2$ and $B=2\delta\cdot R_h-12R_h^2$. For the small perturbations $\delta\varphi_j(t)$ to the phases $\varphi_j^{(0)}$, one obtains:

$$\frac{d}{dt}\begin{pmatrix}\delta\varphi_1(t)\\\delta\varphi_2(t)\\\delta\varphi_3(t)\end{pmatrix}=-2\delta\cdot R_h\underbrace{\begin{pmatrix}1 & 1 & 1\\1 & 1 & 1\\1 & 1 & 1\end{pmatrix}}_{=:S^\varphi}\cdot\begin{pmatrix}\delta\varphi_1(t)\\\delta\varphi_2(t)\\\delta\varphi_3(t)\end{pmatrix}+\alpha\cdot E\begin{pmatrix}\delta\varphi_1(t)-\delta\varphi_1(t-\tau)\\\delta\varphi_2(t)-\delta\varphi_2(t-\tau)\\\delta\varphi_3(t)-\delta\varphi_3(t-\tau)\end{pmatrix}.$$

Our analysis applies for the same reasons as stated for stripes. It is straightforward to show that the matrices S^R and S^φ can be diagonalized and one can calculate the eigenvalues μ_j^R and μ_j^φ, respectively. Again, one obtains a traveling wave bifurcation if any of the $\mu_j^R<0$ is destabilized via TDF, whereas a drift bifurcation sets in when $\mu^\varphi=0$ is destabilized.

A comparison between analytical predictions and direct numerical simulations is shown in Fig. 21.8. Here, green (red) points indicate that the hexagonal solution is stable (unstable), when subjected to TDF. The change of stability is well described by the obtained neutral stability curve $\tau_c(\alpha)$ (21.7). One can see that the first eigenvalue that becomes unstable is $\mu=0$, which corresponds to a drift bifurcation of

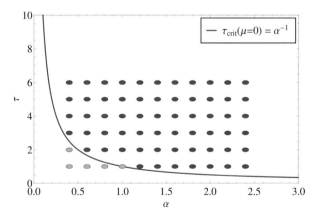

Fig. 21.8 Comparison of analytical predictions with a direct numerical simulation of a destabilization scenario of hexagons. *Green (red)* points indicate that the hexagon steady state is stable (unstable), whereas the neutral stability curve $\tau_c(\alpha)$ (*blue solid line*) mimics the change of stability. Simulation parameters are: $\epsilon=1$, $\delta=1$, TDF turned on at time $t=500$ after the hexagons have fully developed

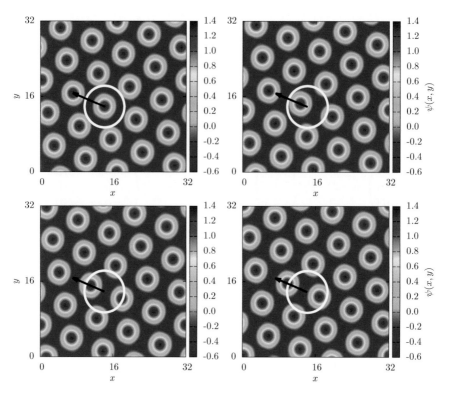

Fig. 21.9 Four snapshots of the time simulation of Eq. (21.2) showing the drifting hexagons. The *white circle* is a reference point, which allows to identify the movement to the *left top*. Parameters are: $\epsilon = 0.1, \delta = 1.0, \alpha = 0.5, \tau = 2.2$, TDF turned on after hexagons relaxed in their steady state

the hexagonal solution (cf. Fig. 21.9, where four snapshots of the time simulation showing the drifting hexagons are presented). Here, the white circle is a reference point, whereas the black arrow illustrates the motion to the left top corner.

21.4 Summary and Outlook

In this chapter, the stability properties of homogeneous and spatially periodic patterns in a Swift-Hohenberg equation subjected to time-delayed feedback were investigated in detail. Starting with the general form of an arbitrary system subjected to TDF, we derived the neutral stability curves at which the stability of a particular stationary solution changes and showed that for real-valued spectra, system states can be destabilized, but not stabilized with the used type of TDF control. Furthermore, we showed that solutions which feature translational invariance lose stability in a drift bifurcation. In the analysis, we assumed that the linear stability problem can be diagonalized. For the Swift-Hohenberg equation with TDF, we applied the results of our linear stability analysis to investigate stability properties of the homogeneous stationary state and of spatially periodic solutions, such as stripes and hexagons. A set of order parameter equations for stripe and hexagonal solutions was derived and analyzed in detail. It was demonstrated that the homogeneous steady state loses stability in a traveling wave bifurcation, whereas stripes and hexagons become unstable via a drift bifurcation, followed by a traveling wave bifurcation for larger values of control parameters α and τ. Direct numerical simulations are also carried out, showing good agreement with analytical predictions based on linear stability analysis and bifurcation theory. However, the explanation of some of the resulting dynamical behavior like drifting zigzag patterns was out of the scope of our current analysis. Indeed, a drift of a zigzag pattern was observed, although the same system parameters would not create a transversal instability of the stripe solution in the system without TDF. Therefore, we conclude that TDF influences the boundaries of the modulation instability of stripes, which provides a good starting point for further analysis. However, this topic is beyond the scope of this paper and is left for future work.

References

1. E. Ott, C. Grebogi, J.A. Yorke, Phys. Rev. Lett. **64**, 1196 (1990). doi:10.1103/PhysRevLett.64. 1196
2. A.S. Mikhailov, K. Showalter, Phys. Rep. **425**(2–3), 79 (2006). doi:10.1016/j.physrep.2005. 11.003
3. E. Schöll, H.G. Schuster, *Handbook of Chaos Control* (Wiley, 2008)
4. K. Pyragas, Phys. Lett. A **170**(6), 421 (1992). doi:10.1016/0375-9601(92)90745-8
5. N. Baba, A. Amann, E. Schöll, W. Just, Phys. Rev. Lett. **89**, 074101 (2002). doi:10.1103/ PhysRevLett.89.074101
6. J. Unkelbach, A. Amann, W. Just, E. Schöll, Phys. Rev. E **68**, 026204 (2003). doi:10.1103/ PhysRevE.68.026204
7. T. Pierre, G. Bonhomme, A. Atipo, Phys. Rev. Lett. **76**, 2290 (1996). doi:10.1103/PhysRevLett. 76.2290
8. H. Friedel, R. Grauer, K.H. Spatschek, Phys. Plasmas **5**(9), 3187 (1998). doi:10.1063/1.872987
9. P.V. Paulau, D. Gomila, T. Ackemann, N.A. Loiko, W.J. Firth, Phys. Rev. E **78**, 016212 (2008)
10. K. Green, B. Krauskopf, F. Marten, D. Lenstra, SIAM, J. Appl. Dyn. Syst. **8**(1), 222 (2009). doi:10.1137/080712982
11. M. Tlidi, A.G. Vladimirov, D. Pieroux, D. Turaev, Phys. Rev. Lett. **103**, 103904 (2009)
12. A. Pimenov, A.G. Vladimirov, S.V. Gurevich, K. Panajotov, G. Huyet, M. Tlidi, Phys. Rev. A **88**, 053830 (2013). doi:10.1103/PhysRevA.88.053830
13. J. Javaloyes, M. Marconi, M. Giudici, Phys. Rev. A **90**, 023838 (2014). doi:10.1103/PhysRevA. 90.023838
14. A.G. Vladimirov, A. Pimenov, S.V. Gurevich, K. Panajotov, E. Averlant, M. Tlidi, Phil. Trans. R. Soc. London A **372**(2027) (2014). doi:10.1098/rsta.2014.0013
15. K. Panajotov, M. Tlidi, Opt. Lett. **39**(16), 4739 (2014). doi:10.1364/OL.39.004739
16. M. Kehrt, P. Hövel, V. Flunkert, M.A. Dahlem, P. Rodin, E. Schöll, EPJ B **68**(4), 557 (2009). doi:10.1140/epjb/e2009-00132-5
17. M.A. Dahlem, F.M. Schneider, E. Schöll, Chaos **18**(2), 026110 (2008)
18. M.A. Dahlem, R. Graf, A.J. Strong, J.P. Dreier, Y.A. Dahlem, M. Sieber, W. Hanke, K. Podoll, E. Schöll, Phys. D **239**(11), 889 (2010)

19. T. Erneux, G. Kozyreff, M. Tlidi, Phil. Trans. R. Soc. London A **368**(1911), 483 (2010)
20. J. Fort, V. Méndez, Phys. Rev. Lett. **89**, 178101 (2002)
21. V. Ortega-Cejas, J. Fort, V. Méndez, Ecology **85**(1), 258 (2004)
22. M. Bestehorn, E.V. Grigorieva, S.A. Kaschenko, Phys. Rev. E **70**, 026202 (2004). doi:10.1103/PhysRevE.70.026202
23. K.A. Montgomery, M. Silber, Nonlinearity **17**(6), 2225 (2004)
24. C.M. Postlethwaite, M. Silber, Phys. D **236**(1), 65 (2007). doi:10.1016/j.physd.2007.07.011
25. M. Stich, A. Casal, C. Beta, Phys. Rev. E **88**, 042910 (2013). doi:10.1103/PhysRevE.88.042910
26. D. Puzyrev, S. Yanchuk, A. Vladimirov, S. Gurevich, SIAM, J. Appl. Dyn. Syst. **13**(2), 986 (2014). doi:10.1137/130944643
27. C. Tian, L. Zhang, Phys. Rev. E **88**, 012713 (2013). doi:10.1103/PhysRevE.88.012713
28. Y.N. Kyrychko, K.B. Blyuss, S.J. Hogan, E. Schöll, Chaos **19**(4), 043126 (2009). doi:10.1063/1.3270048
29. Q.S. Li, H.X. Hu, J. Chem. Phys **127**(15), 154510 (2007). doi:10.1063/1.2792877
30. R. Iri, Y. Tonosaki, K. Shitara, T. Ohta, J. Phys. Soc. Jpn. **83**(2) (2013)
31. M. Tlidi, A. Sonnino, G. Sonnino, Phys. Rev. E **87**, 042918 (2013). doi:10.1103/PhysRevE.87.042918
32. S.V. Gurevich, Phys. Rev. E **87**, 052922 (2013). doi:10.1103/PhysRevE.87.052922
33. S.V. Gurevich, Phil. Trans. R. Soc. London A **372**(2027) (2014). doi:10.1098/rsta.2014.0014
34. J. Swift, P.C. Hohenberg, Phys. Rev. A **15**, 319 (1977). doi:10.1103/PhysRevA.15.319
35. M. Hilali, G. Dewel, P. Borckmans, Phys. Lett. A **217**(45), 263 (1996). doi:10.1016/0375-9601(96)00344-1
36. R. Lefever, N. Barbier, P. Couteron, O. Lejeune, J. Theor. Biol. **261**(2), 194 (2009). doi:10.1016/j.jtbi.2009.07.030
37. M. Tlidi, P. Mandel, R. Lefever, Phys. Rev. Lett. **73**, 640 (1994)
38. G. Kozyreff, S. Chapman, M. Tlidi, Phys. Rev. E **68**(12), 152011 (2003)
39. N. Stoop, R. Lagrange, D. Terwagne, P. Reis, J. Dunkel, Nat. Mater. **14**(3), 337 (2015). doi:10.1038/nmat4202
40. M. Tlidi, A.G. Vladimirov, D. Turaev, G. Kozyreff, D. Pieroux, T. Erneux, Eur. Phys. J. D **59**(1), 59 (2010)
41. S.V. Gurevich, R. Friedrich, Phys. Rev. Lett. **110**, 014101 (2013). doi:10.1103/PhysRevLett.110.014101
42. P. Hövel, E. Schöll, Phys. Rev. E **72**, 046203 (2005). doi:10.1103/PhysRevE.72.046203
43. R.M. Corless, G.H. Gonnet, D.E.G. Hare, D.J. Jeffrey, D.E. Knuth, Adv. Comput. Math. **5**(1), 329 (1996)
44. D.J.B. Lloyd, B. Sandstede, D. Avitabile, A.R. Champneys, SIAM, J. Appl. Dyn. Syst. **7**(3), 1049 (2008). doi:10.1137/070707622
45. D. Avitabile, D.J.B. Lloyd, J. Burke, E. Knobloch, B. Sandstede, SIAM, J. Appl. Dyn. Syst. **9**(3), 704 (2010). doi:10.1137/100782747
46. L.A. Segel, J. Fluid Mech. **38**(01), 203 (1969)
47. M.C. Cross, P.C. Hohenberg, Rev. Mod. Phys. **65**(3), 851 (1993)

Chapter 22
Control of Epidemics on Hospital Networks

Vitaly Belik, Philipp Hövel and Rafael Mikolajczyk

Abstract The spread of hospital-related infections such as antibiotic-resistant pathogens forms a major challenge in public healthcare systems world-wide. One of the driving mechanisms of the pathogen spread are referrals or transfers of patients (hosts) between hospitals or readmissions after their stay in the community, constituting a dynamical network of hospitals. We analyze referral patterns of 1 million patients from one Federal State in Germany over the period of three years. We extract the underlying statistics of relocation patterns and build an agent-based computational model of pathogen spread. We simulate an outbreak of an SIS-type infection (susceptible-infected-susceptible) and evaluate characteristic time scales and prevalence levels. For such recurrent diseases, we finally investigate the effect of control measures based on screening and isolation of incoming patients.

22.1 Introduction

The emergence and transmission of antibiotic-resistant pathogens is an issue of a major challenge for public health on a world-wide scale [1]. Due to the availability of data and computational resources, a number of investigations have been devoted to study the pathogen spread in hospital networks in different countries [2, 3]. It turned out that countries differ in their hospital network structure [2]. Therefore, it is important to analyze healthcare systems in different countries to understand

V. Belik (✉) · P. Hövel
Institut für Theoretische Physik, Technische Universität Berlin,
Hardenbergstraße 36, 10623 Berlin, Germany
e-mail: belik@mailbox.tu-berlin.de

V. Belik · R. Mikolajczyk
Helmholtz Centre for Infection Research, Mascheroder Weg 1,
38124 Braunschweig, Germany
e-mail: Rafael.Mikolajczyk@helmholtz-hzi.de

P. Hövel
Bernstein Center For Computational Neuroscience Berlin,
Philippstraße 13, 10115 Berlin, Germany
e-mail: phoevel@physik.tu-berlin.de

© Springer International Publishing Switzerland 2016   431
E. Schöll et al. (eds.), *Control of Self-Organizing Nonlinear Systems*,
Understanding Complex Systems, DOI 10.1007/978-3-319-28028-8_22

universal features and heterogeneities. Studies that analyze the German healthcare system from the network perspective are very rare [4]. We aim to fill this gap and present an analysis of German data and for the first time model the spread of a pathogen on this network of hospitals in Germany. Specifically, we elaborate on the impact of screening procedures of patients admitted to hospital to reduce the prevalence level of a disease. Additionally we incorporate the possibility of patients carrying the pathogen after their release to the community, which was not considered in the previous studies [2, 3]. Our study should be considered as a proof of concept for approaches combining complex network theory and computational methods in epidemiology.

The rest of this chapter is organized as follows: In Sect. 22.2, we introduce the dataset and provide details of the model. We also present an analysis of the dataset in terms of its network and temporal properties. Section 22.3 contains the main numerical results and discuss the influence of screening procedures of patients upon admittance to a hospital. We finally summarize our findings in Sect. 22.4.

22.2 Dataset and Model

In the presented study, we consider anonymized data on patient referrals, that is, relocations between hospitals or release to/readmission from the community. The dataset contains 1654 hospitals, which are considered as nodes in the network, $9.18 \cdot 10^5$ patients with around 2 million hospital stays over the course of 3 years of data.

The data was obtained from a healthcare provider in a large federal state in Germany. It contains the following information about the referral: day of first admission t_0, number of stays s, duration of each stay τ, and inter-stay time θ. See Fig. 22.1. The color corresponds to different hospitals. Patients can be directly transferred between hospitals or spend some time in the community.

On an average week day there are around 3400 relocation events as shown in Fig. 22.2 (top). These relocations form the set of links in our network. Note that the sequence of links is crucial to ensure the causality of a spreading process. In

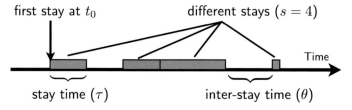

Fig. 22.1 Schematic of the available referral data for an exemplary patient with the day of first admission t_0, number of stays s, duration of stay τ, and inter-stay time θ. The *color* indicates different hospitals

Fig. 22.2 Properties of the time-aggregated, undirected, and non-weighted hospital network with major 176 nodes depicted in Fig. 22.3. *Top* Histogram of the number of all relocations per day. *Middle* Direct relocation between pairs of hospitals. *Bottom* In- and out-degree histograms

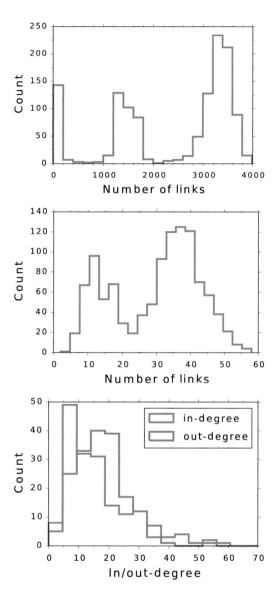

other word, the network under investigation constitutes a temporal network [5, 6]. If we consider only direct relocations between pairs of hospitals without relocations between hospitals and homes (community), we observe around 40 relocations on an average week day. See Fig. 22.2 (middle). The in- and out-degree distributions of the aggregated network of hospitals are presented in Fig. 22.2 (bottom). The in-degree is broader distributed than the out-degree. Note that there is one outlier referring to a node with in-degree 120 (not shown).

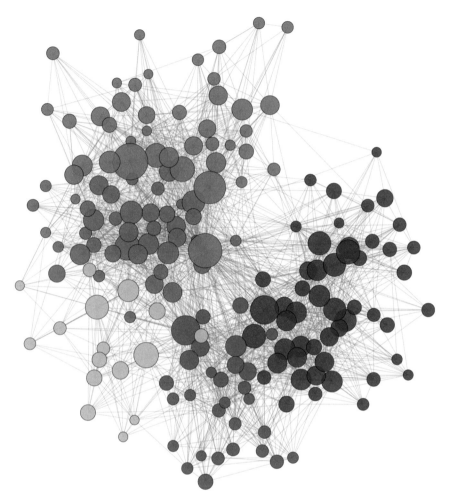

Fig. 22.3 Modular structure of the patient transfer network. 176 nodes (hospitals) can be subdivided into 5 modules indicated by the *color* of the node. The position is chosen according to a spring-embedded layout and thus, corresponds to the topological position in the network and not to the geographical location. Node sizes corresponds to hospital sizes (number of beds) as estimated from the data

The data might also contain information about referrals to hospital outside the federal state under consideration, but does not include full referral records beyond the state borders. We identified the hospitals located in the state under consideration as those hospitals with a maximum number of patients larger than a threshold which was set to 30. This resulted in 176 frequently visited hospitals as depicted in Fig. 22.3. In this reduced network we computed 4 modules using the Louvain algorithm described in Ref. [7]. This method sorts all nodes into different modules by maximizing the so-called modularity Q, which is defined as

$$Q = \frac{1}{2m} \sum_{i,j=1}^{N} \left(a_{ij} - \frac{k_i k_j}{2m} \right) \delta\left(c_i, c_j \right), \tag{22.1}$$

where m denotes the total number of links, $\{a_{ij}\}$ is the adjacency matrix, and k_i and c_i refer to the degree and module of node i, respectively. To obtain Fig. 22.3, we applied this algorithm to the time-aggregated, undirected, and non-weighted network. Information on the modular structure of networks can be used to identify critical links for an effective prevention of epidemics. This is important, for instance, to contain an outbreak locally and prevent spreading across different modules [8].

Due to reasons of privacy we do not have information on geographical coordinates of hospital or access to other types of metadata. However, we are able to estimate hospital sizes from the data. For this purpose we take the maximum number of patients sojourning in the individual hospital as an estimate of its size, i.e. number of beds. To verify our estimates, we compare these with the data from statistical bureau by ranking both numbers.[1] The accuracy of this procedure is shown in Fig. 22.4. We find that the ratio of the ranked hospital size, which are estimated from the data, and the real hospital sizes remains constant around 0.5 for the first 150 nodes. Assuming a uniform distribution of the customers of the health insurance company, which provided us with data, this ratio nicely reflects its market share as it is close to the real value of around 40 %. The agreement can be further improved, if we take into account an occupancy rate of hospitals below 100 %.

Quantifying categories of links in our network, we also computed the numbers of relocations between community and hospitals ($2.974 \cdot 10^6$ or 99 % of all relocations) and direct transfers between different hospitals ($3.3 \cdot 10^4$ or 1 % of all relocations). Therefore, it is to be expected that the role of the community is very important, as the majority of patients are not directly transferred between hospitals, but first stay for some time in the community, potentially carrying the pathogen. In the simulations presented below, we assume no disease spreading in the community for simplicity,

Fig. 22.4 Evaluation of the robustness of hospital size estimations from the data. N_{SB} are the number of beds in hospitals as given from the statistical bureau. N_{data} are the sizes of hospitals as estimated from the data, i.e. the maximal number of patients on any day

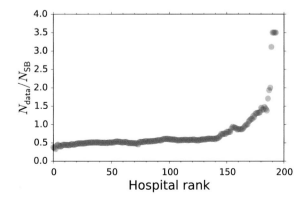

---

[1] See *Landesamt für Statistik Niedersachsen*: http://www.statistik.niedersachsen.de.

because we lack information about the moving patterns except for the times of hospital (re)admission and release. Since we do not have any information about the internal ward structure of the hospitals either, we assume that within a hospital patients are well mixed, and the law of mass action holds.

Fig. 22.5 Statistics of the dataset: histograms of (*top*) number of stays, distribution of (*middle*) duration of stays, (*bottom*) distribution of inter-stay times

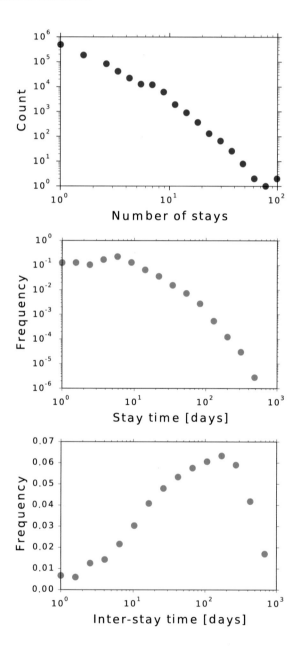

Fig. 22.6 Total number of agents in the system in dependence on time

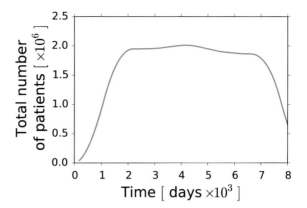

Figure 22.5 depicts basic statistics of the individual referrals in the considered dataset. See also schematics shown in Fig. 22.1. The number of stays and stay times (top and middle panels in Fig. 22.5) are broadly distributed with long tails. The majority of the patients (around 50 %) were admitted just one time to a hospital. The inter-stay time distribution peaks around 110 days (bottom panel in Fig. 22.5).

In order to simulate the whole population of size N, we generate the corresponding number of agents from the data on only 40 % of the patients using the following procedure: (i) we chose N times a patient ID from the dataset with its referral record and assign a random day of its first appearance in the interval $[0, T]$ with $T = 3$ a, (ii) we periodically repeat these records for the intervals $[(n - 1)T, nT]$ with $n = 1, 2, 3$ leading to a total observation time of 12 a; (iii) taking into account mortality rates for the agent, an agent is removed assuming a death rate of 0.007/a and at the same time, a new agent with the same referral profile is added. Following this procedure, we reach a constant population level after an initial transient as shown in Fig. 22.6.

In order to model an endemic prevalence level of resistant pathogens, we consider an SIS (susceptible-infected-susceptible) epidemic model. Given the number I of infected and S of susceptible individuals in one hospital, the dynamics follow a chemical kinetic equations for infectious dynamics:

$$S + I \xrightarrow{\alpha} 2I \tag{22.2a}$$

$$I \xrightarrow{\beta} S \tag{22.2b}$$

where α and β denote the infection and recovery rate per individual. Thus we consider the frequency-dependent model, where the chance of infection is proportional to the product of the number of susceptible and infected individuals in a single population and inversely proportional to its size. Equation (22.2) describes an undetected, free-running spread of pathogens in the absence of control measures.

We use a stochastic agent-based computational epidemic model on a network of hospitals and implement the events according to the empirical data using a priority queue data structure [9] to keep track of single individuals, their infection status and

Fig. 22.7 Histogram of the time-averaged endemic prevalence values in hospitals with a non-zero number of infected agents. Infectious rate $\alpha = 0.1$/day and recovery rate $\beta = 2.7 \cdot 10^{-3}$/day $= 1$/year

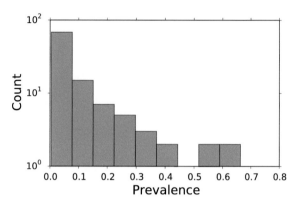

time events (arrival at and release from a hospital, recovery). For the time between two subsequent events, the local node dynamics follow the SIS-model described by the kinetics (22.2). The infectious rate α is chosen to ensure the average prevalence within hospitals around 5 %. We consider a recovery rate $\beta = 1/a = 2.7 \cdot 10^{-3}$/d, which corresponds to typical carriage times of a bacterial pathogen. We configure the system in the following way: we populate the hospitals according to the procedure described above using the empirical transfer profiles of the dataset. As initial condition, we implement 0.99 % of the patients in all hospitals as infected. We find that the dynamics of our network model reach an endemic state after 1000 days (Fig. 22.4).

Figure 22.7 depicts the histogram of the time-averaged endemic node prevalence. The median is 0.02 and the mean value is 0.07. One can see that the prevalence distribution is inhomogeneous and skewed towards low values, indicating a small prevalence for many nodes.

22.3   Results of Simulations and Control by Screening

Extending the model described in the previous section, we additionally implement the following control measure. We randomly screen a fraction ν of patients incoming in every hospital. Assuming a test sensitivity of 100 %, patients that are detected as infected are immediately isolated and cured from the disease.

Figure 22.8 presents the time series of the prevalence after the control is applied at $t = 3000$ d for different screening fractions ν. Note that $\nu = 0$ corresponds to the uncontrolled case. As intuitively expected, screening leads to a reduction of the prevalence level. We observe that the screening rate has to be considerably high in order to achieve significant results. For a 10-fold reduction within 300 days, for instance, a screening of 90 % of incoming patients is required.

Figure 22.9 shows the time required to reduce the prevalence to 50 %, which is known as *half reduction time*, in dependence on the screening fraction ν. This half reduction time is marked in Fig. 22.8 by vertical lines. We find that the half

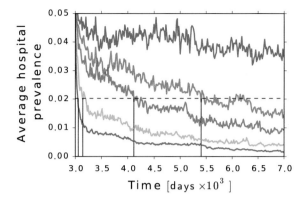

Fig. 22.8 Averaged prevalence for different fraction of incoming patients screened: $\nu = 0$, i.e. no screening, (*blue*), $\nu = 0.1$ (*green*), $\nu = 0.2$ (*red*), $\nu = 0.5$ (*cyan*), $\nu = 0.9$ (*magenta*). The average is computed over all hospitals for each time step. The *dashed horizontal line* indicates the 50% reduction level and the vertical lines mark the half reduction times. System parameters of the SIS model (22.2) as in Fig. 22.7

Fig. 22.9 Time until 50% prevalence reduction from the start of the screening. System parameters of the SIS model (22.2): infectious rate $\alpha = 0.1/$day and recovery rate $\beta = 1\,\text{year}^{-1}$

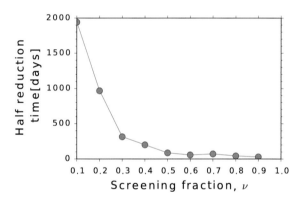

reduction time decreases strongly with screening rates up to $\nu = 30$–40%. For higher values of ν the half reduction time equilibrates around 100 days and does not change significantly. Thus, if the goal is to reduce prevalence to 50%, moderate screening fractions $\nu = 30$–40% are sufficient.

22.4 Conclusion

We have shown how complex network theory and computational methods of agent-based stochastic reaction-diffusion processes can help to control nosocomial infectious diseases, such as antibiotic-resistant pathogens, e.g. Methicillin-resistant Staphylococcus Aureus (MRSA) or *Clostridium Difficile*. We have analyzed patients referral patterns in one federal state in Germany over the period of 3 years. We have extracted the corresponding hospital network of patient relocations and built a computational agent-based stochastic model of disease dynamics including the full history of hospital stays on the single patient level. We have assessed the efficiency of screening a fraction of incoming patients as a potential control measure. This means that in the case of a positive screening test, the patient is isolated and cured before admission to the hospital. For typical values of parameters, we have found that the endemic prevalence can be halved within 100 days for screening fraction around 30–40 %.

Our study represents a proof of concept and opens roads for the future analysis of the potential impact of different epidemic control measures in a network of hospitals.

Acknowledgments All the authors acknowledge the courtesy of the AOK Niedersachsen for providing the anonymized data on patient referrals. VB thanks André Karch and Johannes Horn for fruitful discussions. VB and PH acknowledge funding by the Deutsche Forschungsgemeinschaft in the framework of Collaborative Research Center 910. At the early stage of this study VB was financially supported by the fellowship "Computational sciences" of the VolkswagenStiftung.

References

1. World Health Organization, *Antimicrobial Resistance: Global Report on Surveillance*. WHO, Geneva (2014)
2. T. Donker, J. Wallinga, R. Slack, H. Grundmann, PLoS ONE **7**(4), 1 (2012). doi:10.1371/journal.pone.0035002
3. J.F. Gracia, J.P. Onnela, M.L. Barnett, V.M. Eguíluz, N. Christakis, arXiv preprint arXiv:1504.08343 (2015)
4. M. Ciccolini et al., Int. J. Med. Microbiol. **303**(6–7), 380 (2013). doi:10.1016/j.ijmm.2013.02.003
5. P. Holme, J. Saramäki, Phys. Rep. **519**, 97 (2012). doi:10.1016/j.physrep.2012.03.001
6. A. Casteigts, P. Flocchini, W. Quattrociocchi, N. Santoro, Int. J. Parallel Emergent Distrib. Syst. **27**(5), 387 (2012). doi:10.1080/17445760.2012.668546
7. V.D. Blondel, J.L. Guillaume, R. Lambiotte, E. Lefebvre, J. Stat. Mech. **10**, P10008 (2008)
8. N. Masuda, New J. Phys. **11**(12), 123018 (2009)
9. T.H. Cormen, *Introduction to Algorithms*. MIT Press (2009)

Chapter 23
Intrinsic Control Mechanisms of Neuronal Network Dynamics

Josef Ladenbauer, Moritz Augustin and Klaus Obermayer

Abstract The brain is a complex dynamical system which employs specific mechanisms in a self-organized way to stabilize functionally relevant patterns of activity and switch between them, depending on computational demands. We first provide an overview of control mechanisms that involve delayed feedback of activity, plasticity of synaptic coupling strengths and changes of neuronal adaptation properties, and then focus on the latter, summarizing recent results for different spatial levels, obtained through mathematical bottom-up modeling.

Cognitive processing is linked to the activation of neurons that interact within and across brain areas. Specific cognitive tasks are associated with certain collective dynamics of local (spatially confined) neuronal populations and interareal (spatially distributed) brain networks [1]. Of particular relevance is the concerted spiking activity, which often exhibits oscillatory overall dynamics due to synchronization. For example, invasive studies on rodents and monkeys have shown that in an awake state local cortical networks exhibit largely asynchronous collective spiking activity, in contrast to sleep and drowsiness during which slow oscillations (0.1–5 Hz) are more pronounced [2, 3]. Attentional modulation, however, leads to changes in fast oscillatory activity (30–90 Hz) [3, 4]. In these studies electrodes were inserted into the brain, measuring the (spiking) dynamics of local neuronal populations within up to 1 mm from the electrode contacts. On a larger spatial scale functional networks have been identified—such as the visual, sensory-motor and executive control networks—based on correlations (i.e., weak synchronization) between the fluctuating activities of different brain areas measured by functional magnetic resonance imaging, magnetoencephalography or electroencephalography [5, 6]. Each of these brain areas contains at least several mm^3 of brain tissue with millions of neurons.

J. Ladenbauer (✉) · M. Augustin · K. Obermayer
Department of Software Engineering and Theoretical Computer Science,
Technische Universität Berlin, Marchstraße 23, 10587 Berlin, Germany
e-mail: jl@ni.tu-berlin.de

J. Ladenbauer · M. Augustin · K. Obermayer
Bernstein Center for Computational Neuroscience Berlin, Philippstraße 36,
10623 Berlin, Germany

© Springer International Publishing Switzerland 2016
E. Schöll et al. (eds.), *Control of Self-Organizing Nonlinear Systems*,
Understanding Complex Systems, DOI 10.1007/978-3-319-28028-8_23

441

Furthermore, oscillatory network dynamics have also been related to brain diseases. During epileptic seizures or in Parkinson's disease, for example, strong oscillatory activity in particular frequency bands occurs due to excessive synchrony across large populations of neurons [7, 8]. Patients with schizophrenia, on the other hand, show abnormally weak oscillations in two major frequency classes and reduced synchronization between them [9].

To be able to control the various dynamical states of neuronal networks at the local and interareal levels the brain is equipped with several mechanisms. These include feedback of neuronal spiking activity (short timescale), plasticity of synaptic coupling strengths (long timescale), and changes of the dynamical properties of neurons (intermediate timescale). In the following three Sects. 23.1–23.3 we provide an overview of these mechanisms and their control potential. In Sect. 23.4, we then focus on a mechanism of the latter type that involves neuromodulatory systems which affect the adaptation properties of neurons. We summarize the results from recent computational studies on the capability of that mechanism to control the collective dynamics at different spatial levels, ranging from single neurons to large networks. The bottom-up modeling approach and analyses used in these studies are outlined. Concluding remarks are provided in Sect. 23.5.

23.1 Control Through Feedback of Neuronal Activity

Cortical networks, especially sensory systems, are organized in a hierarchical fashion, where information is processed in subsequent stages and where increasingly abstract representations are formed (cf. [10, 11]). A key feature of this architecture are feedback connections from higher to lower areas in the hierarchy. In the visual system, for example, these feedback projections are characterized by a large degree of convergence and divergence indicating pooling of signals. At the same time, the feedback projection patterns are specific, targeting populations of neurons with similar response properties [12]. Feedback input is always excitatory, but leads to effective inhibition when the local inhibitory population is affected. The feedback time delay due to synaptic transmission can range from only a few up to several tens of milliseconds depending on axonal length and myelination as well as dendritic morphology and postsynaptic receptor kinetics.

The experimental data of these feedback loops and the corresponding flow of signals are reminiscent of time-delayed feedback control schemes that have been extensively investigated in the theoretical physics literature [13–16]. Because time delays arise naturally in neuronal networks, the potential of those control schemes has also been examined using generic models of coupled neurons. For pairs [17] and large ensembles of globally coupled neuronal oscillators [18] synchronization can be enhanced or suppressed by that mechanism, depending on the delay and feedback amplification. Feedback control schemes of this kind, however, have been predominantly studied for the development of stimulation devices that suppress pathological rhythms (see, e.g., [19–23]), rather than understanding brain-intrinsic control.

More specific examples of intrinsic feedback control include interaction loops between cortical and thalamic circuits [24] and well-characterized inhibitory feedback pathways in sensory systems [25]. Using in vitro experiments complemented by computational models it was found that delayed corticothalamic (excitatory) feedback is capable to induce highly synchronous slow (2–4 Hz) oscillations in visual thalamic circuits of ferrets [24]. In vivo recordings from sensory pyramidal neurons[1] and results from an effective circuit model revealed that delayed inhibitory feedback allows the sensory neurons to differentiate between communication-like and prey-like naturalistic stimuli in their spike rate response dynamics (oscillatory and nonoscillatory, respectively) [25].

Delayed local feedback through populations of inhibitory neurons, on the other hand, has been shown to produce rhythmic collective dynamics in local networks, with oscillation frequencies that are largely determined by the delay times—even when the spiking activity of single neurons is irregular and sparse (i.e., cycles are skipped), see [26, 27] for theoretical work, and [28, 29] for experimental results.

On the level of large-scale brain networks the impact of local feedback inhibition on the global dynamics has recently been studied in models constrained by diffusion imaging data of human subjects [30]. It was shown that local inhibitory feedback stabilizing an asynchronous state can—at the global network level—lead to improved prediction of the measured interareal correlations as well as enhanced information capacity and discrimination accuracy [30].

Finally, feedback control schemes have been widely studied in movement neuroscience. An established paradigm considers the integration of sensory feedback state estimation processes which affect motor commands that cause movement [31–33]. Here, a variety of (local) neuronal network states that reflect motor commands need to be controlled [34, 35].

23.2   Control Through Plasticity of Coupling Strengths

The strengths of many types of synapses through which neurons are coupled undergo long-lasting changes in an activity-dependent manner. The effects of such plasticity processes depend on the (correlated) spiking activity of pre- and postsynaptic neurons, potentiating (strengthen) or depressing (weaken) synaptic efficacies, which can last up to several hours [36, 37]. A form of synaptic modification that has received lots of attention is spike-timing-dependent plasticity (STDP) [37–40]. With STDP, repeated presynaptic spike arrival a few milliseconds before postsynaptic spikes leads to long-term potentiation of the synapse, while repeated spike arrival after postsynaptic spikes leads to long-term depression. The study of STDP has been mainly focused at excitatory synapses on excitatory cells but there is also accumulating evidence for STDP at inhibitory synapses and on inhibitory neurons [41].

---

[1]Pyramidal cells are the largest class of excitatory neurons in the brain.

Functionally, STDP is believed to underlie learning and information storage in the brain, as well as the development and refinement of neuronal circuits [39, 42]. Considering its effects on neuronal dynamics, STDP can regulate both, the rate and variability of (postsynaptic) spiking [43, 44]. At the network level, the collective spiking dynamics lead to a slow modification of the coupling strength pattern through STDP, which in turn shapes the spiking activity within the network. Results from computational modeling studies show a variety of network states stabilized by STDP, as reviewed below.

In networks of regular (i.e., constant input driven) neuronal oscillators STDP can lead to stationary locking states, characterized by multiple synchronized clusters where neurons within clusters are largely decoupled [45–47], and slow oscillations between weakly and strongly globally synchronized states [48].

The picture changes when (input) fluctuations are considered which lead to irregular neuronal spiking [1, 49]. In this case, STDP can, for example, stabilize "mixture states" of short-lasting synchronous bursts that interrupt asynchronous spiking activity in networks of excitatory model neurons with delayed coupling [50]. Coupling strengths increased during asynchronous activity, but neurons engaged in synchronous bursts were decoupled due to axonal propagation delays, which is in agreement with decoupling within clusters in oscillator networks. In balanced networks of excitatory and inhibitory neurons, on the other hand, STDP generally leads to stabilization of asynchronous irregular states [51]. There, a slightly modified, weight-dependent STDP rule that includes a dependence on the coupling strength was used to prevent abnormal synaptic strengths, which further led to improved fits of experimental data [51, 52]. Asynchronous irregular network states have also been shown to be stabilized by STDP on inhibitory synapses, by balancing excitatory and inhibitory synaptic currents [53].

When broadly distributed synaptic delays are considered in networks of model neurons driven by fluctuating inputs STDP can lead to locking phenomena where spiking within subnetworks is time locked according to the synaptic delays of the connections between the neurons [54]. This concept has been coined as polychronization [55] and has been further examined in the context of working memory [56]. These spike-timing patterns emerge and reoccur with millisecond precision although spiking within the network appears random and uncorrelated. The number of such polychronous groups that can coexist far exceeds the number of neurons in the network; hence, STDP can cause a large memory capacity of the system in this way.

In summary, STDP seems to facilitate asynchronous irregular states in networks where noise is considered and can—at the same time—promote spike-to-spike locking between neurons, while synchronized activity may only occur transiently due to decoupling of synchronized neurons.

23.3   Control Through Changes of Neuronal Dynamics

The brain is equipped with neuromodulatory systems which regulate the dynamics of neuronal populations by altering the properties of ion channels in their membranes. Several circuits have been identified, comprising executive and sensory cortical areas as well as subcortical neuromodulatory centers, that are involved in regulating brain states through neuromodulators such as acetylcholine (ACh), norepinephrine, or serotonin [57, 58]. Neuromodulation by ACh, for example, which can occur at the timescale of seconds [59], has been experimentally shown to change the neuronal response and synchronization properties [60, 61] often facilitating desynchronized population activity [3, 58], and functionally linked to arousal, vigilance and selective attention [58, 60, 62]. ACh can mediate synaptic plasticity [63–65] and is involved in circuits where feedback control may play an important role [64]; here, however, we focus on its direct impact on the intrinsic dynamical properties of neurons.

ACh affects the neuronal membrane by reducing the conductance of specific types of ion channels, in particular for potassium (K^+), in a type-dependent way [57, 66], see Fig. 23.1a. These K^+ channels produce slowly decaying currents, which rapidly accumulate when the membrane voltage increases. A number of such K^+ channel types with different activation characteristics have been identified [67–69]. Some channels are already activated at membrane voltage values below those leading to a spike (i.e., subthreshold voltage values), while others are activated at higher, suprathreshold values (i.e., driven by spikes). A common feature of these different types of slowly deactivating K^+ channels is that they generate transmembrane currents which lead to spike rate adaptation (see Fig. 23.1b)—a phenomenon shown by many types of neurons. Those K^+ currents are thus also termed adaptation currents. The distinct types of K^+ channels exhibit different sensitivities to ACh. Specifically, ACh has been shown to inhibit the channels which cause a spike-driven adaptation current at a lower concentration than that necessary to inhibit those channels which effectively generate a subthreshold voltage-driven adaptation current [57, 66], cf. Fig. 23.1a. To gain insight into a control mechanism based on the neuromodulator ACh and to assess its potential it is, therefore, important to understand how adaptation currents affect neuronal spiking activity and synchronization phenomena.

23.4   On the Role of Neuronal Adaptation for Controlling the Collective Dynamics: Insights from in Silico studies

Here, we focus on how distinct types of adaptation currents shape neuronal dynamics at different spatial levels and the potential consequences for ACh-based neuromodulatory control (cf. Sect. 23.3). Specifically, we summarize recent results on spike rates and interspike interval (ISI) variability of single neurons (Sect. 23.4.2), spike synchronization and spike-to-spike locking in small networks (Sect. 23.4.3), and the population spike rate dynamics in large networks (Sect. 23.4.4).

These questions can be effectively addressed in silico, using computational models, and through mathematical analyses. Below, we outline a bottom-up approach based on an experimentally validated neuron model of the integrate-and-fire type, which covers spikes, the fast subthreshold membrane voltage and slow adaptation current dynamics. To efficiently describe and analyze the dynamics of these model neurons at the three spatial levels different suitable methods from statistical physics and nonlinear dynamics—including mean-field, phase reduction and master stability function techniques—have recently been extended for that model class. This approach allows to examine the relationships between microscopic interactions (neuron biophysics) and macroscopic features (network dynamics) in a direct way and simplifies bridging scales.

23.4.1 Central Model of Neuronal Activity

Over the last two decades substantial efforts have been exerted to develop single neuron models of reduced complexity that can reproduce a large repertoire of observed neuronal behavior, while being computationally less demanding and, more importantly, easier to understand and analyze than detailed biophysical models. A prominent example that is used in the following is the adaptive exponential integrate-and-fire (aEIF) model [70, 71], which is a single-compartment spiking neuron model given by a two-variable differential equation system with a (discontinuous) reset condition.

Specifically, for each neuron ($i = 1, \ldots, N$) of a population of N neurons, the dynamics of the membrane voltage V_i is described by

$$C\frac{dV_i}{dt} = I_L(V_i) + I_{exp}(V_i) - w_i + I_{syn,i}(V_i, t), \qquad (23.1)$$

where the capacitive current through the membrane with capacitance C equals the sum of three ionic currents and the synaptic current I_{syn}. The ionic currents consist of a linear leak current $I_L(V_i)$, an exponential term $I_{exp}(V_i)$ that approximates the rapidly increasing Na^+ current at spike initiation, and the adaptation current w_i which reflects a slowly deactivating K^+ current. The adaptation current evolves according to

$$\tau_w\frac{dw_i}{dt} = a(V_i - E_w) - w_i, \qquad (23.2)$$

with adaptation time constant τ_w. Its strength depends on the subthreshold membrane voltage via conductance a. E_w denotes its reversal potential. When V_i increases beyond a threshold value, it diverges to infinity in finite time due to the exponentially increasing current $I_{exp}(V_i)$, which defines a spike. In practice, however, the spike is said to occur when V_i reaches a given value V_s—the spike voltage. The downswing of the spike is not explicitly modeled; instead, when $V_i \geq V_s$, the membrane voltage

V_i is instantaneously reset to a lower value V_r. At the same time, the adaptation current w_i is incremented by a value of b, which implements suprathreshold (spike-dependent) activation of the adaptation current. Immediately after the reset, V_i and w_i are clamped (i.e., remain constant) for a short refractory period T_{ref}, and subsequently governed again by (23.1–23.2).

In contrast to higher dimensional smooth neuron models of the Hodgkin-Huxley type [72], integrate-and-fire models describe the actual spike shape in a strongly simplified way for the sake of reduced model complexity. This is justified by the observation that neuronal spike shapes are stereotyped and their duration is very short (about 1 ms). The timing of spikes contains most information, as compared to their shapes, and the aEIF model can accurately reproduce the spike times (up to 96 %) of Hodgkin-Huxley neurons [70].

The aEIF model exhibits rich subthreshold dynamics [73], a variety of biologically relevant spike patterns [74], and it can be easily calibrated using well established electrophysiological measurements [70]. The model parameters are physiologically relevant and—importantly—they can be tuned such that the model reproduces sub-threshold properties [75, 76] and predicts spiking activity of real (cortical) neurons [77–79] to a high degree of accuracy (see Fig. 23.1c). Furthermore, it allows to study the effects of adaptation currents with subthreshold and spike-dependent activation in separation (via parameters a and b). Depending on the values of these parameters the model effectively reflects different types of K^+ channels [80]. The model dynamics in response to an input current step, exhibiting spike rate adaptation for the subthreshold and spike-dependent activation types, are shown in Fig. 23.1e.

To complete the model (23.1) we need to specify the total synaptic current I_{syn} for each cell. It consists of recurrent synaptic inputs I_{rec}, received from neurons (with indices $\{j\}$ and spike times $\{t_j^k\}$) of the network (if a network is considered) and an input current I_{ext} generated from network-external neurons,

$$I_{syn,i}(V_i, t) := I_{rec,i}(V_i, \{t_j^k\}, t) + I_{ext,i}(V_i, t). \qquad (23.3)$$

These currents describe the excitatory or inhibitory postsynaptic effects of chemical synapses—the by far most abundant type of connections between neurons in the vertebrate brain. In particular, the recurrent synaptic current is given by $I_{rec,i} := \sum_j G_{ij} \sum_k s_i(V_i, t - t_j^k - d_{ij})$, where every presynaptic spike triggers a postsynaptic current (weighted by the coupling strength G_{ij}), which causes an excursion of the postsynaptic membrane voltage V_i after a time delay d_{ij} has passed, accounting for the (axonal and dendritic) spike propagation times. The shape of I_{rec} elicited by a presynaptic spike differs slightly across the different spatial levels considered below. In Sect. 23.4.3 a biexponential conductance-based description (cf. [81]), $s_i(V_i, t) \propto [\exp(-t/\tau_d) - \exp(-t/\tau_r)](E_{syn} - V_i)$, and exponentially decaying currents (cf. [82]), $s_i(V_i, t) \propto \exp(-t/\tau_d)$, are used, with rise and decay time constants $\tau_r \leq \tau_d$ and synaptic reversal potential E_{syn}. For both variants, $s_i(V_i, t) := 0$ if $t < 0$. In Sect. 23.4.4 conductance-based membrane voltage jumps are applied (cf. [83]), $s_i(V_i, t) := \delta(t)(E_{syn} - V_i)$. Excitatory and inhibitory effects are implemented by different values for E_{syn} or—for the current-based description—by positive and negative weights G_{ij}.[2] The results summarized below do not depend on a particular choice of synaptic model, the presence (absence) of certain model features, however, simplifies the respective methodology. The external input current, I_{ext}, is described using stochastic processes (Sects. 23.4.2 and 23.4.4) or deterministic (constant and step) functions (Sect. 23.4.3).

23.4.2 Single Neurons: Effects on Threshold, Gain, and Variability of Spiking

How adaptation currents shape the relationship between driving input, spike rate output and ISI variability of single neurons has recently been examined in a probabilistic setting [80]. Considered were aEIF neurons exposed to fluctuating inputs mimicking synaptic bombardment as observed in vivo:

$$I_{\text{syn},i}(V_i, t) = C\left[\mu(t) + \sigma(t)\xi_i(t)\right], \tag{23.4}$$

where $\mu(t)$ and $\sigma(t)$ determine the time-dependent mean and standard deviation of the input, respectively, and $\xi_i(t)$ is a Gaussian white noise process with correlation function $\langle \xi_i(t)\xi_j(t + \tau)\rangle = \delta(\tau)$ for $i = j$ and 0 otherwise. Here, the index i reflects the trial number for a single neuron. Spike rates and ISI distributions were characterized for a wide range of input statistics using an approach based on the Fokker-Planck equation and an adiabatic approximation, where the adaptation current w_i is replaced by its trial average under the reasonable assumption that its timescale is large compared to the membrane time constant ($\tau_w \gg \tau_m := C/g_L$, where g_L denotes the leak conductance of I_L). This leads to the following equations for the probability density $p(V, t)$ and the time-varying mean adaptation current $\langle w \rangle$ (see [80] for details),

$$\frac{\partial p}{\partial t} = -\frac{\partial}{\partial V}q_p(I_L, I_{\text{exp}}, \langle w \rangle, \mu, \sigma), \qquad \frac{d\langle w \rangle}{dt} = \frac{a\left(\langle V \rangle_p - E_w\right) - \langle w \rangle}{\tau_w} + b\,r(t). \tag{23.5}$$

The left (partial differential) equation is supplemented with appropriate boundary conditions (see Fig. 23.1d) accounting for the (discontinuous) reset condition in the aEIF model. The spike rate r is given by the probability flux q_p evaluated at the spike voltage, $r(t) = q_p|_{V=V_s}$. The ISI distribution p_{ISI} can be calculated within this (Fokker-Planck) framework by self-consistently solving the associated first passage time problem [80]. The two output quantities of interest, r and p_{ISI}, were numerically calculated as a function of the input moments $\mu(t)$ and $\sigma(t)$. The results of this method for two aEIF parametrizations, together with simulations of the underlying stochastic system, are shown in Fig. 23.1f.

---

[2]Note, that an excitatory (inhibitory) neuron can only produce excitatory (inhibitory) synaptic effects at its target neurons.

Using this approach it was found [80] that an adaptation current which is primarily driven by the subthreshold membrane voltage ($a > 0$) increases the threshold for spiking and leads to an increase of ISI variability for a broad range of input statistics, by subtracting from the mean input to the neuron (Fig. 23.1g, h; left). A spike-dependent adaptation current ($b > 0$), on the other hand, always reduces the spike rate gain while leaving the threshold for spiking unaffected by predominantly dividing the mean input, and decreases ISI variability for fluctuation-dominated inputs (i.e., when the input mean is substantially smaller in magnitude than its variance). These computational results were supported by analytical expressions derived for the steady-state spike rate and ISI variability as a function of adaptation parameters and synaptic input moments (upon a small model simplification) [80]. Additionally, it was shown that the distinct effects of the two adaptation mechanisms are consistently reproduced by specific types of K^+ currents using a biophysically detailed Hodgkin-Huxley neuron model.

Subthreshold and spike-dependent adaptation currents have also been shown to affect the correlations between subsequent ISIs in different ways. The former type causes positive and the latter type induces negative serial ISI correlations (i.e., short ISIs are typically followed by longer ones and vice versa) [86, 87]. Furthermore, spike-dependent adaptation can produce strong resonance of the spike rate in response to an oscillating mean input, which has been shown using a Fokker-Planck based approach similar to the one outlined above [88].

23.4.3 Small Networks: Effects on Spike Synchronization and Locking

We next turn our focus to small networks of up to hundreds of neurons, for which the precise temporal relationships between spikes are relevant. We cover recent results on how adaptation currents affect spike-to-spike synchronization and locking of synaptically coupled neurons driven to periodic (so-called regular) spiking. This type of spiking activity is exhibited by several types of excitatory and inhibitory neurons [89–91]. For other classes of neurons, whose ISIs show a high degree of variability—even during rhythmic overall network activity [1]—spike synchronization phenomena are preferentially treated in a probabilistic manner (cf. Sect. 23.4.4). We first consider the effects on pairs of coupled aEIF neurons and networks with all-to-all connectivity [81], before we move to networks with different connection patterns [82]. The model dynamics was analyzed using a phase reduction method and the master stability function (MSF) formalism, both of which are outlined in the following. The former is based on a characteristic of single neurons—the phase response curve (PRC)—and allows to effectively predict the synchronization and locking behavior of coupled pairs. The latter can be used to assess the stability of synchrony (and locking states) for many different network topologies at once.

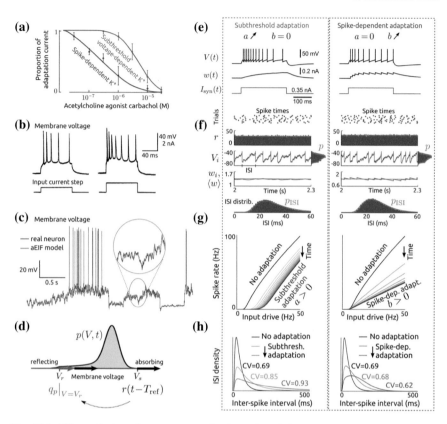

Fig. 23.1 Effects of adaptation currents on the dynamics of single neurons **a** Inhibition of two adaptation current types by carbachol. Data from [66]. **b** Spike rate adaptation observed by in vitro measurements of pyramidal neurons. Adapted from [84]. **c** Overlayed membrane voltage traces of a fast-spiking interneuron measured in vitro and as produced by the aEIF model in response to the same fluctuating input current. Adapted from [85]. **d** Visualization of the boundary conditions for the membrane voltage distribution (reflecting for $V \to -\infty$, absorbing at V_s) and the reinjection of probability flux, $q_p|_{V \searrow V_r} - q_p|_{V \nearrow V_r} = r(t - T_{\mathrm{ref}})$, for the Fokker-Planck (FP) model (23.5). **e** Membrane voltage V and adaptation current w time series of an aEIF neuron, with a purely subthreshold voltage-driven ($a > 0, b = 0$, left) and spike-dependent ($a = 0, b > 0$, right) adaptation mechanism, in response to an input current step. From [81]. **f** Spike times, spike rate histogram (in Hz), sample trace of the membrane voltage (in mV) and the corresponding histogram (attached right), adaptation current (in $\mu A/cm^2$) and ISI histogram of an aEIF neuron subject to a fluctuating input. The corresponding quantities $r(t)$, $p(V, t)$, $\langle w \rangle$ and p_{ISI} were obtained from the FP model (*orange lines*). **g** Adapting spike rate r of an aEIF neuron for varying input intensity (the moments, μ and σ, both increase with increasing drive, see [80] for details). **h** ISI distributions (i.e., p_{ISI}) of aEIF neurons and the corresponding coefficient of variations (CV). **g** and **h** show results from the FP model. **f**–**h** are adapted from [80] where all parameter values are specified

It is convenient to express the network system (23.1)–(23.3), with deterministic external driving input (constant $I_{ext,i}$), in compact form,

$$\frac{dx_i}{dt} = f(x_i) + \sum_{j=1}^{N} G_{ij} h(x_i, x_{j,d}),$$

(23.6)

for $i = 1, \ldots, N$, where $x_i \in \mathbb{R}^m$ contains the state variables V_i and w_i, together with a small number $(m-2)$ additional variables to represent the synaptic effects on neuron i (i.e., the time course of s_i, cf. Sect. 23.4.1) by a differential equation. f and h are piecewise smooth functions with discontinuities due to membrane voltage reset and pulses involved in the model of synaptic interaction. $x_{i,d} := x_i(t - d)$ with axonal plus dendritic propagation delay d (identical delays assumed here for notational simplicity).

Phase reduction method

Under the assumption of weak synaptic interaction (i.e., $|G_{ij}|$ is small) the network model (23.6) can be reduced to a lower dimensional system, where neuron i is represented by its phase $\vartheta_i \in [0, T)$ as follows (see [81] for details):

$$\frac{d\vartheta_i}{dt} = 1 + \sum_{j=1}^{N} \frac{G_{ij}}{T} \int_0^T q(s)^{\mathrm{T}} h\big(\bar{x}(s), \bar{x}_d(s + \vartheta_j - \vartheta_i)\big) \, ds$$

$$=: 1 + \sum_{j=1}^{N} G_{ij} H_d(\vartheta_j - \vartheta_i).$$

(23.7)

$\bar{x}$ is a stable limit cycle solution with period T of the equation $dx/dt = f(x)$, representing the (piecewise smooth) oscillatory dynamics of an isolated neuron. q is the normalized solution of the adjoint variational equation around $\bar{x}$ that determines the PRC [81] and H_d denotes the (so-called) interaction function for the phase network (23.7). The PRC specifies the phase shift of an oscillating neuron caused by a transient input (perturbation) as a function of phase at which the input is received. It can be efficiently calculated from that adjoint equation (see, e.g., [92])—a method, which has recently been extended for aEIF neurons and piecewise smooth systems in general [81].

Applying the phase network model (23.7) synchrony and phase locking of neuronal pairs can be conveniently analyzed. The phase difference $\varphi := \vartheta_2 - \vartheta_1$ then evolves according to the scalar differential equation

$$\frac{d\varphi}{dt} = G_{21} H_d(-\varphi) - G_{12} H_d(\varphi),$$

(23.8)

(assuming no autapses, i.e., $G_{ii} = 0$), whose fixed point solutions correspond to locking states. For symmetric synaptic strengths ($G_{12} = G_{21}$) the stability of synchrony and anti-phase locking can even be "read off" the PRC without additional calculations (see supplementary material of [81]).

Using that approach it was found [81] that a subthreshold voltage-driven adaptation current can change the PRC of an aEIF neuron from type I (only phase advances for excitatory perturbations) to type II (phase advances and delays) (Fig. 23.2a) by altering the rest-spiking transition from a saddle-node to a Hopf bifurcation. A purely spike-dependent adaptation current, on the other hand, does not change the bifurcation type and thus leaves the type of PRC unaffected, but changes its skew. These effects translate, for coupled excitatory neurons, into stabilization of synchrony by subthreshold adaptation, and locking with a small phase difference due to spike-driven adaptation, respectively, as long as the external inputs are weak and the synaptic strengths and delays are small. For inhibitory pairs it was found that synchrony is stable and robust to changes in delay and input strength [81], and (both types of) adaptation currents can mediate bistability of in-phase and anti-phase locking (Fig. 23.2a). The locking behavior which was observed for pairs is reflected by the dynamics of corresponding larger networks with all-to-all connectivity. Stable synchrony in pairs of coupled neurons is a good indicator for stable synchrony in larger networks, and bistability often predicts the emergence of cluster states (Fig. 23.2b). Application of a detailed Hodgkin-Huxley neuron model as an alternative to the (computationally much simpler) aEIF model led to the same conclusions [81].

In the context of neuromodulation by ACh, it has been shown in vitro that the PRC of a cortical pyramidal neuron can be changed from type II to type I and its skew can be decreased by ACh agonist carbachol [61]. This experimental result is consistent with the computational results above and indicates that ACh may have a desynchronizing effect on excitatory neurons by reducing their adaptation currents (cf. Fig. 23.1a).

Master stability function method

To summarize the MSF technique for analyzing network synchrony we consider again the model (23.6). Assuming constant row sum $\bar{g} = \sum_j G_{ij}$ of the coupling matrix G, in the synchronized state every neuron evolves according to the equation $dx/dt = f(x) + \bar{g}h(x, x_d)$. We denote this solution by x_s. To assess the stability of the synchronous solution we linearize (23.6) around $x_i = x_s$ to obtain the (mN-dimensional) variational equation for the evolution of the deviations $\underline{\xi} := (\xi_1, \dots, \xi_N)^T$ from the synchronous solution along its smooth intervals,

$$\frac{d\underline{\xi}}{dt} = \left[I_N \otimes \left(D_x f(x_s) + \bar{g} D_x h(x_s, x_{s,d}) \right) \right] \underline{\xi} + \left[G \otimes D_{x_d} h(x_s, x_{s,d}) \right] \underline{\xi}_d,$$

$$(23.9)$$

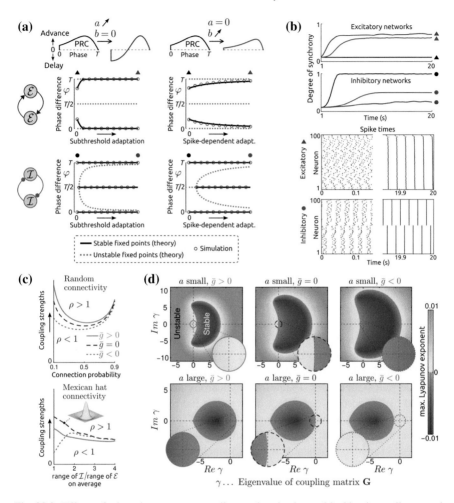

Fig. 23.2 Effects of adaptation currents on spike synchronization and locking in small neuronal networks. **a** Phase response curve of an aEIF neuron (*top*) and phase-locked states for pairs of weakly coupled excitatory ($\mathcal{E}$, *middle*) and inhibitory ($\mathcal{I}$, *bottom*) neurons (i.e., stationary solutions of (23.8)), with different strengths of subthreshold (*left*) and spike-dependent (*right*) adaptation currents. $\varphi = 0$, T and $\varphi = T/2$ correspond to synchrony and anti-phase locking, respectively. Adapted from [81]. **b** Dynamics of simulated networks of $N = 100$ all-to-all coupled aEIF neurons [81]. *Colored icons* indicate the type and strength of adaptation current and the type of coupling (excitatory or inhibitory). **c** Radius ρ of the coupling matrix eigenvalue bulk spectrum for random (*top*) and distance-dependent Mexican hat (*bottom*) connectivity. *Lines* of equal radius $\rho = 1$ are shown. Adapted from [82]. **d** Master stability functions for weak (*top*) and strong (*bottom*) subthreshold adaptation, and for excitation-dominated (*left*), balanced (*center*), and inhibition-dominated (*right*) synaptic interaction. *Circles* indicate the unit disk where stability is predicted for a wide range of different coupling matrices (cf. **c**). Adapted from [82]

where $\otimes$ denotes the Kronecker product, $\boldsymbol{I}_N$ the N-dimensional identity matrix and $D_y := \partial/\partial \boldsymbol{y}$. At the nonsmooth locations (i.e., discontinuities and kinks) of $\boldsymbol{x}_s$ (23.9) is complemented by appropriate linear transition conditions, which are derived via Taylor expansion of (23.6) around $\boldsymbol{x}_s$ in the neighborhood of those locations (see [82] for details). Block-diagonalization of the combined variational system then leads to

$$\frac{d\boldsymbol{\zeta}}{dt} = \left(D_x\boldsymbol{f}(\boldsymbol{x}_s) + \bar{g}D_x\boldsymbol{h}(\boldsymbol{x}_s, \boldsymbol{x}_{s,d})\right)\boldsymbol{\zeta} + \gamma D_{x_d}\boldsymbol{h}(\boldsymbol{x}_s, \boldsymbol{x}_{s,d})\boldsymbol{\zeta}_d, \qquad (23.10)$$

together with linearly transformed transition conditions [82]. $\boldsymbol{\zeta} := \left(z^{\mathrm{T}} \otimes \boldsymbol{I}_m\right)\underline{\boldsymbol{\xi}}$, where z is the normalized eigenvector of $\boldsymbol{G}$ that corresponds to the eigenvalue γ. The transformed variational equation thus separately describes the evolution of deviations from synchrony in the longitudinal direction[3] (for $\gamma = \bar{g}$) and the transverse directions (for the other $N - 1$ eigenvalues of $\boldsymbol{G}$). The MSF is defined by the largest Lyapunov exponent for the solution $\boldsymbol{\zeta} = \boldsymbol{0}$ of that system as a function of arbitrary $\gamma \in \mathbb{C}$. It provides general information about the stability of the synchronous state for many different coupling matrices at once. To assess the stability for a particular matrix $\boldsymbol{G}$ one needs to determine whether or not all of its eigenvalues are contained in the region of stability as indicated by the MSF. One eigenvalue of $\boldsymbol{G}$ is always equal to $\bar{g}$. The other eigenvalues (bulk spectrum) are confined by a circle whose radius ρ may be determined analytically, for example, in case of random connectivity [93, 94], and otherwise numerically (see Fig. 23.2c).

Using this method it has been shown [82] that a subthreshold voltage-driven adaptation current stabilizes or destabilizes synchrony in a homogeneous population of coupled aEIF neurons, depending on whether the recurrent excitatory ($\bar{g} > 0$) or inhibitory ($\bar{g} < 0$) synaptic couplings dominate (Fig. 23.2d). This is consistent with the behavior of neuronal pairs and generalizes that result for larger networks. Synchrony was found to be unstable for homogeneous networks with balanced recurrent synaptic inputs, that is, where excitatory and inhibitory synaptic inputs cancel each other ($\bar{g} = 0$). These synchronization properties are similar for networks with qualitatively different patterns of connections, including random and spatially structured connectivity, as long as the synaptic couplings are not too strong (cf. Fig. 23.2c). By generalizing the MSF method to more heterogeneous networks of excitatory and inhibitory subpopulations with distinct dynamical properties it was further found that adaptation currents provide a mechanism to control the stability of different cluster states, independent of a particular connection pattern [82].

---

[3]The direction of the flow along the synchronous solution.

23.4.4 Large Networks: Effects on Asynchronous States and Sparse Synchronization

Finally, we are interested in the collective dynamics of large networks that consist of at least several thousands of neurons, whose spike times and the temporal relationships between them are subject to variability because of fluctuations in the synaptic inputs.[4] For example, pyramidal neurons have been shown to exhibit highly irregular spiking, even during overall oscillatory network activity (see [1] for a comprehensive review). In this case, the network dynamics is often described by an instantaneous spike rate which quantifies the time-varying mean spiking activity across the population.[5] Oscillations of the spike rate then correspond to a rhythmic modulation of the spiking probability within the population, that is, individual neurons do not participate in every cycle and their spike times are stochastic, as frequently observed [1]. This type of network behavior reflects so-called sparse synchronization in contrast to precise spike-to-spike synchrony as covered in Sect. 23.4.3.

To effectively study how adaptation currents shape asynchronous states and sparse synchronization mean-field methods based on the Fokker-Planck equation have recently been extended to networks of aEIF neurons [83, 96] (based on [26, 97, 98]). This approach allows for convenient analyses of network states while having the properties of single neurons retained by the variables and parameters of the aEIF model. In [83] a large network consisting of two aEIF populations, one excitatory and one inhibitory, was considered. The neurons were sparsely and randomly connected, and exposed to fluctuating inputs. Synaptic delays and connection probabilities were chosen to describe a (generic) local cortical network.[6] From this high-dimensional stochastic network description (multiple thousands of coupled aEIF neurons) a mean-field model can be derived using the Fokker-Planck equation, given by a compact system of coupled partial and ordinary differential equations of the form (23.5) (for each population). Here, the synaptic coupling model involving conductance-based membrane voltage jumps (cf. Sect. 23.4.1) self-consistently carries over into the moments $\mu = \mu(V, r_d^{\mathcal{E}}, r_d^{\mathcal{I}})$ and $\sigma = \sigma(V, r_d^{\mathcal{E}}, r_d^{\mathcal{I}})$ of the input, both of which depend on the membrane voltage V and the delayed population spike rates $r_d^{\mathcal{E}}, r_d^{\mathcal{I}}$ obtained as $r_d(t) := p_d * r(t)$ for a given delay distribution p_d. Approximating presynaptic spike times by Poisson processes (see, e.g., [26]) these moments can be expressed in a straightforward way (see [83] for details). The derived system

---

[4]Synaptic noise is the by far largest-amplitude noise source in neurons of the central nervous system [49].

[5]The dynamics of population-averaged activity measures of this kind relate to those of experimentally widely applied neuronal mass signals (such as local field potentials or electroencephalograms [95]) and are thus of major interest.

[6]Electrophysiological recordings from neuronal pairs have shown that the connection probability in cortical networks is often very low (see, e.g., [99]). Random connectivity is a simplifying assumption that implies negligible noise correlations as measured in vivo [100]. This is exploited in the mean-field reduction below.

allows for an efficient numerical calculation of the network activity in time and thus enables exhaustive explorations of the parameter space.

Using this approach it has been shown [83] that spike-driven adaptation currents together with sufficiently strong recurrent excitatory inputs provide a mechanism to generate slow oscillations (α frequency range and below), see Fig. 23.3a–c. Faster rhythmic activity (β frequency range and above) can be mediated by a feedback loop that involves recurrent excitatory and inhibitory inputs with distinct delay times. Oscillation frequencies increase with the strengths of inhibition and external drive, and adaptation currents (of either type) play a facilitating role in stabilizing these spike rate rhythms (see Fig. 23.3d). In parameter regimes for which the population spike rates are constant (asynchronous state), networks with dominant inhibition show strong resonances when oscillating external inputs are considered. Increased adaptation currents mediate resonance behavior also for networks with stronger excitation by producing amplified population activity within a narrow frequency band [83], in consistency with the resonance effects shown for single neurons (cf. Sect. 23.4.2).

Neuronal adaptation currents have also been found to enable bistability of low and high activity asynchronous states as well as asynchronous and oscillatory states, using Fokker-Planck based mean-field approaches for coupled integrate-and-fire neurons equipped with (only) spike-driven adaptation [101] and for aEIF neurons (Sect. 4.2 of [96]). In these studies, low-dimensional systems of ordinary differential equations describing the collective dynamics were derived from the network models and analyzed.

Differential effects of the two types of adaptation currents on network dynamics have further been observed in the context of self-sustained activity, that is, in the absence of external driving inputs ($I_{ext} \rightarrow 0$). Using numerical simulations it was found that increasing the proportion of neurons with subthreshold adaptation currents promotes self-sustained asynchronous states whereas spike-dependent adaptation can effectively silence the network [102].

To examine the intrinsic controllability of oscillatory and asynchronous network states through ACh-based neuromodulation two experimental findings can be exploited: (i) the distinct effect of the neuromodulator ACh on subthreshold and spike-dependent adaptation (cf. Sect. 23.3 and Fig. 23.1a) and (ii) the existence of synaptic (long range) connections from sensory cortices via prefrontal cortex to populations of cholinergic neurons in the basal forebrain which project (back) to the sensory area, implementing a modality-specific closed circuitry [58, 103, 104]. These physiological evidences can be translated into a closed-loop control system, where the network activity affects the local ACh concentration, which in turn changes the neuronal adaptation current strength, thereby regulating the network dynamics (Fig. 23.3e). Preliminary results from this dynamical system, using a low-dimensional mean-field model derived from a large aEIF network (cf. Sect. 4.2 of [96]), show, that the hypothesized control mechanism can indeed be used to stabilize and switch between a variety of asynchronous and oscillatory states (see Fig. 23.3f).

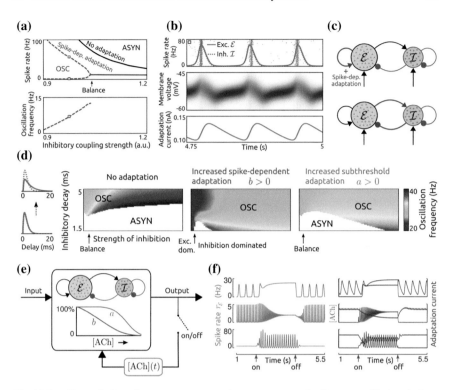

Fig. 23.3 Effects of adaptation currents on asynchronous states and spike rate oscillations in large networks of excitatory ($\mathcal{E}$) and inhibitory ($\mathcal{I}$) aEIF neurons subject to fluctuating inputs. **a** Stable constant spike rate (*solid*) indicating asynchronous states (ASYN) and min./max. of oscillating spike rate (*dashed*, OSC) together with the corresponding frequency. Results from the Fokker-Planck mean-field model [83]. **b** Time series of relevant model variables for the parametrization indicated in **a**. Overlayed are the spike times of individual neurons from a complementary numerical simulation of the original aEIF network ($N = 5 \times 10^4$) [83]. **c** Schematic diagrams visualizing two oscillation mechanisms, based on synaptic excitation and spike-driven adaptation (*top*), and $\mathcal{E}$-$\mathcal{I}$ synaptic interaction (*bottom*). **d** Frequency of stable spike rate oscillations for networks with and without adaptation currents, different strengths of synaptic inhibition and inhibitory delay distributions (*green*) [83]. **e** Closed-loop control circuit that involves local acetylcholine concentration $[ACh] =: z$ described by $\tau_z \, dz/dt = -z + c_z r_{\mathcal{E}}(t - d_z)$ with timescale τ_z, coupling parameter c_z and time delay d_z. "on" condition: Adaptation parameters (a and b) are adjusted according to the indicated sensitivity curves (cf. Fig. 23.1a). "off" condition: values of a and b remain constant. **f** Examples of switching between stable oscillatory (sparsely synchronized) and asynchronous network states via the control loop in **e**. Oscillation mechanisms are based on recurrent excitation and spike-driven adaptation (*top*), recurrent inhibition (*middle*), and recurrent $\mathcal{E}$-$\mathcal{I}$ interplay (*bottom*)

23.5   Conclusion

Control mechanisms employed by the brain to (de)stabilize neuronal network states include feedback of neuronal spiking activity, changes of synaptic coupling strengths, and changes of the dynamical properties of neurons. Here, we reviewed experimental and theoretical results on the control potential of each of these mechanisms, covering the latter in more detail. We focused on the role of neuronal adaptation currents in shaping the collective dynamics and the implications for neuromodulatory (acetylcholine-based) control at different spatial levels. Using an experimentally validated compact model of single neuron activity across these levels and different suitable methods (mean-field, phase reduction, master stability functions) outlined above, it could be shown that neuromodulatory regulation of neuronal adaptation (microscopic property) allows to switch between biologically relevant synchronized and asynchronous network states (macroscopic dynamics) by changing the neuronal spiking characteristics in particular ways. This contribution further demonstrated the benefits of bottom-up modeling and analyses in contributing to our understanding of neuronal dynamics across different scales.

Acknowledgments This work was supported by the Deutsche Forschungsgemeinschaft in the framework of Collaborative Research Center 910.

References

1. X.J. Wang, Physiol. Rev. **90**, 1195 (2010)
2. L. Marshall, H. Helgadóttir, M. Mölle, J. Born, Nature **444**, 610 (2006)
3. K.D. Harris, A. Thiele, Nat. Rev. Neurosci. **12**, 509 (2011)
4. P. Fries, J.H. Reynolds, A.E. Rorie, R. Desimone, Science **291**, 1560 (2001)
5. E. Bullmore, O. Sporns, Nat. Rev. Neurosci. **10**, 186 (2009)
6. M.D. Fox, A.Z. Snyder, J.L. Vincent, M. Corbetta, D.C. Van Essen, M.E. Raichle, Proc. Natl. Acad. Sci. USA **102**, 9673 (2005)
7. M. Zijlmans, J. Jacobs, R. Zelmann, F. Dubeau, J. Grotman, Neurology **72**, 979 (2009)
8. C. Hammond, H. Bergman, P. Brown, Trends Neurosci. **30**, 357 (2007)
9. P.J. Uhlhaas, W. Singer, Nat. Rev. Neurosci. **11**, 100 (2010)
10. D.C. Van Essen, C.H. Anderson, D.J. Felleman, Science **255**, 419 (1992)
11. S. Hochstein, M. Ahissar, Neuron **36**, 791 (2002)
12. A. Angelucci, J.B. Levitt, E.J.S. Walton, J.M. Hupe, J. Bullier, J.S. Lund, J. Neurosci. **22**, 8633 (2002)
13. K. Pyragas, Phys. Lett. A **170**, 421 (1992)
14. W. Just, T. Bernard, M. Ostheimer, E. Reibold, H. Benner, Phys. Rev. Lett. **78**, 203 (1997)
15. P. Hövel, E. Schöll, Phys. Rev. E **72**, 046203 (2005)
16. E. Schöll, H.G. Schuster (eds.), *Handbook of Chaos Control* (Wiley-VCH, Weinheim, 2008)
17. E. Schöll, G. Hiller, P. Hövel, M.A. Dahlem, Phil. Trans. R. Soc. A **367**, 1079 (2009)
18. M. Rosenblum, A. Pikovsky, Phys. Rev. Lett. **92**, 114102 (2004)
19. O. Popovych, C. Hauptmann, P. Tass, Phys. Rev. Lett. **94**, 164102 (2005)
20. C. Hauptmann, J.C. Roulet, J.J. Niederhauser, W. Döll, M.E. Kirlangic, B. Lysyansky, V. Krachkovskyi, M.A. Bhatti, U.B. Barnikol, L. Sasse, C.P. Bührle, E.J. Speckmann, M. Götz, V. Sturm, H.J. Freund, U. Schnell, P.A. Tass, J. Neural Eng. **6**, 066003 (2009)
21. N. Tukhlina, M. Rosenblum, Phys. Rev. E **75**, 011918 (2007)
22. B. Rosin, M. Slovik, R. Mitelman, M. Rivlin-Etzion, S.N. Haber, Z. Israel, E. Vaadia, H. Bergman, Neuron **72**, 370 (2011)
23. C.A.S. Batista, R.L. Viana, F.A.S. Ferrari, S.R. Lopes, A.M. Batista, J.C.P. Coninck, Phys. Rev. E **87**, 042713 (2013)
24. T. Bal, D. Debay, A. Destexhe, J. Neurosci. **20**, 7478 (2000)

25. B. Doiron, M.J. Chacron, L. Maler, A. Longtin, Nature **421**, 539 (2003)
26. N. Brunel, J. Comput. Neurosci. **8**, 183 (2000)
27. N. Brunel, X.J. Wang, J. Neurophysiol. **90**, 415 (2003)
28. P. Tiesinga, T.J. Sejnowski, Neuron **63**, 727 (2009)
29. J.S. Isaacson, M. Scanziani, Neuron **72**, 231 (2011)
30. G. Deco, A. Ponce-Alvarez, P. Hagmann, G.L. Romani, D. Mantini, M. Corbetta, J. Neurosci. **34**, 7886 (2014)
31. E. Todorov, M.I. Jordan, Nat. Neurosci. **5**, 1226 (2002)
32. S.H. Scott, Nat. Rev. Neurosci. **5**, 532 (2004)
33. J. Diedrichsen, R. Shadmehr, R.B. Ivry, Trends Cogn. Sci. **14**, 31 (2010)
34. M.M. Churchland, J.P. Cunningham, M.T. Kaufman, J.D. Foster, P. Nuyujukian, S.I. Ryu, K.V. Shenoy, Nature **487**, 51 (2012)
35. M. Mattia, P. Pani, G. Mirabella, S. Costa, P. Del Giudice, S. Ferraina, J. Neurosci. **33**, 11155 (2013)
36. D.E. Feldman, Annu. Rev. Neurosci. **32**, 33 (2009)
37. D.E. Feldman, Neuron **75**, 556 (2012)
38. H. Markram, J. Lübke, M. Frotscher, B. Sakmann, Science **275**, 213 (1997)
39. G. Bi, M. Poo, Annu. Rev. Neurosci. **24**, 139 (2001)
40. J. Sjöström, W. Gerstner, Scholarpedia **5**, 1362 (2010)
41. D.M. Kullmann, A.W. Moreau, Y. Bakiri, E. Nicholson, Neuron **75**, 951 (2012)
42. H. Ko, L. Cossell, C. Baragli, J. Antolik, C. Clopath, S.B. Hofer, T.D. Mrsic-Flogel, Nature **496**, 96 (2013)
43. R. Kempter, W. Gerstner, J. van Hemmen, Phys. Rev. E **59**, 4498 (1999)
44. S. Song, K.D. Miller, L.F. Abbott, Nat. Neurosci. **3**, 919 (2000)
45. H. Câteau, K. Kitano, T. Fukai, Phys. Rev. E **77**, 051909 (2008)
46. T. Aoki, T. Aoyagi, Phys. Rev. Lett. **102**, 034101 (2009)
47. J. Ladenbauer, L. Shiau, K. Obermayer, Comput. Neurosci. Conf. Abstr. BMC Neurosci. **12**, P240 (2011)
48. K. Mikkelsen, A. Imparato, A. Torcini, Phys. Rev. Lett. **110**, 208101 (2013)
49. A. Destexhe, M. Rudolph-Lilith, *Neuronal Noise* (Springer, New York, 2012)
50. E.V. Lubenov, A.G. Siapas, Neuron **58**, 118 (2008)
51. A. Morrison, A. Aertsen, M. Diesmann, Neural Comput. **19**, 1437 (2007)
52. A. Morrison, M. Diesmann, W. Gerstner, Biol. Cybern. **98**, 459 (2008)
53. T.P. Vogels, H. Sprekeler, F. Zenke, C. Clopath, W. Gerstner, Science **334**, 1569 (2011)
54. E.M. Izhikevich, J.A. Gally, G.M. Edelman, Cereb. Cortex **14**, 933 (2004)
55. E.M. Izhikevich, Neural Comput. **18**, 245 (2006)
56. B. Szatmáry, E.M. Izhikevich, PLOS Comput. Biol. **6**, e1000879 (2010)
57. D.A. McCormick, Progr. Neurobiol. **39**, 337 (1992)
58. S.H. Lee, Y. Dan, Neuron **76**, 209 (2012)
59. V. Parikh, R. Kozak, V. Martinez, M. Sarter, Neuron **56**, 141 (2007)
60. S. Soma, S. Shimegi, H. Osaki, H. Sato, J. Neurophysiol. **107**, 283 (2012)
61. K.M. Stiefel, B.S. Gutkin, T.J. Sejnowski, PLOS One **3**, e3947 (2008)
62. J.L. Herrero, M.J. Roberts, L.S. Delicato, M.A. Gieselmann, P. Dayan, A. Thiele, Nature **454**, 1110 (2008)
63. M.E. Hasselmo, Curr. Opin. Neurobiol. **16**, 710 (2006)
64. J.J. Letzkus, S.B.E. Wolff, E.M.M. Meyer, P. Tovote, J. Courtin, C. Herry, A. Lüthi, Nature **480**, 331 (2011)
65. A.J. Giessel, B.L. Sabatini, Neuron **68**, 936 (2010)
66. D.V. Madison, B. Lancaster, R.A. Nicoll, J. Neurosci. **7**, 733 (1987)
67. P.C. Schwindt, W.J. Spain, W.E. Crill, J. Neurophysiol. **61**, 233 (1989)
68. P.C. Schwindt, W.J. Spain, W.E. Crill, J. Neurophysiol. **67**, 216 (1992)
69. D.A. Brown, P.R. Adams, Nature **283**, 673 (1980)
70. R. Brette, W. Gerstner, J. Neurophysiol. **94**, 3637 (2005)
71. W. Gerstner, R. Brette, Scholarpedia **4**, 8427 (2009)

72. A.L. Hodgkin, A.F. Huxley, J. Physiol. **117**, 500 (1952)
73. J. Touboul, R. Brette, Biol. Cybern. **99**, 319 (2008)
74. R. Naud, N. Marcille, C. Clopath, W. Gerstner, Biol. Cybern. **99**, 335 (2008)
75. L. Badel, S. Lefort, R. Brette, C.C.H. Petersen, W. Gerstner, M.J.E. Richardson, J. Neurophysiol. **99**, 656 (2008)
76. L. Badel, S. Lefort, T.K. Berger, C.C.H. Petersen, W. Gerstner, M.J.E. Richardson, Biol. Cybern. **99**, 361 (2008)
77. R. Jolivet, F. Schürmann, T.K. Berger, R. Naud, W. Gerstner, A. Roth, Biol. Cybern. **99**, 417 (2008)
78. R. Jolivet, R. Kobayashi, A. Rauch, R. Naud, S. Shinomoto, W. Gerstner, J. Neurosci. Meth. **169**, 417 (2008)
79. M. Pospischil, Z. Piwkowska, T. Bal, A. Destexhe, Biol. Cybern. **105**, 167 (2011)
80. J. Ladenbauer, M. Augustin, K. Obermayer, J. Neurophysiol. **111**, 939 (2014)
81. J. Ladenbauer, M. Augustin, L. Shiau, K. Obermayer, PLOS Comput. Biol. **8**, e1002478 (2012)
82. J. Ladenbauer, J. Lehnert, H. Rankoohi, T. Dahms, E. Schöll, K. Obermayer, Phys. Rev. E **88**, 042713 (2013)
83. M. Augustin, J. Ladenbauer, K. Obermayer, Front. Comput. Neurosci. **7**, 1 (2013)
84. D.V. Madison, R.A. Nicoll, J. Physiol. **354**, 319 (1984)
85. R. Naud, The Dynamics of Adapting Neurons. Ph.D. thesis, École Polytechnique Fédérale de Lausanne (2011)
86. T. Schwalger, K. Fisch, J. Benda, B. Lindner, PLOS Comput. Biol. **6**, e1001026 (2010)
87. L. Shiau, T. Schwalger, B. Lindner, J. Comput. Neurosci. **38**, 589 (2015)
88. G. Gigante, P. Del Giudice, M. Mattia, Math. Biosci. **207**, 336 (2007)
89. T. Netoff, M. Banks, A. Dorval, C. Acker, J. Haas, N. Kopell, J.A. White, J. Neurophysiol. **93**, 1197 (2005)
90. T. Klausberger, P.J. Magill, L.F. Márton, J.D.B. Roberts, P.M. Cobden, G. Buzsáki, P. Somogyi, Nature **421**, 844 (2003)
91. E. De Schutter, V. Steuber, Neuroscience **162**, 816 (2009)
92. M.A. Schwemmer, T.J. Lewis, in *Phase Response Curves, in Neuroscience: Theory, Experiment, and Analysis*, ed. by N. Schultheiss, A. Prinz, R. Butera (Springer, New York, 2012), pp. 3–31
93. K. Rajan, L. Abbott, Phys. Rev. Lett. **97**, 188104 (2006)
94. R.T. Gray, P.A. Robinson, J. Comput. Neurosci. **27**, 81 (2009)
95. G. Buzsáki, C.A. Anastassiou, C. Koch, Nat. Rev. Neurosci. **13**, 407 (2012)
96. J. Ladenbauer, The Collective Dynamics of Adaptive Neurons: Insights from Single Cell and Network Models. Ph.D. thesis, Technische Universität Berlin (2015)
97. M. Mattia, P. Del Giudice, Phys. Rev. E **66**, 051917 (2002)
98. S. Ostojic, N. Brunel, PLOS Comput. Biol. **7**, e1001056 (2011)
99. C. Holmgren, T. Harkany, B. Svennenfors, Y. Zilberter, J. Physiol. **551**, 139 (2003)
100. A.S. Ecker, P. Berens, G.A. Keliris, M. Bethge, N.K. Logothetis, A.S. Tolias, Science **327**, 584 (2010)
101. G. Gigante, M. Mattia, P. Giudice, P. Del Giudice, Phys. Rev. Lett. **98**, 148101 (2007)
102. A. Destexhe, J. Comput. Neurosci. **27**, 493 (2009)
103. L. Golmayo, A. Nuñez, L. Zaborszky, Neuroscience **119**, 597 (2003)
104. G.N. Fournier, K. Semba, D.D. Rasmusson, Neuroscience **126**, 257 (2004)

Chapter 24
Evolutionary Dynamics: How Payoffs and Global Feedback Control the Stability

Jens Christian Claussen

Abstract Biological as well as socio-economic populations can exhibit oscillatory dynamics. In the simplest case this can be described by oscillations around a neutral fixed point as in the classical Lotka-Volterra system. In reality, populations are always finite, which can be discussed in a general framework of a finite-size expansion which allows to derive stochastic differential equations of Fokker-Planck type as macroscopic evolutionary dynamics. Important applications of this concept are economic cycles for "cooperate—defect—tit for tat" strategies, mating behavior of lizards, and bacterial population dynamics which can all be described by cyclic games of rock-scissors-paper dynamics. Here one can study explicitly how the stability of coexistence is controlled by payoffs, the specific behavioral model and the population size. Finally, in socio-economic systems one is often interested in the stabilization of coexistence solutions to sustain diversity in an ecosystem or society. Utilizing a diversity measure as dynamical observable, a feedback into the payoff matrix is discussed which stabilizes the steady state of coexistence.

24.1 Game Theory and Evolutionary Dynamics

The roots of game theory [1] date back to the 2nd world war's need to predict strategic operations between nations. Albeit game theory opened an own 'economic' approach to behavioral decision theory, its essential remains limited to the static analysis of fixed points, especially of the type of Nash equilibria [2]. Dynamics was invoked by John Maynard Smith's genious transfer of these ideas to biology: he discretized *biological* behavioral strategies by a finite set of formal strategies [3]. The strategies, e.g. *hawk* and *dove* refer to biological metaphors which are based on qualitative

J.C. Claussen (✉)
Computational Systems Biology Lab, Jacobs University Bremen, Research 2,
Campus Ring 1, 28759 Bremen, Germany
e-mail: j.claussen@jacobs-university.de

© Springer International Publishing Switzerland 2016   461
E. Schöll et al. (eds.), *Control of Self-Organizing Nonlinear Systems*,
Understanding Complex Systems, DOI 10.1007/978-3-319-28028-8_24

observations of outcomes when hawks meet doves, or when hawks and doves meet among each other, respectively. In biology, fitness is defined via the reproductive success and, in principle, fitness is a quantity accessible to experiments. In practice, fitness values are usually estimated or assigned by hand, including plausible parameters like costs and benefits. Subsequently one analyzes the parameter space depending on those parameters to identify stability regimes and bifurcations.

In [3], John Maynard Smith also introduced a dynamical perspective into game theory, borrowed from the motivating analogy of biological evolution. The standard ansatz of evolutionary game theory henceforth assigns to each payoff matrix $A = (a_{kl})$ of payoffs (of strategy k, when played against opponent l), a time-evolution dynamics (for the densities of strategies x_k)

$$\frac{d}{dt}x_k = x_k(\pi_k - \langle \pi \rangle) \tag{24.1}$$

$$= \sum_l x_k x_l(\pi_k - \pi_l) \tag{24.2}$$

usually known as the replicator dynamics [4]. Here $\pi_k := \sum_l a_{kl}x_l$ denotes the payoff received by strategy k, and $\langle \pi \rangle = \sum_l \pi_l x_l$ is the average payoff obtained in the population which, together with the normalization $\sum_k x_k = 1$, allows to rewrite (24.1) in the form of (24.2).

The replicator equation in its payoff difference form (24.2) can be interpreted as a 'chemical' reaction kinetics where (molecular) species k and l meet at rates proportional to each of their densities (mass action kinetics) and react (i.e. change strategy) at a rate modulated by the payoffs of the game. The rationale is that high payoffs (compared to the opponent's) lead to increased reproduction rate, or, in social systems, incentive to change strategy. We should note that, as the payoffs depend on the actual densities x_k, the replicator equations of an evolutionary game are inherently *nonlinear* in the systems variables. For illustration, let us consider the snowdrift game: Two persons in a car get stuck in a snowdrift. Either one (at full cost c) or both (at shared cost $c/2$) can decide to cooperate and shovel, or defect and stay in the car. If at least one of them shovels, they receive the benefit b of escaping the snowdrift. This is equivalent to the payoff matrix [4]

$$A_{SD} = \begin{pmatrix} b - c/2 & b - c \\ b & 0 \end{pmatrix}$$

which has the replicator equation for (x_1, x_2)

$$\dot{x}_1 = x_1[x_1((b - c/2)x_1 - (b - c)x_2) - x_2(bx_1 + 0 \cdot x_2)]$$

where the second equation for x_2 is obtained analogously, but provides only redundant information as we assume both strategies to be normalized, i.e. $x_2 = 1 - x_1$, hence the system is one-dimensional but nonlinear in x_1,

$$\dot{x}_1 = x_1 \left[x_1((b - c/2)x_1 - (b - c)(1 - x_1)) - bx_1(1 - x_1) \right]$$

and exhibits an internal attracting fixed point for suitable parameters [4].

While being the standard approach in evolutionary game theory through three decades, the replicator equation relies on a continuum approximation for the population densities, i.e., the assumption of an infinite number of individuals in the population. Alternative replicator equations have been proposed as the adjusted replicator equation [4] albeit its justification remained unclear.

As the 'reproductive' and 'social' strategy updates suggest, changes of strategies should be considered to follow different microscopic mechanisms and consequently different models should be adopted. Intuitively, the analogy of thermodynamics derived from statistical physics suggested that macroscopic equations of evolutionary dynamics should not depend on fine details of the microscopic interactions of single actions. This viewpoint however neglected that not all microscopic update rules belong to the same universality class in macroscopic dynamics, which will be detailed in the next section.

24.2 Evolutionary Dynamics in Finite Populations

In a finite population of N individuals, population densities x_k have to be replaced by discrete abundances and the dynamics then is defined by a Markov process on $N + 1$ possible states $i \in \{0, 1, \ldots N\}$ and all densities take the form $x = i/N$. In the remainder i thus acts as an index of the Markov states, and at the same time is the dynamical variable of the discrete population.

Among the interaction processes that have been recently introduced are the frequency-dependent Moran process [5–7] that generalizes the Moran process [8] for overlapping generations. In this process, the population size N is kept constant (fixed) restricting to a process where birth and death occur in the same time point, and precisely, reproduction is proportional to fitness (calculated from the payoff) normalized to the average fitness in the population; finally a randomly selected individual is removed by death (Fig. 24.1).

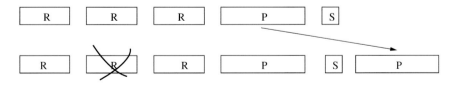

Fig. 24.1 In the frequency-dependent Moran process, the reproduction probability is proportional to each individual payoff, normalized by the average payoff in the population. In the same time step, one randomly selected individual is removed. Here three strategies, rock (*R*), paper (*P*), and scissors (*S*) are shown, and as paper wraps rock, paper receives a high payoff from winning against three players playing *R*, and losing against one player *S* (scissors cut paper). As rock crushes scissors, the *S* player accumulates the lowest payoff (here, the different fitness values are indicated by different lengths of the *boxes*)

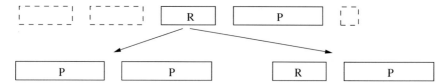

Fig. 24.2 In the pairwise comparison process (Local update) two individuals compete. One focal individual is chosen and its fitness is compared to a second individual. The individual switches strategy with probability larger than 1/2 if the other individual receives a higher payoff. The illustrated population is the same as in Fig. 24.1, with the same payoffs determined by the game

In the frequency-dependent Moran process, the normalization by the average payoff requires global information about received payoff values in the whole population. In most biological systems, such information will not be available and it is more realistic to assume that individuals compete pairwise. Such processes are called *pairwise comparison processes*, and their most basic implementation is the linear pairwise comparison process (or Local Update [9], see Fig. 24.2). The transition probabilities are displayed in Table 24.1.

24.3 From Microscopic Interactions to Macroscopic Processes

In a finite population, strategies as well as abundances are discrete, thus the time evolution is defined by a Markov process. The transition rates between the states depend on the game payoffs and the actual abundances. For instance, if the game consists of two strategies, A and B, the transition probabilites are given by the left column in Table 24.1. The transition rates for, e.g., the Local Update process for a change from state i to state $i \pm 1$ are

$$T^{\pm}(i) = \left(\frac{1}{2} \pm \frac{w}{2} \frac{\pi^A(x) - \pi^B(x)}{\Delta\pi_{\max}} \right) \frac{i}{N} \frac{N-i}{N} \qquad (24.3)$$

where i now is the number of agents in strategy A, and due to normalization to constant population size N the number of agents in strategy B is $N - i$. Here w is a parameter that accounts for the strength of selection; in social systems w can be large (up to 1), whereas the weak selection limit $w \to 0$ is usually considered in biological evolution; fitness differences are usually small and effects observable after many generations. If one identifies w with an inverse temperature, and $(-1) \times$ fitness is interpreted as an energy, the transition rates of the Fermi process corresponds to Boltzmann factors. In this picture, the (biological) weak selection limit is a high-temperature limit, and low temperature ranges are used to model behavioral decisions.

Table 24.1 Microscopic processes with their transition probabilities for an agent to change from strategy B to strategy A, and the corresponding deterministic equations that are obtained in the limit of large populations $N \to \infty$

Microscopic process	Deterministic equation
Moran process $\quad p_{B \to A} = \frac{1-w+w\pi^A(i)}{1-w+w\langle\pi(i)\rangle}$	Adjusted replicator equation $\dot{x} = x(1-x)\frac{1-w+w\pi^A(x)}{\frac{1-w}{w}+w\langle\pi(x)\rangle}$
Local update $\quad p_{B \to A} = \frac{1}{2} + \frac{w}{2}\frac{\pi^A(i)-\pi^B(i)}{\Delta\pi_{\max}}$	(Ordinary) replicator equation $\dot{x} = \frac{w}{\Delta\pi_{\max}}x(1-x)(\pi^A(x)-\pi^B(x))$
Fermi process $\quad p_{B \to A} = \frac{1}{1+e^{-w(\pi^A(i)-\pi^B(i))}}$	(A nonlinear) replicator equation $\dot{x} = x(1-x)\tanh(w(\pi^A(x)-\pi^B(x)))$

Here $\Delta\pi_{\max}$ is the maximal payoff difference and w denotes the strength of selection. In Sects. 24.4.1 and 24.7 we investigate specifically the ordinary replicator equation and in Sect. 24.8 the corresponding Local Update process. For the other processes, the respective replicator equations have to be considered [9, 10]

For three or more strategies, there are accordingly more transitions between the strategies, for the RPS game one has T^{RS}, the backward rate and all cyclic permutations, i.e., 6 transitions [11]. These transition rates then can be inserted into a Master equation for the probability density. While the Master equation often is not solvable analytically, for large populations one can proceed via the Kramers-Moyal expansion to the Fokker-Planck equation for the probability densities for the relative abundances $x := i/N$. In general, one obtains a noise term (the second order term) that scales with $1/\sqrt{N}$ thus vanishes for $N \to \infty$, and a deterministic term containing $a(x) = T^+(x) - T^-(x)$ that is also found in the corresponding (Itô) Langevin equation $\dot{x} = a(x)+b(x)\xi$ (see [9, 11] for a more elaborate derivation). This general scheme can be applied to various microscopic processes, e.g. the Fermi process also listed in Table 24.1 which leads to a nonlinear replicator equation [10] in the limit $N \to \infty$.

24.4   Rock-paper-scissors Game and Non Zero-Sum Payoffs

This classical game, also known as *yanken* in Japan, and under numerous other names worldwide, is comprised of three cyclically outbeating strategies $R \to P \to S \to R$, where the payoffs in direction of the arrows are higher $(+1)$ than in counterdirection $(-s)$,

$$A = \begin{pmatrix} 0 & -s & 1 \\ 1 & 0 & -s \\ -s & 1 & 0 \end{pmatrix} \tag{24.4}$$

where the row and column entries refer to the strategies R (rock), P (paper) and S (scissors), respectively. Such cyclic games appear in various contexts, including an

economic cyclic dilemma between strategies *cooperate*, *defect*, and *Tit-for-Tat* [12].
Here the strategies 'cooperate' (C) and 'defect' (D) follow the Prisoner's Dilemma
with the payoff matrix $\begin{pmatrix} 3 & 0 \\ 5 & 1 \end{pmatrix}$, and a memory-one strategy Tit-for-Tat (TFT) is introduced playing C in the first round, and subsequently imitating the other's strategy. If
played iteratively, these strategies can be shown to outcompete each other cyclically
[12].

In the rock-paper-sciccors game, the standard setting of the game $s = 1$ is a
zero-sum game (i.e., $a_{lk} = -a_{kl}$) referring to the rule that the losing player passes
one coin (of value 1) to the winner. However, in most realistic situations benefits
and costs for both players need not sum up to zero (similar as in the Prisoner's
Dilemma). E.g., for mating lizards [13, 14] the experimentally observed fitnesses
indicate payoffs corresponding to that the bank loses with the game. In well-mixed
populations of different strains of colicin-producing bacteria [15, 16] the metabolic
costs of establishing resistance mechanism and poison production lead, in sum, to a
payoff loss by the bacteria, corresponding to the game situation that 'the bank wins'.
The parameter s accounts for this important distinction. In [11], it has been shown
that in the 'bank loses' case there is always a reversal of stochastic stability, i.e.,
for every $s > 1$ there is a critical population size N_c above which the coexistence
fixed point is metastable, i.e. extinction of one (or more) strategies is observed only
through extensively long transients.

It should be pointed out that here we fully neglect the influence of spatial domain,
which in ecology, especially for territorial species, is well known to account for
niches. The joint effect of finiteness of the population and spatial structure of the
population has beautifully been elucidated by Csaran, Hoekstra and Paigie [17]. In
general, species or agent interactions can follow complex networks of connectivities,
see [18, 19] for recent reviews into this eleborate field.

24.4.1 The Replicator Equations for Non Zero-Sum RPS

Let x, y, z again denote the relative densities of the three strategies R, P, S. Then we
observe the payoffs

$$\pi_R = z - sy$$
$$\pi_P = x - sz$$
$$\pi_S = y - sx$$

and the average payoff is given by $\langle \pi \rangle = (1 - s)(xy + xz + yz)$. Utilizing the
normalization $z = 1 - x - y$ we arrive at the replicator equations

$$\frac{\mathrm{d}}{\mathrm{d}t}x = x\left[1 - x - y - sy - (1 - s)(xy + (x + y)(1 - x - y))\right]$$

$$\frac{\mathrm{d}}{\mathrm{d}t}y = y\left[x - s(1 - x - y) - (1 - s)(xy + (x + y)(1 - x - y))\right] \quad (24.5)$$

Besides three fixed points at the borders, which correspond to states where two species are extinct, there is an interior fixed point $x = y = 1/3$ at the intersection of the nullclines.

24.5   Observables for (bio)diversity

It is widely assumed in global ecological contexts that holding up a high state of diversity is a primary or essential goal for sustainability of the biosphere. In ecological context, spatial abundances are often cast into spatial diversity measures. However, detailed spatial information is often lacking. In genetic contexts, genetic diversity is assumed to be highly beneficial for the adaptability of an evolving system [20] but it remains nontrivial how much diversity in fact is needed, as for pure adaptability the (exponentially increasing) number of combinatorial possibilities even of quite short genomes would offer sufficient optimization spaces. On the population ecology level with a large number of species the role of biodiversity is likewise unclear, but even more, it is ambiguous how one should define, quantify or observe diversity. A naive (and even most unbiased) definition is to let only the number of species account for the diversity, i.e. populations with largest number of abundant species are classified as of highest diversity.

In contrary to statistical physics, where the scaling of variables is either intensive (independent of system size, as temperature), or extensive (scaling linearly with systems size, as energy or mass or volume), here it is apparent that it is not a priori clear how the complexity (or biodiversity index) of a population should scale with (or depend on) the chosen phylogenetic level (i.e. the taxonomic scale of coarse-graining biological species).

Leaving this general question unsolved, in light of this discussion we now define the population diversity observable for a three-species ecosystem as

$$H = 27 \cdot x \cdot y \cdot (1 - x - y) \quad (24.6)$$

which is a normalized variant of what has been used in other context as a conserved quantity for evolutionary game dynamics characterized by a cyclically symmetric zero-sum game [11].

This definition may be difficult to generalize in a large ecosystem, but is sufficient in our context to provide a conceptual minimal model of feedback of the diversity state into the dynamics.

24.6   Introduction of a Feedback Term

In our definition (24.6), $H = 1$ refers to the equi-abundant case whereas decreasing H refers to getting closer to extinction, in which case $H = 0$ is reached. Consequently H can be utilized as a feedback variable to influence the payoffs in the game. In the eco-econo context such may be interpreted as a tax that is raised if there is a loss of biodiversity observed. In our case, we define the feedback

$$s(t) = s_0 + \varepsilon \cdot (1 - H(t)). \tag{24.7}$$
$$= s_0 + \varepsilon \cdot (1 - 27 \cdot x(t) \cdot y(t) \cdot (1 - x(t) - y(t))). \tag{24.8}$$

In an iterated dynamics, the observed state would be of delayed knowledge, i.e. $H(t - 1)$ but a delay can be taken into account even in the time-continuous case. The term $(1 - H(t))$ vanishes in the "biodiversity fixed point", thus the control scheme Eq. (24.8) implements a *noninvasive control*, and ε depicts the strength of control (or feedback). Similar feedback schemes have been introduced and investigated widely to study the stabilization of fixed points and unstable periodic orbits embedded in chaotic attractors [21, 22].

24.7   Replicator Equations Including Control

The feedback term defined by (24.8) modifies the actual payoff parameter s in the replicator equations (24.5). Again, we utilize that the normalization of the population renders the dynamics within a two-dimensional simplex. Hence the replicator equations for the RPS game with applied control read explicitly

$$\frac{d}{dt}x = x(1 - x - y - (xy + (x + y)(1 - x - y)))$$
$$+ [s_0 + \varepsilon(1 - 27xy(1 - x - y))]$$
$$\times x((xy + (x + y)(1 - x - y)) - y)$$
$$\frac{d}{dt}y = y(x - (xy + (x + y)(1 - x - y)))$$
$$+ [s_0 + \varepsilon(1 - 27xy(1 - x - y))]$$
$$\times y((xy + (x + y)(1 - x - y)) - (1 - x - y)) \tag{24.9}$$

As the fixed points of (24.5) are independent on s, they are also fixed points of (24.9), and the biodiversity fixed point $x = y = 1/3$ (which is neutrally stable for $s_0 = 1$ and $\varepsilon = 0$) is stabilized when control is switched on ($\varepsilon > 0$).

We note that this replicator equation dynamics is valid only in the limit of an infinite population size. In a finite population of N individuals however, even the $s_0 = 1$ case of conservative oscillations, becomes stochastic which includes the risk of extinction of one or more species when trajectories approach the boundaries. The next section verifies that the control scheme is applicable in a finite population.

24.8 Numerical Investigation of the RPS Process Under Feedback Control

Finally we illustrate that the feedback control scheme is able to stabilize the biodiversity fixed point not only in the limit of infinite populations, but also in a finite population.

Here we have to specify a microscopic process, and as we analyzed beforehand the classical replicator equation, it is natural to consider an interaction process belonging to its universality class. In the Local Update, or pairwise comparison process, pairs of agents are selected at random and an agent playing B adopts the strategy of an agent playing A with the probability [9]

$$p_{B \to A} = \frac{1}{2} + w \frac{\pi_A - \pi_B}{2 \Delta \pi_{\max}}. \tag{24.10}$$

Here, $\Delta \pi_{\max}$ is the maximal payoff difference that can occur between two agents. This is necessary here to keep probabilities within the interval $[0, 1]$. In our case, the control may lead to temporal change of the possible maximal payoff difference, depending on s_0 and ε.

Figure 24.3 displays a typical simulation of the stochastic process defined by the Local Update (24.10) where the payoff s follows the feedback rule (24.8). The initial condition has been chosen far away from the biodiversity fixed point (=low biodi-

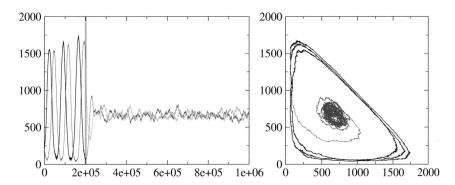

Fig. 24.3 RPS dynamics in a population of $N = 2000$ agents, $w = 0.9$, $s_0 = 1.0$. At time $t = 200,000$, control $\varepsilon = 10$ is switched on to stabilize coexistence. *Left panel* Time series of densities of R (*black/red*) and P (*blue/green*). The orange bar indicates the time when control is switched on. *Right panel* Phase plot (R, P) where the trajectory (same data) firstly is plotted in *black*, and thereafter in *red* after control was switched on

versity). Before control is switched on, trajectories stay close to a neutral oscillation at large amplitudes. After control is switched on, trajectories are attracted towards the biodiversity fixed point and perform low amplitude stochastic oscillations around the fixed point.

24.9   Conclusions and Outlook

Evolutionary dynamics based on game-theoretical models can be formulated in a wide range of contexts from microbe biology to social and economic strategic decision behavior. The resulting models provide qualitative insight into the phase diagram, and a rich variety of dynamical phenomena as fixed points, cycles and chaos can be observed. Here we have investigated, how the observation of an abstract global population state (specifically, a diversity measure) can be used for a feedback term to stabilize the coexistence of strategies, which would be stochastically fragile due to demographic fluctuations naturally occuring in finite populations. While here we have considered one of the most simple cases of stabilization of a biodiversity (or coexistence) fixed point, there is significant potential to transfer the ideas of control of chaos and oscillations to relevant biological and socio-economic models.

References

1. J. von Neumann, O. Morgenstern, *Theory of Games and Economic Behavior*, (Princeton, 1944)
2. J. Nash, Proc. Natl. Acad. Sci. **36**, 4849 (1950)
3. J.M. Smith, G.R. Price, Nature **246**, 15–18 (1973)
4. J. Hofbauer, K. Sigmund, *Evolutionary Games and Population Dynamics* (Cambridge University Press, Cambridge, England, 1998)
5. M.A. Nowak, A. Sasaki, C. Taylor, D. Fudenberg, Nature **428**, 646 (2004)
6. C. Taylor, D. Fudenberg, A. Sasaki, M.A. Nowak, Bull. Math. Biol. **66**, 1621 (2004)
7. M.A. Nowak, *Evolutionary Dynamics* (Cambridge, MA, 2006)
8. P.A.P. Moran, *The Statistical Processes of Evolutionary Theory* (Oxford, 1962)
9. A. Traulsen, J.C. Claussen, C. Hauert, Phys. Rev. Lett. **95**, 238701 (2005)
10. A. Traulsen, M.A. Nowak, J.M. Pacheco, Phys. Rev. E **74**, 011909 (2006)
11. J.C. Claussen, A. Traulsen, Phys. Rev. Lett. **100**, 058104 (2008)
12. L. Imhof, D. Fudenberg, M.A. Nowak, Proc. Natl. Acad. Sci. **102**, 10797 (2005)
13. B. Sinervo, C.M. Lively, Nature **380**, 240 (1996)
14. K.R. Zamudio, B. Sinervo, Proc. Natl. Acad. Sci. **97**, 14427 (2000)
15. B. Kerr, M.A. Riley, M.W. Feldman, B.J.M. Bohannan, Nature **418**, 171 (2002)
16. B.C. Kirkup, M.A. Riley, Nature **428**, 412 (2004)
17. T. Czárán, R.F. Hoekstra, L. Pagie, Proc. Natl. Acad. Sci. **99**, 786 (2002)
18. G. Szabó, G. Fáth, Phys. Rep. **446**, 97 (2007)
19. M. Perc, J. Gómez-Gardeñes, A. Szolnoki, L.M. Floría, Y. Moreno, J.R. Soc, Interface **10**, 20120997 (2013)
20. G.M. Mace et al., Biodiversity targets after 2010. Current Opinion in Environmental Sustainability **2**, 3 (2010)
21. E. Ott, C. Grebogi, J.A. Yorke, Phys. Rev. Lett. **64**, 1196 (1990)
22. E. Schöll, H.G. Schuster, *Handbook of Chaos Control*, 2nd edn. (Wiley-VCH, Weinheim, 2007)

Index

© Springer International Publishing Switzerland 2016
E. Schöll et al. (eds.), *Control of Self-Organizing Nonlinear Systems*,
Understanding Complex Systems, DOI 10.1007/978-3-319-28028-8

H
Hard spheres, 386
Hard spring, 124
Harmonic, 112, 113, 121–123
Harmonic oscillator, damped classical, 317
Harmonic oscillator, damped quantum, 291, 317
Heterogeneous, 267
Hodgkin-Huxley model, 449
Homoclinic explosion, 85
Homogenization, 245
Hopf bifurcation, 49, 50, 66, 112–114, 116, 117, 119, 124, 125, 361
Hopf normal form, 49, 83
Hysteresis branches, 215
Hysteresis topology, 222
Hysteretic discontinuity, 212
Hysteretic transition, 28, 31

I
Ideal control, 284
Incoherent dynamics, 366
Incoherent region, 369
Inertia, 26, 29, 34
Inhibitory neuronal feedback, 443
Initial conditions, 7
In-phase synchronization, 360, 368
Integrate-and-fire model, 446
Interspike interval, 445
Ion channel, 445
Irregular neuronal spiking, 444
Itô calculus, 294
Itô's formula, 256

K
Kinetic equation, 257
Kirkwood closure, 266
Klein 4-group, 111
Kolmogorov equation, 256, 260
Kramers rate, 379
Kuramoto model with inertia, 26, 27

L
Lambert-W function, 416
Landauer-Büttiker theory, 283
Langevin equation, 309
Lang-Kobayashi equations, 357
Laplace transform, 257
Laser networks, 357
Limit cycle, 339, 360
Linear stability analysis, 41, 55, 359, 415

Linear system, 150
  stochastic, 154
Linewidth enhancement factor , 358
Liouville operator, 318–321, 324
Liouvillian, 277
Local neuronal networks, 441
Locked oscillators, 32, 34
Lognormal distribution, 262
Lorenz equations, 84
Lyapunov vector, 42

M
Martingale, 256
Master equation, 277, 291, 317, 318, 320, 321, 324
Master equation, stochastic, 294
Master stability function, 148, 454
Mathieu-Duffing oscillator, 179
Maximal Lyapunov exponent, 28, 41
Maximum entropy, 263, 267
Maxwellian density, 258, 263
Maxwell's demon, 281
Mean, 255
Mean-field, 264
Mean-field description, 455
Mean-field theory, 29
Measurement-based control, 276
Microscopic closure, 264
Modularity, 434
Moment closure, 253, 260
Moment density, 262
Moment equation, 256
Moment generating function, 257
Multicluster chimera, 7
Multiple scaling, 149, 155
Multirhythmicity, 339
Multiscale analysis, 236
Multistability, 5, 16, 131, 364, 367, 370
Mutually coupled lasers, 359

N
Network, 3, 105–109, 124, 259, 267
Neuromodulator, 445
Neuromodulatory system, 445
Neuronal membrane voltage, 446
Neuronal network dynamics, 441
Neuronal oscillations, 441
Neuronal spike, 446
Neuronal spike rate, 448
Neutral delay system, 151
Noise, 11, 133, 172
Noise-induced bifurcation, 179